This book is a comprehensive treatment of fine particle magnetism and the magnetic properties of rocks. Starting from atomic magnetism and magnetostatic principles, the authors explain why domains and micromagnetic structures form in ferrimagnetic crystals and how these lead to magnetic memory in the form of thermal, chemical and other remanent magnetizations.

Such magnetizations are phenomenally stable and provide a record of the plate tectonic motions of continents and oceans over millions of years. Thermal activation theory, in which changes in magnetic domains and domain walls are blocked below a critical temperature or above a critical grain size, explains the high stability. One chapter is devoted to practical tests of domain state and paleomagnetic stability. Another deals with pseudo-single-domain magnetism, that is particles that contain domain walls but behave like a single domain. The final four chapters place magnetism in the context of igneous, sedimentary, metamorphic, and extraterrestrial rocks.

This book will be of value to graduate students and researchers in geophysics and geology, particularly in paleomagnetism and rock magnetism, as well as physicists and electrical engineers interested in fine-particle magnetism and magnetic recording.

Rock Magnetism

Cambridge Studies in Magnetism

EDITED BY

David Edwards
Department of Mathematics,
Imperial College of Science, Technology and Medicine

Rock Magnetism

Fundamentals and frontiers

David J. Dunlop
Geophysics, Department of Physics, University of Toronto

Özden Özdemir
Geophysics, Department of Physics, University of Toronto

CAMBRIDGE
UNIVERSITY PRESS

PUBLISHED BY THE PRESS SYNDICATE OF THE UNIVERSITY OF CAMBRIDGE
The Pitt Building, Trumpington Street, Cambridge, CB2 1RP United Kingdom

CAMBRIDGE UNIVERSITY PRESS
The Edinburgh Building, Cambridge CB2 2RU, United Kingdom
40 West 20th Street, New York, NY 10011-4211, USA
10 Stamford Road, Oakleigh, Melbourne 3166, Australia

First published 1997

Printed in the United Kingdom at the University Press, Cambridge

Typeset in Times $10\frac{1}{4}/13\frac{1}{2}$ pt

A catalogue record for this book is available from the British Library

Library of Congress Cataloguing in Publication data

Dunlop, David J.
  Rock magnetism: fundamentals and frontiers / by David J. Dunlop
  and Özden Özdemir.
    p.   cm. – (Cambridge studies in magnetism)
  ISBN 0 521 32514 5 (hc)
    1. Rocks – Magnetic properties.   I. Özdemir, Özden, 1946–   .
  II. Title. III. Series.
  QE431.6.M3D86   1997
  552'.06–dc20   96-31562 CIP

ISBN 0 521 32514 5 hardback

To those who have gone before us, the pioneers of rock magnetism, and to our students and colleagues, from whom we have learned so much.

Contents

Preface

The magnetic compass was one of mankind's first high-technology devices. Possession of the compass gave the Islamic world an early edge in navigation and led to the rapid eastward spread, by sea, of their trade, religion and civilization. But man was a comparative latecomer in magnetically aided navigation. Birds, fish, insects, and even bacteria had evolved efficient compasses millions of years earlier.

Magnetic memory, whether of a compass needle, a lava flow, or a computer diskette, is a remarkable physical phenomenon. The magnetic moment is permanent. It requires no expenditure of energy to sustain. Yet it can be partly or completely overprinted with a new signal. Nowhere is this more strikingly demonstrated than in rocks. A single hand sample can record generations of past magnetic events. This family tree can be decoded in the laboratory by stripping away successive layers of the magnetic signal.

Paleomagnetism is the science of reading and interpreting the magnetic signal of rocks. Rock magnetism is more concerned with the writing or recording process. The principles are no different from those of fine-particle magnetism as applied in permanent magnet and magnetic recording technology. But the physical parameters are rather different. Weak magnetic fields are involved, on the order of the present geomagnetic field (0.3–0.6 G or 30–60 µT), much less than the switching fields of the magnetic particles. Temperatures may be high: thermoremanent magnetization of igneous rocks is acquired during cooling from the melt. Times are long, typically millions of years.

Rock magnetism and paleomagnetism trace their origins to the mid-nineteenth century, but they really came into prominence in the 1950's and

1960's because of two daring questions that shook and ultimately revolutionized earth science: Does the earth's magnetic field reverse itself? And do the continents drift? Because rocks record in their magnetizations the polarity and direction of past geomagnetic fields, paleomagnetism was able to answer both questions.

Rocks of the same age from around the globe always recorded the same geomagnetic polarity. Their ancient compasses pointed north (normal polarity) during certain time intervals and south at other times. Most strikingly, strips of seafloor on either side of mid-ocean ridges (the birthplace of new seafloor) were unmistakably striped magnetically: either normal or reverse in response to the prevailing geomagnetic polarity as they formed. The earth's field does indeed reverse.

The same internal compasses showed that continents or subcontinents (now recognized as sections of lithospheric plates, containing ocean floor as well as continent) had rotated away from present-day north and had changed latitude during geological history. But prior to 175 Ma ago, their compass bearings coincided. They were originally assembled in a single supercontinent.

These findings shook earth science to its foundations and led to its rebuilding around a new guiding principle, plate tectonics. The revolution was not quite as straightforward as we have implied. The multiple generations of magnetization in rocks clouded the issue until methods were developed to strip away all but the most ancient. The stability of this ancestral magnetization came under scrutiny. It seemed to many inconceivable that rocks could preserve an unchanging magnetic memory for 175 Ma when the best products of human technology could be rather easily remagnetized by extraneous fields or stresses.

Such is the exciting history of rock magnetism. In this book, we will show how rocks manage to achieve a fidelity of magnetic memory that is beyond human experience. After developing the principles of ferromagnetism in Chapters 1–5, we will see in Chapter 6 how ferromagnetic domains appear under the microscope and in Chapter 7 what new micromagnetic structures are currently being predicted in grains too small to observe optically. Chapters 8, 9 and 10 reveal how the joint influences of temperature or time and magnetic fields permit the writing of a magnetic signal that cannot be erased by subsequent geomagnetic field changes. These are the fundamentals.

Chapters 11, 12 and 13 deal with some of the frontiers in understanding rock magnetic recording. As well as developing laboratory parameters and techniques that can predict stability on geological time scales (Chapter 11), we will look at how rather large particles can achieve stable memory that rivals that of submicroscopic particles (the pseudo-single-domain effect: Chapter 12) and how chemical changes in minerals degrade or enhance magnetic memory (Chapter 13).

Paleomagnetists sometimes complain that the 'rock' is frequently left out of 'rock magnetism'. Rock magnetic research looks too much like magnetic materials research. We have tried to answer that criticism in Chapters 14–17 by looking at magnetic minerals and their magnetic signals in the real world of igneous, sedimentary, metamorphic, and extraterrestrial rocks. This is a whole subject in itself and no one person can claim to be expert in all aspects, the present authors included. We have tried to convey the flavour of current thinking and research, rather than serving up an overwhelming banquet.

What background do you need to appreciate this book? A grounding in electricity and magnetism at junior undergraduate level is a help. Those with more geological background may wish to skip over the mathematical details. The key results can stand without them, and we have tried to maintain the story line uninterrupted wherever possible. Similarly a knowledge of basic earth science is helpful, but those with a physical science or engineering background should not turn away because they can't tell a hyperbyssal rock from a descending plate. This knowledge too is usually peripheral to the main message of the book.

We enjoyed writing this book and hope you will enjoy reading it. It would certainly never have been completed without a lot of help from our friends. We would especially like to mention those who taught us and passed on so much of their knowledge: Subir Banerjee, Ken Creer, Zdenek Hauptmann, Ted Irving, Takesi Nagata, Bill O'Reilly, Minoru and Mituko Ozima, Frank Stacey, David Strangway, Emile Thellier and Gordon West. With our colleagues Susan Halgedahl, Ron Merrill, Bruce Moskowitz, Michel Prévot, Valera Shcherbakov, Wyn Williams and Song Xu, we have passed countless hours of pleasurable discussion. Among our former students, we want to mention in particular Ken Buchan, Randy Enkin, Franz Heider and Andrew Newell; their work has had a central influence on the ideas in the book. Sherman Grommé, Ted Irving, Ed Larson, Michel Prévot and Naoji Sugiura kindly read and commented on early versions of some of the chapters.

We are grateful to Jennifer Wiszniewski, Carolyn Moon and Li Guo for patiently shepherding the manuscript through its many incarnations. Thanks to Khader Khan and Raul Cunha, different generations of figures were skillfully drawn and transformed through successive changes in technology. The photography was done with dedication and skill by Alison Dias, Steve Jaunzems and Judith Kostilek. Unlike some authors' families, ours did not eagerly await the appearance of each new chapter, but they did have a collective sigh of relief when the last page was written and life returned to normal.

Mississauga, Canada
April 1996

Chapter 1

Magnetism in nature

Magnetism has fascinated mankind since the invention of compasses that could track invisible magnetic field lines over the earth's surface. Much later came the discovery that rocks can fossilize a record of ancient magnetic fields. Unravelling this record – the 'archeology' of magnetism – is the science of paleomagnetism, and understanding how the microscopic fossil 'compasses' in rocks behave has come to be known as rock magnetism.

Rock magnetism is both a basic and an applied science. Its fundamentals concern ferromagnetism and magnetic domains and were developed most authoritatively by Néel. Its applications continue to expand, giving impetus to new research into the mechanisms and fidelity of rock magnetic recording. Some of the history and applications are described in this chapter.

1.1 A brief history

1.1.1 Earth magnetism

Compasses were used in China and the Arab world for centuries before Petrus Peregrinus in 1269 gave the first European description of a working compass. The earliest compasses were lodestones, naturally occurring ores of magnetite (Fe_3O_4). Particular areas, or poles, of one lodestone would attract or repel the poles of another lodestone. This magnetic polarization is the key to their use as compasses in navigation. A suspended lodestone will rotate until its axis of magnetization or polarization, joining north and south poles of the lodestone, lines up with imaginary field lines joining the north and south geomagnetic poles. In modern terminology, the magnetization M aligns with the field H.

Although the analogy between the polarization of a magnet and the polarization of the earth's magnetic field had been known at least since the time of Petrus Peregrinus (and probably much earlier), it was not until 1600 that William Gilbert in his celebrated treatise *De Magnete* documented the similarity between the field lines around a lodestone cut in the form of a sphere (a terella) and the field lines around the earth. He concluded: the earth is a great magnet. At the 'north' pole of the terella (actually a south magnetic pole by modern definition), marked A in Fig. 1.1, a compass needle points down, and at the 'south' pole B, it points up, just as for the earth's field. At the equator, the field is horizontal. At intermediate latitudes on the terella, the field is shown as flat-lying, whereas on the earth the field is inclined 60° or more to the horizontal. Nevertheless, Gilbert correctly recognized the essential similarity between the field of a uniformly magnetized sphere and the earth's field. Both are dipole fields (Chapter 2).

In the early nineteenth century, geomagnetic observatories were established around the world, although the coverage was rather uneven in both latitude and longitude (the same remains true today). The accumulated data on the inclination, declination (deviation from north) and intensity of the earth's field

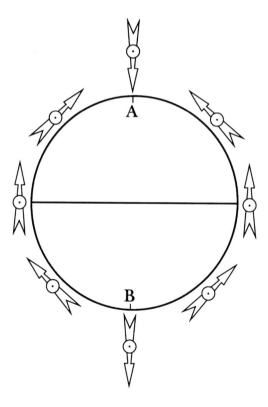

Figure 1.1   Sketch of a terella (a sphere of magnetite) as shown by Gilbert (1600) in *De Magnete*. The inclinations of a compass needle around the terella approximately match those around the earth.

were sufficient for Gauss (1839) to make a pioneering spherical harmonic analysis. He found that the bulk of the field originates within the earth, as Gilbert had surmised, and that it is predominantly a dipole field, again confirming Gilbert's conjecture. Gauss was able to calculate several higher multipole terms in the field with sufficient accuracy to show that the best-fitting dipole is inclined 11.5° to the earth's rotation axis. That is, the magnetic north and south poles do not coincide with the geographic poles.

Gauss and his contemporaries observed that the earth's field is not static. The most striking change in the field between Gauss' time and our own is a >5% decrease in the earth's dipole moment. Similar changes in the paleofield intensity are well documented in rocks. Sometimes they presage a reversal of the field, in which the magnetic north pole becomes the new south pole and vice versa. The question of whether the entire field reverses or only the main dipole is still hotly debated, but the reality of polarity reversals is not in question. Earlier this century, however, it was not clear whether the field had actually reversed or whether some rocks could become magnetized antiparallel to the field ('self-reversal').

Brunhes (1906) and David (1904) were the first to measure natural magnetizations that were antiparallel to the local geomagnetic field. They observed that both lavas and the clays they had baked were reversely magnetized. Since the two materials have very different mineralogies, it is unlikely that both would exhibit self-reversal. Matuyama (1929) showed that younger Quaternary lavas, now known to be <0.78 Ma in age, were all magnetized normally (parallel to the present geomagnetic field) whereas older lavas were reversely magnetized. The association of normal and reverse polarities with different geological times leaves no doubt that the earth's field has reversed. It also forms the basis for the geomagnetic polarity time scale. The most recent polarity epochs, covering the past 4 Ma of earth history, are named in honour of Brunhes (and David), Matuyama, Gauss and Gilbert (see Fig. 1.4).

1.1.2 Ferromagnetism and magnetic domains

Weiss (1907) was intrigued by the fact that iron and other 'soft' ferromagnetic materials with small permanent magnetizations become strongly magnetized when exposed to quite weak magnetic fields. Weiss proposed that external fields play a minor role compared to a hypothesized internal 'molecular field' which aligns the magnetic moments of individual atoms, producing a spontaneous magnetization, M_s. In the absence of any external field, the magnetic moments of regions with different directions of M_s (now called domains) cancel almost perfectly, but even a small applied field will either rotate domains or enlarge some at the expense of others. Weiss' theory is the starting point for modern ideas about ferromagnetism (Chapter 2) and ferromagnetic domains (Chapter 5).

The origin of Weiss' molecular field was explained by Heisenberg (1928). Electrostatic 'exchange coupling' of 3d orbitals in neighbouring atoms results in the alignment of the magnetic moments of their electrons. Néel (1948) returned to Weiss' molecular field description in his Nobel-prize-winning treatment of exchange coupling between magnetic sublattices in antiferromagnetic and ferrimagnetic substances.

It was not until the 1930's that F. Bitter developed a method of observing magnetic domains. In the Bitter technique, one observes a smooth (usually polished) surface of a ferromagnetic grain or crystal under the microscope after it has been coated with a thin colloidal suspension of ultrafine ($\ll 1$ μm) magnetite particles. The magnetic colloid gathers at domain boundaries, where flux leaks out of the crystal, and moves with the boundaries in response to applied fields. A photograph of simple domain boundaries on a {110} crystal face of magnetite appears in Fig. 1.2.

Without the benefit of any substantial data base, Landau and Lifschitz (1935) correctly predicted the form and dimensions of magnetic domains and domain walls (the boundaries between domains). Reasoning that domain boundaries would be positioned so as to maximize internal flux closure and eliminate 'magnetic charges' on both internal boundaries and the crystal surface, they postulated exactly the structure illustrated in Fig. 1.2. The plate-like main or body domains (seen here edge on) are terminated by wedge-shaped surface or closure domains which provide a continuous flux path between neighbouring body domains.

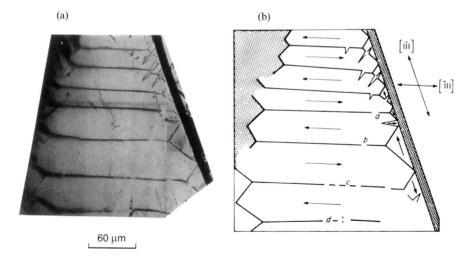

(a)   (b)

60 μm

Figure 1.2 Body and surface (closure) domains observed on a polished {110} surface of a large natural magnetite crystal. The arrows are inferred directions of M_s within each domain. The black lines are domain walls, made visible by means of a magnetic colloid. [After Özdemir and Dunlop (1993a) © American Geophysical Union, with the permission of the authors and the publisher.]

The final development of classic domain theory came with the work of Kittel (1949), whose calculations of wall and domain widths extended and refined Landau and Lifschitz's estimates, and Rhodes and Rowlands (1954) and Amar (1958), who first calculated the magnetic self-energy of a rectangular block, representing a domain wall. Although the self-energy (or demagnetizing energy) had long been known to control the size and shape of domains, its importance in determining domain wall energy and size had not been recognized. In small (< 1 μm) magnetite crystals, walls become broad and their demagnetizing energy increases accordingly. Rhodes and Rowlands' results are also the starting point for most micromagnetic calculations of fine-scale magnetization structure (Chapter 7).

1.1.3   Rock magnetism

Earth magnetism and ferromagnetism developed in mutual isolation until Koenigsberger (1938), Thellier (1938) and Nagata (1943) brought earth magnetism into the laboratory. They attempted to reproduce and understand the process by which igneous rocks are magnetized in nature. The new science was given a name with the publication in 1953 of Nagata's classic book, *Rock Magnetism.*

Koenigsberger, Thellier and Nagata gave new magnetizations to their rocks (lavas and archeological materials like bricks and baked clays) by heating them to high temperatures and cooling them in a weak field. This **thermoremanent magnetization** (TRM for short) was always parallel to the field in which it was acquired and its intensity was proportional to the strength of the field. TRM was therefore an accurate recorder of magnetic field directions, as geomagnetists had assumed, and was also potentially a means of determining paleofield intensities.

TRM in the laboratory had a series of remarkable properties which were most clearly described by Thellier. When a sample was given a **partial TRM**, by exposing it to a field only in a narrow cooling interval (between T_1 and T_2 say), and was subsequently reheated to high temperatures in zero field, the TRM remained unchanged below T_2, but completely disappeared between T_2 and T_1.

Néel (1949) explained this observation as a consequence of **blocking** of TRM during cooling at a single blocking temperature T_B determined by the size and shape of a particular magnetic grain. The TRM is unblocked when reheated through T_B. The wide spectrum of grain sizes and shapes in a rock leads to a continuous distribution of blocking temperatures of partial TRM, but each individual grain has one and only one value of T_B.

This individuality of blocking temperatures illuminates other experimental TRM 'laws'. Partial TRM's acquired in different temperature intervals are mutually independent. For example, if two partial TRM's are added vectorially in a single cooling run, by rotating the field by 90° after the first partial TRM is produced, the two partial TRM's are observed to demagnetize independently of each other during zero-field reheating. Each partial TRM disappears over its own blocking temperature interval and the total magnetization vector retraces the exact pattern it followed during cooling.

The 'additivity law' is another consequence of the one grain/one blocking temperature principle. In this case, we combine results of separate experiments in which partial TRM's are produced over different temperature intervals. The intervals, taken together, cover the entire temperature range from the Curie point (the ferromagnetic ordering temperature, usually several hundred °C) to room temperature. The sum of all partial TRM intensities is found to be equal to the intensity of total TRM produced in a single cooling. Since each partial TRM contains a unique part of the T_B spectrum of the total TRM, this law is easy to understand also. Figure 1.3 illustrates how total TRM is built up from a spectrum of partial TRM's.

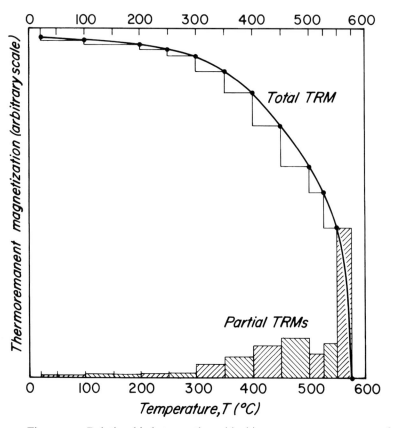

Figure 1.3 Relationship between the unblocking temperature spectrum of total TRM (solid curve, with steps indicating increments over 25–100 °C heating intervals) and the blocking temperature spectrum of partial TRM's (hatched). The partial TRM gained over a particular temperature interval equals the incremental TRM lost over the same interval (the reciprocity law) and the sum of all partial TRM's equals the intensity of total TRM (additivity law).

Koenigsberger, Thellier and Nagata were fortunate in their choice of fine-grained rocks containing an abundance of single-domain magnetite and hematite (αFe_2O_3) grains. **Single-domain** grains are too small to accommodate a domain wall and must change their magnetizations by rotation. Only single-domain grains exhibit a unique TRM blocking temperature and obey the Thellier laws (see Chapter 8). The TRM of larger **multidomain** grains (Chapter 9) is complicated by the mobility of domain walls and is less obviously suitable for recording the paleomagnetic field. Stacey (1962, 1963) proposed a new view: multidomain grains that have aspects of single-domain behaviour. Pseudo-single-domain (PSD) models have inspired and shaped rock magnetic research ever since, and form the subject of Chapter 12.

1.2   How rock magnetism is applied

1.2.1   Magnetic lineations: the seafloor record of reversals

Lithospheric plate motions result in a smooth spreading of newly erupted seafloor lavas away from mid-ocean ridges. The lavas contain fine-grained titanomagnetite ($Fe_{2.4}Ti_{0.6}O_4$) which acquires an intense TRM on cooling below its Curie temperature. The geomagnetic field which produces the TRM reverses at irregular intervals. The overall result (Fig. 1.4) is a set of normally and reversely magnetized bands of seafloor, symmetrical about mid-ocean ridges and increasing in age with distance from the ridge.

The magnetized seafloor produces a magnetic field (the 'anomaly' field) which is added to or subtracted from the local geomagnetic field. The polarized strips of seafloor are typically 10 km (Atlantic) to 80 km (Pacific) wide. Usually the field is measured at or just above the sea surface, 1–2 km above the seafloor. Because of this fortunate source geometry, fine details of the seafloor field are filtered out and the anomaly field is quite flat except over the boundaries between magnetized bands of seafloor, where it reverses sign. When contoured over the ocean, the anomaly field reproduces the lineated pattern of seafloor TRM that underlies it. These lineated anomalies are often called **magnetic stripes**. The boundaries between stripes mark times of field reversals.

Magnetic stripes are the most tangible evidence of seafloor spreading and plate tectonics. Historically they were decisive in converting geologists to a mobilist view of the earth. They are the main means of determining plate velocities, and they provide the most complete record of geomagnetic reversal history during the last 175 Ma. In view of this importance, and the exhaustive research that has been carried out since Vine and Matthews first proposed their model in 1963, it might be thought that very little remains to be learned. Surprisingly, this is not so. Using a tape recorder as an analogy, the tape drive (the spreading

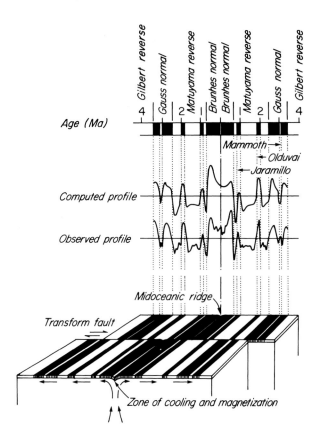

Figure 1.4 The Vine and Matthews' (1963) model of seafloor spreading (bottom) and linear magnetic anomalies over the oceans. The anomaly profile (centre) is a spatial replica of the time sequence of geomagnetic field reversals (top). [After Takeuchi *et al.* (1970) Fig. 8.4, with the permission of the authors.]

seafloor and the tectonic forces that drive it) and the time series of geomagnetic reversals that are recorded have been thoroughly looked into, but the recording medium itself – the seafloor and its magnetic minerals – have not.

The first surprise is that much of the seafloor carries no TRM at all. Before new seafloor has spread more than 20 km from its parent ridge, seawater has penetrated deep along fissures and oxidized the primary titanomagnetites, changing both their chemistry and their magnetization. A second major question is how much of the magnetic stripe signal originates in the surface layer of seafloor lavas and how much has a deeper source in rocks to which we have no direct access. How these problems are being tackled is described in Chapters 13 and 14.

1.2.2   **Other magnetic anomalies**

Magnetic stripes are not the only interesting anomalies in the geomagnetic field. Anomalies of local and regional extent are used by geologists as tools in mapping lithologies, structures and deformational styles, and delineating metamorphic terranes. They may also have direct economic interest if they pinpoint mineable concentrations of magnetite or hematite or if they trace iron formations outlining the structure of greenstone belts in which gold and other precious metals are concentrated.

The exact connection between anomaly fields and the magnetic petrology of the source body or region is not always clear, usually because the source cannot be sampled directly. The interpretation of an anomaly also depends on whether the magnetization is mainly **induced** by the present geomagnetic field and is therefore in the present field direction, or is mainly **remanent** (permanent) and parallel to some ancient field, which because of plate movements and field reversals may be at a large angle to the present field direction. The Koenigsberger Q_n ratio of remanent to induced magnetization is therefore an important parameter.

Some major questions have been raised by long-wavelength (regional to global scale) anomalies of the earth and moon revealed by satellite and spacecraft magnetometers. Lunar anomalies are often too large to be explained by typical magnetizations of returned Apollo samples (Chapter 17). Closer to home and just as vexing are broad anomalies over both continents and oceans whose amplitudes demand a substantial source of magnetization in the lower crust or upper mantle. Exotic magnetic minerals seen in crystal xenoliths have been favoured by some (Wasilewski and Mayhew, 1982; Haggerty and Toft, 1985) and an enhancement of magnetization by high temperatures at depth by others (see §10.2.2). The question is far from settled (Toft and Haggerty, 1988).

1.2.3 Records of geomagnetic field variation

Reversals are only one (albeit the most spectacular one) of many geomagnetic field variations. Magnetostratigraphy, the dating and correlation of geological strata based on the magnetic reversals they record, has become a major science. For example, the Cretaceous–Tertiary boundary at which the dinosaurs became extinct has been dated very precisely at 66 Ma using magnetostratiography (Lowrie and Alvarez, 1981). Reversal records are recovered both from sediments cored in the oceans and from ancient sedimentary sections preserved on land. For times before the mid-Jurassic (>175 Ma) there are no surviving ocean sediments or seafloor and our only record of reversals comes from sedimentary rocks. Most sections include times when sedimentation was slow or non-existent and the quality of the magnetic recording may vary considerably down the section. Deciphering the reversal record requires much time, labour, insight, comparison of sections – and rock magnetic tests (Chapter 11).

The behaviour of the geomagnetic field during a polarity transition and during excursions (which may be aborted transitions) is of vital importance in understanding the core dynamo. A reversal lasts only a few thousand years. The best magnetic records of reversals come from lavas which are closely spaced in time and from rapidly deposited ocean sediments. Figure 1.5 is a high quality record from Ocean Drilling Program cores in the equatorial Pacific. The gradual decline

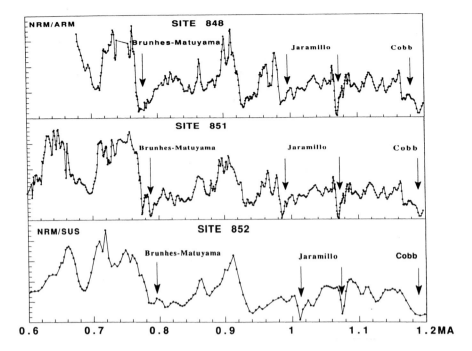

Figure 1.5   A high quality record of geomagnetic intensity variations recorded in oceanic sediment cores. The Brunhes–Matuyama polarity transition (see Fig. 1.4) and the Jaramillo and Cobb events (aborted reversals) are indicated. [After Meynadier *et al.* (1995), with the permission of the authors and the publisher, The Ocean Drilling Program, College Station, TX.]

in field intensity preceding a transition or excursion, the rapid recovery of intensity after the event, and the correlation between the amount of intensity recovery and the time until the next event are clear in the record.

The geomagnetic field changes direction and intensity even at times of constant polarity. The decline in dipole moment in the past 150 years was mentioned earlier. The field declination and inclination also vary. Lake sediments, soils, and loess (windblown sediment) record similar behaviour of the earth's field in the past, known as paleosecular variation (PSV). PSV records, if of high quality, are a powerful tool for dating and correlating sediments and soils over the past few thousand years. The correlation cannot be made over large distances because secular variation records fairly local variations in the field.

The main problem in paleofield variation records is fidelity of the recording. Both sediments and lavas may record a magnetization whose inclination is systematically shallower than the field inclination. Why inclination errors occur and what can be done to recognize and overcome them will be considered in Chapters 14 and 15. Random noise often obscures the details of the signal in sediments. Some of this noise is introduced in the coring process, but much is inherent to the magnetic grains themselves. Effective ways of 'cleaning' the records to remove the effects of the noisier magnetic grains (generally the larger grains) are still being developed (Chapters 11, 12 and 15). Chemical alteration of magnetic minerals is a problem with both lavas and sediments. This question is dealt with in Chapter 13.

1.2.4   Paleointensity determination

Archeomagnetism, the dating of archeological materials by magnetic means, also deals with geomagnetic field variations, but more emphasis is put on careful determinations of paleofield intensity. In fact, Thellier's study of TRM and partial TRM's was part of a lifelong program of determining reliable paleointensities from archeological pottery and bricks. Figure 1.6 illustrates how the Thellier laws form the basis for a rigorous method of paleointensity determination, described in detail in Chapter 8. Since partial TRM acquired at T_B during cooling is erased at T_B during heating, one can replace the natural remanent magnetization (NRM) of an igneous or baked rock in stepwise fashion with laboratory TRM in a series of double heatings (the first in zero field and the second in a known laboratory field) to increasing temperatures. The Thellier method is effective because the ratio between NRM and laboratory TRM intensities is the same for each fraction of the blocking temperature spectrum. Since TRM is proportional to the field strength, the same ratio gives replicate determinations of

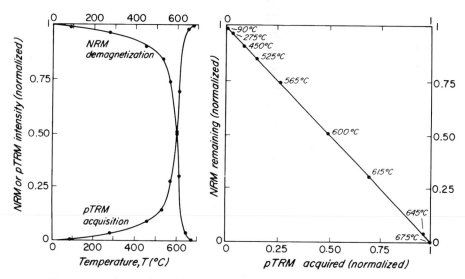

Figure 1.6   Illustration of the Thellier method of paleointensity determination, using data for hematite (αFe_2O_3). Since NRM demagnetization and partial TRM acquisition are reciprocal processes (left), a plot of NRM remaining versus partial TRM acquired after heating to a particular temperature is linear (right). The slope of the plot is proportional to paleofield intensity. [After Roquet and Thellier (1946), with permission of the publisher, L'Institut de France, Paris, France.]

paleofield/laboratory field intensity. Deviations from ideality may result from decay of part of the NRM in nature (affecting the low-T_B fraction) or chemical alteration in laboratory heatings (affecting the high-T_B fraction).

The intensity of the earth's field has also been studied for very old rocks. The earth's field existed at least 3500 Ma ago (Hale and Dunlop, 1984). We have only a fragmentary knowledge of how its strength has varied with time, however. Rocks which behave ideally in the Thellier experiment (single-domain size grains which do not alter in repeated heatings) are unfortunately few and far between. Recently there have been attempts to extend the range of suitable rocks; submarine volcanic glasses are particularly promising (Pick and Tauxe, 1993).

The Thellier method determines an absolute paleofield intensity. *Relative* paleointensities for a suite of related samples should be much easier to determine. It would be naive to assume that even closely related samples, e.g., sediments of similar mineralogy deposited under similar conditions, have an unvarying fraction of fine-grained magnetite. Various rock magnetic techniques, such as partial demagnetization of the NRM and normalization to synthetic magnetizations, are used to correct for the presence of softer, less reliable magnetic grains. These methods are described in Chapters 11 and 15. Ultimately one is still making inter-sample comparisons of total remanences; the subtlety of replicate determinations on various magnetization fractions of the same sample is lacking.

1.2.5   Paleomagnetism and plate motions

Plate motions during the past 175 Ma are most precisely tracked using magnetic stripes recorded in the oceanic portions of the plates. There are limitations in this approach, the most obvious being that one is restricted to the final 5% of earth history. Another problem is the lack of a trustworthy absolute reference frame. Paleomagnetism provides a reference axis – the earth's rotation axis – if measured declination and inclination of NRM of a sample are compared to the expected (long-term average) dipole field of the earth. Declinations deviating from north record rotations of a plate, or of smaller blocks within a plate. Anomalies in expected inclination record either changes in paleolatitude or tilting of blocks. Paleolongitude is not recorded because a dipole field is symmetric about the axis.

The paleomagnetic evidence that continents have moved about the earth was ignored for a long time by the geological community. Why did this happen, and why was the magnetic evidence from the spreading seafloor felt to be so much more compelling a decade later? At least part of the explanation is rock magnetic. Young seafloor lavas seemed to be simple magnetic recorders (more recent evidence to the contrary has been largely ignored) and their TRM record is binary: normal or reverse polarity. Continental rocks comprise a wide range of lithologies and ages. Once demagnetization techniques had developed, it became obvious that only part of the NRM was stable and by implication reliable. Furthermore, the information recorded was a vector, and small errors in directions of the vector components could cause large differences in the inferred paleopositions of the continents.

Paleomagnetically based plate reconstructions are now generally accepted without comment if they fit geological and paleoclimatological constraints, but the problem of assessing the reliability of individual results remains. Paleomagnetists increasingly have come to recognize that the older the rock, the more intricate (and interesting) the mixture of different NRM's it is likely to contain. Figure 1.7 is an example of two NRM's of very different ages carried by separate minerals (magnetite and hematite) in an Appalachian rock. The older primary

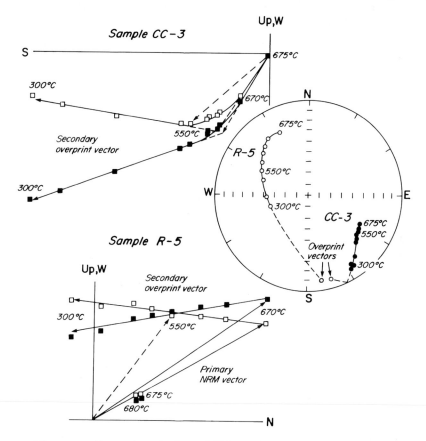

Figure 1.7   Separation of primary NRM carried by hematite with unblocking temperatures up to 675 °C and a later overprint (secondary NRM) carried mainly by magnetite (sample CC-3) or both magnetite and hematite (sample R-5). The two NRM components have distinct linear trajectories on vector diagrams. The stereonet shows that the primary NRM has opposite polarities in the two samples: downwards (solid circles) in CC-3 and upwards (open circles) in R-5. [After Kent and Opdyke (1985) © American Geophysical Union, with the permission of the authors and the publisher.]

NRM in this rock (and similar ones in other Appalachian rocks) escaped detection for many years, until careful thermal cleaning (Chapter 8) above the magnetite Curie point revealed its presence. The dominant lower unblocking temperature NRM is a late Carboniferous (≈ 300 Ma) chemical remagnetization found on both sides of the Atlantic and associated with the Hercynian/Alleghenian orogeny.

Not many remagnetizations have proven as elusive as this one. The message is clear, however. The more searching the rock magnetic tests and techniques applied to a rock, the more trustworthy the paleomagnetic result and any tectonic interpretations based on that result.

1.2.6   Biomagnetism

Many organisms, man included, contain intracelluar magnetite. Some species of neutral-density bacteria living in muddy water use their magnetite to sense geomagnetic field lines and thereby swim down to feed or to optimize their microaerobic environment. South of the magnetic equator, where the inclination is upward (Fig. 1.1), these magnetotactic bacteria swim in a direction opposite to the field. Honeybees, homing pigeons, migratory birds and dolphins all contain magnetite sensors which may be (in some cases, definitely are) utilized as compasses. An account of this fascinating subject is given by Kirschvink *et al.* (1985).

Magnetotactic bacteria have evolved efficient compass 'needles' consisting of chains of single-domain magnetite crystals polarized along the chain axis. Rock magnetic tests for single-domain behaviour and for interactions between crystals in each chain (Chapter 11) are widely used by biomagnetists to categorize bacterial and other biogenic magnetites (Moskowitz *et al.*, 1988a, 1993).

1.2.7   Environmental magnetism

One of many pressing environmental questions of our day is global warming. Soil profiles are a rich record of past climatic changes, but the record is usually fragmentary. The most continuous record is preserved in the loess plateau of China. Alternating horizons of loess and paleosol were laid down in dry and moist climates, respectively. Magnetism is used indirectly, in the dating and correlation of horizons by their recorded PSV. It may also be possible to use magnetism directly as a climate indicator. Loess is relatively barren, containing mainly wind-transported magnetic particles from distant sources. Soils, on the other hand, develop magnetic minerals in situ and the particular minerals that form are a good guide to moisture, pH and redox conditions.

There is a wealth of environmental applications of magnetism and magnetic particles: as tracers of pollutants, as indicators of the provenance and distance of transport of soils during erosion, as an aid to determining sediment load and velocity in streams, and so on. These and many other applications are outlined by Thompson and Oldfield (1986), who devote much space to the rock magnetic tests used by environmental scientists.

1.3   Plan of the book

We have seen how rock magnetism originated and have reviewed some of its practical applications, which nowadays extend well beyond rocks in the narrow sense. In the rest of the book, we will first outline the fundamentals of ferromagnetism (Chapter 2), magnetic mineralogy (Chapter 3), magnetostatic principles (Chapter 4) and magnetic domain structure and hysteresis (Chapter 5). Then we will turn to observations and calculations of micromagnetic structure in magnetite and other common magnetic minerals (Chapters 6 and 7). Next we will summarize the basic experimental properties of various remanent magnetizations that are important in nature: TRM in Chapters 8 and 9, viscous remanent magnetization (VRM) in Chapter 10, and crystallization or chemical remanent magnetization (CRM) in Chapter 13. In these chapters, we also review our present understanding of the underlying mechanisms of TRM, VRM and CRM. Chapters 11 and 12 deal with more specialized topics, isothermal rock magnetic tests and techniques, and pseudo-single-domain effects. Finally, we examine rock magnetism and magnetic mineralogy in specific geological environments: igneous and baked rocks (Chapter 14), sediments and sedimentary rocks (Chapter 15), altered and metamorphosed rocks (Chapter 16) and lunar rocks and meteorites (Chapter 17).

There are many other books, old and new, to which the reader may turn for additional information. For ferromagnetism, Bozorth (1951), Jiles (1959), Smit and Wijn (1961), Kneller (1962, 1969), Chikazumi (1964), Morrish (1965), Cullity (1972), and Craik (1995) are excellent sources. Among the older books, Bozorth (1951) and Kneller (1962) are encyclopedic in their coverage and therefore indispensable. Magnetic domain observations and calculations are covered in classic works by Kittel (1949), Brown (1962, 1963a) and Craik and Tebble (1965). The mainstream texts in rock magnetism are Nagata (1961), Stacey and Banerjee (1974), and O'Reilly (1984). Similar cornerstones of paleomagnetism, geomagnetism and planetary magnetism are Irving (1964), McElhinny (1973), Merrill and McElhinny (1983), and Butler (1992). Finally, we have found four specialized books very useful: Collinson (1983) for magnetic techniques and instrumentation; Kirschvink *et al.* (1985) for biomagnetism; Thompson and Oldfield (1986) for environmental magnetism; and Tarling and Hrouda (1993) for magnetic anisotropy.

Chapter 2

Fundamentals of magnetism

2.1   Introduction

There are two views of how magnetism originates: microscopic current loops and magnetic dipoles. The latter view is actually the older and is the basis of magnetostatics, the analog of electrostatics. It was discredited when it was demonstrated that magnetic dipoles cannot be separated into isolated + and − magnetic charges ('monopoles'). Yet the criticism is hardly damning. The component charges of electric dipoles (nuclei and electron clouds, respectively, expressing the polarization of atoms by an electric field) cannot readily be isolated either. Admittedly there is no analog of the free or conduction electron in magnetism, but in many ways magnetically polarized or polarizable materials are close analogs of electrically polarizable materials (dielectrics) and we shall make considerable use of this analogy. Ultimately neither current loops nor charge pairs can explain ferromagnetism. Ferromagnetic moments arise from a non-classical phenomenon, electron spin (§2.3).

A current loop or a dipole produces a magnetic field H or B. What is this magnetic field? H and B are defined not by their cause but by their effect, the forces they exert on physical objects. A magnetic field exerts a torque on a compass needle or pivoted bar magnet (macroscopic dipoles) tending to align their axes with H or B; this is the magnetostatic definition. A magnetic field also exerts a Lorentz force on a moving charged particle, either in free space or when channeled through a conductor as current I, at right angles to both the field and the particle velocity or the current flow; this is the electrodynamic definition. Neither definition gives us a feeling for what the source of H or B is. This comes from Coulomb's law or the Biot–Savart law (below) which tell us how to calculate H from a specified set of sources (either magnetic charges or current elements).

B is an augmented field which includes not only H but also the macroscopic magnetization M:

$$B = \mu_0(H + M) \text{ in SI}, \quad \text{or } B = H + 4\pi M \text{ in cgs emu}. \qquad (2.1)$$

Microscopic dipole moments μ are associated with individual atoms, but M, the dipole moment per unit volume, is an average over a region containing many atoms. For this reason, M, and therefore also B, can only be defined in a mesoscopic or macroscopic region. H, on the other hand, can be defined and calculated at any scale, including the atomic.

B is sometimes claimed to be more 'fundamental' than H. In view of the discussion of the last paragraph, this is clearly not so. It is fair to say that B is of greater practical and technological importance. Magnetic induction, for example, depends on the rate of change of magnetic flux, which is the integrated effect of lines of B cutting a particular area (B is in fact often called the magnetic induction or flux density). The sum of H and M is important in these phenomena, with M greatly outweighing H inside ferromagnetic materials (although not inside rocks, which are very dilute dispersions of ferromagnetic grains).

Outside a magnetic material, where $M = 0$, B and H are parallel and their strengths are in the ratio $B/H = \mu_0$, the permeability of free space. In SI, $\mu_0 = 4\pi \times 10^{-7}$ H/m so that numerical values of B (in tesla) and H (in A/m) are very different for the same field. In cgs electromagnetic units (emu), $\mu_0 = 1$ and the values of B (in gauss) and H (in oersted) are numerically equal.

In rock magnetism, both B and H, and both systems of units, are widely used. Conversion factors appear in Table 2.1, which also defines other quantities and ratios of practical interest. Very often rock magnetists quote values of $\mu_0 H$ in T as though they were values of H. This situation has come about because the numerical conversion of fields is particularly easy for B. For example, a field of 1 Oe or G ($\approx$ the earth's field) equals 10^{-4} T or 100 μT ($\mu_0 H$) but equals 79.6 kA/m (H). Note, however, that all formulas involving H require numerical values in A/m.

2.2   Magnetic moments of dipoles and current loops

Figure 2.1(a) illustrates a magnetic dipole: charges $\pm Q_m$ a distance d apart along the z axis. The **magnetic moment** of the dipole is $\mu = Q_m d\hat{k}$. At a point P, separated from $+Q_m$ by r_1, from $-Q_m$ by r_2, and from the centre of the dipole by r, where r_1, r_2 and r are all $\gg d$, the magnetic field is given by the magnetic equivalent to Coulomb's law:

Table 2.1   Definitions and units for magnetic quantities

Quantity	Definition and units	
	SI	cgs
Magnetic moment, μ	$A\,m^2$	emu
Magnetization, M	A/m	'emu/cm^3'
($\equiv$ magnetic moment/volume)		
Magnetic field, H	A/m	oersted (Oe)
Flux density or magnetic induction, B	$\boldsymbol{B} = \mu_0(\boldsymbol{H} + \boldsymbol{M})$	$\boldsymbol{B} = \boldsymbol{H} + 4\pi\boldsymbol{M}$
	tesla (T)	gauss (G)
Magnetic susceptibility, χ	$\boldsymbol{M} = \chi\boldsymbol{H}$	$\boldsymbol{M} = \chi\boldsymbol{H}$
	dimensionless	dimensionless

N.B. $\chi_{SI} = 4\pi\chi_{cgs}$

Some useful conversions

Magnetic field

1 gamma (γ) field	$10^{-9}\,T\,(1\,nT)$	$10^{-5}\,G$ or Oe
1 Oe or 1 G field	$10^{-4}\,T\,(100\,\mu T)$ or	1 G or 1 Oe
($\approx$ earth's field)	$10^3/4\pi = 79.6\,A/m$	

Magnetizations

'Typical' NRM – sediment	$10^{-3}\,A/m$	$10^{-6}\,emu/cm^3$
'Typical' NRM – volcanic rock	$1\,A/m$	$10^{-3}\,emu/cm^3$
Susceptibility of multidomain magnetite	$\chi_0 \approx 2.5$	$\chi_0 \approx 0.2$

$$\boldsymbol{H} = \frac{\boldsymbol{B}}{\mu_0} = \frac{1}{\mu_0}\frac{\mu_0}{4\pi}\left(\frac{Q_m}{r_1^2}\hat{r}_1 - \frac{Q_m}{r_2^2}\hat{r}_2\right) = \frac{\mu}{4\pi r^3}(2\cos\theta\hat{r} + \sin\theta\hat{\theta}). \tag{2.2}$$

(In the cgs version of eqns. (2.2)–(2.5), the factor $1/4\pi$ is omitted.) Along the axis of $\boldsymbol{\mu}$ (the z axis), $r = z$, $\hat{r} = \hat{k}$, and $\theta = 0$. Then

$$\boldsymbol{H}_{\text{axial}} = \frac{2\mu}{4\pi z^3}\hat{k}. \tag{2.3}$$

In the case of a circular current loop (Fig. 2.1b) of radius a carrying current I, $\boldsymbol{H}$ is prescribed by the Biot–Savart law. For a point P on the axis of the loop, as shown,

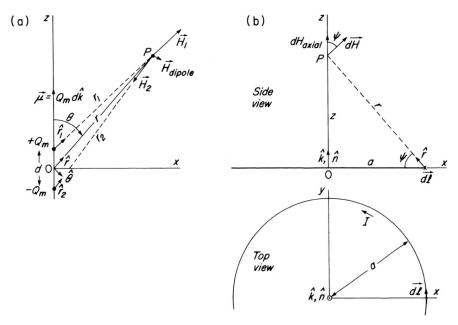

Figure 2.1 (a) Sketch of a magnetic dipole: magnetic charges $\pm Q_m$ separated by a distance d have a dipole moment $\mu = Q_m d\hat{k}$ and produce a field H_{dipole} at P. (b) Side and top views of a circular current loop. The loop has an equivalent dipole moment $I\pi a^2 \hat{n}$ and produces a field equivalent to H_{dipole} at distances $\gg a$.

$$H_{axial} = \oint \frac{Idl \times \hat{r}}{4\pi r^2} = \oint dH = \frac{I\cos\psi}{4\pi(a^2 + z^2)} \oint dl\hat{k}$$

$$= \frac{2I\pi a^2}{4\pi(a^2 + z^2)^{3/2}}\hat{k}, \tag{2.4}$$

which at large distances ($z \gg a$) reduces to

$$H_{axial} = \frac{2I\pi a^2}{4\pi z^3}\hat{k} = \frac{2\mu}{4\pi z^3}\hat{n}. \tag{2.5}$$

Equation (2.5) is identical to (2.3) if dipole moment μ is $I\pi a^2\hat{n}$ (current $I \times$ area $A \times$ outward normal $\hat{n}$, using the right-hand rule). Likewise (2.4) gives the same result as (2.2) for other points $P(r)$ if $\mu = IA\hat{n}$ for the current loop.

In the Bohr model of the atom, an s electron has a circular orbit. Its orbital magnetic moment, $IA\hat{n} = [e/(2\pi/\omega)]\pi r^2\hat{n}$, is always parallel to its orbital angular momentum vector, $L = m_e r^2 \omega\hat{n}$, where e and m_e are electronic charge and mass and ω is angular velocity. In a magnetic field H, the torque $\mu_0\mu \times H$ causes L, and with it μ, to precess like a gyroscope around H. The time-average L and μ are the observed components L_z and μ_z along H (redefining the z axis to lie along H, not along μ as in Fig. 2.1). We next turn to a quantum picture of the atom, in which L, L_z and μ_z are quantized and L and μ can only occupy certain specified directions relative to H.

2.3   Magnetic moments of atoms and ions

In quantum mechanics, the orbital energy equation for an electron transforms into Schrödinger's wave equation. Solving the wave equation for specified boundary conditions in a spherical region, such as an atom, we find a set of eigenfunctions Ψ_{lmn}, which are analogs of standing waves or normal modes of a classical system. These **wave functions** describe the probability of finding the electron at point (r, θ, ϕ) and involve spherical harmonics:

$$\Psi_{lmn}(r, \theta, \phi) = A_{lmn} P_l^m(\cos \theta) \cos, \sin(m\phi) f_n(r). \tag{2.6}$$

Here A_{lmn} is a probability amplitude, P_l^m are Legendre polynomials and the $f_n(r)$ are functions of distance r from the nucleus.

The eigenvalues l, m and n, as well as another number s, are **quantum numbers**. Pauli's exclusion principle states that each electron has a different set of quantum numbers, representing distinct values of energy, angular momentum, magnetic moment, and spin:

n specifies the shell ($n = 1 \rightarrow$ K shell, etc.)

l specifies the orbital angular momentum $L(l = 0 \rightarrow$ s, $l = 1 \rightarrow$ p, $l = 2 \rightarrow$ d, etc.). N.B. $0 \leq l \leq n - 1$: n values

m (or m_l or l_z) specifies the component of orbital angular momentum L_z in the direction of an applied magnetic field B or H (see Fig. 2.2). N.B. $-l \leq m \leq +l$: $2l + 1$ values

s (or m_s or s_z) specifies the spin angular momentum S in the direction of H (see Fig. 2.2). N.B. $s = \pm\frac{1}{2}$: 2 values.

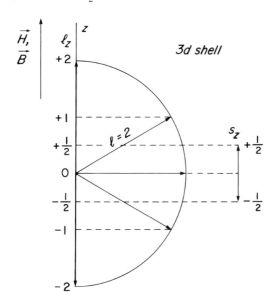

Figure 2.2 Quantized orientations, relative to the direction of a magnetic field H or B, of the orbital angular momentum vector for a 3d electron ($l = 2$), and allowed values of the time-average orbital and spin angular momenta, l_z and s_z, due to precession about H.

The elementary unit or quantum of magnetic moment is the Bohr magneton. A classical picture of an electron in a circular orbit of radius r around the nucleus gives the correct answer. The orbital magnetic moment μ_l is the product of the effective current $I = e\omega/2\pi$ and the orbital area πr^2. The angular momentum $L = m_e r^2 \omega$ is quantized in multiples of Planck's constant $\hbar$. Since most electron orbits have their axes at an angle to the z axis (the direction of H), it is L_z that matters (see §2.2). L_z takes on values $m\hbar = 0, \pm\hbar, \ldots, \pm l\hbar$. Thus we have

$$\mu_l = \frac{er^2\omega}{2} = \frac{e}{2m_e}L_z = m\frac{e\hbar}{2m_e} = m\mu_B, \tag{2.7}$$

where $\mu_B = e\hbar/2m_e$ is the Bohr magneton. Substituting numerical values, we find $\mu_B = 9.274 \times 10^{-24}$ A m^2 or 9.274×10^{-21} emu.

Like macroscopic orbiting bodies, electrons have spin as well as orbital angular momentum, and a corresponding spin magnetic moment

$$\mu_s = 2s\mu_B = \mu_B. \tag{2.8}$$

The difference of a factor 2 between eqns. (2.7) and (2.8) is conveniently incorporated in the Landé g factor:

$$\mu_l = g_l m\mu_B, \quad \mu_s = g_s s\mu_B, \tag{2.9}$$

with $g_l = 1, g_s = 2$. When all the electrons in an atom or ion are considered, the Landé factor is more complex.

The atoms of most elements have only one unpaired electron (odd atomic numbers) or no unpaired electrons (even atomic numbers). Their permanent magnetic moments are small or non-existent. Atoms of the transition metals Cr, Mn, Fe, Co and Ni have larger moments which arise from unpaired 3d ($n = 3, l = 2$) electrons. All these elements have filled 4s ($n = 4, l = 0$) valence shells but only partly filled 3d shells. Compounds of these metals may also have large moments because the 4s electrons are removed first in ionic bonding.

Let us take as an example iron, Fe^0. The 26 electrons have the distribution $1s^2 2s^2 2p^6 3s^2 3p^6 3d^6 4s^2$. All shells except 3d are filled. A filled shell has zero net moment, since it contains paired $s = \pm\frac{1}{2}$ electrons. In filling the 3d shell, the vector sum $S = \Sigma s_i$ is maximized (Hund's first rule), so that the first five electrons have unpaired spins ($s = +\frac{1}{2}$) and parallel spin magnetic moments. Thus $S = (5 \times \frac{1}{2}) - \frac{1}{2} = 2$. Next $L = \Sigma l_i$ (or $L = \Sigma m_i$) is maximized (Hund's second rule), giving $L = (2 + 1 + 0 - 1 - 2) + 2 = 2$. Notice that there was no choice in filling the half shell with spins parallel (numbers in parentheses); L is always zero for such a half-shell. Finally the total angular momentum J is determined. $J = |L - S|$ if the shell is less than half-filled, $J = L + S$ if the shell is more than half-filled, and $J = S$ if the shell is half-filled (since $L = 0$). For Fe^0, $J = 2 + 2 = 4$.

Experimentally, the transition metals and their compounds behave as though $J = S$, not $|L \pm S|$. The orbital contribution to the magnetic moment is said to be quenched, and to a first approximation the magnetic moment is due to electron spin only. The paramagnetic moments per atom for the important cations $Fe^{2+}(3d^6 4s^0, S = 2, J = 4)$ and $Fe^{3+}(3d^5 4s^0, S = 5/2, J = 5/2)$ can then be calculated from

$$\mu_J = g_J[J(J+1)]^{1/2}\mu_B = g_s[S(S+1)]^{1/2}\mu_B. \tag{2.10}$$

They are about 5 (actually 5.4) and 6 Bohr magnetons, respectively.

Orbital angular momentum is quenched because the 3d electrons, which occupy the outermost orbitals, experience an electrostatic 'crystal field' due to neighbouring ions in the crystal which outweighs the electrostatic L–S coupling within the atom. Kittel (1976, p. 444) gives an enlightening discussion. The crystal field also establishes preferred directions for L in the crystal lattice and residual L–S coupling makes these preferred directions for μ as well. This is the origin of magnetocrystalline anisotropy (§2.8).

2.4   Diamagnetism and paramagnetism

Diamagnetism is a property of all materials. A moving charge, such as an electron in orbit, experiences a Lorentz force $e\boldsymbol{v} \times \boldsymbol{B}$ in a magnetic field $\boldsymbol{B}$ ($\boldsymbol{v}$ is electron velocity). The Lorentz force has a gyroscopic effect, causing all electron orbits to precess about $\boldsymbol{B}$. This Larmor precession of Z electrons (Z is atomic number) is equivalent to an electric current and produces a negative induced magnetic moment

$$\mu_d = -\frac{Ze^2\langle\rho^2\rangle}{4m_e}B. \tag{2.11}$$

In (2.11), $\langle\rho^2\rangle = \langle x^2 + y^2\rangle$ is the mean square distance of electrons from the z axis (i.e., $\boldsymbol{B}$) and is related to $\langle r^2\rangle = \langle x^2 + y^2 + z^2\rangle$ by $\langle\rho^2\rangle = (2/3)\langle r^2\rangle$. The diamagnetic susceptibility is then

$$\chi_d = \frac{M}{H} = \frac{n\mu_d}{B/\mu_0} = -\frac{\mu_0 nZe^2\langle r^2\rangle}{6m_e}, \tag{2.12}$$

in which n is the number of atoms per unit volume. This is the classical Langevin expression for diamagnetic susceptibility.

Most non-iron-bearing minerals like quartz, calcite and feldspars are purely diamagnetic (see Table 2.2). Their susceptibilities are small but temperature independent. Thus at high temperatures if the amount of paramagnetic and ferromagnetic minerals is small, the negative moment of a quartz sample holder or the silicate matrix of a rock may become appreciable.

Paramagnetism is the partial alignment of permanent atomic magnetic moments in the direction of an applied field $\boldsymbol{B}$. Although the degree of alignment is slight at ordinary temperatures, it is sufficient to outweigh diamagnetism. The

Table 2.2 Susceptibilities of some diamagnetic and paramagnetic minerals

Mineral	Magnetic susceptibility (per unit mass)	
	$(10^{-6} \, \text{emu/g Oe})$	$(10^{-8} \, \text{m}^3/\text{kg})$
Diamagnetic		
Quartz (SiO_2)	−0.50	−0.62
Orthoclase feldspar ($KAlSi_3O_8$)	−0.46	−0.58
Calcite ($CaCO_3$)	−0.38	−0.48
Forsterite (Mg_2SiO_4)	−0.31	−0.39
Water (H_2O)	−0.72	−0.90
Paramagnetic		
Pyrite (FeS_2)	+24	+30
Siderite ($FeCO_3$)	+98	+123
Ilmenite ($FeTiO_3$)	+80–90	+100–113
Orthopyroxenes ((Fe,Mg)SiO_3)	+34–73	+43–92
Fayalite (Fe_2SiO_4)	+100	+126
Intermediate olivine ((Fe,Mg)$_2SiO_4$)	+29	+36
Serpentinite ($Mg_3Si_2O_5(OH)_4$)	$\geq 95^*$	$\geq 120^*$
Amphiboles	+13–75	+16–94
Biotites	+53–78	+67–98
Illite (clay)	+12	+15
Montmorillonite (clay)	+11	+14

*Higher values are due to precipitated magnetite (Fe_3O_4).

moments $\boldsymbol{\mu}_J$ ($\boldsymbol{\mu}$ for short) of the atoms are assumed to be non-interacting (this is not strictly true in a crystal lattice, where atoms interact both magnetically and electrostatically). The torque of $\boldsymbol{B}$ on each atomic moment is expressed in the magnetic field (or Zeeman) energy

$$E_{\text{m}} = -\boldsymbol{\mu} \cdot \boldsymbol{B} = -\mu B \cos \phi. \tag{2.13}$$

Only thermal perturbations prevent perfect alignment of the moments with $\boldsymbol{B}$.

Suppose there are only two possible states, $\boldsymbol{\mu}$ parallel to $\boldsymbol{B}$ ($\phi_1 = 0$) or antiparallel to $\boldsymbol{B}$ ($\phi_2 = 180°$). This is the situation for transition metals in which $\boldsymbol{L}$ is quenched and $\boldsymbol{J} = \boldsymbol{S}$. The Boltzmann probabilities of the states are

$$\text{Pr}(\phi_i) = \frac{1}{Z'} \exp\left(-\frac{E_{\text{m}}(\phi_i)}{kT}\right) = \frac{1}{Z'} \exp\left(\pm\frac{\mu B}{kT}\right), \tag{2.14}$$

in which Z' is the partition function or sum of exponential factors over all states and k is Boltzmann's constant. We can now calculate the net moment and the magnetization of the ensemble:

$$M(B, T) = n\frac{\mu e^{\alpha} - \mu e^{-\alpha}}{e^{\alpha} + e^{-\alpha}} = n\mu \tanh\left(\frac{\mu B}{kT}\right), \tag{2.15}$$

where we have used α to stand for $\mu B/kT$.

In the classical Langevin calculation, all orientations ϕ of $\boldsymbol{\mu}$ relative to $\boldsymbol{B}$ are permitted. The result is

$$M(B, T) = n\mu L(\alpha) = n\mu\left(\coth(\alpha) - \frac{1}{\alpha}\right). \tag{2.16}$$

$L(\alpha)$ is called the Langevin function. The essential results predicted by (2.15) or (2.16) are that M is positive (i.e., M is parallel to $\boldsymbol{B}$) and decreases with increasing temperature.

Because individual atomic moments μ are small, for most fields and temperatures $\alpha \ll 1$. The alignment of moments is therefore slight and M is a small fraction of its saturation value $n\mu$. From a series expansion of $\tanh(\alpha)$ in (2.15) or $L(\alpha)$ in (2.16) for small α, it is easy to show that the paramagnetic susceptibility is

$$\chi_{\mathrm{p}} = \frac{M}{H} = \frac{1}{3}\frac{n\mu^2\mu_0}{k}\frac{1}{T} = \frac{C}{T}. \tag{2.17}$$

This is Curie's law and C is the Curie constant. The factor $\frac{1}{3}$ is included if (2.16) holds and omitted if (2.15) holds. In either case, the suceptibility is inversely proportional to T.

In the corresponding quantum mechanical calculation, there are $2J + 1$ states with energies $E_{\mathrm{m}} = -m_J\mu_{\mathrm{B}}B$, where the quantum number m_J takes on values $-J, \ldots, J$. The result is

$$M = ng_J J\mu_{\mathrm{B}}B(x) = ng_J J\mu_{\mathrm{B}}\left\{\frac{2J + 1}{2J}\coth\left[\frac{(2J + 1)x}{2J}\right] - \frac{1}{2J}\coth\left(\frac{x}{2J}\right)\right\}. \tag{2.18}$$

$B(x)$ is the Brillouin function and x stands for $g_J J\mu_{\mathrm{B}}B/kT$. For most B and T, $x \ll 1$ and we find for the paramagnetic susceptibility

$$\chi_{\mathrm{p}} = \frac{ng_J^2 J(J + 1)\mu_{\mathrm{B}}^2\mu_0}{3k}\frac{1}{T} = \frac{C}{T}. \tag{2.19}$$

In the limits $J = S = \frac{1}{2}$ (a single spin) and J very large, (2.19) reduces to the two alternatives of (2.17), since by (2.10), $\mu_J = g_J[J(J + 1)]^{1/2}\mu_{\mathrm{B}}$.

Most iron-bearing sulphides, carbonates and silicates are paramagnetic, with susceptibilities usually 10–100 times larger than those of diamagnetic minerals. Some typical values are listed in Table 2.2.

2.5   **Ferromagnetism**

The magnetization of paramagnetic and diamagnetic substances is induced or temporary: M disappears when B is removed. Furthermore at ordinary temperatures extremely large fields, on the order of 10^6 G or 100 T, are required to align the permanent moments of paramagnetic atoms or ions and yield a saturation magnetization $n\mu_J$. Ferromagnetic materials like metallic iron and nickel behave quite differently. Fields of 100 G or so are enough to saturate the magnetization of a soft iron ring, and when the field is suppressed, a measurable permanent or remanent magnetization M_r remains. Weiss (1907) explained these puzzling observations by postulating the existence of an internal 'molecular field' so strong that it brings atomic moments into nearly saturation alignment, producing a **spontaneous magnetization M_s** (Fig. 2.3). Different domains with different directions of M_s have mutually compensating moments, but relatively small fields could serve to enlarge domains in which M_s vectors happen to be close to the direction of B.

   In reality, the molecular field must depend on the local environment of a particular atom. As a first approximation, however, Weiss proposed that the local field has the same magnitude throughout a domain and is proportional to the mean magnetization of the domain:

$$H_m = \lambda M. \tag{2.20}$$

This mean-field approximation must break down at high temperatures, when local thermal fluctuations in M become too large to ignore. However, because the moment of each atom is exchange coupled to its neighbours (this is the real

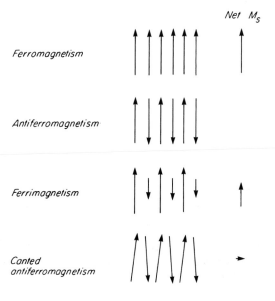

Ferromagnetism

Antiferromagnetism

Ferrimagnetism

Canted antiferromagnetism

Net M_S

Figure 2.3   Different possible exchange-coupled spin structures and the net spontaneous magnetization M_s of each.

origin of the molecular field: see §2.6), local or short-range order is preserved up to and even above the Curie point, the temperature at which M_s and long-range ordering of moments disappear.

Equation (2.20) describes a cooperative phenomenon, in which any increase in M causes H_m to increase, and vice versa. (The reverse is also true. Decreases in M, due to thermal fluctuations at high temperature, cause H_m to decrease, thus provoking further decreases in M. For this reason, M_s remains close to its saturation value $n\mu$ at low temperatures, but drops rapidly to zero near the Curie point (Fig. 2.4b)). If there is no externally applied field, H_m is the only field acting on μ and the energy equation (2.13) becomes

$$E_m = -\mu_0 \boldsymbol{\mu} \cdot \boldsymbol{H}_m = -\mu_0 \mu \lambda M \cos \phi. \tag{2.21}$$

Then (2.15) or (2.16) become

$$M = n\mu \tanh(\alpha) \quad \text{or} \quad n\mu L(\alpha) \tag{2.22}$$

with

$$\alpha = \frac{\mu_0 \mu \lambda M}{kT}, \quad \text{i.e., } M = \frac{kT\alpha}{\mu_0 \mu \lambda}. \tag{2.23}$$

Since (2.22) and (2.23) must be simultaneously satisfied, M is easily found by plotting $M(\alpha)$ for each separately and taking the intersection between the two curves (Fig. 2.4a). It is clear from the diagram that at low temperatures, the two curves intersect at a value of M_s close to the saturation magnetization $n\mu$. As T increases, the short-range and long-range alignment of moments is increasingly perturbed and M_s decreases. At the Curie temperature T_C, (2.23) becomes tangent to (2.22) and $M_s \to 0$. The ferromagnetic–paramagnetic phase transition is of second order, since there is no discontinuity in M at T_C.

This simple adaptation of the Langevin theory of paramagnetism gives a first-order description of measured $M_s(T)$ curves for iron and nickel (Fig. 2.4b). $M_s(T)$ curves for magnetite and other iron oxides require a more elaborate molecular field theory (§2.7).

We can estimate the strength of the molecular field by equating the initial slopes $dM/d\alpha$ of (2.22) and (2.23) at $T = T_C$. The result is

$$\lambda = (3)\frac{kT_C}{n\mu^2\mu_0} = \frac{T_C}{C}, \tag{2.24}$$

where C is the Curie constant from (2.17). Substituting numerical values for iron ($T_C = 765\,°C = 1038\,K$, $n = 3.3 \times 10^{23}\,cm^{-3}$, $k = 1.38 \times 10^{-16}\,erg/K$ and $\mu_B = 9.274 \times 10^{-21}\,emu$), we find $\lambda \approx 5000$. Since $M_s = 1715\,emu/cm^3$ at room temperature, $H_m \approx 8.5 \times 10^6\,Oe$. This is such a strong field that it cannot possibly be due to magnetic interaction between neighbouring atoms. A magnetic ion produces a dipole field (see §2.2) of order $\mu_B/a^3 = n\mu_B \approx 3000\,Oe$ at neighbouring lattice sites (a is lattice spacing).

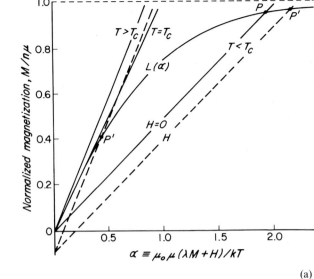

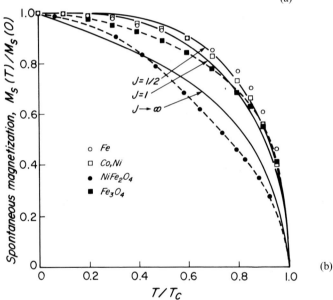

Figure 2.4 The Weiss molecular field theory of ferromagnetism. (a) The intersection P between $L(\alpha)$ (eqn. (2.16)) and one of the solid lines (the Weiss molecular field relation: eqn (2.23)) gives the spontaneous magnetization M_s. Intersections like P′ give M produced by an external field H, even above the Curie temperature T_C. (b) Predicted $M_s(T)$ curves for different values of J compared with measured $M_s(T)$ for some ferromagnetic metals and ferrimagnetic oxides.

Above T_C, there is no spontaneous magnetization in the absence of an applied field. However, adding an external field H to the molecular field will produce a magnetization M (Fig. 2.4a). In this case, solving (2.22) with $\alpha = \mu_0\mu(H + \lambda M)/kT$ for α small and using (2.24), we have

$$\chi_p = \frac{M}{H} = \frac{C}{T - T_C}. \tag{2.25}$$

This is the Curie–Weiss law. The magnetization is paramagnetic in that thermally disordered atomic moments are partly aligned by an external field, but the individual moments are now (exchange) coupled through the molecular field rather than mutually independent.

It is often convenient for theoretical calculations to have an analytic expression for $M_s(T)$. Let us make a critical-point expansion of (2.22) in the region of T_C, where α is small. Rewriting (2.22) and (2.23) in terms of reduced variables $M^* = M/n\mu$, $T^* = T/T_C$ and expanding $\tanh(\alpha)$ in powers of α, we obtain $M^* = \alpha T^* = \tanh(M^*/T^*) = \tanh(\alpha) \approx \alpha(1 - \alpha^2/3)$, or $\alpha^2 \approx 3(1 - T^*)$. From this it follows that $\alpha \propto (1 - T/T_C)^{1/2}$, or that

$$M_s(T) \propto (T_C - T)^\gamma, \tag{2.26}$$

with $\gamma = \frac{1}{2}$ in the molecular field approximation. Experimentally an expression like (2.26) fits the $M_s(T)$ data of ferromagnetic and ferrimagnetic substances quite well, not only near T_C but at ordinary temperatures as well. For magnetite, $\gamma = 0.43$ gives the best fit (see Fig. 3.5), while for iron, $\gamma = 0.34$.

2.6   Exchange coupling and exchange energy

We have seen that magnetic interactions between neighbouring atoms are inadequate by many orders of magnitude to explain the strength of the molecular field $H_m(\approx 10^7 \, \text{Oe or} \approx 10^9 \, \text{A/m})$. Electrostatic Coulomb interactions, which are responsible for chemical bonding, are the other likely possibility. Within an atom, there is considerable overlap of orbitals. It is reasonable that electrons should interact electrostatically at such close quarters. This coupling is the underlying reason why s and l vectors of different electrons couple in the way described in §2.3 to give S, L and J for the whole atom. What is surprising is that 3d electrons of *neighbouring* atoms can have a strong spin-dependent interaction. This must be the case because ferromagnetism (or ferrimagnetism: see §2.7) is a property of electrically neutral atoms (e.g., Fe^0) as well as ions (e.g., Fe^{2+} and Fe^{3+} in Fe_3O_4). The 3d rather than the 4s valence electrons must be the key. (In the band theory of ferromagnetism, there is a mixing of states, with electrons in overlapping 3d and 4s orbitals spending part of their time in either state; good accounts are found in Kittel (1976), Cullity (1972) and Chikazumi (1964).)

Heisenberg (1928) explained ferromagnetic exchange coupling as analogous to spin-dependent hydrogen bonding. We shall give a simple outline, following Stacey and Banerjee (1974). Consider 3d electrons 1 and 2 associated with atoms a and b. Because the electrons are indistinguishable and we assume the 3d orbitals of a and b overlap significantly, 1 will sometimes be associated with a and sometimes with b: in effect, electrons 1 and 2 can be exchanged between atoms a and b. Electrons 1 and 2 have the same quantum numbers l, m and n, and the orbital wave functions (eqn. (2.6)) Ψ_a and Ψ_b have the same mathematical form (except that they refer to different origins, the nuclei of a and b respectively). Electrons 1 and 2 can have either the same values of s (in which their spins are parallel) or different values (spins antiparallel). We want to show that the Coulomb energy of interaction between them is lower in one case than in the other.

Since the 3d orbitals overlap significantly, we can gauge the interaction from the wave function Ψ of the *pair* of electrons, which is the product of their individual wave functions. The product wave function Ψ expresses the joint probability of finding electron 1 at position r_1 and simultaneously electron 2 at position r_2. Consider first the spin parallel case. If both 1 and 2 have $s = +\frac{1}{2}$, their spin wave functions will be named α and their combined spin wave function is

$$\Psi_s(r_1, r_2) = \alpha(r_1) \cdot \alpha(r_2). \tag{2.27}$$

If 1 and 2 both have $s = -\frac{1}{2}$, with spin wave functions β,

$$\Psi_s(r_1, r_2) = \beta(r_1) \cdot \beta(r_2). \tag{2.28}$$

These wave functions are symmetric with respect to exchange of electrons 1 and 2. That is, if $r_1 \rightarrow r_2$ and $r_2 \rightarrow r_1$, Ψ_s is unchanged. However, electrons and other fermions have the property that the overall wave function Ψ, which is the product of Ψ_s and the orbital wave function Ψ_l of the pair of electrons, must be antisymmetric. Thus Ψ_l must be antisymmetric:

$$\Psi_l(r_1, r_2) = \Psi_a(r_1) \cdot \Psi_b(r_2) - \Psi_b(r_1) \cdot \Psi_a(r_2). \tag{2.29}$$

Notice that if $r_1 = r_2$, $\Psi_l = 0$. There is zero probability of electrons with spins parallel being in the same location at the same time. If they were, both could be associated with the same atom and would have the same set of quantum numbers n, l, m and s, in violation of Pauli's exclusion principle.

However, if Ψ_l is symmetric:

$$\Psi_l(r_1, r_2) = \Psi_a(r_1) \cdot \Psi_b(r_2) + \Psi_b(r_1) \cdot \Psi_a(r_2), \tag{2.30}$$

so that $\Psi_l \neq 0$ when $r_1 = r_2$ and electrons 1 and 2 can simultaneously be in the same vicinity, then we require Ψ_s to be antisymmetric:

$$\Psi_s(r_1, r_2) = \alpha(r_1) \cdot \beta(r_2) - \beta(r_1) \cdot \alpha(r_2). \tag{2.31}$$

In this case electrons 1 and 2 always have opposite spins ($s_1 = +\frac{1}{2}$, $s_2 = -\frac{1}{2}$, or vice versa when they exchange atoms), and there is no violation of Pauli's principle.

For s_1 parallel to s_2, the overall wave function $\Psi_{\parallel}$ is the product of Ψ_s from either (2.27) or (2.28) and Ψ_l from (2.29). If s_1 is antiparallel to s_2, the overall wave function is $\Psi_{-\parallel}$, the product of Ψ_s from (2.31) and Ψ_l from (2.30). $\Psi_{\parallel}$ is quite a different spatial function from $\Psi_{-\parallel}$. As a result, the Coulomb potential energy of interaction between electrons 1 and 2,

$$E_{el} = \int\int \frac{e^2 \Psi^2}{|r_1 - r_2|} \, dr_1 \, dr_2, \tag{2.32}$$

has different values for spins parallel and antiparallel. The difference is called the exchange integral J_e.

Whether parallel or antiparallel coupling is favoured (i.e., whether the exchange integral is positive or negative) depends crucially on the degree of overlap of 3d orbitals. For small values of a/r_{3d}, where a is interatomic spacing and r_{3d} is mean 3d orbital radius, antiferromagnetic or negative exchange coupling (§2.7) is favoured. For larger a/r_{3d}, i.e., less overlap, there is ferromagnetic or positive exchange coupling. At still larger a/r_{3d}, exchange coupling is lost and the neighbouring atoms behave paramagnetically. Among the transition metals, in order of increasing a/r_{3d}, Cr and Mn are antiferromagnetic (at least at low temperatures – the strength of exchange coupling determines the ordering temperature), Fe, Co and Ni are ferromagnetic, and Cu is paramagnetic.

In a solid, we are no longer concerned with isolated interactions between pairs of neighbouring 3d electrons, but with simultaneous exchange coupling between spins in all neighbouring atoms. The situation is analogous to the coupling by which lattice vibrations transmit elastic waves through a crystal. Standing waves in the lattice (phonons) have their counterpart in quantized spin waves (magnons). These normal modes are the fundamental thermal excitations of a ferromagnetic system. They also have particle properties, and can be excited during neutron bombardment of a magnetic crystal.

Spin waves (Fig. 2.5) are a long-range manifestation of the short-range (basically nearest neighbour) exchange force. Instead of each spin having only two choices, parallel or antiparallel to a field H (Fig. 2.2), neighbouring spins trace out a cone, each with the same component s_z along H. Because large numbers of atoms participate collectively, quantization requires only that the value of s_z

Side view

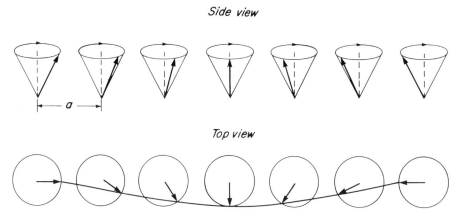

Top view

Figure 2.5 Sketch of a spin wave in a line of exchange-coupled spins, showing one half wavelength. [Redrawn after Kittel (1976) *Introduction to Solid State Physics*, 5th edition. Copyright © Wiley 1976. Reprinted by permission of John Wiley & Sons, Inc.]

characterizing the entire spin wave change by $\hbar/2$ from one mode to another. In a line of m coupled spins across a crystal, therefore, the angle between neighbouring atomic magnetic moments can be as small as $180°/m$. The exchange energy per pair of spins in a coupled lattice is then

$$E_{\text{ex}} = -2J_e S_i \cdot S_j = -2J_e S_i S_j \cos\theta_{ij}, \tag{2.33}$$

where S_i, S_j are the spins of neighbouring atoms i and j and θ_{ij} is the angle between them. Terms in the Coulomb energy which are independent of θ_{ij} have been ignored. Although quantum mechanical in origin, (2.33) can be interpreted and used as a classical energy in the applications we shall discuss.

Why does the Weiss molecular field theory work so well? The basic reason is that long-range coupling of atomic moments through spin waves ensures that the local value of M approximates to M elsewhere in the crystal, at least for the easily excited long wavelength modes. We can relate the Weiss and Heisenberg models quantitatively as follows. Suppose all atomic moments μ are aligned, so that $M = n\mu$. Then using (2.21) and (2.24), $E_m = -\mu_0\mu\lambda M = -kT_C$ for a particular atom. The molecular field is due to exchange coupling with the z nearest neighbours of this atom and so the exchange energy from (2.33) is $E_{\text{ex}} = -2zJ_e S^2$. Equating these energies, we have

$$J_e = \frac{kT_C}{2zS^2}. \tag{2.34}$$

The Curie temperature is therefore a direct measure of the strength of the exchange interaction.

2.7   Ferrimagnetism and antiferromagnetism

2.7.1   Magnetic sublattices

In terrestrial rocks, we are concerned with iron oxides, hydroxides and sulphides (Chapter 3). In these compounds, iron atoms are never close enough for direct exchange interaction. Instead, there is indirect exchange (or superexchange) interaction of 3d electrons through overlap of their orbitals with 2p orbitals of intervening oxygen ions or 3p orbitals of sulphur ions. O^{2-} ($1s^2 2s^2 2p^6$) and S^{2-} ($1s^2 2s^2 2p^6 3s^2 3p^6$) have filled 2p or 3p shells. A pair of 2p (or 3p) electrons with opposite spins is involved in the exchange coupling. One 'exchanges' its electron with a 3d electron of a neighbouring Fe^{2+} or Fe^{3+} ion and the other does the same with a 3d electron from another neighbouring ion. The 3d–2p (or 3p) coupling is negative in both cases, and so the effective coupling between the 3d electrons is the same as that between the 2p electrons – negative.

This mechanism describes the dominant negative interaction which establishes two magnetic sublattices A and B of oppositely directed spins and magnetic moments. The coupling, described by a molecular field coefficient λ_{AB}, is strongest if the Fe–O bonds make an angle close to $180°$ and weak if the angle is close to $90°$. In the cubic spinel lattice of magnetite (Fig. 3.4), the Fe–O bonds for cations in A (tetrahedrally coordinated) sites make angles of $125.2°$ with bonds for B or octahedral site cations. There are also weak negative AA and BB interactions between cations in the same sublattice. The bond angles are $79.6°$ for AA and $90°$ for BB.

Many years before the existence of magnetic sublattices was proven by neutron diffraction, Néel (1948) postulated their existence and explained the properties of what he named antiferromagnetic and ferrimagnetic substances (Fig. 2.3) with an adaptation of Weiss' molecular field model. In Néel's theory, sublattices A and B have oppositely directed magnetizations M_A and M_B, and each sublattice experiences its own molecular field:

$$H_A = -\lambda_{AA}M_A - \lambda_{AB}M_B, \quad H_B = -\lambda_{AB}M_A - \lambda_{BB}M_B. \tag{2.35}$$

Notice that the λ are positive numbers; the negative interactions are indicated by $-$ signs. A full mathematical treatment is beyond the scope of this book. The following simplified treatment due to Kittel (1976) gives considerable insight.

2.7.2   Ferrimagnetism

Consider first a ferrimagnetic substance, in which the A and B sublattices contain different numbers and kinds of cations. In magnetite the ionic distribution is $Fe^{3+}[Fe^{2+}Fe^{3+}]O_4^{2-}$, where octahedral or B-site cations are in brackets. The numbers and moments of Fe^{3+} ions on the two sublattices balance; the net magnetization $(M_A + M_B)$ is due to B-site Fe^{2+} ions. (However the exchange coupling and magnetic properties depend on all the cations, not just Fe^{2+}.) By analogy with eqn. (2.21), we can write the magnetic field energy density e_m (the energy E_m per unit volume) as

$$\begin{aligned} e_m &= -\tfrac{1}{2}\mu_0(M_A \cdot H_A + M_B \cdot H_B) \\ &= \tfrac{1}{2}\mu_0(\lambda_{AA}M_A^2 + 2\lambda_{AB}M_A \cdot M_B + \lambda_{BB}M_B^2). \end{aligned} \tag{2.36}$$

The factor $\tfrac{1}{2}$ appears because each interaction is being counted twice. Under what conditions is the antiparallel arrangement of M_A and M_B stable? Clearly if λ_{AA} becomes too large compared to λ_{AB}, the moments of A site cations will become disordered or mutually antiparallel and $M_A \to 0$. Likewise if λ_{BB} is large, $M_B \to 0$. Thus we need to know when e_m is less than zero, which is its value if $M_A, M_B \to 0$. For M_A antiparallel to M_B, we have

$$\tfrac{1}{2}\mu_0(\lambda_{AA}M_A^2 - 2\lambda_{AB}|M_A||M_B| + \lambda_{BB}M_B^2) < 0$$

or

$$\tag{2.37}$$

$$\lambda_{AB} > \frac{\lambda_{AA}M_A^2 + \lambda_{BB}M_B^2}{2|M_A||M_B|}.$$

Since λ_{AB} is $\gg$ either λ_{AA} or λ_{BB} for all ferrimagnetic oxides and sulphides we shall be concerned with, condition (2.37) is met. We can make a considerable simplification in what follows by ignoring the AA and BB interactions and considering only the AB interaction.

First let us look at the paramagnetism above, and in the limit at, the ferrimagnetic Curie point T_C. Atoms on the A and B sublattices have moments μ_A, μ_B, with numbers per unit volume n_A, n_B, respectively. Each sublattice follows a separate Curie law with Curie constants $C_A = n_A\mu_A^2\mu_0/k$, $C_B = n_B\mu_B^2\mu_0/k$:

$$M_A = \frac{C_A}{T}(H - \lambda_{AB}M_B), \quad M_B = \frac{C_B}{T}(H - \lambda_{AB}M_A). \tag{2.38}$$

As we approach any magnetic ordering temperature from above, χ_p becomes very large. In effect, as $T \to T_C$, we require finite sublattice magnetizations M_A, M_B even though the applied field $H \to 0$. Putting $H = 0$ and $T = T_C$ in (2.38) gives the simultaneous equations

$$M_A + \frac{C_A\lambda_{AB}}{T_C}M_B = 0, \quad \frac{C_B\lambda_{AB}}{T_C}M_A + M_B = 0, \tag{2.39}$$

which have non-zero solutions for M_A and M_B if and only if the determinant of the coefficients equals 0. Thus

$$T_C = \lambda_{AB}(C_A C_B)^{1/2} \tag{2.40}$$

is the ferrimagnetic Curie temperature. For non-zero H at $T > T_C$, we can solve the simultaneous eqns. (2.38) to find

$$\chi_p = \frac{M_A + M_B}{H} = \frac{(C_A + C_B)T - 2C_A C_B\lambda_{AB}}{T^2 - T_C^2}. \tag{2.41}$$

Unlike the situation for true paramagnetism, ferromagnetism and antiferromagnetism, inverse susceptibility χ^{-1} above the magnetic ordering temperature is not linear in T for ferrimagnetism (see Fig. 2.6).

Below T_C, there is a spontaneous magnetization M_s given by the sum of the sublattice magnetizations M_A and M_B. Because the A and B sites are not equivalent, $M_A(T)$ is not a mirror image of $M_B(T)$. Different combinations of n_A, n_B, μ_A, μ_B, λ_{AA} and λ_{BB} give rise to $M_s(T)$ variations of several types (Fig. 2.7). Q-type curves resemble those of true ferromagnets. The AB interaction is dominant and $M_A(T)$ is similar to $M_B(T)$. Magnetite is of this type (Figs. 2.4b, 3.5). R-type curves are found for many ferrites (e.g., $NiFe_2O_4$, Fig. 2.4b). P-type behaviour results from very different $M_A(T)$ and $M_B(T)$ variations. $M_s(T)$ has a hump, often at low temperature. (Apparent P-type behaviour can result from non-saturation of the magnetization by laboratory fields at low temperatures in hard, i.e., high-coercive-force, materials.) N-type behaviour is an extreme situation in which $M_A(T)$ and $M_B(T)$ are so different that M_s is in the direction of M_A at low temperatures but in the direction of M_B at high temperatures, or vice versa. Thus M_s 'self-reverses' in the course of heating or cooling. Certain

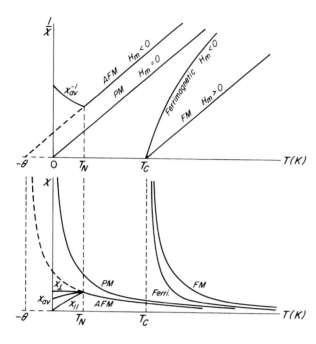

Figure 2.6 The temperature dependence of susceptibility χ and inverse susceptibility $1/\chi$ for antiferromagnetic (AFM), paramagnetic (PM), ferrimagnetic, and ferromagnetic (FM) materials. Positive and negative values of the molecular field H_m correspond to positive and negative exchange coupling.

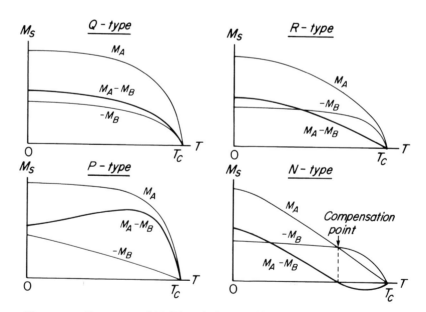

Figure 2.7 Four types of $M_\mathrm{s}(T)$ variation resulting from different temperature variations of the sublattice magnetizations M_A, M_B in ferrimagnetic materials.

natural chromites behave in this way, but their reversal temperatures are much below room temperature.

2.7.3   Antiferromagnetism

We conclude by considering antiferromagnetic substances. An antiferromagnet is a ferrimagnet in which $M_A = -M_B$ at all temperatures. Thus there is no observable M_s. Many natural magnetic minerals, among them hematite (αFe_2O_3), ilmenite ($FeTiO_3$), goethite ($\alpha FeOOH$), and pyrrhotite (Fe_7S_8) are perfect or imperfect (slightly ferrimagnetic) antiferromagnets. Although the sublattice magnetizations balance perfectly in the absence of an applied field, antiferromagnetic substances do have an induced magnetization in the presence of H.

The ordering temperature for antiferromagnets, called the Néel temperature T_N, is given by (2.40) with $C_A = C_B = C/2$:

$$T_N = (C/2)\lambda_{AB}, \tag{2.42}$$

again neglecting AA and BB interactions. Equation (2.41) then gives

$$\chi_p = \frac{C}{T + \Theta} \tag{2.43}$$

with $\Theta = T_N$ if $\lambda_{AA} = \lambda_{BB} = 0$. The difference between experimental values of Θ and T_N gives a measure of the importance of AA and BB interactions.

Below T_N, applying a field H perpendicular to the sublattice magnetizations deflects M_A and M_B so that each has a small component along H (Fig. 2.8). Magnetocrystalline anisotropy (§2.8) ensures that the deflection angle ϕ is small for all reasonable fields. Adapting (2.36), and again ignoring λ_{AA} and λ_{BB}, we have

$$\begin{aligned}
e_m &= -\mu_o[-\lambda_{AB}M_A \cdot M_B + (M_A + M_B) \cdot H] \\
&= -\mu_0(\lambda_{AB}M^2 \cos 2\phi + 2MH \sin \phi) \\
&\approx -\mu_0 M[\lambda_{AB}M(1 - 2\phi^2) + 2H\phi] \text{ for small } \phi, \tag{2.44}
\end{aligned}$$

with $|M_A| = |M_B| = M$. At equilibrium, $de_m/d\phi = 0$, giving $\phi_{eq} = H/2\lambda_{AB}M$. The perpendicular susceptibility is therefore

$$\chi_\perp = \frac{2M \sin \phi}{H} = \frac{1}{\lambda_{AB}}. \tag{2.45}$$

Notice that (2.42) and (2.43) give $\chi_p(T_N) = C/2T_N = 1/\lambda_{AB}$, which is equal to the (temperature-independent) value of $\chi_\perp$ below T_N (see Fig. 2.6).

Applying H parallel to one of the sublattice magnetizations applies zero torque to either M_A or M_B and should in principle induce no magnetization. This is true at $0\,K$, but at higher temperatures, local thermal rotations of M_A and M_B result in some torque and a small susceptibility which increases as T increases. In very strong fields, the two spin sublattices may rotate through $90°$, if permitted by the crystalline anisotropy, to become perpendicular to H. This phenomenon is called spin flopping.

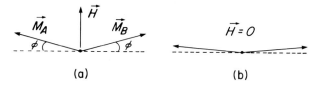

(a)   (b)

Figure 2.8   (a) Perpendicular susceptibility in an antiferromagnet: the sublattice magnetizations M_A and M_B are deflected through an angle ϕ by the field H. (b) Canted antiferromagnetism: permanent deflections of the sublattice magnetizations when $H = 0$ (see also Fig. 2.3).

Spin flopping of a different sort may occur spontaneously even when $H = 0$ because different directions in the lattice ('easy axes') for M_A, M_B are favoured by magnetocrystalline anisotropy at different temperatures (see §2.8). The Morin transition just below room temperature in hematite is such a spin-flop transition (see Figs. 3.20, 3.21). Above T_{Morin}, hematite has a weak 'parasitic' ferromagnetism, which arises from **spin-canting** (Dzyaloshinsky, 1958). The sublattice magnetizations are permanently deflected through a small angle $\phi(\approx 0.2°)$ in the absence of any external field H, producing a small M_s vector perpendicular to M_A and M_B. Although the ferromagnetism of hematite is weak, it is extremely important in recording ancient geomagnetic fields.

2.8   Magnetocrystalline anisotropy and magnetostriction

2.8.1   Macroscopic description of magnetocrystalline anisotropy

Certain directions in a crystal, called easy axes, are preferred by M_s (or by M_A and M_B in ferrimagnetic and antiferromagnetic materials). Initial magnetization curves (Fig. 2.9) show that it is considerably more difficult to saturate the magnetization of a magnetite crystal along $\langle 100 \rangle$ than along $\langle 111 \rangle$. The magnetic potential energy required to set up a field H with magnetization M in an initially unmagnetized sample is

$$\int H \cdot dB = \int \mu_0 H \, dH + \int \mu_0 H \cdot dM = \tfrac{1}{2}\mu_0 H^2 + \mu_0 \int H \, dM, \qquad (2.46)$$

assuming M is parallel to H (M is an average over all domains). The first term exists in all space, but the second is specific to the magnetized material. It equals the area between either of the magnetization curves in Fig. 2.9 and the M axis. This energy is clearly larger for the hard $\langle 100 \rangle$ axis than for the easy $\langle 111 \rangle$ axis. The difference between the two energies when the magnetization is taken to saturation (the area between the curves) is a measure of the **magnetocrystalline anisotropy**.

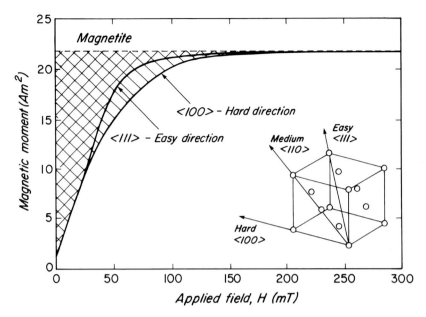

Figure 2.9 Magnetization curves for magnetite (Fe_3O_4) along $\langle 111 \rangle$ (easy) and $\langle 100 \rangle$ (hard) directions. The area between the two curves is the difference in magnetic potential energy between the hard- and easy-direction magnetization functions (hatched regions). It equals the magnetocrystalline anisotropy energy. [After Moskowitz (1992), with the permission of the author.]

Magnetocrystalline anisotropy energy E_K is usually much less than its maximum value because M_s is usually close to an easy axis. In a cubic crystal,

$$E_K = K_1 V(\alpha_1^2\alpha_2^2 + \alpha_2^2\alpha_3^2 + \alpha_3^2\alpha_1^2) + K_2 V\alpha_1^2\alpha_2^2\alpha_3^2, \qquad (2.47)$$

where K_1 and K_2 are anisotropy constants, V is sample volume, and the α_i are direction cosines of M_s with respect to the $\langle 100 \rangle$ axes. Equation (2.47) applies within any region of uniform magnetization, e.g., within individual domains even when the sample magnetization M is far from saturation. There is no second-order term in (2.47) because $\alpha_1^2 + \alpha_2^2 + \alpha_3^2 = 1$, independent of orientation. If $K_1 > 0$, as for iron and high-Ti titanomagnetites, the minimum E_K is 0, for M_s along a $\langle 100 \rangle$ (easy) axis (e.g., $\alpha_1 = 1$, $\alpha_2 = \alpha_3 = 0$). The maximum E_K is $K_1 V/3 + K_2 V/27$, for M_s along a $\langle 111 \rangle$ hard axis ($|\alpha_1| = |\alpha_2| = |\alpha_3| = 1/\sqrt{3}$). The difference between them (cf. Fig. 2.9) is $K_1 V/3 + K_2 V/27$. If $K_1 < 0$, as for magnetite and low-Ti titanomagnetites, the easy axes are $\langle 111 \rangle$, with $E_K = -|K_1|V/3 + K_2 V/27$, and the hard axes are $\langle 100 \rangle$, with $E_K = 0$. K_1 and K_2 are usually determined by Fourier analysis of the angular dependence of the torque exerted by a strong applied field H on a magnetized sample, or else from ferromagnetic resonance.

Some substances have only one easy axis. Their **uniaxial anisotropy** is described by

$$E_K = K_{u1} V \sin^2 \theta + K_{u2} V \sin^4 \theta, \tag{2.48}$$

where $\theta = 0$ or $180°$ are equivalent directions along the easy axis. Non-magneto-crystalline sources of uniaxial anisotropy include magnetoelastic anisotropy (§2.8.4) and shape anisotropy of strongly magnetic minerals in elongated grains (§4.2.4). For simplicity, (2.48) or its equivalent is sometimes used in micromagnetic modelling (Chapter 7) of cubic minerals instead of (2.47).

Above T_{Morin}, a strong anisotropy confines the magnetic sublattices of hematite (αFe_2O_3) to the rhombohedral basal plane, perpendicular to the (hard) c-axis. The parasitic ferromagnetism M_s, perpendicular to the sublattices, is also confined to the basal plane. There is a weaker anisotropy with hexagonal symmetry within the basal plane, defining three easy axes for M_A and M_B. The directions perpendicular to these axes are in effect the easy directions for M_s. We have for E_K (Dunlop, 1971)

$$E_K = KV \sin^2(3\theta), \tag{2.49}$$

$\theta = 0$ being a reference easy direction of M_s.

2.8.2   Microscopic view of magnetocrystalline anisotropy

Thus far, we have taken a phenomenological view, based on macroscopic experimental results. Microscopically crystalline anisotropy has two sources, often referred to as single-ion anisotropy and dipolar anisotropy. In §2.3, we saw that $L = 0$ for a half-filled 3d shell (such as $Fe^{3+}:3d^5$), while $J = L + S$ (i.e., L is coupled parallel to S) if the 3d shell is more than half-filled (e.g., $Fe^{2+}:3d^6$). Although L is said to be quenched, there is some residual L–S coupling. L is coupled to the crystal lattice through ionic bonding, the magnetic moments are controlled mainly by S (actually it is individual electron spins and moments that are exchange coupled), and the weak L–S coupling provides the link between moments and preferred directions in the lattice. This effect is called single-ion anisotropy because an ion like Fe^{2+} will experience an equivalent lattice coupling even if its surroundings change (different neighbouring anions and cations, different lattices). The essential link is the L–S coupling within the single Fe^{2+} ion. Fe^{3+} has little or no single-ion anisotropy because $L = 0$.

Dipolar anisotropy does not actually arise from magnetic interaction of atomic dipoles (this coupling is too weak), but the anisotropic exchange interactions that are responsible have an angular dependence that is pseudo-dipolar or pseudo-quadrupolar. Consider the atomic dipoles shown in Fig. 2.10. A central dipole is surrounded by three of its six nearest neighbours on a simple cubic lattice. All moments μ are deflected within the y–z plane by an angle θ from the z-axis. The dipole–dipole interaction energy between the central dipole 0 and dipole 1 at distance a along the x axis is

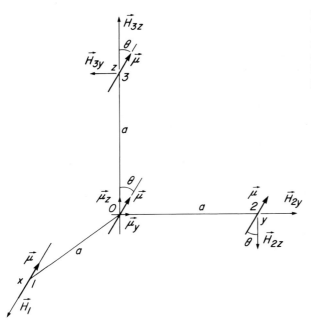

Figure 2.10 Illustration of dipolar anisotropy: a magnetic dipole and three of its six nearest neighbours on a simple cubic lattice.

$$E_{01} = -\mu_0 \boldsymbol{\mu} \cdot \boldsymbol{H}_1 = -\frac{\mu_0 \mu^2}{4\pi a^3} \sin 180° = +\frac{\mu_0 \mu^2}{4\pi a^3}, \tag{2.50}$$

since dipoles 0 and 1 are parallel and separated by a line perpendicular to either of them (cf. Fig. 2.1 for the calculation of the field $\boldsymbol{H}_1$ due to dipole 0 at the location of dipole 1). To calculate interactions between dipole 0 and dipoles 2 and 3, it is easiest to treat the central dipole as the vector sum of two dipoles $\boldsymbol{\mu}_z = \mu \cos \theta \hat{k}$, $\boldsymbol{\mu}_y = \mu \sin \theta \hat{j}$, producing fields $\boldsymbol{H}_{2z}, \boldsymbol{H}_{2y}$ at the site of dipole 2 and $\boldsymbol{H}_{3z}, \boldsymbol{H}_{3y}$ at dipole 3 (see Fig. 2.10). Using eqn. (2.2), we have

$$E_{02} = -\mu_0 \boldsymbol{\mu} \cdot (\boldsymbol{H}_{2y} + \boldsymbol{H}_{2z}) = -\frac{\mu_0 \mu^2}{4\pi a^3}(2\sin^2 \theta - \cos^2 \theta) \tag{2.51}$$

$$E_{03} = -\mu_0 \boldsymbol{\mu} \cdot (\boldsymbol{H}_{3z} + \boldsymbol{H}_{3y}) = -\frac{\mu_0 \mu^2}{4\pi a^3}(2\cos^2 \theta - \sin^2 \theta). \tag{2.52}$$

Summing,

$$E_{01} + E_{02} + E_{03} = -\frac{\mu_0 \mu^2}{4\pi a^3}(\sin^2 \theta + \cos^2 \theta - 1) = 0. \tag{2.53}$$

The other three nearest neighbour interactions similarly sum to zero. This expresses a general result: parallel dipoles within spherical shells in a cubic lattice have interactions with a central dipole that sum to zero. Since there is no dependence on angle θ, such interactions cannot be a source of anisotropy and there is no second-order term in the macroscopic anisotropy energy (eqn. (2.47)), as we concluded earlier.

If the lattice is other than cubic, the interactions do not sum to zero and the energy has a dependence on θ. The same is true if the dipoles are not perfectly parallel (sometimes called **configurational anisotropy**).

Notice that (2.50), (2.51) and (2.52) are all of the form $-(2\mu_0\mu^2/4\pi a^3)\frac{1}{2}(3\cos^2\phi - 1)$ where ϕ is the angle between either dipole in a pair and the line joining their centres (for the pair $0 - 1$, $\phi = 90°$; for pair $0 - 2$, $\phi = 90° - \theta$; for pair $0 - 3$, $\phi = \theta$). We recognize $\frac{1}{2}(3\cos^2\phi - 1)$ as the degree $l = 2$ Legendre polynomial P_2^0. Thus interactions between parallel dipoles can be described as a function of the local pair angle ϕ by a quadrupolar Legendre polynomial. Similarly magnetic quadrupole interactions give rise to an $l = 4$ polynomial, $P_4^0(\cos\phi)$.

Van Vleck (1937) developed a pair model of anisotropic exchange coupling in which the pair interaction energy has exactly this form:

$$E_{ij} = E_{ex} + c_2 P_2^0(\cos\phi) + c_4 P_4^0(\cos\phi). \tag{2.54}$$

E_{ex} is the isotropic exchange energy of eqn. (2.33). As expected, when (2.54) is summed over nearest neighbours in simple cubic, body-centred cubic or face-centred cubic lattices, $c_2 = 0$. The dominant anisotropy energy is the pseudo-quadrupolar term $c_4 P_4^0$, which accounts for the term in K_1 in eqn. (2.47). In lattices of lower symmetry, the pseudo-dipolar ($l = 2$) term is much larger than the $l = 4$ term. For this reason, non-cubic magnetic minerals generally have larger magnetocrystalline anisotropy than cubic minerals (Table 3.1).

2.8.3 Temperature dependence of magnetocrystalline anisotropy

Magnetocrystalline anisotropy constants usually change much more rapidly with temperature than does M_s, particularly in cubic minerals (Fig. 2.11). There is a good reason for this. Changes in M_s, at ordinary temperatures anyway, are caused by small changes in dipole orientation from one atom to the next, specified by the local direction cosines β_1, β_2, β_3. Crystalline anisotropy depends on correlated changes in local angles (since two atoms are involved in anisotropic exchange) and for cubic crystals the anisotropy energy, to lowest order, at temperature T is

$$\begin{aligned} E_K(T) &= K_1(0)\langle\beta_1^2\beta_2^2 + \beta_2^2\beta_3^2 + \beta_3^2\beta_1^2\rangle \\ &= K_1(0)\langle P_4^0(\cos\psi)\rangle(\alpha_1^2\alpha_2^2 + \alpha_2^2\alpha_3^2 + \alpha_3^2\alpha_1^2), \end{aligned} \tag{2.55}$$

in which 0 refers to $T = 0$ K, $\langle\ \rangle$ indicates average, the α_i are the long-range average values of the β_i over the crystal, and ψ is the local angle between a moment $\boldsymbol{\mu}$ and the average $\boldsymbol{M}_s$ vector (i.e., the angle between the directions specified by β_i and α_i). Comparing (2.55) to (2.47), we see that $K_1(T) = K_1(0)\langle P_4^0(\cos\psi)\rangle = K_1(0)\langle(35\cos^4\theta - 30\cos^2\theta + 3)/8\rangle$.

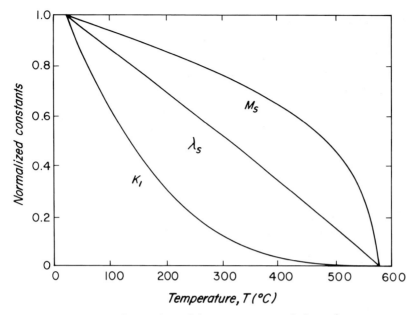

Figure 2.11 Comparison of the temperature variations of spontaneous magnetization M_s, magnetostriction constant λ_s, and magnetocrystalline anisotropy constant K_1 for magnetite. [After Banerjee (1991), with the permission of the author and the publisher, the Mineralogical Society of America, Washington, DC.]

The local deviations ψ from long-range parallelism of moments are due to thermal excitations. A simple classical approach (Akulov, 1936) is to assume that the ψ are distributed like independent paramagnetic moments, so that the probability that ψ falls between values ψ and $\psi + d\psi$ is proportional to $\exp(\alpha \cos \psi) \sin \psi \, d\psi$, with $\alpha \equiv \mu B / kT$ as in (2.15) and (2.16). (This is clearly wrong since the atomic moments are not independent but exchange coupled into spin waves, but fortunately the essential result does not depend on the distribution used in the averaging.) The result for large values of α (low temperatures) is

$$K_1(T)/K_1(0) = \langle P_4^0(\cos \psi) \rangle = 1 - 10/\alpha + 45/\alpha^2 + \dots \qquad (2.56)$$

In the same approximation, $M_s(T)$ is prescribed by the Langevin function (cf. (2.16)):

$$\frac{M_s(T)}{M_s(0)} = L(\alpha) = \coth(\alpha) - \frac{1}{\alpha} \approx 1 - \frac{1}{\alpha}. \qquad (2.57)$$

Comparing (2.56) and (2.57), we see that to first order in $1/\alpha$,

$$\frac{K_1(T)}{K_1(0)} = \left[\frac{M_s(T)}{M_s(0)}\right]^{10}. \qquad (2.58)$$

This 'law' is approximately obeyed at high as well as low temperatures. For magnetite, the experimentally determined power is between 8 and 9 (Fletcher and O'Reilly, 1974; Fletcher, 1975).

Callen and Callen (1966) review all the theories of the temperature dependence of K_1 and show that the general 'law' is

$$\frac{K_1(T)}{K_1(0)} = \left[\frac{M_s(T)}{M_s(0)}\right]^{l(l+1)/2}. \tag{2.59}$$

For cubic minerals, $l = 4$ is dominant in the anisotropy energy (eqn. (2.54)). Then we have $l(l+1)/2 = 10$, as in (2.58). For non-cubic minerals, $l = 2$ is dominant in (2.54) and $l(l+1)/2 = 3$, a much less radical temperature dependence. These 'laws' apply whether the source of K_1 is single-ion anisotropy or two-ion anisotropy (anisotropy exchange).

2.8.4   Magnetostriction and magnetoelastic anisotropy

Magnetostriction is a spontaneous change in the dimensions of a ferromagnetic or ferrimagnetic crystal when it is magnetized. If the crystal expands in the direction of magnetization, the magnetostriction is positive. In magnetite at room temperature, the magnetostriction depends crucially on direction, being positive for magnetization along a $\langle 111 \rangle$ easy axis but negative for magnetization along a $\langle 100 \rangle$ hard axis. Magnetocrystalline anisotropy is also very different for these two directions. Since both single-ion and dipolar two-ion anisotropies depend on interionic distance, the former by reason of the electrostatic crystal field due to neighbouring bonded anions and the latter because of exchange interaction, it is logical that a change in distances should cause a change in macroscopic magnetization (magnetoelastic effect) and vice versa.

Magnetoelasticity is fundamentally the strain dependence of crystalline anisotropy. Macroscopically, the saturation magnetostriction λ is the uniform and spontaneous longitudinal strain $\Delta L/L$ in a direction specified by direction cosines γ_i when the magnetization M (direction cosines α_i) changes from 0 to M_s. In cubic crystal, the anisotropy of λ is described by (e.g., Kittel, 1949, pp. 556–8)

$$\lambda = \tfrac{3}{2}\lambda_{100}(\alpha_1^2\gamma_1^2 + \alpha_2^2\gamma_2^2 + \alpha_3^2\gamma_3^2 - \tfrac{2}{3})$$
$$+ 3\lambda_{111}(\alpha_1\alpha_2\gamma_1\gamma_2 + \alpha_2\alpha_3\gamma_2\gamma_3 + \alpha_3\alpha_1\gamma_3\gamma_1), \tag{2.60}$$

where $\lambda_{100}(\lambda_{111})$ is the saturation magnetostriction measured in the $\langle 100 \rangle (\langle 111 \rangle)$ direction for M in the same direction. Averaged over randomly oriented crystals, e.g., in a rock, the 'isotropic' magnetostriction λ_s is

$$\lambda_s = \tfrac{2}{5}\lambda_{100} + \tfrac{3}{5}\lambda_{111}. \tag{2.61}$$

When a crystal deforms, whether spontaneously as a result of changing M or because of stress σ (either externally applied or internal), both the elastic strain energy E_{el} (described by elastic constants c_{ij}) and the magnetoelastic strain energy E_{me} (described by magnetoelastic coupling constants B_1, B_2) change. Kittel (1949) and Ye *et al.* (1994) show that the magnetostriction constants are related to these more fundamental constants by

$$\lambda_{100} = -\frac{2B_1}{3(c_{11} - c_{12})}, \quad \lambda_{111} = -\frac{B_2}{3c_{44}} \tag{2.62}$$

and that, for zero stress σ, the total magnetocrystalline plus magnetoelastic energy is given by

$$E'_{K} = E_K + E_{el} + E_{me} = K'_1 V (\alpha_1^2 \alpha_2^2 + \alpha_2^2 \alpha_3^2 + \alpha_3^2 \alpha_1^2), \tag{2.63}$$

where

$$K'_1 - K_1 = K_\lambda = \tfrac{9}{4}[(c_{11} - c_{12})\lambda_{100}^2 - 2c_{44}\lambda_{111}^2]. \tag{2.64}$$

Equation (2.63) has the same form as eqn. (2.47), but the crystalline anisotropy K_1 is augmented by a magnetoelastic or magnetostrictive anisotropy K_λ. Since torque curves are measured in a zero-stress (atmospheric pressure) rather than a zero-strain condition, K'_1 rather than K_1 is the quantity determined (Ye *et al.*, 1994). On the other hand, ferromagnetic resonance (FMR) probably responds to K_1. The difference is about 20% for magnetite: $K'_1 = -1.35 \times 10^4 \, \text{J/m}^3$ according to torque curves (Table 3.1) but $K_1 = -1.12 \times 10^4 \, \text{J/m}^3$ from FMR (Bickford, 1950).

However, in the case of TM60, Sahu and Moskowitz (1995) find that $K_\lambda > K_1$ at most temperatures. The strain-dependent part of crystalline anisotropy is more important than the strain-free part in this mineral. This finding has important consequences. Because the strained lattice is no longer perfectly cubic, $l = 2$ terms become significant. For this reason, magnetostriction constants in all minerals, and K'_1 in minerals where K_λ is large, change more gradually between room temperature and the Curie point than K_1 does. In the example of Fig. 2.11, λ_s of magnetite goes as $M_s(T)$ to the power 2.5 approximately (Klapel and Shive, 1974; Moskowitz, 1993a), whereas K_1 decreases with T as M_s to the power 8–9.

In TM60, λ_{100}, λ_{111} and λ_s all decrease as $M_s^3(T)$ approximately and K'_1 decreases as $M_s^6(T)$ (Moskowitz, 1993a; Sahu and Moskowitz, 1995; see Fig. 3.13). Any additional strain, due to internal or external stress, will make magnetoelastic anisotropy even more important compared to K_1. For this reason, domain patterns in TM60 are frequently stress-controlled rather than intrinsic, i.e., controlled by K_1 (see Chapter 6).

Both K_1 and K_λ are positive in TM60, but the sign of K_1 becomes negative in titanomagnetites with composition around $x = 0.55$. It is then quite possible that the different temperature variations of K_1 and K_λ may result in a change in sign of K'_1 between T_0 and T_C. That is, in some compositions of titanomagnetite, there may be an isotropic point, like that in magnetite (§3.2.2) but *above room temperature*, with devastating consequences for TRM production.

Very often we are interested in the change of magnetization M that results from an imposed strain, due to stress σ, rather than vice versa. If the strain is uniform, due to a uniform applied stress or to magnetostrictive mismatch between adjacent domains with different magnetization directions, e.g., between closure and body domains (Fig. 1.2), and if we make the approximation of isotropic magnetostriction, eqn. (2.60) reduces to

$$\lambda = \tfrac{3}{2}\lambda_s[(\alpha_1\gamma_1 + \alpha_2\gamma_2 + \alpha_3\gamma_3)^2 - \tfrac{1}{3}] = \tfrac{3}{2}\lambda_s(\cos^2\theta - \tfrac{1}{3}), \qquad (2.65)$$

where θ is the angle between M and the direction in which $\Delta L/L$ is measured. Then if a uniform tension σ is applied in the same direction, the stress-dependent part of E'_K is given by

$$E_\sigma = \lambda\sigma V = \tfrac{3}{2}\lambda_s\sigma V \sin^2\theta, \qquad (2.66)$$

apart from a constant term.

Equation (2.66) has the same form as eqn. (2.48). E_σ is a uniaxial anisotropy, with an equivalent anisotropy constant $K_{ul} = (3/2)\lambda_s\sigma$, and an easy axis $\theta = 0$ corresponding to the axis of tension in materials with positive λ_s. The assumption of isotropic magnetostriction is not a good one, in magnetite or most other minerals, but the problem is otherwise almost intractable.

Frequently we are interested in local strains and magnetization deflections due to the non-uniform stress fields around dislocations or other crystal defects. These cannot be treated by macroscopic equations like (2.66). In Chapters 9 and 11, we will consider how to address these problems.

Chapter 3

Terrestrial magnetic minerals

3.1 Introduction

The most important terrestrial magnetic minerals are oxides of iron and titanium. Their compositions are conveniently represented on a Ti^{4+}–Fe^{2+}–Fe^{3+} ternary diagram (Fig. 3.1). The **titanomagnetites** (TM for short) are cubic minerals with inverse spinel structure (see §3.2). The mole per cent of Ti^{4+} is measured by the composition parameter x. TM0 (i.e., titanomagnetite with $x = 0$) is magnetite and TM100 is ulvöspinel. The **titanomaghemites** are also spinel minerals. They are cation-deficient oxidation products of titanomagnetites. The degree of oxidation is measured by the oxidation parameter z. The **titanohematites** (often called hemoilmenites after the end-member minerals hematite and ilmenite) are also oxidized equivalents of the titanomagnetites, but their crystal structure is rhombohedral. Notice that minerals of the same composition but different structures occupy the same point on the ternary diagram. For example, maghemite (cubic or γFe_2O_3) and hematite (rhombohedral or αFe_2O_3) both plot in the lower right corner.

The titanomagnetites form a complete solid-solution series for all values of x at very high temperatures (see Lindsley (1976, 1991) for an equilibrium phase diagram) but intermediate compositions can only be preserved as single-phase minerals at ordinary temperatures if they cooled very rapidly. Submarine pillow lavas, for example, have been quenched by extrusion into seawater. Their primary magnetic oxides are single-phase TM60 grains. If the same basaltic magma cools more slowly in an oxygen-poor setting, the primary oxide will not be a single-phase mineral but an exsolution intergrowth of low-x (near-magnetite) and high-x (near ulvöspinel) cubic minerals. Similarly titanohematites of

45

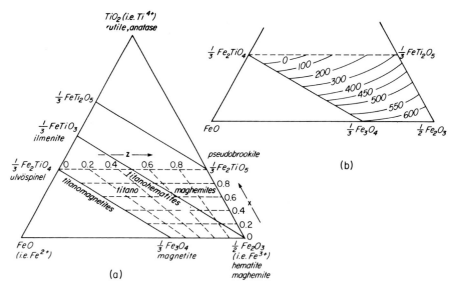

Figure 3.1 (a) The TiO_2–FeO–Fe_2O_3 ternary diagram, showing the titanomagnetite and titanohematite solid-solution lines and the titanomaghemite field. During either low-temperature or high-temperature oxidation of titanomagnetites, the bulk composition follows the horizontal dashed lines, although the oxidation product in the latter case is not a single-phase spinel. (b) Curie temperature contours for synthetic titanomaghemites. [After O'Reilly (1976) and Readman and O'Reilly (1972), with the permission of the authors and the publishers, Institute of Physics Publishing, Bristol, UK and Journal of Geomagnetism and Geoelectricity, Tokyo, Japan.]

intermediate composition tend to exsolve into intergrown iron-rich (near-hematite) and titanium-rich (near ilmenite) rhombohedral phases.

Magnetite-ulvöspinel intergrowths are rather uncommon in nature because there is usually enough oxygen in the melt to oxidize the titanomagnetites. This process is **high-temperature oxidation**, also called **deuteric oxidation** if it occurred during initial cooling of the rock rather than in later high-temperature metamorphism. During both high-temperature oxidation and **low-temperature oxidation** or **maghemitization** (the conversion of titanomagnetite to titanomaghemite, usually as a result of secondary weathering at ordinary temperatures), the bulk compositions of titanomagnetite grains follow the same oxidation lines (Fig. 3.1), but the resulting phase assemblages are entirely different. Low-temperature oxidation converts a single-phase spinel to another single-phase spinel with a different lattice parameter. High-temperature oxidation results in intergrown spinel (near-magnetite) and rhombohedral (near-ilmenite) phases (Fig. 3.2). The intergrowths mimic the exsolution texture of intergrown phases with the same crystal structure, and the process is often called **oxyexsolution**.

Figure 3.2   Electron micrograph of trellis-like oxyexsolution intergrowth texture in high-temperature oxidized titanomagnetite. Magnification is ×10 400. [After Gapeyev and Tsel'movich (1983) © American Geophysical Union, with the permission of the authors and the publisher.]

Oxyexsolution of titanomagnetite is one of many **topotactic transformations** of iron oxides and hydroxides. In a topotactic transformation, one phase converts to another while preserving some of the original crystalline planes and directions. The common iron oxides and hydroxides are all composed of different stackings of close-packed oxygen/hydroxyl sheets, with various arrangements of the iron (and titanium) ions in octahedral and tetrahedral (cf. §2.7, 3.2) interstitial sites (Bernal *et al.*, 1959). For example, orthorhombic goethite (αFeOOH) transforms to rhombohedral αFe_2O_3 by removing the hydroxyl sheets and some of the oxygen in strips parallel to the c-axis to form water. The [100], [010] and [001] axes of goethite become the [111], [110] and [112] axes of hematite (Bate, 1980). The low-temperature oxidation of Fe_3O_4 to γFe_2O_3 is topotactic, as is the subsequent inversion of γFe_2O_3 to αFe_2O_3. Figure 3.3 illustrates the crystallographic orientation of growing hematite platelets relative to the {111} planes of a maghemite or magnetite octahedron. The hematite [0001] axis forms

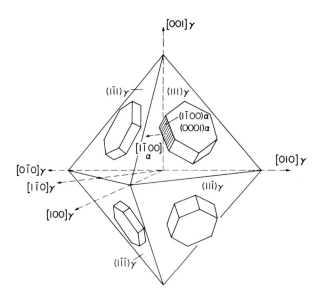

Figure 3.3 An example of a topotactic transformation: hexagonal hematite crystallites forming with their *c*-planes in {111} planes of maghemite. [After Kachi *et al.* (1963), with the permission of the publisher, The Physical Society of Japan, Tokyo, Japan.]

parallel to the spinel [111] axis, and the rhombohedral [1100] and spinel [110] axes coincide (Bate, 1980).

The remainder of this chapter discusses the properties of single-phase iron oxides, hydroxides and sulphides. Table 3.1 provides a summary of their basic material and magnetic properties (see also Hunt *et al.*, 1995a). The reader should remember, however, that multiphase assemblages and phase transformations are common in nature. These have fascinating properties and consequences, which we will focus on later (e.g., interactions, §11.7; self-reversal, §14.5; crystallization remanent magnetization or CRM, Chapter 13).

3.2   Magnetite

3.2.1   Crystal structure and magnetic sublattices

Magnetite ($Fe^{2+}Fe_2^{3+}O_4$), or an exsolved phase close to TM0 in composition, is the single most important magnetic mineral on earth. It occurs on the continents and in the ocean crust as a primary or secondary mineral in igneous, sedimentary, and low- and high-grade metamorphic rocks. For reviews of the crystal chemistry and magnetic properties of magnetite and related spinels, see Gorter (1954), Blasse (1964) and Banerjee and Moskowitz (1985).

Magnetite is a cubic mineral with spinel structure (Fig. 3.4). Oxygen anions form a face-centred cubic lattice, with Fe^{2+} and Fe^{3+} cations in interstitial sites. A unit cell, with lattice constant $a = 8.396$ Å, consists of four units like the one illustrated. The unit cell contains 8 cations on A sites, each surrounded by $4\,O^{2-}$ anions at the corners of a tetrahedron (e.g., site A1 in Fig. 3.4), and 16 cations on

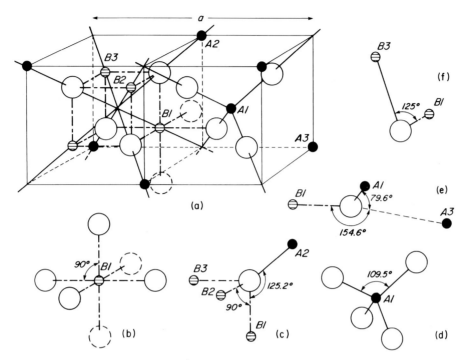

Figure 3.4 (a) Sketch of $\frac{1}{4}$ of a unit cell of magnetite. The lattice parameter is a. Solid and hatched circles represent cations in tetrahedral (A-site) and octahedral (B-site) coordination, respectively, with O^{2-} ions (large open circles). (b)–(f) Bond angles for specific cation pairs in (a). Bond angles near 90° are unfavourable for superexchange coupling. [After Gorter (1954), with the permission of the publisher, Philips Research, Eindhoven, The Netherlands.]

B sites, each surrounded by 6 O^{2-} ions at the corners of an octahedron (e.g., site B1 in Fig. 3.4). Tetrahedral and octahedral bonds are in $\langle 111 \rangle$ and $\langle 100 \rangle$ directions, respectively (Fig. 3.4b,d). Therefore the nearest-neighbour $B - O^{2-} - B$ and $A - O^{2-} - B$ bond angles should be 90° and 125.2°, respectively (Fig. 3.4c).

In the ideal cubic close-packed arrangement of Fig. 3.4, the tetrahedral interstices between oxygen ions are smaller than the octahedral interstices, too small in fact to accommodate Fe^{2+} or Fe^{3+} cations. By displacing the oxygen ions slightly, the lattice can accommodate Fe^{3+} (ionic radius 0.67 Å) in A sites, while the larger Fe^{2+} (ionic radius 0.83 Å) and the remaining Fe^{3+} enter B sites. This cation distribution, in which divalent cations are confined to octahedral sites while trivalent cations populate A and B sites equally, defines an **inverse spinel**. At 0 K, the net magnetic moment per formula weight of Fe_3O_4 is $4.1\mu_B$, close to the moment of Fe^{2+} ($S = 2, \mu = 4\mu_B$). This is as expected since Fe^{3+} ($S = 5/2, \mu = 5\mu_B$) occurs in equal numbers on both sublattices.

The bond angles change only slightly when the oxygen lattice distorts. The strongest exchange interaction is the negative AB coupling at 125.2° (e.g.,

A2 — O^{2-} — B1, Fig. 3.4c). This defines the A and B magnetic sublattices and gives magnetite its ferrimagnetic character. The BB interaction (e.g., B1 — O^{2-} — B2, Fig. 3.4c) is less effective because of the unfavourable 90° bond angle. Interactions that include one next-nearest-neighbour anion-cation pair are weak because the 3d-2p orbital overlap for the longer bonds is too small for effective exchange coupling. These interactions (in order of decreasing exchange coupling) and their bond angles are AA at 79.6° (e.g., A1 — O^{2-} — A3, Fig. 3.4e), AB at 154.6° (e.g., A3 — O^{2-} — B1, Fig. 3.4e) and BB at 125.0° (e.g., B1 — O^{2-} — B3, Fig. 3.4f).

Because BB_{90} is negative while $AA_{79.6}$ is weakly positive, the B sublattice magnetization M_B decreases somewhat more rapidly than M_A as the temperature rises (Fig. 3.5). $AB_{125.2}$ is much larger than either, however. The net magnetization $M_s(T)$ is of Q-type, similar to that of true ferromagnets (Fig. 2.4b). At room temperature, M_s is 480 kA/m or emu/cm^3. The Curie temperature T_C is 580 °C (Table 3.1).

Below about 120 K, there is an ordered arrangement of Fe^{2+} and Fe^{3+} ions on the octahedral sublattice and the unit cell is distorted very slightly from cubic to monoclinic symmetry. Above ≈120 K, electron hopping from Fe^{2+} to neighbouring B-site Fe^{3+}, converting the Fe^{2+} to Fe^{3+} and vice versa, destroys the cation ordering. All $\langle 100 \rangle$ directions are then equivalent and the lattice is perfectly cubic. At the **Verwey transition** between these states, the increased electron mobility converts magnetite from an electrical insulator to a semiconductor.

3.2.2   Magnetocrystalline anisotropy

The change in lattice at the Verwey transition temperature, $T_V \approx 120$ K, is too small to significantly affect the AB interaction and there is little if any change in M_s. However the Fe^{2+} single-ion anisotropy, which is the source of a relatively large positive K_1 below T_V, is greatly affected. Above T_V, because of electron hopping, there are no longer localized Fe^{2+} ions. The B sublattice can be thought of as being populated by $Fe^{2.5+}$ ions: each lattice site is occupied about half the time by Fe^{2+} and about half the time by Fe^{3+}. The crystalline anisotropy is much reduced and has contributions from both octahedral and tetrahedral cations. The net K_1 is negative (the room-temperature value is -1.35×10^4 J/m^3: see Table 3.1).

Magnetic properties which depend on crystalline anisotropy, such as remanent magnetization, susceptibility and coercive force in multidomain or equidimensional single-domain grains, change abruptly around T_V (Özdemir *et al.*, 1993; Fig. 3.6a). This property is a useful means of detecting magnetite in rocks. It is also the basis of selective **low-temperature demagnetization** of remanence carried by such grains (§12.4) by cycling them through T_V in zero field. Elongated single-domain particles in principle should not exhibit a transition in their

Table 3.1 Magnetic and X-ray properties of some common minerals

Mineral	Composition	M_s (kA/m or emu/cm³)	T_C (°C)	K_1 (J/m³)	$\lambda \times 10^{-6}$	a (Å)	Density (g/cm³)
Iron	αFe	1715	765	4.8×10^4 $\pm 0.5 \times 10^4$ (K_2)	-7 -21 (λ_{111}) 21 (λ_{100})	2.886	7.874
Magnetite	Fe_3O_4	480	580	-1.35×10^4 $-0.28 \times 10^4 (K_2)$	35.8 72.6 (λ_{111}) -19.5 (λ_{100})	8.396	5.197
Maghemite	γFe_2O_3	380	590–675 (Table 3.2)	-4.65×10^3	-8.9	8.337 (a_0) 24.99 (c_0)	5.074
Titanomagnetite	$Fe_{2.4}Ti_{0.6}O_4$	125	150	2.02×10^3	114 142.5 (λ_{100}) 95.4 (λ_{111})	8.482	4.939
Hematite	αFe_2O_3	≈2.5	675	1.2×10^6 (c-axis)	8	5.034 (a_0) 13.749 (c_0)	5.271
Goethite	$\alpha FeOOH$	≈2	120			4.596 (a_0) 9.957 (b_0) 3.021 (c_0)	4.264
Pyrrhotite	Fe_7S_8	≈80	320	$\approx 10^4$ (c-plane)	< 10	6.885 (a_0) 28.679 (c_0)	4.662
Greigite	Fe_3S_4	≈125	≈330	$\approx -10^3$		9.881	4.079

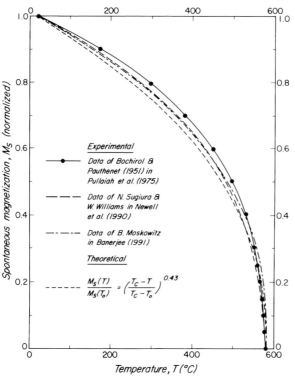

Figure 3.5 (a) Experimental data on the variation of spontaneous magnetization M_s with temperature in magnetite and a theoretical fit. (b) Sublattice magnetizations M_A, M_B as a function of temperature in magnetite. [After Stacey and Banerjee (1974), with the permission of the authors.]

(a)

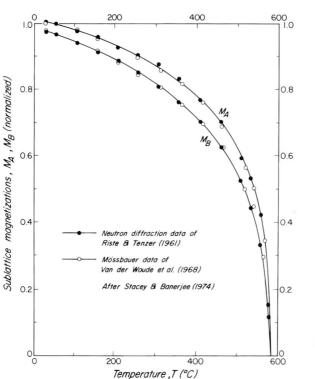

(b)

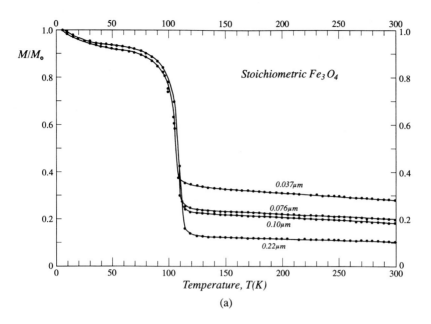

(a)

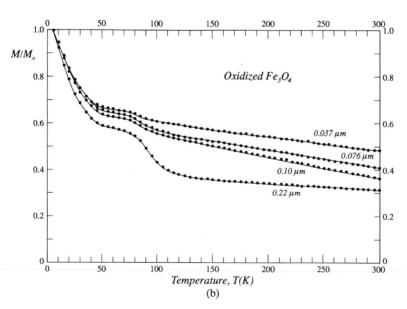

(b)

Figure 3.6 Saturation remanence acquired at 10 K monitored during zero-field heating to room temperature. The Verwey transition is expressed through the large drop in remanence around 110 K in stoichiometric magnetites, but is largely suppressed in non-stoichiometric (partially oxidized) magnetites. [After Özdemir *et al.* (1993) © American Geophysical Union, with the permission of the authors and the publisher.]

remanence or susceptibility because their anisotropy is determined by particle shape (§4.2.4) although the acicular magnetites used in magnetic recording, which contain chains of equidimensional crystallites, do exhibit a remanence transition. Some fraction of the initial remanence is always recovered after low-temperature cycling, whatever the size and shape of the magnetite grains. This fraction is called the **low-temperature memory**. Memory is largest for shape-controlled SD grains and decreases as grain size increases, but even mm-size magnetite crystals have some memory (Heider *et al.*, 1992).

The isotropic point T_I at which K_1 momentarily becomes zero as it changes from negative to positive during cooling (or vice versa during heating) is ≈ 135 K (Fig. 3.7a; Syono, 1965), about 15 °C above T_V as determined by changes in electrical conductivity or lattice structure. Furthermore K_2 does not change sign at the same temperature as K_1. The latter observation is not surprising since the anisotropies represented by K_1 and K_2 have different symmetries and origins (§2.8). The former observation implies that the onset of electron hopping is gradual rather than abrupt, and that a greater degree of electron mobility is required to suppress Fe^{2+} anisotropy (T_I) than to amplify conductivity (T_V).

There has been controversy about whether remanence transitions like those shown in Fig. 3.6 pinpoint T_V or T_I. Logically they should occur at or near T_I, since the pinning of remanence is much reduced when K_1 and $K_2 \rightarrow 0$ (the pinning would virtually disappear if K_1 and K_2 vanished at the same temperature). Most reported remanence transitions occur below 135 K, however. Some of the confusion may result from the variability of transition temperatures. Substitution of Ti^{4+} and other cations in the lattice inhibits Fe^{2+}–Fe^{3+} electron hopping and lowers T_I (Fig. 3.7a). Non-stoichiometry has a similar effect, because of the lattice vacancies introduced during low-temperature oxidation. Oxidation depresses, broadens, and eventually (beyond $z = 0.3$) eliminates the remanence transitions (Özdemir *et al.*, 1993; Fig. 3.6b), which are sharp in near-stoichiometric ($z < 0.1$) magnetites (Fig. 3.6a). This property may prove useful in detecting maghemitization of magnetite, for example in soils where the small quantities and fine particle sizes of the magnetic material make other methods difficult to apply. Grain size itself does not appear to be a factor. Large crystals of magnetite have remanence transitions very similar to those in Fig. 3.6 (except that their memories are much lower).

3.2.3   Magnetostriction

Magnetostriction makes a minor contribution to coercive force in magnetite. At room temperature, $\lambda_{111} = +72.6 \times 10^{-6}$, $\lambda_{100} = -19.5 \times 10^{-6}$ and $\lambda_s = (3\lambda_{111} + 2\lambda_{100})/5 = +35.8 \times 10^{-6}$ (Bickford *et al.*, 1955). The constant λ_s is the appropriate average for randomly oriented crystals (e.g., Fig. 2.11). Even a uniaxial stress $\sigma = 10$ MPa ($= 0.1$ kb or 10^8 dyne/cm^2), which is close to the

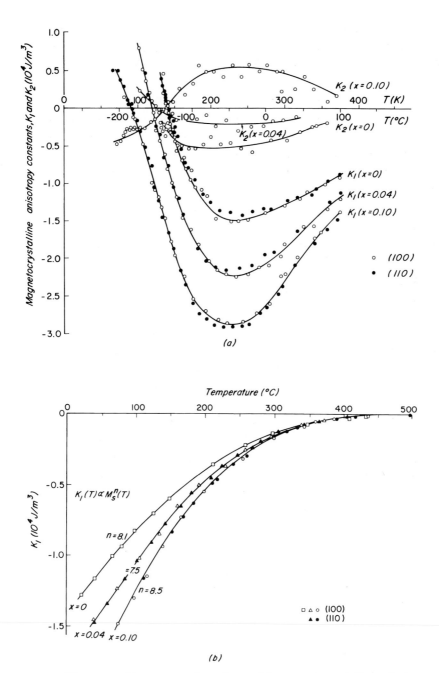

Figure 3.7   Temperature dependence of the magnetocrystalline anisotropy
constants K_1 and K_2 in magnetite ($x = 0$) and titanomagnetites near magnetite in
composition. Notice the isotropic point at which $K_1 = 0$ at $T_I = 135\,K(x = 0)$ to
$90\,K(x = 0.1)$. [After Syono and Ishikawa (1963a) and Fletcher and O'Reilly
(1974), with the permission of the authors and the publishers, The Physical Society
of Japan, Tokyo, Japan and Institute of Physics Publishing, Bristol, UK.]

breaking strength, only produces a critical field for single-domain magnetization rotation (§5.7.1) of $\mu_0 H_c = 3\lambda_s \sigma / M_s \approx 2\,\text{mT}$ or $20\,\text{G}$. The corresponding critical field due to magnetocrystalline anisotropy is approximately $0.5|K_1|/M_s \approx 15\,\text{mT}$ or $150\,\text{G}$.

The magnetostriction constants decrease in magnitude with heating above room temperature, less rapidly than K_1 but more rapidly than M_s (Figs. 2.11, 3.7b, 3.8b). Thus stress-induced coercivities will decrease with heating, although not as rapidly as those due to crystalline anisotropy. Below room temperature, λ_{100} is almost temperature independent while λ_{111} begins to decrease sharply below $\approx 200\,\text{K}$ (Fig. 3.8a). The sign of λ_{111} does not change at or near T_V, however. Any remanence pinned by stress, for example the local internal stresses around dislocations and other crystal defects, will be insensitive to cycling through T_V. Nuclei of original magnetization localized in highly stressed regions may be the key to the partial recovery of remanence in low-temperature demagnetization.

Hydrostatic or undirected stress has only a weak effect on the material constants of magnetite. According to Nagata and Kinoshita (1967), $|K_1|$ and $|K_2|$ decrease at a rate of $-5\%/\text{kb}$ while λ_{111} and $|\lambda_{100}|$ increase at a rate of $+15\%/\text{kb}$. The Curie temperature changes by only $\approx 2\,°\text{C}/\text{kb}$ (Schult, 1970).

Euhedral magnetite crystals in the form of cubes, octahedra or dodecahedra are often used in laboratory rock magnetic experiments and assumed in micromagnetic calculations (Chapter 7). In nature, less regular forms are common. Magnetites in lavas quenched from the melt may have skeletal, dendritic or other irregular shapes. Secondary magnetites in sedimentary and metamorphic rocks conform to the space available for their growth. Magnetites forming as oxyexsolution intergrowths (Fig. 3.2, §14.4.3) or within silicate host minerals (§14.6.3) may have tabular or rod-like forms. Any such irregularity of form will result in shape anisotropy that outweighs crystalline and magnetoelastic anisotropies.

3.3   Maghemite

3.3.1   Structure and saturation magnetization

Maghemite ($\gamma\text{Fe}_2\text{O}_3$) is the fully oxidized equivalent of magnetite. To emphasize its inverse spinel structure, we can write its formula as $\text{Fe}^{3+}[\text{Fe}^{3+}_{5/3}\square_{1/3}]\text{O}^{2-}_4$, where octahedral sites are enclosed by the brackets and $\square$ stands for a lattice vacancy. The vacancies are ordered on a tetragonal superlattice with $c = 3a$. Maghemite can be identified by the appearance of X-ray superlattice lines and by its smaller unit cell edge $a = 8.337\,\text{Å}$. Its Mössbauer spectrum is also distinctively different from that of magnetite.

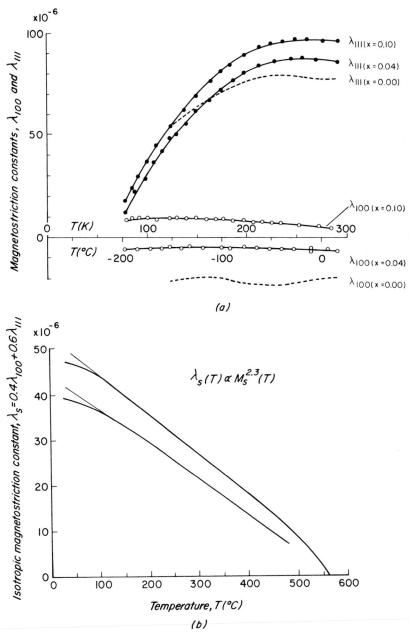

$$\lambda_s(T) \propto M_s^{2.3}(T)$$

(a)

(b)

Figure 3.8   Temperature dependence of the magnetostriction constants (a) of titanomagnetites near magnetite in composition below room temperature, and (b) of magnetite above room temperature. In (b), $\lambda_s(T)$ measurements were repeated 12 times on the same polycrystalline sample. The magnetostriction increased irreversibly between the first 8 runs (lower curve) and the final 4 runs (upper curve). [After Syono and Ishikawa (1963b) and Moskowitz (1993a) © American Geophysical Union, with the permission of the authors and the publishers, the Physical Society of Japan, Tokyo, Japan and The American Geophysical Union, Washington, DC.]

The ferrimagnetism of γFe_2O_3 is the result of an average excess per formula weight $Fe_{8/3}O_4$ of $\frac{2}{3}$ of an Fe^{3+} ion on the B sublattice. Per formula weight Fe_2O_3, this is equivalent to an average excess of $\frac{1}{2} Fe^{3+}$ and a net moment of $\frac{1}{2} \times 5\mu_B = 2.5\mu_B$. The measured moment at 0 K is $(2.36-2.38)\mu_B$ (Weiss and Forrer, 1929; Henry and Boehm, 1956). The 0 K saturation magnetization is $\sigma_s = (5585/M)n_B = 82.5 \, Am^2/kg$ or emu/g (the molecular weight M is 159.7 and the number of Bohr magnetons is $n_B = 2.37$). Takei and Chiba (1966) reported a 0 K moment of $2.9\mu_B$ from FMR measurements on epitaxial single-crystal films (Table 3.2). This high value may reflect film–substrate mismatch or a different vacancy distribution in films compared to particles.

At room temperature, the saturation magnetization of γFe_2O_3 is $\sigma_s = 74.3 \, Am^2/kg$ or $M_s = 380 \, kA/m$, compared to $480 \, kA/m$ for Fe_3O_4. The $M_s(T)$ curve of maghemite is similar in shape to that of magnetite. The Curie temperature T_C is $\approx 645 \, °C$ (Özdemir and Banerjee, 1984), higher than that of magnetite.

Maghemite is the ultimate low-temperature oxidation or weathering product of magnetite and is common in both subaerial and submarine environments. The presence of maghemite is particularly important in soil science because it gives information about the soil formation process and the nature of subsoil (Le Borgne, 1955). In paleomagnetism, the presence of γFe_2O_3 indicates an NRM that is mainly a CRM in origin.

Table 3.2 Spontaneous moment at 0 K (as n_B, the number of Bohr magnetons μ_B per formula weight) and Curie temperature T_C of maghemite

n_B (Bohr magnetons per molecule)	T_C (°C)	Authors
2.38		Weiss and Forrer (1929)
	675	Michel and Chaudron (1935)
	675	Maxwell et al. (1949)
	591	Michel et al. (1951)
2.36		Henry and Boehm (1956)
	590	Aharoni et al. (1962)
2.90	470	Takei and Chiba (1966)
	675	Frölich and Vollstädt (1967)
	695	Readman and O'Reilly (1972)
2.43		Mollard et al. (1977)
	645	Özdemir and Banerjee (1984)

3.3.2 Inversion and Curie temperatures

Cubic γFe_2O_3 is metastable and inverts to weakly magnetic rhombohedral αFe_2O_3 (hematite) when heated in vacuum or in air. The transformation is topotactic (§3.1), occurring by restacking of atomic planes rather than by wholesale recrystallization. Maghemite inversion consists of restacking of close-packed O^2-layers accompanied by displacement of interstitial Fe^{3+} ions (Kachi et al., 1963, 1971). The relationship between growing crystallites of hematite and the maghemite host is shown in Fig. 3.3.

The inversion temperature T_{inv} has been variously reported as occurring at 250 °C to $\geq$750 °C (Table 3.3). Naturally occurring maghemites have higher, often much higher, T_{inv} values than synthetic maghemites (e.g., Wilson, 1961; Oades and Townsend, 1963; Özdemir, 1990). Grain size, the degree of oxidation (z), and incorporation of impurity ions into the lattice (Wilson, 1961) may influence the inversion temperature. It was once believed that H^+ or $(OH)^-$ were responsible for stabilizing maghemite. It is now known that neither ion is incorporated in the maghemite structure, although an aqueous environment does promote maghemitization rather than direct oxidation of magnetite to hematite (Elder, 1965; Sakamoto et al., 1968).

The Curie temperature of γFe_3O_3 has generally been determined indirectly because T_C often lies above T_{inv}. For example, Michel and Chaudron (1935) determined T_C values for Na-substituted maghemites and extrapolated to zero Na concentration to obtain $T_C = 675$ °C. Maxwell et al. (1949) measured order-

Table 3.3 Inversion temperature T_{inv} ($\gamma Fe_2O_3 \rightarrow \alpha Fe_2O_3$)

T_{inv} (°C)	Authors
250	Verwey (1935)
250	Bernal et al. (1957)
700–850	Wilson (1961)
>568	Oades and Townsend (1963)
525–650	Gustard and Schuele (1966)
545	Fröhlich and Vollstädt (1967)
350–900	Sato et al. (1967)
250, 540	Imaoka (1968)
400	Kachi et al. (1971)
550	Özdemir and Dunlop (1988)
750	Özdemir (1990)

ing temperatures for partially oxidized cation-deficient magnetites (see §3.4) and extrapolated to $z = 1$ to obtain $T_C = 675\,°C$. Aluminium and other impurities in the lattice stabilize the maghemite structure and have allowed a few direct determinations of T_C. When 7% Al was incorporated, Michel *et al.* (1951) measured $T_C = 591\,°C$, while Özdemir and Banerjee (1984) found $T_C = 645\,°C$ for a maghemite with smaller amounts of impurities. The latter value is the best present estimate of T_C for γFe_2O_3 (see Table 3.2 for a complete listing).

3.3.3   Magnetocrystalline anisotropy and magnetostriction

Takei and Chiba (1966) measured $K_1 = -4.6 \times 10^3\,J/m^3$ on epitaxially grown single-crystal films of γFe_2O_3. This is one-third the value of K_1 for magnetite. The disagreement of saturation magnetization values measured on Takei and Chiba's films with values determined for bulk material has been mentioned earlier. However, their K_1 value agrees closely with the value $|K_1| = 4.7 \times 10^3\,J/m^3$ calculated by Birks (1950) based on measurements of initial permeability (the ratio of B to H) for polycrystalline γFe_2O_3.

Kaneoka (1980) measured saturation magnetostriction λ_s throughout the magnetite-maghemite solid-solution series (see next section). For γFe_2O_3, he obtained $\lambda_s = -8.9 \times 10^{-6}$, which is less than one-third the value for magnetite and of opposite sign. The measured values of λ_s increased with decreasing z. For Fe_3O_4, he found 23×10^{-6}, which is considerably lower than the accepted value of 36×10^{-6} (§3.2.3).

3.4   Magnetite–maghemite solid-solution series

A partially oxidized or cation-deficient magnetite has the structural formula $Fe^{3+}_{2+2z/3}Fe^{2+}_{1-z}\square_{z/3}O^{2-}_4$, where z is the oxidation parameter. Saturation moment decreases steadily with increasing z, from $4\mu_B$ for Fe_3O_4 to $\approx 3\mu_B$ when $z = 0.95$ (Fig. 3.9). The X-ray cell edge a also decreases with oxidation, as a result of the change in Goldschmidt radius from 0.83 to 0.67 Å as Fe^{2+} is converted to Fe^{3+}. Cracking of maghemitized regions in the crystal may result (Fig. 3.10).

Maghemitization is a low-temperature ($\leq 200\,°C$) process occurring mainly at the crystal surface or in fissures (Fig. 3.10). The mechanism is diffusion of Fe^{2+} from the crystal interior to a free surface, where it is converted to Fe^{3+}. Oxidation is therefore slow, being controlled by Fe^{2+} diffusion rates and distance to the surface. The interiors of large crystals may remain unoxidized indefinitely (Fig. 3.10; Cui *et al.*, 1994). On the other hand, oxidation may be substantially complete even at room temperature T_0 in fine grains with a large surface/volume ratio. For example, ≈ 100-Å size particles were found to be 95% oxidized to γFe_2O_3 after 50 days at T_0 (Haneda and Morrish, 1977). Factors promoting oxi-

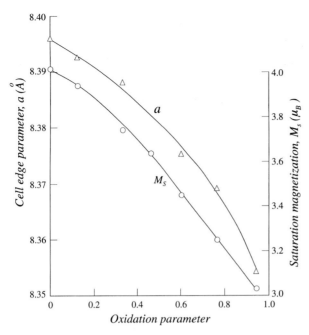

Figure 3.9 Unit cell edge *a* and saturation moment in number of Bohr magnetons μ_B per molecule at 0 K as a function of oxidation parameter *z* in partially maghemitized magnetites. [After Readman and O'Reilly (1972), with the permission of the authors and the publisher, Journal of Geomagnetism and Geoelectricity, Tokyo, Japan.]

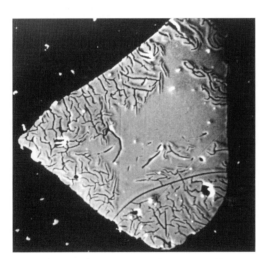

Figure 3.10 Maghemitization and surface cracking in a magnetite crystal $\approx 30\,\mu\text{m}$ in size. The electron micrograph shows that maghemitization is not uniform throughout the crystal. The oxidized surface layer, with a smaller lattice parameter than the unoxidized magnetite core, stretches and cracks. [After Gapeyev and Tsel'movich (1988), with the permission of the authors.]

dation are crystalline imperfection (Colombo *et al.*, 1964), a large specific surface area (Gallagher *et al.*, 1968), and adsorbed water.

3.5 Titanomagnetites

Single-phase titanomagnetites, $Fe_{3-x}Ti_xO_4$, are inverse spinels. Increasing substitution of Ti^{4+} weakens the AB exchange coupling and the Curie point falls almost linearly with increasing x (Fig. 3.11). A Curie point of $\approx$150–200 °C is characteristic of stoichiometric $Fe_{2.4}Ti_{0.6}O_4$ or TM60, which is the primary titanomagnetite in rapidly cooled basaltic lavas. An independent way of determining composition is from the X-ray unit cell edge, which increases steadily from 8.396 Å in Fe_3O_4 to about 8.54 Å in ulvöspinel (Fe_2TiO_4) (O'Reilly, 1984, Fig. 2.4). If x values deduced from T_C and from X-ray data disagree, the titanomagnetite is probably oxidized or contains impurities such as Al and Mg.

The distribution of cations on A and B sublattices is uncertain for intermediate titanomagnetites. The predicted variations of magnetic moment with composition for several different models are shown in Fig. 3.11 and compared with experimental observations. Part of the problem in testing cation distribution models stems from the difficulty of preparing stoichiometric single-phase spinels. The observations all support a regular decrease in magnetic moment from $4\mu_B$ for Fe_3O_4 to 0 for Fe_2TiO_4. At room temperature, TM60 has an M_s value of $\approx$125 kA/m, about one-quarter that of magnetite. Natural intermediate titanomagnetites may contain up to 10% Mg and Al, further lowering their M_s and T_C values. Thermomagnetic or $M_s(T)$ curves for pure and Al-substituted TM60 are illustrated in Fig. 3.12(a).

Electrical conductivity falls rapidly with Ti^{4+} substitution and the Verwey transition is suppressed for $x > 0.1$. However, the isotropic point does not disappear. To preserve charge balance, when Ti^{4+} is substituted for Fe^{3+}, another Fe^{3+} must be converted to Fe^{2+}: $2Fe^{3+} \rightarrow Ti^{4+} + Fe^{2+}$. The additional anisotropy of the Fe^{2+} ions causes K_1 to increase to a peak value of -2.50×10^4 J/m^3

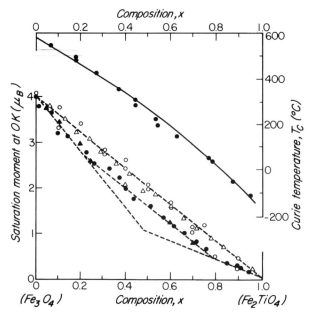

Figure 3.11 Saturation moment at 0 K and Curie temperature of titanomagnetite as a function of titanium content x. Saturation moment data: solid circles, Akimoto (1962); solid triangles, O'Reilly and Banerjee (1965); open circles, Stephenson (1969); open triangles, Bleil (1971). The dashed lines are theoretical predictions based on the Akimoto and Néel–Chevallier models of sublattice cation distributions. [After O'Reilly (1976), with the permission of the author and the publisher, Institute of Physics Publishing, Bristol, UK.]

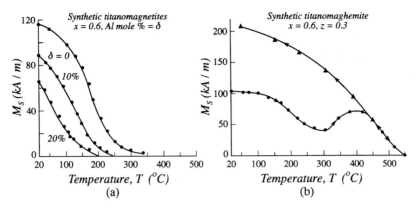

Figure 3.12 Vacuum thermomagnetic curves of synthetic $x = 0.6$ titanomagnetites and titanomaghemite. (a) The $M_s(T)$ curves of the titanomagnetites are reversible. (b) The titanomaghemite inverts during heating to a phase assemblage that includes magnetite, giving rise to an irreversible $M_s(T)$ curve. [Reprinted from Özdemir and O'Reilly (1981, 1982a) with kind permission of the authors and Elsevier Science – NL, Sara Burgerhartstraat 25, 1055 KV Amsterdam, The Netherlands.]

around $x = 0.1$, at which point T_I is ≈ 90 K (Fig. 3.7a). With further Ti^{4+} substitution, K_1 begins to decrease in value, eventually becoming positive around $x = 0.55$. In TM60, the $\langle 100 \rangle$ directions, not the $\langle 111 \rangle$ directions, are the crystalline easy axes at room temperature.

For TM61, $K_1' = 2.02 \times 10^3$ J/m³ (Sahu and Moskowitz, 1995). The critical field for magnetization rotation in single-domain grains with no shape anisotropy is then $\mu_0 H_c \approx 0.5 K_1'/M_s = 8$ mT or 80 G, about half the critical field for equant grains of magnetite. At higher temperatures in TM60, $K_1'(T) \propto M_s^6(T)$ (Fig. 3.13), a less extreme decrease than in magnetite (see §2.8.4, Fig. 3.7b).

As x increases from 0 (i.e., Fe_3O_4), λ_{111} at first increases slightly, then remains constant with a room-temperature value $\approx 100 \times 10^{-6}$ (Fig. 3.8a). However, λ_{100} changes sign by $x = 0.1$ and its value at T_0 increases to $+170 \times 10^{-6}$ when $x = 0.56$. Klerk et al. (1977) measured λ_{100} and λ_{111} over very broad ranges of both composition and temperature (Fig. 3.14). Both magnetostriction constants increase steadily with increasing x (at fixed T/T_C) over the entire range $0 \le x \le 1$.

For $x = 0.61$, Sahu and Moskowitz (1995) found $\lambda_{100} = 142.5 \times 10^{-6}$ and $\lambda_{111} = 95.4 \times 10^{-6}$ at T_0, giving $\lambda_s = 114 \times 10^{-6}$, more than three times λ_s for magnetite. A uniaxial stress of 10 MPa or 0.1 kb then produces $\mu_0 H_c = 3\lambda_s\sigma/M_s \approx 30$ mT or 300 G, 15 times larger than the corresponding critical field for magnetite. Magnetoelastic anisotropy is thus a major contributor to coercivity in TM60. H_c due to crystalline anisotropy was estimated above to be only ≈ 8 mT, while critical fields due to shape anisotropy are proportional to M_s (§4.2.4, 5.7.1), which is four times weaker in TM60 than in magnetite.

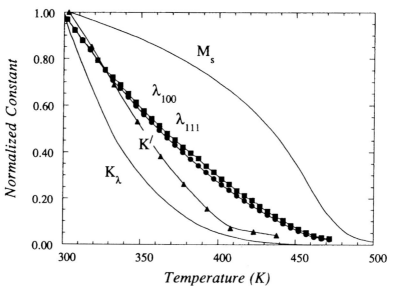

Figure 3.13 Temperature dependences of M_s, crystalline anisotropy constant K_1', and magnetostriction constants λ_{100}, λ_{111} measured for a single crystal of TM61. The magnetoelastic portion, K_λ, of K_1' was calculated from the data according to eqn. (2.64). [After Sahu and Moskowitz (1995) © American Geophysical Union, with the permission of the authors and the publisher.]

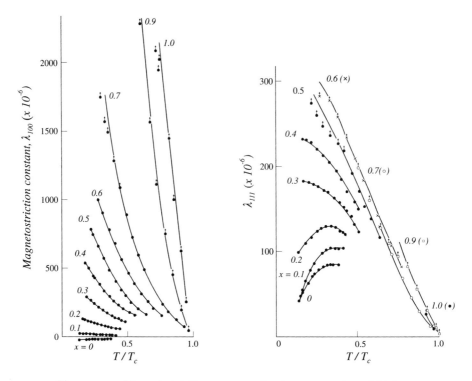

Figure 3.14 Magnetostriction constants for titanomagnetites with Ti contents x ranging from 0 to 1 as a function of reduced temperature T/T_C. [After Klerk *et al.* (1977), with the permission of the authors and the publisher, Les Editions de Physique, Les Ulis, France.]

Sahu and Moskowitz (1995) found that $\lambda(T) \propto M_s^2(T)$ approximately for TM61, a weaker dependence than that of K_1' (Fig. 3.13). Thus magnetoelastic anisotropy will increase in importance relative to crystalline anisotropy at high temperatures, for example in the TRM blocking temperature range.

The magnetostriction is also strongly dependent on the presence of impurities, for example Al^{3+} which is the most frequent impurity cation in natural TM60 in submarine basalts (Chapter 14). λ_s decreases linearly with increasing Al content (Özdemir and Moskowitz, 1992; Fig. 3.15).

Figure 3.16 compares predicted and measured coercive forces for titanomagnetites of different x values. Finely ground synthetic titanomagnetites have such high internal stresses that their coercive forces are stress-controlled for most values of x. Larger and less stressed single-domain grains have crystalline-controlled coercive forces for most compositions. Very large multidomain grains have much lower coercive forces whose average values are probably also determined by crystalline anisotropy. However, local variations in stress become important when large grains of TM60 are polished for domain observations (§6.1, 6.4). As a result of the high magnetostriction of TM60, residual surface stresses produce intricate 'maze' patterns which obscure the underlying fundamental domain structure. Even careful annealing often fails to remove the

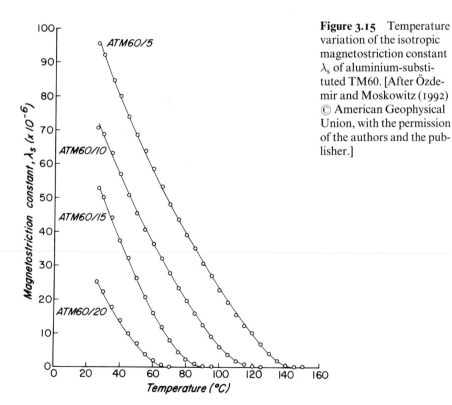

Figure 3.15 Temperature variation of the isotropic magnetostriction constant λ_s of aluminium-substituted TM60. [After Özdemir and Moskowitz (1992) © American Geophysical Union, with the permission of the authors and the publisher.]

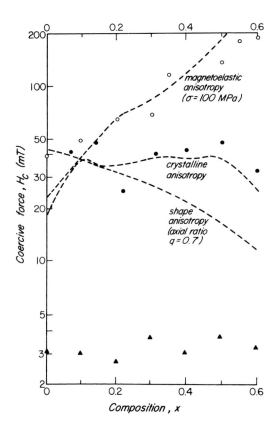

Figure 3.16 Measured bulk coercive force H_c as a function of Ti content x. Open circles, finely ground single-domain grains ($\approx 0.1\,\mu m$); solid circles, $\approx 1.5\,\mu m$ grains; triangles, large ($> 150\,\mu m$) multi-domain grains. Dashed curves are theoretical dependences of critical fields in single-domain grains resulting from pure magnetoelastic, magneto-crystalline, and shape anisotropies, respectively. [After Day (1977), with the permission of the author and the publisher, Journal of Geomagnetism and Geoelectricity, Tokyo, Japan.]

unwanted patterns. For this reason, domain observations on TM60 are more challenging than on magnetite.

3.6   Titanomaghemites

A partially oxidized titanomagnetite is usually referred to as **titanomaghemite**. This is a different usage than in the titanium-free mineral, where maghemite (§3.3) is the *fully oxidized* end-member of magnetite oxidation. A titanomaghemite has the formula $Fe^{3+}_{[2-2x+z(1+x)]R}Fe^{2+}_{(1+x)(1-z)R}Ti^{4+}_{xR}\square_{3(1-R)}O^{2-}_4$, where $R = 8/[8 + z(1 + x)]$ (O'Reilly, 1984, p. 14), provided oxidation proceeds by the addition of oxygen, building new unit cells at the surface (see Fig. 14.6). (In the alternative iron removal mechanism, which occurs under submarine conditions, oxygen is not present at the surface and iron is leached by the seawater. In this case, the Fe/Ti ratio falls as oxidation proceeds and the oxidation lines on the Ti^{4+}–Fe^{2+}–Fe^{3+} ternary diagram (Fig. 3.1) slope upward to the right.) During the low-temperature oxidation of titanomagnetites, the face-centred cubic oxygen lattice is preserved (§3.2) and Fe^{2+} cations migrate to the crystal surface.

Cation diffusion is a slow process at ordinary temperatures. The crystal surface oxidizes much more quickly than the interior. As a result, there is a strong gradient in cation deficiency (i.e., vacancy concentration) and oxidation parameter z in partially oxidized spinels. In fine grains, the gradient region covers much of the crystal and X-ray lines may be broadened and diffuse.

Because the lattice parameters of the stoichiometric and cation-deficient spinels are so different, the oxidized surface is strained and frequently cracks (Fig. 3.10). Thus partially oxidized titanomagnetite grains, although they are single-phase in the sense of preserving a spinel lattice, are not homogeneous in their structure or magnetic properties. The smearing out of the Verwey transition and the large drop in remanence between 5 and 20 K in partially oxidized magnetites (Fig. 3.6b) are manifestations of this inhomogeneity.

One difference between low-temperature oxidation of magnetite and of titanomagnetite is that in the latter case, the removal of iron from the crystal leads to a falling Fe/Ti ratio in the crystal interior as oxidation proceeds. If the mechanism of oxidation were addition of oxygen diffusing to the crystal interior, as originally proposed by Readman and O'Reilly (1972), the Fe/Ti ratio would not change.

M_s and unit cell edge decrease and T_C rises with oxidation in titanomaghemites of all compositions (Figs. 3.1b, 3.17, 3.18; Hauptmann, 1974; Nishitani and Kono, 1983). The difficulty of synthesizing stoichiometric titanomagnetites and variations in cation distribution depending on the method of preparation probably account for differences between the results of different experimenters (Moskowitz, 1987).

Coercive force H_c and initial susceptibility χ_0 vary in approximately reciprocal fashion with z in synthetic single-domain titanomaghemites (Fig. 3.19). The

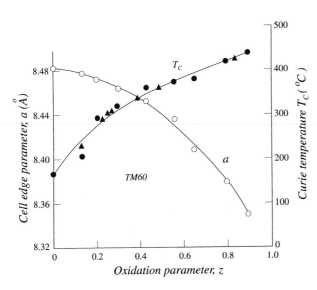

Figure 3.17 Unit cell edge a and Curie temperature T_C as a function of oxidation parameter z in titanomaghemites of composition $x = 0.6$. [After Özdemir and O'Reilly (1982a) (solid and open circles) and Brown and O'Reilly (1988) (solid triangles), with kind permission of the authors and Elsevier Science – NL, Sara Burgerhartstraat 25, 1055 KV Amsterdam, The Netherlands.]

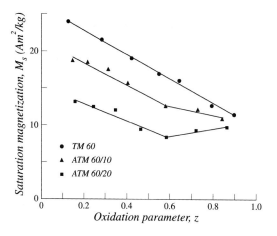

Figure 3.18 Variation of room-temperature saturation magnetization M_{s0} with oxidation parameter z in Al-substituted titanomaghemites with Ti content $x = 0.6$. [Reprinted from Özdemir and O'Reilly (1982a) with kind permission of the authors and Elsevier Science – NL, Sara Burgerhartstraat 25, 1055 KV Amsterdam, The Netherlands.]

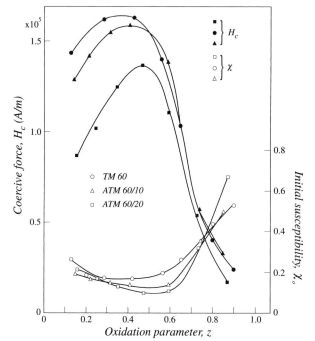

Figure 3.19 Variation of room-temperature coercive force H_c and initial susceptibility χ_0 with oxidation parameter z in Al-substituted titanomaghemites with Ti content $x = 0.6$. [Reprinted from Özdemir and O'Reilly (1982a) with kind permission of the authors and Elsevier Science – NL, Sara Burgerhartstraat 25, 1055 KV Amsterdam, The Netherlands.]

starting titanomagnetites were highly strained during their preparation; H_c and χ_0 are therefore probably magnetoelastic in origin, and their variations with z may give an indication of the corresponding variation in λ. No direct measurements of λ have yet been reported.

Curie temperatures and other high-temperature properties of titanomaghemites are difficult to measure because the spinel structure becomes unstable during heating and changes to rhombohedral, or to intergrown spinel and rhombohedral phases. This transformation, like the one in maghemite, is known as *inversion*. Figure 3.12(b) illustrates the typical irreversible behaviour of an oxidized titanomagnetite when it is heated in vacuum. M_s at first decreases, but around 250 °C, before the titanomaghemite reaches its Curie temperature, inversion begins. The phases which develop are a titanium-rich rhombohedral phase (near ilmenite) and a titanium-poor spinel, with composition and Curie temperature near magnetite. The compositions and textures of the intergrown phases are very similar to those of the oxyexsolution products that result when a titanomagnetite of the same x value undergoes high-temperature oxidation (Özdemir, 1987; Fig. 3.2). In subsequent heatings and coolings, the phase assemblage is stable and the $M_s(T)$ curve is reversible.

Maghemitization of primary titanomagnetites and inversion of the resulting titanomaghemites occur in the upper oceanic crust (Chapter 14). Maghemitization reduces the 'magnetic stripe' signal (§1.2.1) but pushes the Curie point isotherm deeper into the crust. Inversion produces magnetite and pushes the Curie point isotherm much deeper, perhaps into the upper mantle.

The cation distribution in titanomaghemites with general x and z values is not known with certainty. In some cation distribution models, with sufficient oxidation the tetrahedral sublattice magnetization M_A can outweigh the octahedral sublattice magnetization M_B, causing a self-reversal of M_s. Ozima and Sakamoto (1971) produced experimental single and even double reversals of magnetization in extreme low-temperature oxidation of high-titanium titanomagnetites. However, the oxidation products were multiphase intergrowths (inverted titanomaghemites) and the reversals may have resulted from exchange coupling of the phases (Chapter 13) rather than self-reversal of $M_A + M_B$ in a single phase.

3.7 Hematite

Rhombohedral hematite (αFe_2O_3) is an antiferromagnetic mineral with a Néel temperature $T_N = 675$ °C. Between the Morin transition just below room temperature and T_N, the sublattice magnetizations M_A and M_B are canted by a fraction of a degree out of antiparallelism (§2.7.3 and Fig. 2.8), giving rise to a weak spontaneous magnetization M_s lying in the basal plane perpendicular to M_A and M_B (Fig. 2.3). The magnitude of M_s is about 2.5 kA/m, $\approx 0.5\%$ that of magnetite.

The **spin-canting** phenomenon is often called **parasitic ferromagnetism** because the ferromagnetism originates in, and is a small fraction of, the antiferromagnetism. M_s changes very slowly with temperature until just below T_C, which coincides with T_N. The 'blocky' $M_s(T)$ curve of hematite (Fig. 3.20) is distinctive and quite different from $M_A(T)$ and $M_B(T)$. It results in a concentration of remanence blocking temperatures just below 675 °C. It is usually easy to resolve hematite and magnetite remanences in a rock by thermal demagnetiza-

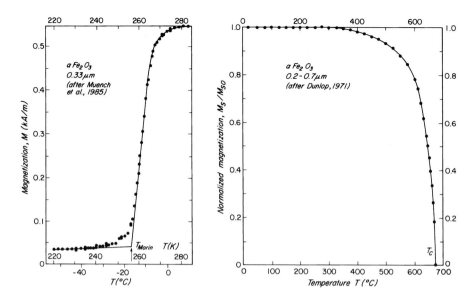

Figure 3.20   The variation of high-field magnetization of hematite below and above room temperature. Note the change of horizontal and vertical scales between the two diagrams. The $M_s(T)$ variation in hematite is 'blocky', being almost temperature independent except close to the Morin transition and the Curie point. [After Muench *et al.* (1985) and Dunlop (1971), with the permission of the authors and the publishers, Akademie Verlag GmbH, Berlin, Germany and Gauthier-Villars, Montrouge, France.]

tion because their blocking temperature spectra normally do not overlap. Although hematite remanence is often weak compared to magnetite remanence, it cannot be ignored. Being the highest blocking-temperature remanence in a rock, it may survive heating during metamorphism that resets magnetite remanences (§16.3). It also has higher coercivities than magnetite (see below) and is the ultimate remanence isolated in alternating field (AF) as well as in thermal demagnetization.

The Morin transition is a discontinuity in magnetic and other properties around $-15\,°C$ (Liebermann and Banerjee, 1971; Figs. 3.20, 3.21). Small amounts of titanium substitution lower T_{Morin} and eventually suppress the transition. Decreasing particle size has the same effect. The transition vanishes in particles $\leq 0.1\,\mu m$ in size (Bando *et al.*, 1965).

In hematite, dipolar and single-ion magnetocrystalline anisotropies are both important but have opposite signs. Their relative magnitudes change with temperature and become equal at T_{Morin}. With the change in sign of the net anisotropy, the crystallographic easy axis changes and there is an antiferromagnetic spin flop. When $T < T_{\text{Morin}}$, $\boldsymbol{M}_A$ and $\boldsymbol{M}_B$ are pinned along the c-axis and there is no spin-canting. Above T_{Morin}, $\boldsymbol{M}_A$, $\boldsymbol{M}_B$, and their canted resultant $\boldsymbol{M}_s$ are pinned perpendicular to the c-axis (Fig. 3.22a) but have relative freedom of

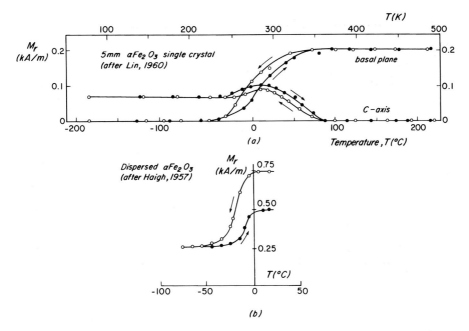

(a)

(b)

Figure 3.21 The temperature variation of strong-field remanence of hematite, (a) for a large single crystal with magnetizing fields applied in the basal plane and along the *c*-axis, respectively, and (b) for randomly oriented fine particles. A large fraction of the remanence is lost in cooling through the Morin transition, but part is recovered on reheating through the transition (memory fraction). [After Lin (1960) and Haigh (1957), with the permission of the publishers, The American Institute of Physics, Woodbury, NY and Taylor & Francis, Basingstoke, UK.]

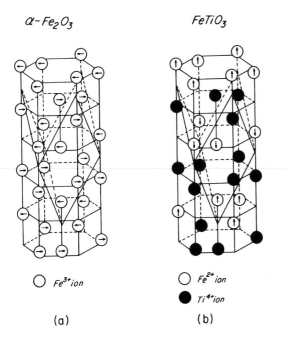

(a)   *(b)*

Figure 3.22 The magnetic structures of hematite (αFe_2O_3) and ilmenite ($FeTiO_3$). Ilmenite has an ordered cation distribution, with Fe^{2+} and Ti^{4+} ions on alternate basal planes. [After Nagata (1961) Fig. 3.25, with the permission of the publisher, Maruzen Co. Ltd., Tokyo, Japan.]

rotation within the basal plane. There is an intrinsic six-fold anisotropy within the basal plane (eqn. (2.49)), but more important in practice is a uniaxial magnetoelastic anisotropy governed by the stress due to crystal twinning.

In addition to the anisotropic spin-canted ferromagnetism that vanishes below T_{Morin}, hematite possesses an underlying **isotropic ferromagnetism** that is unaffected by the Morin transition (Fig. 3.21b). Since its origin seems to lie in chemical or lattice defects, it is often called the **defect moment**. Defect moments may be a general feature of magnetic minerals but only in hematite is the intrinsic ferromagnetism weak enough to reveal the underlying defect component.

As in the case of magnetite remanence cycled in zero field through the Verwey transition (§3.2.2), hematite remanence cycled through the Morin transition decreases to low values in the cooling half of the cycle but recovers a fraction of the initial remanence, the **remanence memory**, during reheating to room temperature (Fig. 3.21b). The original direction of remanence is also recovered. The origin of memory in hematite is something of a mystery. Spin-canted remanence should vanish in cooling through T_{Morin}, along with the anisotropy that pins it, unlike the situation in magnetite. Cycling across the Morin transition is a chemically non-destructive alternative to thermomagnetic analysis for detecting hematite (Fuller and Kobayashi, 1964).

Coercivity due to shape anisotropy is proportional to M_s and is negligible in hematite. Coercive forces due to crystalline anistropy and magnetostriction vary from ≈ 1 mT or 10 Oe in large multidomain grains to many hundreds of mT in single-domain hematite. The triaxial crystalline anisotropy in the basal plane is weak, certainly not more than $10–100$ J/m^3, compared to $> 10^4$ J/m^3 in magnetite. Since $H_c \propto K/M_s$, both minerals in single-domain grains have coercivities of $10–20$ mT due to magnetocrystalline anisotropy. The source of the observed high coercivities in single-domain hematite must be magnetoelastic anisotropy due to internal stress. The magnetostriction of hematite is $\lambda_s \approx 8 \times 10^{-6}$, comparable to that of magnetite or maghemite, but the small value of M_s in hematite means that $H_c = 3\lambda_s\sigma/2\mu_0 M_s \approx 500$ mT or 5000 Oe for an internal stress of 100 MPa or 1 kb.

Hematite can form by primary high-temperature oxidation of titanomagnetite during cooling from the melt or as a secondary inversion product of titanomaghemite during later reheating of a rock. Its plate-like crystals or crystallites, which are elongated in the basal {0001} plane, grow in spinel {100} planes (Fig. 3.3). Hematite also forms as the end product of prolonged oxidation of magnetite at ordinary temperatures. In this form, pseudomorphing the original magnetite crystals, it is called martite. Other important secondary processes that generate hematite are inversion of maghemite (§3.3.2), dehydration of weathering products like goethite (αFeOOH), and precipitation of ultra-fine-grained hematite cement from iron-rich solutions in the pore spaces of clastic sediments. This last type of hematite gives red beds their distinctive colour.

3.8   Titanohematites

The rhombohedral titanohematites or hemoilmenites ($Fe_y^{2+}Fe_{2-2y}^{3+}Ti_y^{4+}O_3^{2-}$) have compositions on the tie-line between the endmembers hematite and ilmenite (Fig. 3.1). As with the titanomagnetites, solid solution is complete only at high temperatures. Approximately single-phase titanohematites of intermediate composition ($y \approx 0.5$–0.7) can be preserved by rapid chilling of dacitic pyroclastic rocks, and have the fascinating property of acquiring a self-reversed thermoremanent magnetization (Uyeda, 1958; Heller *et al.*, 1986; Hoffman, 1992). This self-reversing property will be discussed in detail later (§14.5.4). It does not result from self-reversal of M_s of a single phase but from negative coupling between M_s vectors in microscopically exsolved ferrimagnetic and parasitic ferromagnetic phases. Some titanohematites of relatively low y value exhibit self-reversal of partial thermoremanent magnetizations (§1.1.3) acquired in certain temperature ranges (Carmichael, 1961).

Titanohematites have a corundum structure, with oxygen anions in a hexagonal-close-packed lattice and Fe and Ti cations occupying two-thirds of the interstices. The cations are arranged very nearly in basal {0001} planes (Fig. 3.22) with O^{2-} in intervening layers. Neighbouring Fe^{2+} or Fe^{3+} ions in adjacent cation planes are negatively exchange coupled, but exchange coupling within each plane may be either positive or negative. Only for certain compositions do alternate cation planes correspond to magnetic sublattices.

For $0 \leq y \leq 0.5$, the cation distribution is disordered. Ti^{4+} and Fe^{2+} are equally distributed among all c-planes. Titanohematite in this compositional range is essentially antiferromagnetic with a weak parasitic ferromagnetism, like hematite. In the range $0.5 \leq y < 1$, titanohematite is ferrimagnetic as a result of an ordered (in reality, partially ordered) cation distribution. In the fully ordered distribution, Ti^{4+} is confined to alternate cation planes and Fe^{2+} to the intervening planes. The peak magnetic moment occurs around $y = 0.7$ (Fig. 3.23). The theoretical spontaneous moment for this composition is $0.7 \times 4\mu_B = 2.8\mu_B$ per formula weight $Fe_{1.3}Ti_{0.7}O_3$, corresponding to $\sigma_s = (5585/M)n_B = 101.5\,Am^2/kg$ or emu/g ($M = 154.1$, $n_B = 2.8$) at $0\,K$. At ordinary temperatures, the spontaneous magnetizations are considerably reduced, since for $0.5 \leq y \leq 0.7$, T_C is between 20 and $200\,°C$ (Fig. 3.23) and for $y > 0.7$, T_C is below room temperature. An approximate value for $y \approx 0.5$ is $M_{s0} \approx 100\,kA/m$ at $T_0 = 20\,°C$.

Since homogeneous titanohematite of intermediate composition is not an equilibrium phase at ordinary temperatures, it will slowly exsolve on a microscale towards an equilibrium intergrowth of Ti-rich (ferrimagnetic) and Ti-poor (parasitic ferromagnetic) phases. This microintergrowth of ordered and disordered phases is at the root of most reported self-reversals of TRM (Hoffman, 1992).

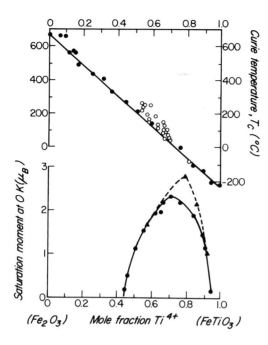

Figure 3.23 Curie temperature T_C and saturation moment at 0 K (in μ_B) for titanohematite, $Fe_{2-y}Ti_yO_3$, as a function of the composition parameter or Ti content y. For $0.5 < y < 1$, titanohematite is ferrimagnetic, but T_C is below T_0 if $y > 0.7$. [After Nagata (1961) Figs. 3.29 and 3.30, with the permission of the publisher, Maruzen Co. Ltd., Tokyo, Japan.]

3.9   Iron oxyhydroxides

Hydrous iron oxides form as microcrystalline weathering products in nature and are often collectively called limonite. The most important mineral among them is orthorhombic goethite (αFeOOH), which is a common constituent of soils and sediments. Goethite is antiferromagnetic with a Néel temperature of 120 °C (Fig. 3.24), but small amounts of impurity ions lower T_N considerably. Like hematite, with which it is often intergrown, goethite has a superimposed ferromagnetism whose Curie temperature coincides with T_N (Özdemir and Dunlop, 1996a; Fig. 3.25). However, the ferromagnetism is weaker than that of hematite. It seems to be a defect moment, perhaps due to unbalanced numbers of spins in the very fine crystallites which are typical of goethite. Although the intrinsic anisotropy is probably not unusually high, the extremely weak M_s means that high magnetic fields must be applied to deflect the moments. The resulting coercive forces are often many hundreds of mT (thousands of G or Oe), higher even than those of fine-particle hematite. As a result, goethite remanences cannot be erased by AF demagnetization. They are efficiently removed by thermal demagnetization (Fig. 3.25; §15.3.3).

During routine thermal cleaning to erase the remanences of minerals with higher T_C's, goethite dehydrates in the range 250–400 °C to form hematite. The needle-like form of the goethite crystals is preserved but each crystal contains many hematite crystallites. The effective hematite grain size is therefore extre-

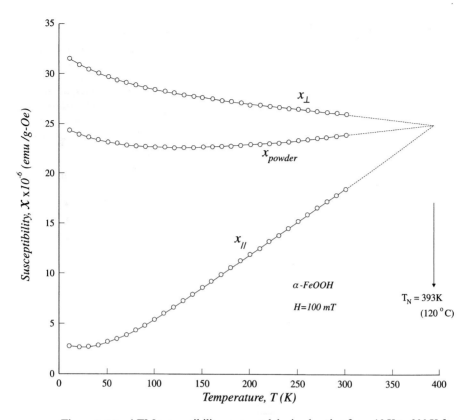

Figure 3.24 AFM susceptibility measured during heating from 10 K to 300 K for a natural goethite crystal. $\chi_\parallel$, $\chi_\perp$ were measured parallel and perpendicular to the goethite c-axis, respectively. χ_{powder} was measured for randomly oriented crystallites in a powdered sample. Notice the similarity to the predicted AFM susceptibilities in Fig. 2.6. The intersection of linear extrapolations of the $\chi_\parallel$, $\chi_\perp$ and χ_{powder} data pinpoints the Néel temperature, $T_N = 120\,°C$. [After Özdemir and Dunlop (1966a) © American Geophysical Union, with the permission of the authors and the publisher.]

mely fine. Hematite that has formed by the dehydration of goethite is generally superparamagnetic (§5.5.2, 5.7.2) and carries no remanence.

Since goethite at 20 °C is close to its Curie temperature, its remanence sometimes shows a striking increase during zero-field cooling to liquid nitrogen temperature (77 K) and a corresponding decrease in reheating to T_0, arising from the reversible change in $M_s(T)$. There is very little irreversible decrease in remanence in such a cooling–heating cycle because most goethite remanence has unblocking temperatures well above T_0 (Fig. 3.25). Dekkers (1989a) suggests using this property as a non-destructive method of detecting goethite.

Lepidocrocite (γFeOOH) is a minor constituent of soils and sediments. The antiferromagnetic Néel point is far below T_0 and so lepidocrocite at ordinary temperatures has no remanence-carrying potential. Its importance stems from

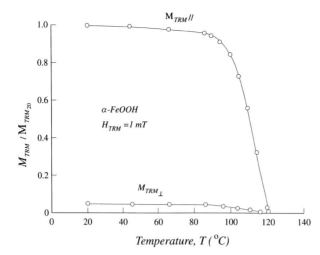

Figure 3.25 Stepwise thermal demagnetization of thermoremanent magnetizations (TRM's) produced parallel and perpendicular to the c-axis of the goethite crystal of Fig. 3.24. Both TRM's are unaffected by zero-field heating to 90 °C but demagnetize completely when heated to the Curie point, $T_C = 120$ °C. [After Özdemir and Dunlop (1996a) © American Geophysical Union, with the permission of the authors and the publisher.]

the fact that it dehydrates to strongly magnetic maghemite when heated above 250 °C. The chemical remanence acquired if dehydration takes place in a magnetic field has peculiar behaviour when the maghemite inverts to hematite (around 400 °C). The hematite remanence may form at a large angle to the maghemite remanence, in some cases 180° (§13.4.6). Although this process is unlikely to be significant in nature because of the temperatures involved, the mechanism is intriguing.

3.10   Iron sulphides

Greigite (Fe_3S_4), which was formerly thought to be a rare mineral, occurs quite commonly in sediments formed under anoxic, e.g., sulphate reducing, conditions (Roberts, 1995; §15.3.1). Magnetotactic bacteria (§3.13) can also biomineralize Fe_3S_4 (Mann *et al.*, 1990; Fassbinder and Stanjek, 1994). Greigite is the sulphide counterpart of magnetite and has the same inverse spinel structure. It is ferrimagnetic but M_s is only $\approx \frac{1}{4}$ that of magnetite (Hoffmann, 1992). T_C is around 330 °C, which is very similar to the Néel point of troilite (FeS) and the Néel and Curie point of pyrrhotite (320 °C, Table 3.1). K_1 is estimated to be $\sim 10^3 \text{ J/m}^3$, about an order of magnitude less than that of magnetite. Nevertheless coercivities of single-domain greigite (Diaz Ricci and Kirschvink, 1992) are sufficient to ensure a stable remanence.

Pyrrhotite ($Fe_{1-x}S$) is a common accessory mineral in igneous, metamorphic and sedimentary rocks, although it seldom dominates the remanence. Natural pyrrhotite is actually a mixture of the monoclinic mineral Fe_7S_8, which is ferrimagnetic, and antiferromagnetic hexagonal phases like Fe_9S_{10} and $Fe_{11}S_{12}$. The deficiency of Fe^{2+} is accounted for by vacancies in the lattice, and ordering of these vacancies on superlattices gives rise in Fe_7S_8 to different numbers of Fe^{2+} ions on the two magnetic sublattices. Thus monoclinic pyrrhotite is like maghemite in owing its ferrimagnetism to cation deficiency and vacancy ordering. The stoichiometric mineral troilite (FeS) is common in meteorites and lunar rocks (Chapter 17), but not on earth.

Pyrrhotite has a NiAs crystal structure, with a c-axis of hexagonal symmetry and Fe and S confined to alternate c-planes (Fig. 3.26). Fe cations within a particular layer are ferromagnetically coupled, while Fe cations in adjacent layers are negatively exchange coupled via intervening S^{2-} ions. Thus alternate Fe planes form the two magnetic sublattices. In Fe_7S_8, the preferential arrangement of vacancies on one of the sublattices (Fig. 3.26) distorts the hexagonal lattice to a monoclinic one and also gives rise to a fairly strong ferrimagnetic moment. M_{s0} is approximately 80 kA/m, although a range of values has been reported. The Curie point is 320 °C (Fig. 3.27a). Grain-size dependent properties have been studied in detail by Clark (1984) and Dekkers (1988, 1989b).

As with hematite, the c-axis is hard and M_s and the sublattice magnetizations are confined to the c-plane. However the triaxial anisotropy within the c-plane is quite strong, much larger than that due to magnetostriction. As a result, domain patterns are easily viewed on a {0001} surface of Fe_7S_8 without elaborate precautions for removing the stressed layer resulting from polishing.

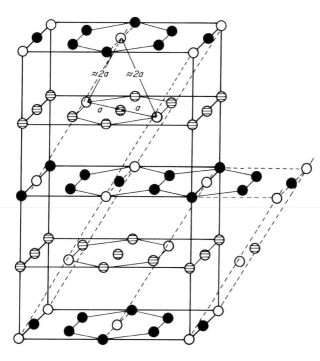

Figure 3.26 The magnetic structure and distribution of vacancies (open circles) and Fe^{2+} ions (black and hatched circles, with oppositely directed moments) in pyrrhotite (Fe_7O_8). The structure is basically hexagonal, with magnetic sublattices corresponding to alternate basal planes. Vacancies are arranged on a monoclinic superlattice (dashed lines) and are located preferentially on one of the two magnetic sublattices, giving rise to ferrimagnetism. [After Stacy and Banerjee (1974), with the permission of the authors.]

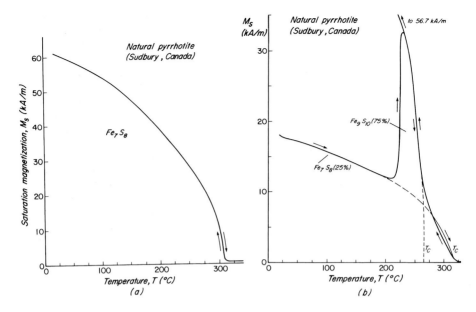

Figure 3.27   (a) Reversible thermomagnetic curve of ferrimagnetic pyrrhotite (Fe_7S_8). (b) Irreversible thermomagnetic curve of natural pyrrhotite containing a mixture of Fe_7S_8 and Fe_9S_{10}. Fe_9S_{10} is ferrimagnetic only between $\approx 200\,°C$ and its Curie temperature of $265\,°C$; this is the cause of the 'lambda' transition in the heating curve. Fe_9S_{10} breaks down to Fe_7S_8 in the example shown and only the latter phase, with its $320\,°C$ Curie point, is expressed in the cooling curve. [After Schwarz (1975), courtesy of the Geological Survey of Canada, Ottawa, Canada.]

Hexagonal pyrrhotite (Fe_9S_{10}) is also ferrimagnetic over a restricted temperature range between the so-called λ transition around $200\,°C$, at which thermally activated vacancy ordering occurs, and the Curie point of $\approx 265\,°C$ (Schwarz and Vaughan, 1972; Fig. 3.27b). The λ transition is distinctive and diagnostic of hexagonal pyrrhotite, although in very rapid cooling, the ferrimagnetic phase and its vacancy ordering can be preserved metastably well below $200\,°C$.

Monoclinic Fe_7S_8 has no λ transition, but it does have a low-temperature transition in remanence and coercive force at 30–$35\,K$ by which it may be recognized (Dekkers *et al.*, 1989; Rochette *et al.*, 1990; Fig. 3.28). This is probably not a spin-flop transition, at which the sublattice magnetizations rotate from the *c*-plane to the *c*-axis, since M_s is almost continuous. It must be an isotropic point for the magnetocrystalline anisotropy, however, because multidomain grains lose more of their remanence in cooling through the transition than do single-domain grains, and they also have much less memory of their initial remanence when reheated through the transition (Fig. 3.28).

During thermal demagnetization of a rock above $500\,°C$, pyrrhotite transforms irreversibly, usually to magnetite (Bina and Daly, 1994). If the ambient field is not perfectly nulled, a chemical remanence can result. The breakdown of

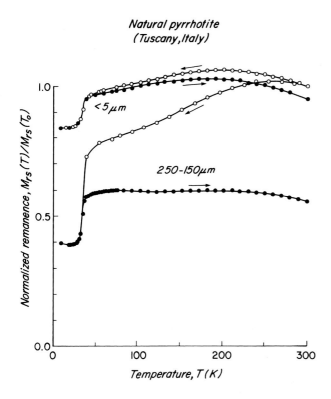

Natural pyrrhotite
(Tuscany, Italy)

Figure 3.28 Low-temperature remanence transition around 35 K in natural pyrrhotites. As in cycling across the Verwey transition in magnetite, fine single-domain grains ($< 5\,\mu$m) lose less remanence in cooling through the transition and recover more remanence in reheating (memory) than large multidomain grains. [After Dekkers *et al.* (1989) © American Geophysical Union, with the permission of the authors and the publisher.]

pyrrhotite is evidenced in strong-field thermomagnetic analysis by a large irreversible increase in magnetization. At higher temperatures, pyrrhotite transforms to hematite, either directly or by oxidation of magnetite (Dekkers, 1990).

3.11 Other magnetic minerals

Fe–Cr spinels or chromites ($FeCr_2O_4$–Fe_3O_4, often containing also Mg^{2+}, Mn^{2+}, Ni^{2+}, Al^{3+} or Ti^{4+}) are common in submarine gabbros and peridotites (Dunlop and Prévot, 1982), in marine sediments derived from these rocks, and in lunar igneous rocks. Depending on composition they may be ferrimagnetic, antiferromagnetic or paramagnetic at ordinary temperatures (Robbins *et al.*, 1971; Schmidbauer, 1971). The Curie temperature $T_C = -185\,°C$ for $FeCr_2O_4$ and rises with decreasing Cr content. Some naturally occurring chromites can carry remanence (Kumar and Bhalla, 1984), but they are seldom major contributors to the total NRM.

Siderite ($FeCO_3$) is common in carbonate sediments and rocks (Ellwood *et al.*, 1986, 1988). It is paramagnetic at ordinary temperatures and therefore carries no natural remanence. Even at room temperature, natural siderites oxidize by measurable amounts when exposed to air for periods of weeks to months, acquiring chemical remanence (CRM) in the process (Hus, 1990). When heated, rapid oxidation to magnetite or maghemite and ultimately to hematite occurs above 300 °C. In even a small stray field, the magnetite chemical remanence so produced tends to overshadow the generally weak NRM of the limestone or other carbonate sample being studied, effectively terminating the thermal demagnetization run (e.g., Özdemir and Deutsch, 1984).

3.12   Magnetism in silicates

Many silicate minerals behave paramagnetically because of the Fe^{2+}, Fe^{3+} or Mn^{2+} they contain. Examples are olivines, pyroxenes, amphiboles like hornblende and actinolite, iron-rich micas like biotite, garnets and cordierites. Measured paramagnetic susceptibilities (Table 2.2; Hunt *et al.*, 1995a) are in excellent agreement with estimates based on the paramagnetic moments calculated from eqn. (2.10): $5.4\mu_B$ (theoretically $4.9\mu_B$) for Fe^{2+} and $5.9\mu_B$ for Fe^{3+} and Mn^{2+}.

In addition, some silicates exhibit ferromagnetic behaviour which is quite variable from one sample or locality to another. An example is shown in Fig. 3.29. In some cases minerals which have no intrinsic paramagnetism, such as plagioclase feldspar, exhibit magnetic hysteresis and can carry substantial remanence. The ferromagnetism is clearly extrinsic and in a number of cases has been traced to magnetite impurities (see §14.6.2, 14.6.3). In biotites, the magnetite can be quite abundant, filling cracks and planar voids between the mica sheets. In plagioclases and pyroxenes, the magnetite seems to have exsolved from the host mineral and is crystallographically oriented, often as parallel needles within the host (Murthy *et al.*, 1971; Davis, 1981). If the needles are of single-domain size, they are capable of carrying extremely stable remanence due to their shape anisotropy (§4.2.5). We will return to this important topic in Chapter 14.

3.13   Biogenic magnetic minerals

The new science of biomagnetism (Kirschvink *et al.*, 1985) is concerned with how living organisms produce magnetic fields or use the geomagnetic field as a navigational aid. The presence of biochemically precipitated magnetite, greigite and iron oxyhydroxides in bacteria, molluscs, insects and higher animals is well established (see review by Lowenstam and Kirschvink, 1985), although it is not clear that the magnetic crystals are useful in all cases as field sensors. Lowenstam and Kirschvink distinguish between 'biologically induced' mineralization, in which the organism exercises minimal control over the precipitates, and 'organic

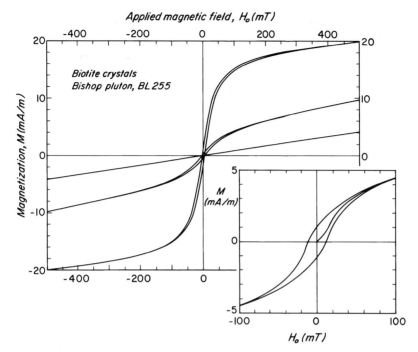

Figure 3.29 Hysteresis (M–H) curves of biotite single crystals, showing behaviour ranging from purely paramagnetic to strongly ferrimagnetic. Inset: expanded view of the middle curve between ± 100 mT. The ferrimagnetism is due to variable amounts of magnetite in the biotites.

matrix-mediated' mineralization, in which the precipitates are monomineralic, narrowly sized and crystallographically aligned. In addition, the crystals may have unusual, magnetically advantageous forms. For example, in magnetotactic bacteria (Blakemore, 1975; Mann, 1985) magnetite crystallites line up in chains parallel to their $\langle 111 \rangle$ easy axes, thereby maximizing both magnetic moments and shape anisotropy (Moskowitz $et\ al.$, 1988a).

Biological organisms are capable of 'product control', in terms of strict confinement of particle size and shape to the single-domain range (§5.5), that humans cannot yet achieve. For this reason, biogenic minerals, magnetite in particular, are of considerable interest as raw materials for experimentation. They are also being used in improved magnetic colloids to reveal details of domain structures (Futschik $et\ al.$, 1989; Funaki $et\ al.$, 1992) and are being proposed as ideal components for probe tips in magnetic force microscopes (§6.9).

In a paleomagnetic context, judging by the wide variety of organisms that are able to biomineralize magnetite or other minerals, it is reasonable to assume that this capability evolved in geologically ancient times. If so, ancient biogenic magnetites should be fossilized in sediments and the rocks that formed from them. Kirschvink and Lowenstam (1979) suggest that biogenic magnetites could be the main NRM carriers in deep-sea sediments, which are largely isolated from sources of detrital magnetite. Petersen *et al.* (1986) and Vali *et al.* (1987) have reported striking scanning electron micrographs of what appear to be fossil biomagnetic crystals and chains of crystals in Atlantic and Pacific deep-sea sediments 5–50 Ma in age and in 150-Ma-old marine limestones.

Chapter 4

Magnetostatic fields and energies

4.1   Introduction

Chapters 2 and 3 were concerned mainly with material properties like spontaneous magnetization, Curie temperature, magnetocrystalline anisotropy and
magnetostriction. If there were no other factors to consider, exchange coupling
would cause ferromagnets and ferrimagnets to be magnetized to saturation
along a magnetocrystalline or magnetoelastic easy axis throughout their
volumes, apart from thermal disordering at high temperatures. Such **single-
domain (SD)** grains do exist, but in most magnetic minerals they are quite small,
often $< 1 \mu m$ in size. Larger **multidomain (MD)** grains spontaneously subdivide
themselves into two or more domains (Fig. 1.2). M_s is uniform within each
domain (at least on a macroscopic viewing scale) but M_s vectors are in different
directions in different domains.

Why do domains exist? Landau and Lifschitz (1935) recognized that the long-
range effect of dipole–dipole interactions between atomic moments is to generate
a **magnetostatic** or **demagnetizing** energy which eventually outweighs the tendency of exchange forces and magnetocrystalline anisotropy to produce a uniform magnetization. Without the guidance of domain observations, Landau
and Lifschitz predicted the basic pattern of body and closure domains which
were later (in the 1950's) observed experimentally.

Because dipole–dipole interactions are a long-range effect, involving each pair
of dipoles in a crystal, magnetostatic calculations can be computationally gruel-
ling. Visualizing the fields involved is not easy, even for quite simple magnetization distributions. Ultimately the goal is to generate magnetization structures
which resemble the domains actually observed in large grains by starting from

first principles, without imposing initial constraints. This subject is called **micro-magnetism** and will be described fully in Chapter 7. Luckily, we can make a simple beginning which gives considerable insight and leads to the structures Landau and Lifschitz predicted. We do this by using our knowledge of electrostatics, which has close analogies with magnetostatics.

4.2   Self-demagnetization and the internal demagnetizing field

4.2.1   The parallel-plate capacitor: an electrostatic analog

Figure 4.1 illustrates the electric polarization of a rectangular slab of a dielectric (a material containing electric dipoles or charge pairs – the analog of a paramagnet). In (a), the parallel metal plates of a capacitor are given charge densities $\pm\sigma_f$ by applying a difference of potential V between them. An electric field E_0 (the vector force on a unit test charge released between the plates) is created, mainly between the plates but also fringing out at the top and bottom edges. Each field line of E_0 begins on a positive charge and follows the path a positive test charge would trace out, ending on a negative charge.

In Fig. 4.1(b), E_0 **polarizes** a sheet of dielectric which exactly fills the space between the plates (the plates themselves are not shown). That is, positive charges in atoms or molecules move in the direction of E_0 while negative charges move antiparallel to E_0. Since the positive and negative charges are bound together within atoms or molecules, the actual amount of movement is minute, but it creates atomic-scale dipole moments, all aligned (apart from thermal disordering) with E_0. The electric dipole moment per unit volume is the electric polarization P.

Imagine now taking a microscopic view in which we look first at the positive charge distribution within the dielectric (represented by the hatching downward to the right) and then at the negative volume charge distribution (hatching downward to the left). The distributions are shifted with respect to each other by the same amount that the nucleus and the centre of mass of the electron cloud within an individual atom are offset to form a dipole moment. The bulk of the dielectric is still electrically neutral, but there are two strips with atomic-scale thicknesses at the left and right surfaces which are negatively and positively charged, respectively.

The matching sheets of surface charges are called **bound charges** because they are not free to move like the conduction electrons that are responsible for the free charge density $\pm\sigma_f$ on the capacitor plates. Instead they are bound to particular near-surface atoms and hence to the dielectric itself. Their surface density

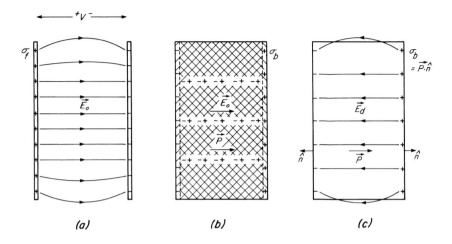

Figure 4.1 (a) Electric field lines in a parallel-plate capacitor. (b) Polarization of a dielectric between the capacitor planes, producing surface bound charges with density σ_b. (c) The depolarizing field E_d caused by the surface charges. Notice the fringing of the field lines out of the dielectric at the top and bottom.

$\pm\sigma_b$ is equal to $\boldsymbol{P} \cdot \hat{n}$ (Fig. 4.1c), where $\hat{n}$ is the outward normal to the surface. Thus there is positive charge on the right surface, where $\boldsymbol{P}$ is $\|\hat{n}$, negative charge on the left surface, and almost no charge on the top, bottom, front and back surfaces, where $\boldsymbol{P}$ is almost $\perp\hat{n}$.

The surface charges generate their own electric field E_d. Each line of E_d begins on a positive surface charge and ends on the matching negative charge on the left surface. Because E_d is opposite in direction to $\boldsymbol{P}$ (which creates it) and tends to reduce $\boldsymbol{P}$ (since it partially cancels E_0, which creates $\boldsymbol{P}$), E_d is called the **depolarizing field**. In the interior, E_d is exactly antiparallel to $\boldsymbol{P}$. However near the top and bottom edges (and the front and back edges as well), even if $\boldsymbol{P}$ is parallel to the surface, E_d fringes out of the dielectric.

4.2.2 Rectangular prism of magnetic material

We next turn to a rectangular slab of magnetic material. Because of exchange coupling, the magnetic polarization M and resulting magnetic depolarizing or **demagnetizing field** H_d in a ferromagnetic or ferrimagnetic substance are much more powerful than the corresponding dielectric polarization and depolarizing field. Another difference is that in dielectrics, an applied field E_0 is necessary to induce a polarization $\boldsymbol{P}$. (Permanently polarized dielectrics, called ferroelectrics, do exist but they are rare.) The magnetization M, on the other hand, is a sum of induced (temporary) and permanent parts:

$$M = \chi H_0 + M_r \tag{4.1}$$

where χ is magnetic susceptibility, H_0 is the externally applied magnetic field, and M_r is remanent magnetization, or remanence for short. There is no fundamental distinction between remanence and induced magnetization at the atomic level. Both are due to the same atomic moments. But macroscopically (4.1) is approximately true for very weak fields like the earth's field ($\approx$ 40 A/m or 0.5 Oe) because the 'soft' moments that produce reversible induced magnetization are localized in different grains, or different parts of grains, than the 'hard' moments that result in irreversible magnetic behaviour or remanence.

Apart from these minor differences, demagnetizing fields are the exact analog of depolarizing fields. In Fig. 4.2, an applied field H_0 (usually produced by a set of current-carrying coils) produces a net magnetization M. (The units of M are A/m, since M is the net dipole moment in A m^2 per unit volume in m^3. The processes by which M is created and the resulting magnetization distribution or domain structure will be considered in Chapter 5. The details need not concern us at this stage, provided M is reasonably uniform when viewed on a macroscopic scale.) Alternatively M may be a remanence M_r, which requires no applied field to maintain it. In either case, we can think of M as creating two sheets of matching ± surface 'charges' on the left and right faces of the crystal. These are the exact analog of dielectric bound charges. We will refer to them as **magnetic charges** or **magnetic poles**.

Magnetic poles ('monopoles') cannot be separated or extracted from a magnetic material, any more than bound charges can be separated from a dielectric. They always occur in dipole pairs. But like bound charges they are very useful in calculating internal fields and the energy that results from these fields. We should

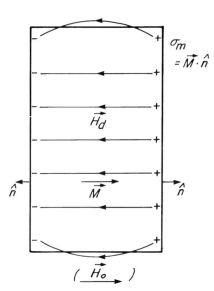

Figure 4.2 The demagnetizing field H_d in a magnetically polarized material. The polarization, the surface charge densities, and the fields are exact analogs to the electrostatic case (Fig. 4.1c).

note that magnetic 'charge' (or pole strength) Q_m and the magnetic 'charge density' or surface pole density,

$$\sigma_m = M \cdot \hat{n}, \tag{4.2}$$

have different units than electrical charge and charge density. The 'magnetic Coulomb' or unit of Q_m is A m (it is seldom used) and the unit of σ_m is A m/m^2 or A/m, as for M and H.

The **internal demagnetizing field** H_d has field lines that begin on positive magnetic charges and end on negative charges. It is antiparallel to M, or nearly so, throughout the magnetized material. A positive 'test pole' would follow a line of H_d from the right face to the left face of the model prism in Fig. 4.2, while a negative pole would move in the opposite direction. Since the two are bound together, it is clear that a dipole, with moment μ, will rotate until its positive pole points to the left and its negative pole to the right. That is, μ will tend to align with H_d, like a miniature compass needle. The torque on μ is

$$\tau = \mu \times B, \tag{4.3}$$

which leads to the expression for magnetostatic potential energy,

$$E_m = -\mu \cdot B, \tag{4.4}$$

given previously in eqn. (2.13).

B includes both the external field H_0 (or $B_0 = \mu_0 H_0$) and the demagnetizing field H_d. The total internal field inside the magnetized material is

$$H_i = H_0 + H_d = H_0 - \mathbf{N} \cdot M, \tag{4.5}$$

where $\mathbf{N}$ is the **demagnetizing tensor**, defined by

$$H_d = -\mathbf{N} \cdot M. \tag{4.6}$$

This is a logical definition. In the interior of the prism, the lines of H_d do not spread out and so H_d is constant. Also the direction of H_d is exactly antiparallel to M. In this region, $H_d = -NM$, where N is a scalar constant, the **demagnetizing factor**. Towards the edges of the crystal, where the lines of H_d fringe outward, weakening $|H_d|$ and causing its direction to diverge from M, $\mathbf{N}$ is the analogous tensor relating components of the non-parallel vectors H_d and M.

Working from the definition (4.5), we find from (4.4) that

$$\begin{aligned} E_m &= -\mu_0 V M \cdot H_0 + \tfrac{1}{2}\mu_0 V M \cdot \mathbf{N} \cdot M \\ &= -\mu_0 V M \cdot H_0 + \tfrac{1}{2}\mu_0 V N M^2 \quad \text{if } H_d \| - M. \end{aligned} \tag{4.7}$$

The factor $\tfrac{1}{2}$ occurs because calculating the magnetostatic energy due to the internal demagnetizing field is really a way of evaluating dipole–dipole interactions throughout the whole prism, and in such an enumeration, each dipole pair ends up being counted twice. The dipole moment μ has been replaced by VM, since M has been assumed uniform throughout the crystal. $H_d = -NM$ is *not* uniform, but the regions of fringing fields are often small enough that $\mathbf{N}$ can be replaced by N, giving the second alternative of (4.7). This is an approximation, but a useful one.

The two terms in (4.7) are usually separated and given different names. The **magnetic field energy** or **Zeeman energy** is

$$E_H = -\mu_0 V \boldsymbol{M} \cdot \boldsymbol{H}_0. \qquad (4.8)$$

The **magnetostatic self-energy** or **demagnetizing energy** is

$$E_d = \tfrac{1}{2}\mu_0 V \boldsymbol{M} \cdot \mathbf{N} \cdot \boldsymbol{M} \rightarrow \tfrac{1}{2}\mu_0 V N M^2 \quad \text{if } \boldsymbol{H}_d \| -\boldsymbol{M}. \qquad (4.9)$$

These are two of the basic energy terms needed to study domain structures.

4.2.3   Uniformly magnetized sphere

The sphere of radius R in Fig. 4.3 has a uniform magnetization $\boldsymbol{M}$ along the z axis and a surface pole density $\sigma_m = \boldsymbol{M} \cdot \hat{n} = M \cos \theta$. Poles are concentrated at the top and bottom surfaces where $\boldsymbol{M}$ cuts the surface at a large angle. There are no poles along the 'equator' of the sphere where $\boldsymbol{M}$ is tangential to the surface.

In order to calculate the magnitude of the internal demagnetizing field $\boldsymbol{H}_d$, we will make use of a result of elementary potential theory. Just as the external grav-

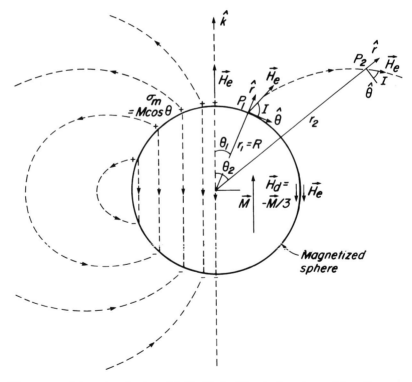

Figure 4.3   Internal and external fields (dashed lines) of a sphere with uniform magnetization $\boldsymbol{M}$ along the z-axis. Surface poles or magnetic charges are indicated by $+$ and $-$.

itational field of a sphere of uniform density is identical to that of a monopole of the same mass at the centre of the sphere, the external magnetic field H_e of a uniformly polarized sphere is identical to that of a central point dipole with the same dipole moment, $\mu = VM$. We can match H_d and H_e at the boundary of the sphere by using the conditions (from Maxwell's equations for electromagnetic fields) that the tangential component H_t of H and the normal component B_n of B are continuous across any interface. M can be regarded as a uniform remanent magnetization M_r, or an applied field H_0 can be added to induce M. In the latter case, H_0 forms a uniform background added to both H_d and H_e and does not affect the result below.

At the equator of the sphere, $H_d = -NM$ and $H_e = -[(4\pi/3)R^3 M /4\pi R^3]\,\hat{k} = -\frac{1}{3}M$, substituting $\mu = VM = (4\pi/3)R^3 M$, $r = R$ and $\theta = 90°$ in eqn. (2.2). Both fields are tangential to the sphere, and so

$$H_d = -\tfrac{1}{3}M, \qquad N = \tfrac{1}{3} \quad \text{(in SI)},$$
$$H_d = -\frac{4\pi}{3}M, \qquad N = \frac{4\pi}{3} \quad \text{(in cgs)}. \tag{4.10}$$

Notice that in cgs emu, demagnetizing fields and factors are multiplied by 4π compared to SI. At the top of the sphere on the z-axis ($r = R$, $\theta = 0$), the fields are normal to the surface. $B_i = \mu_0(H_d + M) = \mu_0(1 - N)M$ and $B_e = +\mu_0\frac{2}{3}M$, using (2.2). Equating, we obtain the same results as in (4.10). Notice that there is no source or sink of B at the surface poles. The continuity of B shows that the surface poles are not isolated monopoles. However, the surface poles *are* sources (positive poles) or sinks (negative poles) for both H_d and H_e. For this reason, in what follows we will always calculate H as our primary field, and derive B from it using $B = \mu_0(H + M)$. There are also fundamental reasons for this approach (see §2.1).

Equations (4.10) show that for two particular surface points on a magnetized sphere, the internal field is antiparallel to M (this is dictated by symmetry for these two special points) and has a magnitude $\frac{1}{3}M$. That is, for these points, the demagnetizing factor N is $\frac{1}{3}$ (or $4\pi/3$ in cgs). In §4.3.2, we will show that this result is perfectly general and that $H_d = -\frac{1}{3}M$ everywhere on the surace and in the interior of a single-domain sphere.

4.2.4   Uniformly magnetized spheroid and shape anisotropy

Figure 4.4 illustrates surface poles and demagnetizing fields for a uniformly magnetized prolate spheroid (an ellipsoid of revolution). In (a), the spheroid is magnetized parallel to its semi-major axis a. The $\pm$ pairs of surface poles are farther apart than they would be in a sphere of the same volume. By analogy with electrostatics, fields decrease with distance (they would decrease as $1/r^2$ for an isolated magnetic charge). Therefore the internal field H_d arising from these poles is less than in the case of a sphere: $N_a < \frac{1}{3}$. On the other hand, in (b) where the spheroid

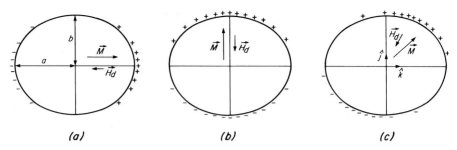

Figure 4.4 Surface poles and internal demagnetizing field H_d for a prolate spheroid with uniform magnetization M (a) along the major axis, (b) along the minor axis, and (c) oblique to the major axis. H_d is uniform in all cases, but in (c), it is not antiparallel to M.

is magnetized parallel to its semi-minor axis b, the surface poles are brought closer together and $N_b > \frac{1}{3}$.

In Fig. 4.4(c), M is at a general angle θ to the positive semi-major axis z. H_d is still uniform, but it is no longer anti-parallel to M. Let us write M in component form: $M = M \sin\theta \hat{j} + M \cos\theta \hat{k}$. Each component of M is in a principal direction and has an associated antiparallel demagnetizing field. Therefore we have that $H_d = -N_b M \sin\theta \hat{j} - N_a M \cos\theta \hat{k}$. The same result is given directly by eqn. (4.6), $H_d = -N \cdot M$, since N is a diagonal tensor with principal values (N_b, N_b, N_a).

Now we can calculate $E_d = \frac{1}{2}\mu_0 V M \cdot H_d$ for M at angle θ to the z axis:

$$E_d(\theta) = \frac{1}{2}\mu_0 V (N_a \cos^2\theta + N_b \sin^2\theta) M^2$$
$$= \frac{1}{2}\mu_0 V [N_a + (N_b - N_a)\sin^2\theta] M^2. \tag{4.11}$$

Notice that E_d is in the form of the second alternative of (4.7) or (4.9), even though H_d is not $\| - M$. However, we cannot interpret $N_a \cos^2\theta + N_b \sin^2\theta$ as an angle-dependent demagnetizing factor $N(\theta)$. Unless M is along a principal axis, N is a tensor and H_d must be found from $N \cdot M$.

N_a and N_b are linked by a general relationship among the principal demagnetizing factors N_x, N_y, N_z for M along the principal axes x, y, z, respectively:

$$N_x + N_y + N_z = 1 \quad (\text{or } 4\pi \text{ in cgs emu}). \tag{4.12}$$

For the prolate spheroid, we have $N_a + 2N_b = 1$. Figure 4.5 plots N_a for a prolate spheroid and N_b for an oblate spheroid as a function of axial ratio b/a.

The difference $N_b - N_a$ is often more important than N_a or N_b separately. Equation (4.11) shows that E_d is anisotropic, not with respect to crystallographic axes but with respect to the exterior form of the crystal, which determines the separation of surface poles for different orientations of M. The anisotropy is uniaxial for a prolate spheroid, with the easy axis of magnetization ($\theta = 0$ or $180°$) parallel to the major axis. The effect is called **shape anisotropy**. It usually over-

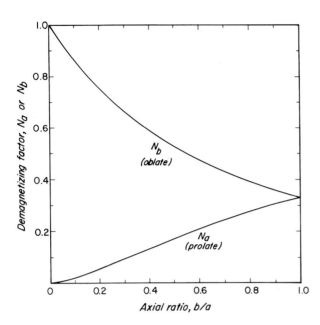

Figure 4.5 Demagnetizing factors along the axis of revolution of spheroids (the long axis for prolate spheroids or the short axis for oblate spheroids) for different elongations b/a. [After Stacey and Banerjee (1974), with the permission of the authors.]

shadows magnetocrystalline and magnetoelastic anisotropies in single-domain grains of strongly magnetic minerals like magnetite, maghemite and iron. Oblate spheroids also have shape anisotropy, but in this case M can rotate freely in the plane of the major axis. A sphere has no shape anisotropy, since $N = \frac{1}{3}$ for any direction of M.

4.2.5 Other bodies with uniform or nearly uniform magnetization

The concept of shape anisotropy is quite wide-ranging. Single-domain grains with shapes other than ellipsoids do not have uniform fields (nor is M uniform in magnitude and direction on a microscale, so they are not strictly speaking single domains – see §7.5.2). The components of H_d and M are related by

$$(H_d)_j = -N_{jk}M_k, \quad j, k = 1, 2, 3, \tag{4.13}$$

which is equivalent to (4.6). M and H_d now vary from point to point in the crystal and are related by the demagnetizing tensor N_{jk}, which also varies with position. However, even in bodies with sharp corners (i.e., real crystals), M, H_d and N tend to be quite uniform in much of the interior. Only at the corners are there large changes in the magnitude and direction of H_d (cf. Figure 4.9), and these tend to average out when the crystal is viewed as a whole.

Except for very irregular grain shapes, a demagnetizing tensor N_{av} can be found that gives a good description of the average demagnetizing field. By diagonalizing this tensor, we reduce N_{jk} to its principal values N_x, N_y, N_z. The principal axes will coincide with crystallographic symmetry axes (and thus in some cases, although not typically, the easy axis of shape anisotropy may coincide with a magnetocrystalline easy axis). A good approximation is to take for the principal demagnetizing factors the corresponding values for an ellipsoid with the same dimensional ratios $a : b : c$.

Applying this principle, a cube has the same principal demagnetizing factors as a sphere: $N_x = N_y = N_z = \frac{1}{3}$ (or $4\pi/3$ in cgs). In fact, $N = \frac{1}{3}$ for a uniformly magnetized cube no matter what the direction of M. Like a sphere, it has no shape anisotropy. In practice, this is not quite true because M is *not* uniform inside a cube, and the non-uniformity is different if M is directed along a $\langle 111 \rangle$ axis, say, than if M is along $\langle 100 \rangle$. The demagnetizing fields and factors also vary depending on the orientation of M, giving rise to a small angular dependence of $\mathbf{N}$, called **configurational anisotropy**.

When a long cylinder or needle (the limiting case of a prolate spheroid) is magnetized along its length (Fig. 4.6a), the $\pm$ pairs of surface poles at either end are separated by such a large distance that $H_d \to 0$. Thus $N_a \approx 0$. Using (4.12), we find that the transverse demagnetizing factor N_b, when M is transverse to the long axis and the $\pm$ poles are brought close together (Fig. 4.6b), is $N_b \approx \frac{1}{2}$ (2π in cgs). Single-domain magnetite grains in needle-like form, such as those that occur in some silicate minerals (§3.12) or in oxidized titanomagnetite crystals subdivided by lamellae of ilmenite (§3.1), have very large shape anisotropy. The coercivity of such grains is determined by the field that must be supplied to rotate M from one easy axis $(\theta = 0)$ to the other $(\theta = 180°)$ through the hard axis $(\theta = 90°)$:

$$H_c = H_d(90°) - H_d(0) = (N_b - N_a)M_s, \tag{4.14}$$

which gives $\frac{1}{2} \times 480 \times 10^3 = 240 \text{ kA/m}$ or 3000 Oe.

The transverse demagnetizing field in a thin sheet (the limiting case of an oblate spheroid; Fig. 4.6b) is even stronger. Since $\pm$ poles are separated by great distances when M lies in the plane of the sheet (Fig. 4.6a), $H_d \to 0$ and $N_a \approx 0$. Then from (4.12), $N_b \approx 1$ (4π in cgs). Notice that in this case, we predict $B_i = \mu_0(-NM + M) = 0$. B vanishes inside the sheet. B_e outside the sheet is also zero because fringing fields have an insignificant effect when the sheet is very large compared to its thickness (cf. Fig. 4.2). The boundary condition B_n continuous is obeyed at both transverse interfaces, as it should be. This agree-

(a)   (b)

Figure 4.6   Surface poles and internal demagnetizing field H_d for a long rod or a thin sheet magnetized (a) along the axis of the rod or in the plane of the sheet, and (b) transversely.

ment, plus the result $N = \frac{1}{3}$ for a sphere, constitute a proof of the important general principle $N_x + N_y + N_z = 1$ in two special cases.

Demagnetizing factors for simple grain shapes are summarized in Table 4.1. Detailed tabulations and graphs of demagnetizing factors for ellipsoids, cylinders and prisms appear in Osborn (1945), Stoner (1945), Nagata (1961, pp.70–71) and Joseph (1976).

4.3 General methods for finding internal fields

4.3.1 The magnetostatic potential

From Maxwell's equations, in a current-free region with static electric and magnetic fields,

$$\nabla \times \boldsymbol{H} = 0, \quad \nabla \cdot \boldsymbol{B} = \mu_0 \nabla \cdot (\boldsymbol{H} + \boldsymbol{M}) = 0. \tag{4.15}$$

It is helpful to write $\boldsymbol{H} = -\nabla \Phi$; this is permissible because $\nabla \times \nabla \Phi \equiv 0$. Φ, the **magnetostatic potential**, physically represents the potential energy E_d (cf. eqn. (4.9)) per unit magnetic charge. Using the second equation of (4.15), Φ is specified by the differential equation

$$\begin{aligned} \nabla^2 \Phi &= \nabla \cdot \boldsymbol{M}, \\ &= 0 \end{aligned} \quad \text{if} \quad \begin{aligned} \rho_\mathrm{m} &= -\nabla \cdot \boldsymbol{M} \\ \rho_\mathrm{m} &= 0, \end{aligned} \tag{4.16}$$

where $\rho_\mathrm{m} = -\nabla \cdot \boldsymbol{M}$ is the magnetic pole density inside a magnetized material (the electrical analog is the volume bound charge density ρ_b). In uniformly mag-

Table 4.1 Demagnetizing factors for uniformly magnetized grains

Grain shape and direction of M	N (SI)	N (cgs)
Sphere (or cube), any direction	$\frac{1}{3}$	$\dfrac{4\pi}{3}$
Prolate spheroid, long axis (N_a)	$< \frac{1}{3}$	$< \dfrac{4\pi}{3}$
Prolate spheroid, short axis (N_b)	$> \frac{1}{3}$	$> \dfrac{4\pi}{3}$
(for exact values, see Fig. 4.5)		
Cylinder, axial M	see eqns. (4.28), (4.29)	
Needle, axial M	≈ 0	≈ 0
Needle, transverse M	$\approx \frac{1}{2}$	$\approx 2\pi$
Thin disc or sheet, in-plane M	≈ 0	≈ 0
Thin disc or sheet, transverse M	≈ 1	$\approx 4\pi$

netized grains, such as we have been studying in §4.2, or within uniformly magnetized domains and some types of domain walls (Bloch walls; §5.3), there are no volume poles. Then we use the second alternative of (4.16), which is Laplace's equation. Notice that finding Φ, and hence H_d and H_e, by solving Laplace's equation is equivalent to assuming that the only sources of H_d and H_e are surface poles.

The general solution to the first alternative of (4.16) (Poisson's equation) is

$$\Phi(r) = -\frac{1}{4\pi} \int_{V'} M \cdot \nabla \frac{1}{|r - r'|} dV', \tag{4.17}$$

in which r specifies the location of the field point where Φ (and ultimately H_d or H_e) are being evaluated and r' is the location of a source point within the magnetized region V' bounded by the surface S'. In cgs, the $1/4\pi$ is omitted. Evaluating the integrals for a sufficient number of points to give a reasonable picture of the spatial pattern of H_d or H_e is extremely laborious unless the distribution of M is simple. Equation (4.17) is therefore not recommended as an actual prescription for calculating Φ.

However, we can transform (4.17) in a way that gives considerable insight. Using Green's theorem for two general scalar potentials Ψ_1, Ψ_2:

$$\int_{V'} \nabla \Psi_1 \cdot \nabla \Psi_2 \ dV' + \int_{V'} \Psi_1 \nabla^2 \Psi_2 \ dV' = \int_{S'} \Psi_1 \frac{\partial \Psi_2}{\partial n} dS' \tag{4.18}$$

and putting $\Psi_1 = 1/4\pi|r - r'|$, $\nabla \Psi_2 = M$, the first term in (4.18) becomes the r.h.s. of (4.17).

Rearranging we find

$$\Phi(r) = \int_{V'} \frac{1}{4\pi|r - r'|} \nabla \cdot M \ dV' - \int_{S'} \frac{1}{4\pi|r - r'|} M \cdot \hat{n} \ dS'$$

$$= -\int_{V'} \frac{\rho_m}{4\pi|r - r'|} dV' - \int_{S'} \frac{\sigma_m}{4\pi|r - r'|} dS'. \tag{4.19}$$

Equation (4.19) explicitly demonstrates that Φ has two sources, volume poles and surface poles. Furthermore, we recognize the integrands as Coulomb potentials of magnetic charges in small regions in the interior and on the surface of the magnetized source region. This is the underlying justification for the approach taken in §4.2.

One simple case in which (4.17) can be evaluated directly is that of a magnetic dipole (Fig. 2.1a) oriented along the positive z axis, so that $\mu = \mu \hat{k}$. V' shrinks to a single point, the origin ($r' = 0$), so that (4.17) becomes

$$4\pi \Phi(r) = -\mu \cdot \nabla \frac{1}{|r|} = -\mu \frac{\partial}{\partial z}\left(\frac{1}{r}\right) = +\mu \frac{z}{r^3} = \mu \frac{\cos \theta}{r^2}. \tag{4.20}$$

$H_e = -\nabla \Phi$ then correctly gives the dipole field of eqn. (2.2).

4.3.2   Internal and external fields of a uniformly magnetized sphere

Let us return to the uniformly magnetized sphere of radius R (Fig. 4.3). Since $\boldsymbol{M} = M\hat{k}$ is uniform and ∇ does not involve the source region coordinates $\boldsymbol{r}'$, we can write (4.17) as

$$
\begin{aligned}
\Phi(\boldsymbol{r}) &= -\boldsymbol{M} \cdot \nabla \int_{V'} \frac{1}{4\pi|\boldsymbol{r} - \boldsymbol{r}'|} \mathrm{d}V' \\
&= -M \frac{4\pi}{3} R^3 \frac{\partial}{\partial z}\left(\frac{1}{4\pi r}\right) = \mu \frac{\cos\theta}{4\pi r^2} \quad \text{if } r > R \\
&= -\frac{M}{6} \frac{\partial}{\partial z}(3R^2 - r^2) = \tfrac{1}{3}Mz \quad \text{if } r \le R.
\end{aligned}
\tag{4.21}
$$

To arrive at the alternative solutions in (4.21), we have used results from elementary potential theory (e.g., the gravitational potential of a homogeneous sphere; Serway, 1986, pp. 302–303). The first alternative shows that the external magnetic field $\boldsymbol{H}_e$ of a uniformly magnetized sphere is the same as that of a dipole of moment $\mu = V\boldsymbol{M}$ at the centre of the sphere, as we stated in §4.2.3. From the second alternative, we find

$$
\boldsymbol{H}_i = \boldsymbol{H}_d = -\nabla \Phi = -\frac{\partial \Phi}{\partial z}\hat{k} = -\tfrac{1}{3}M\hat{k} = -\tfrac{1}{3}\boldsymbol{M}.
\tag{4.22}
$$

Thus a sphere has a uniform demagnetizing field $-\tfrac{1}{3}\boldsymbol{M}$ everywhere in its interior.

An alternative approach is to solve the differential equation for Φ as a boundary value problem. Since $\boldsymbol{M}$ is uniform, there are no volume poles and the second alternative of (4.16) is to be solved. The solutions of Laplace's equation for a source region with a spherical boundary are spherical harmonics:

$$
\Phi_i = \sum A_l r^l P_l(\cos\theta), \qquad \Phi_e = \sum B_l r^{-l-1} P_l(\cos\theta).
\tag{4.23}
$$

With $\boldsymbol{M}$ along the z axis, the symmetry dictates that there is no dependence on azimuthal angle ϕ. No $l = 0$ terms are possible because then Φ_i would be constant and there would be no internal field. We can begin with the $l = 1$ harmonics (and in fact these will prove to be sufficient). We have

$$
\Phi_i = A_1 r \cos\theta = A_1 z, \qquad \Phi_e = B_1 \frac{\cos\theta}{r^2}.
\tag{4.24}
$$

At the equator of the sphere ($r = R$, $\theta = 90°$), $\boldsymbol{H}_i = -A_1\hat{k}$ is tangential to the surface and so is $\boldsymbol{H}_e = B_1 (\sin\theta/R^3)\hat{\theta} = -(B_1/R^3)\hat{k}$. By the continuity of H_t we have that $B_1 = A_1 R^3$. Now on the z-axis at the surface of the sphere ($r = R$, $\theta = 0$), both fields are normal. Applying continuity of B_n, we have that $-A_1 + M = 2B_1/R^3 = 2A_1$, and so $A_1 = M/3$. Thus we have

$$
\Phi_i = \frac{M}{3}z, \qquad \boldsymbol{H}_i = \boldsymbol{H}_d = -\frac{1}{3}\boldsymbol{M},
\tag{4.25}
$$

and

$$\Phi_{\mathrm{e}} = \frac{MR^3}{3}\frac{\cos\theta}{r^2} = \mu\frac{\cos\theta}{4\pi r^2}, \tag{4.26}$$

where $\mu = (4\pi/3)R^3 M$ is the dipole moment of the sphere. These are the same results we found in (4.21) and (4.22).

4.3.3   Internal field of a uniformly magnetized cylinder

In Fig. 4.7, the field point P is on the axis of a circular cylinder of length $2L$ and radius R at a distance z from the centre. The cylinder has uniform magnetization M along its axis, resulting in positive magnetic poles on the right end of the cylinder and negative poles on the left end, with densities $\sigma_{\mathrm{m}} = \pm M$. There are no volume poles.

Working directly from (4.19), it is easy to show that

$$\Phi(z) = M\{z + \tfrac{1}{2}[R^2 + (L-z)^2]^{1/2} - \tfrac{1}{2}[R^2 + (L+z)^2]^{1/2}\}. \tag{4.27}$$

Then

$$\boldsymbol{H}_{\mathrm{d}} = -\frac{\mathrm{d}\Phi}{\mathrm{d}z}\hat{k} = -\boldsymbol{M}\left\{1 - \frac{L-z}{2[R^2 + (L-z)^2]^{1/2}} - \frac{L+z}{2[R^2 + (L+z)^2]^{1/2}}\right\}. \tag{4.28}$$

At the centre of the cylinder ($z = 0$), this becomes

$$\boldsymbol{H}_{\mathrm{d}} = -\boldsymbol{M}\left[1 - \frac{L}{(R^2 + L^2)^{1/2}}\right]. \tag{4.29}$$

The limiting cases are

$$\boldsymbol{H}_{\mathrm{d}} \to 0 \quad\text{as}\quad L/R \to \infty \quad\text{(needle, axial }\boldsymbol{M}\text{)} \tag{4.30}$$

and

$$\boldsymbol{H}_{\mathrm{d}} = -\boldsymbol{M} \quad\text{as}\quad L/R \to 0 \quad\text{(thin disc, transverse }\boldsymbol{M}\text{)} \tag{4.31}$$

These results confirm the conclusions of §4.2.5.

Macroscopic bodies as well as microscopic grains can have shape anisotropy. Paleomagnetic samples, for example, are normally cylindrical cores with a fairly uniform average magnetization. In order to avoid deflection of the NRM vector

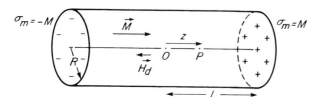

Figure 4.7 Surface poles and internal demagnetizing field $\boldsymbol{H}_{\mathrm{d}}$ for a cylinder with uniform axial magnetization $\boldsymbol{M}$.

by internal demagnetizing fields or to avoid introducing a non-intrinsic source of anisotropy in magnetic susceptibility anisotropy measurements (§16.4.1), we need to eliminate as nearly as possible the sample's shape anisotropy by making the transverse and longitudinal demagnetizing factors of the cylinder the same. This is approximately the case if the length to diameter ratio, $L/2R$, of the sample is about 0.9 (see Stacey and Banerjee, 1974, p.45 for a discussion).

4.3.4   Internal fields in cubes with uniform and non-uniform M

In a single-domain cube (Fig. 4.8) with $M = M_s$, along $+x$ as shown, positive surface poles appear on the $+x$ face of the cube and negative poles on the $-x$ face, with densities $\sigma_m = \pm M_s$. The direct approach to finding H_d would be to sum the vector dipole fields (from eqn. (2.2)) of all atoms at the location $P(r)$. This is too large a task for even a supercomputer! A reasonable approach is to divide the model crystal into $n \times n \times n$ cubic cells, approximate H_e of each cell by a dipole field, and calculate the resultant vector field at P of the $n^3 - 1$ cells other than the one containing P. To map out the field, this calculation has to be repeated for the field point P at the centre of each of the other cells. In all,

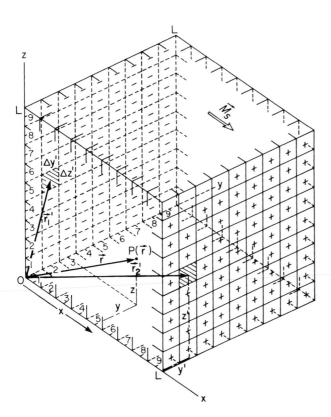

Figure 4.8   Example of subdivision of a model grain into cells for a micro-magnetic calculation of internal fields. The grain in this example is a single-domain cube with M_s along the x-axis. Therefore poles appear only on matching faces of surface cells (hatched) at opposite ends of the grain.

$3n^3(n^3 - 1)$ vector components must be computed. For the cube in Fig. 4.8, which has been subdivided into 9 cells on a side, just under 1.6 million individual calculations are required.

There are two problems with this approach. First, it is not clear a priori how fine a discretization or subdivision into cells is needed to obtain a close approximation to the true field of the (macroscopically) continuous magnetized medium. Trial and error is the easiest way of testing the convergence of H_d (see Williams *et al.*, 1988). This adds considerably to the computation time. The second problem is that although the external field of a sphere is a dipole field at any distance outside the sphere (even, in the limit, at the sphere's surface), this is not true for a cube. Thus a special approach is needed to deal with the fields of near-neighbour cells.

Another approach is to calculate numerically the resultant at P of all the vector Coulomb fields of the 9×9 positive surface charges of the $+x$ face and the matching 9×9 negative poles of the $-x$ face. This is computationally more efficient, since only $6n^5$ vector components are to be calculated (about 135 000, $< 1/10$ as many as in the dipole field calculation). There is also an order of magnitude reduction in the number of calculations needed to test the discretization for convergence. The near-neighbour problem is also less severe, except for field points near the $\pm x$ faces.

An analytical solution is also possible, working directly from (4.19) to find $\Phi(x, y, z)$ at P and then differentiating the result with respect to x, y and z to find the components of H_d. The equations are so cumbersome that there is not much reduction in computation time (Dunlop *et al.*, 1990a), but H_d is exact and covers any possible point P inside the cube (or H_e at any point outside the cube). The only discretization is in approximating a non-uniform M by cells with different uniform M vectors. Figures 4.9 and 4.10 show the results of Dunlop *et al.* for the single domain cube of Fig. 4.8 and for a non-uniform M, a two-domain cube with a Bloch wall.

In much of the interior of a single-domain cube, the internal field due to M is a demagnetizing field (i.e., antiparallel to M). Its strength varies but the volume-averaged field is exactly $-\frac{1}{3}M$. This is as it should be because the demagnetizing energy E_d is $\frac{1}{2}\mu_0 V \frac{1}{3}M_s^2$ for a SD cube as for a SD sphere (this is the basis for the statement in §4.2.5 that $N = \frac{1}{3}$ for a uniformly magnetized cube). H_d is also exactly $-\frac{1}{3}M$ at the centre of the cube. Elsewhere, especially in the corners, H_d is deflected from $-x$ and has transverse (y and z) components. Note that the field lines fringe out of the cube at the sides and edges, exactly as for a capacitor (Figure 4.1). The minimum magnitudes of H_d are at the sides, where the field lines are most widely spaced. H_d is maximum near the $\pm x$ pole faces where the field lines converge and end on surface poles.

Inside a two-domain grain, the fields within domains are considerably reduced. This is because each domain has an axial ratio $b:a$ of $\approx \frac{1}{2}$ and M is in an

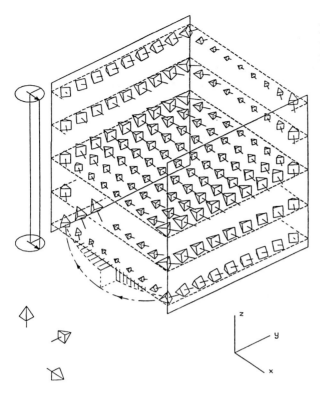

Figure 4.9 Calculated internal demagnetizing field H_d at the centres of selected cells in the uniformly magnetized single-domain cube of Fig. 4.8. The size of each arrow is proportional to the strength of H_d. The x, y, z arrowheads at the lower left each have magnitude $0.5M_s$. H_d at the centre of the grain has a value $(1/3)M_s$. Notice the fringing of field lines out of the corners and edges of the grain. [After Dunlop *et al.* (1990a) © American Geophysical Union, with the permission of the authors and the publisher.]

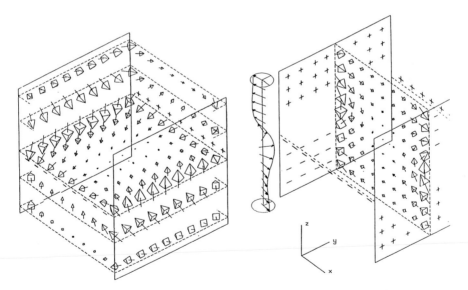

Figure 4.10 Calculated internal demagnetizing field H_d vectors on selected planes and lines inside a model grain with two domains separated by a Bloch wall. A profile of M_s parallel to z, crossing the wall and both domains, is shown at centre. The demagnetizing fields are much smaller than in the single-domain grain (Fig. 4.9), except where the wall intersects the $\pm x$ faces. Here tight flux linkages between adjacent $+$ and $-$ pole sheets lead to strong local fields (right-hand diagram). [After Dunlop *et al.* (1990a) © American Geophysical Union, with the permission of the authors and the publisher.]

easy direction of shape anisotropy. The strongest fields are within the domain wall, particularly where the wall meets the $\pm x$ faces. Here there are short flux links between adjacent $+$ and $-$ poles on the same face, due to the two domains, and a high flux density B. Note that similar tight flux links pass *outside* the crystal. These will attract magnetic colloid particles, which make the flux lines visible (Fig. 1.2), just as iron fillings spread over a bar magnet reveal its field lines.

Numerical computations of the demagnetizing tensor N are now very efficient (Newell *et al.* 1993a). The general formulation of N was developed as a means of calculating demagnetizing energies $E_d = \frac{1}{2}\mu_0 V M \cdot N \cdot M$ required in micromagnetic calculations of the minimum-energy magnetization distribution M (see next section and Chapter 7). Finding the internal field $-N \cdot M$ is one step in these calculations.

4.4 The demagnetizing energy or magnetostatic self-energy

In all the calculations we have considered so far, M was prescribed and the resulting H_d and E_d were then found. In reality, the distribution of M is usually a more interesting and difficult problem than the distribution of H_d. Finding the magnetization structure requires minimizing the total energy, of which E_d is one part, and by far the more laborious part to calculate.

If no initial constraints are being applied, one subdivides the model crystal into cells, usually cubes, square rods or rectangular sheets. Within each cell, $M = M_s$ (taking account of short-range exchange coupling) but the direction of M is allowed to vary from cell to cell. Many iterations of possible mutual orientations of M vectors are required to converge on a minimum-energy structure.

The most direct approach to finding E_d would be to calculate the dipole–dipole interaction energy between the moments of all pairs of cells. As in §4.3.4, there is a problem with near-neighbour interactions, but this can be handled using numerical results for a uniformly magnetized cube (e.g., Fig. 4.9). The energy of interaction of two dipoles, whose moments μ_1, μ_2 make angles θ_1, θ_2 with the line r_{12} joining them, is

$$E_{12} = \mu_0 \frac{\mu_1\mu_2}{4\pi r_{12}^3} [\cos(\theta_1 - \theta_2) - 3\cos\theta_1\cos\theta_2]. \tag{4.32}$$

The main problem with this approach is that it takes a large amount of computing time. With an $n \times n \times n$ array of cells, there are $\frac{1}{2}n^3(n^3 - 1)$ pair interactions to be evaluated in each iteration of the model. If $n = 9$ as in Fig. 4.8, this is about 38 000 interactions per iteration. Because M varies in direction for each cell in each iteration, all the (θ_1, θ_2) change at each step and there is no easy way of economizing on the computational labour.

A much more efficient approach, developed by Rhodes and Rowlands (1954), makes use of the transformation of the magnetostatic potential Φ into a sum or integral of Coulomb potentials (eqn. (4.19)). Since M is uniform within each cell, $\nabla \cdot M = 0$ and volume poles are eliminated from all model structures. Each cell has surface pole densities $\sigma_m = M_s \cdot \hat{n}$ on its six faces. We can think of the poles on these faces as six discrete magnetic charges Q_m. The demagnetizing energy is then the sum of all Coulomb pair interaction energies:

$$E_d = \frac{1}{2}\frac{\mu_0}{4\pi}\sum \frac{Q_j Q_k}{r_{jk}}, \quad j,\, k = 1, \ldots, 6n^p, \tag{4.33}$$

where $p = 3$ for an array of $n \times n \times n$ cells (3-dimensional micromagnetic calculation), $p = 2$ for a 2-dimensional calculation with $n \times n$ cells and $p = 1$ for a 1-dimensional calculation. Rectangular model crystals with $l \times m \times n$ cells would require calculating $\frac{1}{2} \times (6lmn)^2$ interactions.

An example of an array of n cells suitable for a 1-dimensional magnetization minimization is shown in Fig. 4.11. This is a fairly crude discretization in which domain walls are subdivided into only two cells or slabs. Since M_s is constrained to lie in the plane of each slab (this is appropriate for lamellar domains separated by 180° Bloch walls), only four (and in many cases only two) of the six faces of each cell are 'charged'. The number of interactions is thereby much reduced.

As in internal field calculations, there is a near-neighbour problem if charged rectangles are approximated by single charges. Self-energies ($j = k$) and mutual energies of charges on faces shared by two cells cannot be approximated by interactions of discrete charges. Rhodes and Rowlands dealt with this difficulty by solving the general problem of interactions between uniformly charged

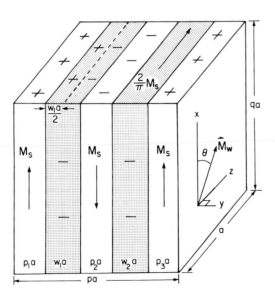

Figure 4.11 Coarse subdivision of a model rectangular grain into slabs of uniform magnetization (M_s in the domains; $(2/\pi)M_s$ in the walls, representing an average of M_w across each wall). Such a discretization is suitable for low-resolution one-dimensional modelling of magnetization structure (see Chapter 7). [After Argyle and Dunlop (1984) © American Geophysical Union, with the permission of the authors and the publisher.]

rectangular sheets of identical size, with their planes either parallel or perpendicular. Their analytic function (the Rhodes and Rowlands' or RR function) is therefore exact for interactions at all spacings.

As a simple example, the per-unit-volume Coulomb interaction energy e_{ij} of domains i and j, bounded at their tops and bottoms by pole sheets of width $p_i a$ and $p_j a$, and with separation r_{ij} (Fig. 4.11) is (Dunlop, 1983a)

$$e_{ij} = (-1)^{i+j} 2\sigma_{mi}\sigma_{mj}[\Delta F(p_i + p_j + r_{ij}, \; q) + \Delta F(r_{ij}, \; q)$$
$$-\Delta F(p_i + r_{ij}, \; q) - \Delta F(p_j + r_{ij}, \; q)]. \tag{4.34}$$

Here σ_{mi}, σ_{mj} are surface pole densities and $\Delta F(p, \; q) = F(p, \; 0) - F(p, \; q)$, where the RR function $F(p, \; q)$ is

$$F(p, \; q) = (p^2 - q^2)\sinh^{-1}\left[\frac{1}{(p^2 + q^2)^{1/2}}\right]$$
$$+ p(1 - q^2)\sinh^{-1}\left[\frac{p}{(1 + q^2)^{1/2}}\right]$$
$$+ pq^2\sinh^{-1}\left(\frac{p}{q}\right) - \pi pq + q^2\sinh^{-1}\left(\frac{1}{q}\right)$$
$$+ 2pq\tan^{-1}\left[\frac{q(1 + p^2 + q^2)^{1/2}}{p}\right]$$
$$- \tfrac{1}{3}(1 + p^2 - 2q^2)(1 + p^2 + q^2)^{1/2}$$
$$+ \tfrac{1}{3}(1 - 2q^2)(1 + q^2)^{1/2}$$
$$+ \tfrac{1}{3}(p^2 - 2q^2)(p^2 + q^2)^{1/2} + \tfrac{2}{3}q^3. \tag{4.35}$$

The great advantage of the RR approach is that all geometrical factors – the spacings of the rectangular pole sheets and their angles of offset – remain fixed. The geometrical part of each RR coefficient is calculated only once, and stored for use in each successive iteration. Varying the magnetization, i.e., the set of orientation angles of M_s for each cell, merely produces a different set of scalar σ_{mi}, σ_{mj} values on the faces of the cells, which are quickly multiplied by the stored geometrical coefficients to complete an iteration of energy. The saving in computing time is substantial.

Still more efficient numerical methods of calculating E_d are actively under pursuit. For example, Fourier transform methods (Yuan and Bertram, 1992; Berkov et al., 1993; Fabian et al., 1996) can reduce the number of computations in each iteration significantly. The model crystal can then be more finely subdivided to increase the resolution of its theoretical minimum-energy magnetization structure. We will return to micromagnetic model calculations in Chapter 7, after discussing simple domain structures based on elementary versions of the energy terms.

Chapter 5

Elementary domain structures and hysteresis

5.1   Introduction

Landau and Lifschitz (1935) predicted the existence of ferromagnetic domains as a means of reducing magnetostatic energy. As they put it, 'Elementary regions of nearly saturation magnetization in a ferromagnetic crystal result from the demagnetizing effect of the surface of the body'. Their predicted domain configurations (Fig. 5.1a) agree quite closely with the patterns observed on a polished {110} surface of magnetite (e.g., Figs. 1.2, 5.8b). Landau and Lifschitz proposed a layered structure: lamellar **body domains**, with alternating directions of magnetization M_{s1}, M_{s2}, in the interior of the crystal; wedge-shaped surface **closure domains**, with M_{sc} parallel or nearly parallel to the surface, providing flux closure paths between adjacent domains; and narrow **domain walls** in which M_s rotates through 90° or 180° between neighbouring domains. In magnetite, M_s rotates through 70.5° or 109.5° (the angles between adjacent $\langle 111 \rangle$ easy directions) rather than 90° across walls between closure and body domains (Fig. 5.1b).

The Landau–Lifschitz (LL) structure is remarkably effective in reducing demagnetizing energy. M_s is uniform within body and closure domains, and so there are no volume poles. M_s is parallel to 180° walls, and so $\sigma_m = M_s \cdot \hat{n} = 0$ for these walls. The 90° (or 70.5° and 109.5°) walls bisect the angle between M_s vectors in neighbouring body and closure domains. Thus σ_m due to the body domain exactly cancels σ_m due to the closure domain, and no surface poles appear on any interior boundaries.

For an idealized rectangular prismatic crystal with $\langle 100 \rangle$ easy axes, e.g., iron or TM60, there are no poles at the crystal surface because M_s vectors in both body and closure domains are surface parallel (Fig. 5.1a). In a magnetite prism,

103

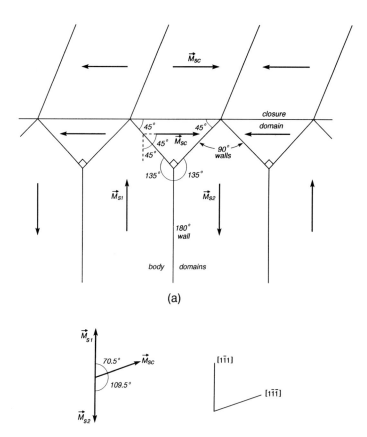

(a)

(b)

Figure 5.1 (a) The Landau–Lifschitz structure, with body and closure domains separated by 90° walls, appropriate for a material with positive K_1 and $\langle 100 \rangle$ easy axes. (b) A possible domain structure, with $\boldsymbol{M}_s$ vectors along $\langle 111 \rangle$ easy axes and 71° and 109° walls separating closure and body domains. This structure produces poles (+ and − signs) where $\boldsymbol{M}_{sc}$ cuts the crystal surface. [After Özdemir *et al.* (1995) © American Geophysical Union, with the permission of the authors and the publisher.]

the $\langle 111 \rangle$ easy axes are oblique to the crystal faces, resulting in some surface poles (Fig. 5.1b). However, the demagnetizing energy is so important compared to the crystalline anisotropy energy that M_s vectors close to the crystal boundary will be deflected away from $\langle 111 \rangle$ so as to be nearly surface-parallel, greatly reducing σ_m. This is sometimes called the μ^* effect.

The only remaining sources of magnetostatic energy are domain walls themselves. In the interior of large crystals, the walls are narrow relative to their lateral dimensions. On a perpendicular traverse across the wall, M_s rotates on a cone about the wall normal (a **Bloch wall**; see Fig. 5.9). For a 180° wall, the 180° cone is a circle in the plane of the wall. 70.5°, 90° and 109.5° cones have both a component of M_s in the plane of the wall ($M_{s\perp}$, perpendicular to the cone axis or wall normal $\hat{n}$; Fig. 5.9) and a component $M_{s\|}$ parallel to $\hat{n}$. $M_{s\|}$ is constant across the wall; therefore $\nabla \cdot M_s = 0$ and no volume poles appear within the wall (cf. eqn. (4.16)). The in-plane component $M_{s\|}$ does produce surface poles where the domain walls intersect the crystal surface but since $N \approx 0$ for a thin sheet with a planar magnetization (§4.2.5, Table 4.1), the demagnetizing energy of Bloch walls in large crystals is small.

Domains may be simple in form, like the ones in Fig. 5.1, or quite complex. It is the domain walls, within which moments rotate in a wave-like fashion from one atom to the next and M_s changes rapidly over short distances (typically 0.1–0.5 μm), which are actually viewed in most methods of domain observation. Because M_s in domains tends to be nearly parallel to the surface, there is little or no leakage of magnetic flux (i.e., lines of B) out of the crystal from the domains themselves, unless closure domains are absent (as in materials with strong uniaxial magnetocrystalline anisotropy). In Bloch walls, however, M_s is oblique to the surface and a concentration of flux lines emerges from or reenters the crystal along the surface intersection of the wall (Fig. 4.10). A colloidal suspension of ultrafine magnetite particles will be attracted to these locations; this is the basis of the Bitter (1931) technique of domain observation.

In domain observations by the Bitter method, M_s directions within domains must be inferred from the response of their bounding walls to applied fields (§5.6). Not all walls are equally visible. Visibility is greater if M_s has a large surface-normal component where the wall cuts the polished viewing surface. This is the case for 180° Bloch walls bounding body domains, but less so for the 70.5°, 90° or 109.5° walls bounding closure domains (because $M_{s\perp}$ is smaller for these walls), and not at all for near-surface **Néel walls**, in which M_s rotates in the plane of the surface (this amounts to a μ^* effect for walls; see Fig. 7.8). Narrow walls may be easier to see than broad ones, because the colloid is more concentrated. Fine structures, in which M_s rotates gradually within a nearly uniform domain, are not easy to detect.

A problem with almost all methods of domain observation, including the Bitter technique, is that the sample must be polished to a smooth surface. One has then modified the shapes and structures of the magnetic grains being observed. Even if the surface is unpolished (e.g., Moskowitz *et al.*, 1988b), only near-surface structures can be viewed. These may not be representative of body domains in the grain's interior.

5.2   Simple domain structures and their energies

5.2.1   Alternative domain states of a particle

Domains have simple and predictable forms if the grain geometry is simple. Figure 5.2 shows four alternative minimum-energy states for the same particle. In the single-domain (SD) state (Fig. 5.2a), there is a high density of surface

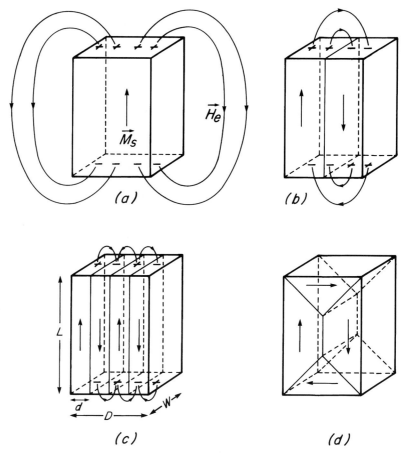

Figure 5.2   Alternative domain structures for the same grain. (a) single-domain structure with widely separated + and − poles, linked by large external flux loops. (b) Two-domain structure, with less pole separation and more localized flux linkages. (c) Four-domain state. (d) Two-domain state with closure domains.

poles and a large magnetostatic energy associated with both the internal field H_d and the external (or stray) field H_e arising from these poles. In the two-domain (2D) state (Fig. 5.2b), the surface pole densities have not changed, but the domains are long and narrow compared to the SD state. Their demagnetizing factors and demagnetizing energy (§5.2.2) are about one-half of SD values. On the other hand, a wall has been created at the cost of some added exchange and crystalline anisotropy energy (§5.3).

Notice that it is usually energetically favourable to subdivide the grain by a wall parallel to the long axis (see Fig. 5.3). The reduction in E_d compared to the same grain with a shorter transverse wall outweighs the extra wall energy. Any rotation of the domains away from the long axis results in shape anisotropy, just as in SD particles.

In the four-domain (4D) state (Fig. 5.2c), the particle has long plate-like domains with small N and E_d. Equivalently, the external flux loops of H_e or B_e linking $+$ and $-$ surface poles are now very short and tight, and generate little stray-field energy. However, the wall energy has tripled.

Another way of reducing or eliminating E_d is by adding closure domains to the two body domains (Fig. 5.2d). There are now no surface poles and no external stray fields; the flux closure is all internal. The wall energy is also reduced compared to the 4D state. However, the closure domains generate additional magnetoelastic energy. Because λ_{111} is positive in magnetite while λ_{100} is negative (Fig. 3.8a), domains magnetized in or near $\langle 111 \rangle$ easy directions will tend to spontaneously expand along the direction of M_s and contract in transverse directions. Domains with parallel M_s vectors (e.g., Fig. 5.2a–c) can expand compatibly, but domains with perpendicular M_s vectors cannot. The body and closure domains in Fig. 5.2(d) are strained. Both are shorter than they would be in isolation. The stress that results from incompatible magnetostrictive strains of body and closure domains gives rise to magnetoelastic energy (§2.8.4).

All four structures shown in Fig. 5.2 are possible minimum-energy states of the same particle. That is, it takes energy to transform one into another. Which of the alternative structures is the **equilibrium** state, with the lowest absolute energy? In order to answer this question, we shall develop expressions for demagnetizing and wall energies in the next two sections.

5.2.2 Demagnetizing energy of lamellar domain structures

In Chapter 4, we determined demagnetizing energies of uniformly or almost uniformly magnetized grains of different shapes. Such single-domain grains respond to an applied field H_0 by a rotation of M_s. Now we are concerned with multidomain grains which have strong and uniform magnetizations within individual domains, but zero or near-zero net magnetization overall in the absence of an applied field. Domain rotation is minor compared to the displacement of domain

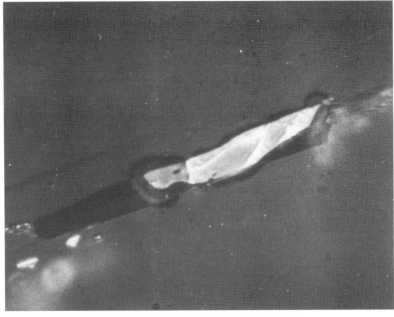

(a)

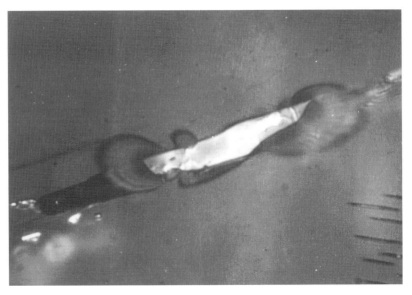

(b)

Figure 5.3   Elongated magnetite grain in a crack in a biotite crystal, with several domains parallel to the long axis. (a) In approximately zero field, a small external flux loop linking poles on two adjacent domains forms at one end of the grain. (b) in an 8-mT (80-Oe) field applied subparallel to the grain axis (the series of dark lines indicate the field direction), walls are driven to the edge of the grain and larger external flux loops develop. [After Özdemir and Dunlop (1992) © American Geophysical Union, with the permission of the authors and the publisher.]

walls, enlarging domains whose M_s vectors are favourably oriented (parallel or nearly parallel to H_0).

During this process of **wall displacement**, the structure and energy of a wall (§5.3) do not change unless crystal defects impede wall motion ('pinning'). Ignoring both wall pinning and internal wall structure for the moment, the only source of energy in lamellar domain structures (Fig. 5.2a–c) is magnetostatic. As an approximation, we can write the demagnetizing energy as

$$E_d = E_{d0} + \tfrac{1}{2}\mu_0 VNM^2 = E_{d0} + 2\mu_0 nLWN(x)M_s^2 x^2/d, \tag{5.1}$$

in which n and d are the number and width of domains in the demagnetized state, L and W are grain length and width (Fig. 5.2c), and x is the distance each wall is displaced in response to H_0. Equation (5.1) is modelled on the corresponding SD expression (4.9), but the anology between the two expressions is superficial. Even in the 'demagnetized' state ($x = 0$ and $M = 0$), the individual domains are strongly magnetized and have a substantial demagnetizing energy E_{d0}. When walls move, unfavourably oriented domains grow narrower and their individual demagnetizing factors and energies decrease (§4.2.5). On the other hand, favourably oriented domains widen and their N and E_d values increase. The increases outweigh the decreases, with the result that E_d increases with increasing x.

E_d and the effective demagnetizing factor for the grain as a whole,

$$N = \frac{2(E_d - E_{d0})}{\mu_0 VM^2}, \tag{5.2}$$

are plotted in Fig. 5.4 for MD cubes ($L = W = D$) as functions of normalized wall displacement or magnetization r. To a first approximation, E_d increases quadratically with x or M and the MD demagnetizing factor N is a constant. This fact justifies the form of eqn. (5.1). The calculations on which Fig. 5.4 is based (Dunlop, 1983a) used the RR formulation of E_d (§4.4) and assumed walls of negligible width and energy with equal displacements. Xu and Merrill (1987) allowed the walls to have varying displacements and obtained very similar results. A full micromagnetic calculation, with wall structure and closure domains or other deviations from one-dimensional magnetization structure included, has yet to be carried out.

Figure 5.5(a) shows how E_{d0}, the demagnetizing energy when $H_0 = 0$, decreases as the number n of domains increases. (If n is even, the grain has zero net magnetization when $H_0 = 0$, but if n is odd, e.g., the three-domain curve in Fig. 5.4a, there is one unbalanced domain and a small net magnetization in zero applied field.) A useful approximation,

$$E_{d0} \approx \frac{1}{n}(E_{d0})_{SD}, \tag{5.3}$$

is plotted in Fig. 5.5(a). Another widely used approxiation (Kittel, 1949, eqn. (2.4.4); Rhodes and Rowlands, 1954, eqn. (3.18)) is

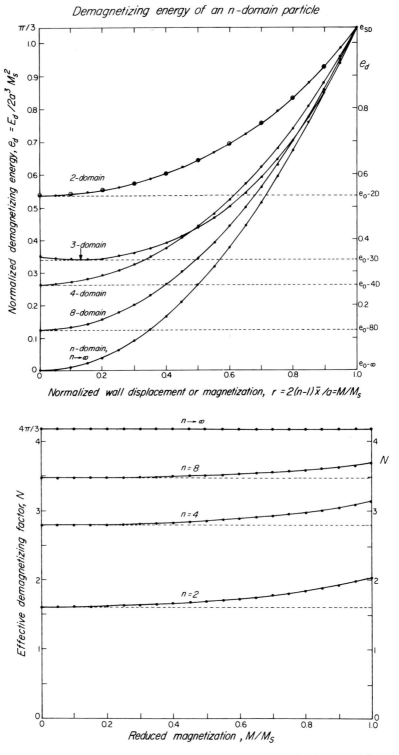

Figure 5.4 Calculated demagnetizing energy E_d and effective demagnetizing factor N of an n-domain particle as a function of domain-wall displacement. [After Dunlop (1983a) © American Geophysical Union, with the permission of the author and the publisher.]

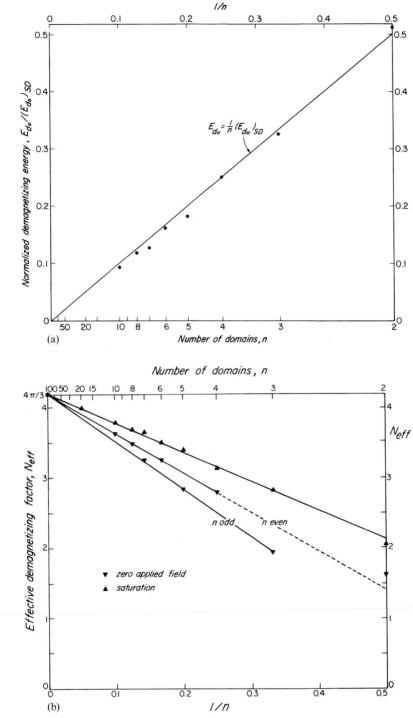

Figure 5.5 Calculated demagnetizing energy E_{d0} in the remanent state and effective demagnetizing factor N in low-field and high-field states as a function of the number of domains, n. [Unpublished calculations by D. J. Dunlop.]

$$E_{d0} \approx 0.8142 \frac{1}{n} (E_{d0})_{SD}. \tag{5.4}$$

This expression considers only interactions between poles on the same face of the cube and ignores interactions between opposite faces. It is exact in the limit $n \to \infty$, but seriously underestimates E_{d0} when n is ≤ 20. For example, for $n = 10$ domains, $E_{d0}/(E_{d0})_{SD} = 0.093$ (Fig. 5.5a). The approximate estimates are 0.100 from (5.3) and 0.081 from (5.4).

N is also an approximately linear function of $1/n$ (Fig. 5.5b). When n is very large, the grain is so finely subdivided into domains on a scale length L or D that it behaves magnetostatically as though it had a uniform magnetization equal to its net magnetization M. N then approaches the value for a SD cube, $N_{SD} = \frac{1}{3}$ ($4\pi/3$ in cgs). As n becomes smaller, the discreteness of the domain magnetizations becomes more and more important. A 'demagnetized' 2D cube has $N_{2D} = 0.127$ (1.6 in cgs), less than one-half of N_{SD}.

For MD grains approaching saturation ($M \to M_s$), eqns. (4.9), (5.2) and (5.3) yield

$$N_{sat} \approx \left(1 - \frac{1}{n}\right) N_{SD}. \tag{5.5}$$

The demagnetizing factor N_0 when walls are near their zero-field positions is significantly less than N_{sat} for the same n. N_0 depends linearly on $1/n$, but in different ways for n even or odd (Fig. 5.5b). As a practical guideline, N_{MD} is within 10% of N_{SD} for any magnetization state if $n \geq 17$.

The average internal field, $\langle H_d \rangle = -N_{2D} M$, inside a 2D cube is greatly reduced compared to that inside a SD cube. However, H_d can be large in local regions (Fig. 4.10). The dispersion in directions of H_d results in a small vector average. There is a similar dispersion on a fine scale inside MD grains with a large number of domains, but for many purposes they can be regarded as a continuum with uniform magnetization M. Then the treatments of §4.2 and 4.3 will correctly predict the average internal field and the part of E_d, $\frac{1}{2}\mu_0 VNM^2$, that varies with wall displacement. Shape anisotropy due to domain rotation requires some modification because E_{d0} is shape dependent as well as N. In other problems, for example the nucleation of new domains, local variations in the magnitude and direction of H_d are of paramount importance.

Generally speaking, E_{d0} is the part of E_d that must be considered in the total energy budget of a grain when comparing alternative domain structures. This is the point of view we shall take in determining the equilibrium number of lamellar domains as a function of grain size in §5.4. When modelling changes in the magnetization state of a grain with a constant number of domains, the term $\frac{1}{2}\mu_0 VNM^2$ is relevant and we must know N. Calculations of this type are the subject of §9.2.

5.3   Width and energy of domain walls

5.3.1   180° Bloch walls

Within the 180° Bloch wall separating two body domains, atomic magnetic moments rotate in the plane of the wall through small angles across ≈ 200–300 adjacent unit cells (Fig. 5.6a). This gradual transition between domain magnetization directions greatly reduces the exchange energy compared to an abrupt reversal of M_s. On the other hand, many of the moments in the wall are rotated out of easy $\langle 111 \rangle$ directions. The wider the wall, the greater the magnetocrystalline anisotropy energy becomes. The competition between E_{ex}, favouring broad walls, and E_K, favouring narrow walls, results in wall structures like those shown in Fig. 5.6(b).

In the LL wall (Landau and Lifschitz, 1935), the angle between adjacent moments is largest in the centre and decreases towards the edges of the wall. This model is appropriate for materials with uniaxial anisotropy, in which moments midway through the wall are in hard directions and have large E_K, but not for magnetite, in which moments in the mid-region of the wall are near $\langle 111 \rangle$ easy axes.

One-dimensional micromagnetic calculations (Moon and Merrill, 1984; Enkin and Dunlop, 1987) that assume uniaxial anisotropy give $\theta(x)$ profiles that resemble the LL structure in the wall interior but overshoot 0 and 180° at either edge (Fig. 5.6b). The 'skirts' of overrotated moments partially offset the net magnetization of the main part of the wall, thus reducing the wall's demagnetizing energy E_d. E_d is ignored in classical calculations of wall structure because the walls are assumed to be thin sheets with negligible demagnetizing factors. Amar (1958) first pointed out that in small particles with few domains, walls have widths comparable to their lateral dimensions. E_d is then not negligible. Indeed it may be the largest term in the total wall energy. Another effect that is important in fine grains because of their large surface/volume ratios is the tendency for near-surface moments in the wall to rotate subparallel to the surface, rather than in the plane of the wall. This reduces pole densities on the crystal surface at wall terminations at the expense of creating some additional volume

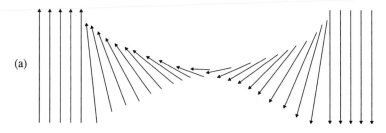

Figure 5.6   (a) Sketch of the rotation of neighbouring spins or atomic magnetic moments across a 180° Bloch wall.

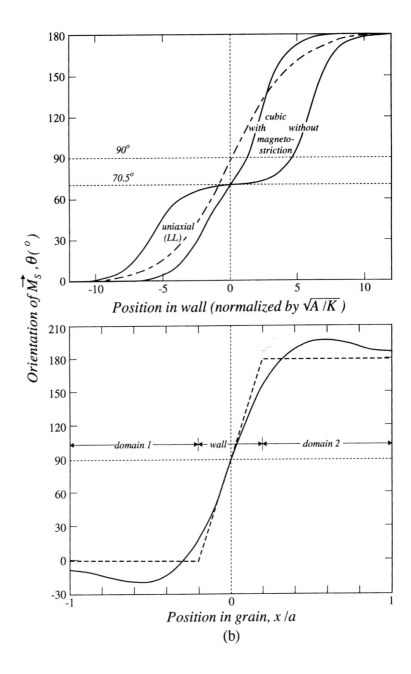

Figure 5.6 (b) Profile of spin orientations, according to different calculations: Landau and Liftschitz (1935) for an unbounded medium; Lilley (1950), for a wall plane containing an intermediate ⟨111⟩ easy axis, so that the 180° wall subdivides into a 70.5° wall and a 109.5° wall; Enkin and Dunlop (1987), showing overshoot of orientations beyond 0 and 180°, forming wall skirts that reduce E_d. [Figure from Enkin and Dunlop (1987) © American Geophysical Union reproduced with the permission of the authors and the publisher.]

poles within the wall. Néel-like walls near the crystal surface are predicted by two- and three-dimensional micromagnetic models (§7.5.4).

We now turn to approximate calculations valid for large crystals containing many domains and walls. We ignore both E_d and near-surface changes in the structure of a wall, and further assume equal angles between adjacent moments within the wall and no overrotations beyond 0 or 180°. The problem is reduced to a one-dimensional minimization of the sum of E_{ex} and E_K for a line of $m+1$ moments, each at an angle $\theta = \pi/m$ with respect to its neighbours. The exchange energy for the m pairs of neighbouring moments, from eqn. (2.33), is

$$E_{ex} = -2mJ_e S_i \cdot S_j = -2mJ_e S^2 \cos\theta \approx -2mJ_e S^2 (1 - \tfrac{1}{2}\theta^2)$$
$$= -2mJ_e S^2 + J_e S^2 \pi^2 \, \frac{1}{m}. \tag{5.6}$$

The approximation for $\cos\theta$ is valid because m is large and so θ is small. The first term in (5.6) is the exchange energy of m pairs of parallel spins. The second term is angle-dependent and is minimized by making m as large as possible.

The anisotropy energy E_K is usually stated to be on the order of the anisotropy constant K times the volume of the wall (e.g., Kittel, 1949, eqn. (3.23); Cullity, 1972, eqn. (9.6)). This is a gross overestimate because very few spins are in or near hard directions (Fig. 5.7a). The spins will rotate from [111] to the opposite direction, [$\bar{1}\bar{1}\bar{1}$], in the easiest possible plane. (The spins are confined to a *plane* (the plane of the wall) in order to avoid producing poles and increasing E_d.) A

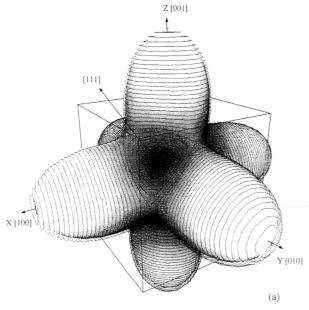

(a)

Figure 5.7   (a) Three-dimensional energy surface of E_K for a cube of magnetite. The $\langle 100 \rangle$ directions are hard axes with high energy. [After Williams and Dunlop (1995) © American Geophysical Union, with the permission of the authors and the publisher.]

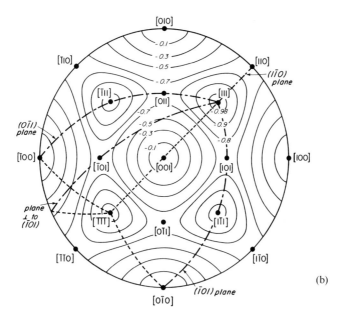

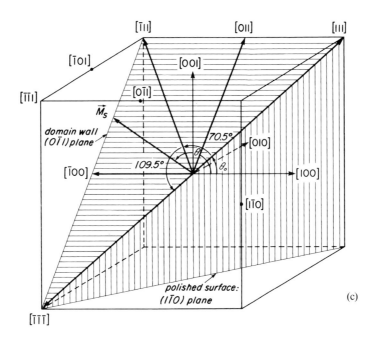

Figure 5.7 (b) A contour map of E_K over half the unit sphere (stereographic projection), with possible wall planes and polished viewing surfaces indicated by dot-dash (upper hemisphere) and dotted (lower hemisphere) lines. (c) The geometry of the $(0\bar{1}1)$ wall plane, containing intermediate $[011]$, $[\bar{1}11]$ and $[\bar{1}00]$ axes, and the $(1\bar{1}0)$ polished surface, relative to a magnetite cube with $\langle 100 \rangle$ principal axes.

relatively easy plane is $(0\bar{1}1)$, which contains an intermediate easy axis, $[\bar{1}11]$, but also the $[\bar{1}00]$ hard axis. This plane is sketched in Fig. 5.7(b), in which E_K is contoured over the unit sphere, and in Fig. 5.7(c) relative to a cube of magnetite.

The *easiest* wall plane is the one shown passing through $[\bar{1}01]$ in Fig. 5.7(b), which is perpendicular to $(\bar{1}01)$. It avoids the $\langle 100 \rangle$ hard axes entirely. Now, in order to best view domain structures in magnetite, the crystal must be sectioned and polished along a $\{110\}$ plane, which contains four $\langle 111 \rangle$ easy directions. Only then can a full set of body and closure domains form with $\boldsymbol{M}_s$ vectors in the plane of view, so as not to produce surface poles. If the viewing plane is $(\bar{1}01)$, the plane of the Bloch wall is perpendicular to the polished surface.

To calculate wall thickness and energy, we need to determine the average value $\langle E_K \rangle$ of E_K for equal-angle rotations of spins in the easiest plane. Exact solutions have been given by Néel (1944a) and Lilley (1950), but we will follow the approach of Stacey and Banerjee (1974), which leads to a simple analytic calculation of $\langle E_K \rangle$. Returning to our earlier relatively easy wall plane, suppose that the spins rotate from [111] to $[\bar{1}1\bar{1}]$ via $(0\bar{1}1)$. (In this case, the plane of the wall is inclined at 60°, rather than 90°, to a polished surface like $(1\bar{1}0)$: see Fig. 5.7c.) During the first 70.5° of rotation, spins pass the [011] axis of intermediate hardness and reach the $[\bar{1}11]$ easy axis (Figs. 5.7b,c). In the final 109.5° of rotation, the spins must pass the $[\bar{1}00]$ hard axis before reaching $[\bar{1}1\bar{1}]$. The anisotropy energy within the $(0\bar{1}1)$ plane at angle θ to the [100] axis (see Fig. 5.7c) can be found from eqn. (2.47) by setting $\alpha_1 = \cos\theta$, $\alpha_2^2 = \alpha_3^2 = \frac{1}{2}(1 - \alpha_1^2) = \frac{1}{2}\sin^2\theta$, giving

$$E_K = K_1 V(\sin^2\theta - 0.75\sin^4\theta) + 0.25 K_2 V(\sin^4\theta - \sin^6\theta). \qquad (5.7)$$

Since K_1 and K_2 are both negative for magnetite above T_V (Fig. 3.7a), the minimum energy is $K_1/3 + K_2/27$ for moments along $\langle 111 \rangle$, an intermediate level of energy is $K_1/4$ for moments along $\langle 110 \rangle$, and the maximum energy is 0 for moments along $\langle 100 \rangle$. The maximum energy difference, between $\langle 100 \rangle$ and $\langle 111 \rangle$, amounts to $|K_1|/3 + |K_2|/27$. The average energy difference of a spin in the wall is only about $\frac{1}{3}$ of this, however. In our equal-angle model of spin rotation, this average energy is well approximated by integrating (5.7) over the 180° between [111] and $[\bar{1}1\bar{1}]$, i.e., from θ_0 to $\pi + \theta_0$ (Fig. 5.7c), dividing the result by π, and subtracting the $\langle 111 \rangle$-direction minimum energy (Stacey and Banerjee, 1974, eqn. (3.28)). The result is

$$\Delta E_K = KV = 0.1146|K_1|V + 0.0214|K_2|V. \qquad (5.8)$$

Using the values $K_1 = -1.35 \times 10^4 \, \text{J/m}^3$, $K_2 = -0.28 \times 10^4 \, \text{J/m}^3$ for magnetite (Table 3.1), we find $K = 1.64 \times 10^3 \, \text{J/m}^3$. By inspection of the E_K contours in Fig. 5.7(b), it is clear that K will be even smaller if the spins rotate in the easiest plane.

A line of $m + 1$ spins occupies a volume $V = ma^3$, where a is the spacing between spins. Its energy is

$$E_w = E_{ex} + \Delta E_K = J_e S^2 \pi^2 \frac{1}{m} + Ka^3 m \qquad (5.9)$$

(the angle-independent part of E_{ex} has been omitted). At equilibrium, $dE_w/dm = 0$, giving a wall width

$$\delta_w = m_{eq}a = \pi \left(\frac{A}{K}\right)^{1/2}, \qquad (5.10)$$

and a wall energy per unit wall area

$$\gamma_w = 2\pi (AK)^{1/2}. \qquad (5.11)$$

The exchange constant A, defined as

$$A = J_e S^2/a, \qquad (5.12)$$

has a value $A = 1.33 \times 10^{-11}$ J/m for magnetite at room temperature (Heider and Williams, 1988). Equations (5.10) and (5.11) are identical, apart from minor numerical differences, to the results of Landau and Lifschitz (1935) and Lilley (1950).

Substituting the values of A and K for magnetite, we find $\delta_w = 0.28\,\mu m$ or about 300 lattice spacings. An experimental determination of δ_w for magnetite is $0.18\,\mu m$ (Moskowitz et al., 1988b).

The specific wall energy for magnetite from (5.11) is 0.93×10^{-3} J/m^2 or 0.93 erg/cm^2. Adapting the values in Lilley (1950, Table 6.1) for Ni (which has $\langle 111 \rangle$ easy axes like Fe_3O_4), one obtains an alternative estimate of 0.85 erg/cm^2. Experimentally, from observations of domains like those in Fig. 1.2, Özdemir and Dunlop (1993a) found $\gamma_w = 0.91$ erg/cm^2. Thus the theoretical estimates are reasonable.

5.3.2   70.5° and 109.5° Bloch walls

70.5° and 109.5° walls occur in the interior of magnetite crystals, separating body domains with M_s vectors along different $\langle 111 \rangle$ axes (Fig. 5.8a). They also form the boundaries of surface closure domains (Fig. 5.8b). The 70.5° and 109.5° angles between M_{sc} in the closure domain and M_{s1} or M_{s2} in the body domains are not directly measurable. It is the angles between the walls themselves which are directly observable in Bitter patterns. For a $\{110\}$ viewing plane, these angles are 125° and 145° (Fig. 5.9a).

What is the structure of 70.5° and 109.5° walls? They differ from 180° Bloch walls in that there is always a component of M_s normal to the wall. However, the in-plane component of M_s behaves just like the spins in a 180° wall and follows the easiest plane described in the previous section. Therefore the planes of 70.5° and 109.5° walls, like the planes of 180° walls, are perpendicular to a

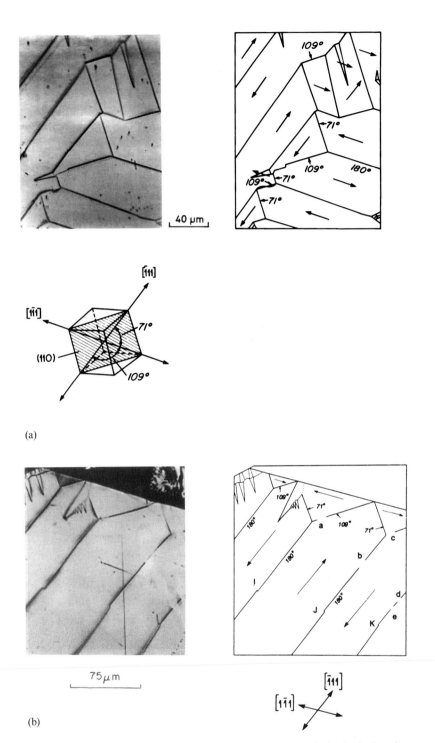

Figure 5.8 (a) Body domains with 71°, 109° and 180° boundaries in the interior of a large magnetite crystal ({110} viewing plane). (b) Body, closure and spike domains at a deep crack in the same crystal. [After Özdemir *et al.* (1995) © American Geophysical Union, with the permission of the authors and the publisher.]

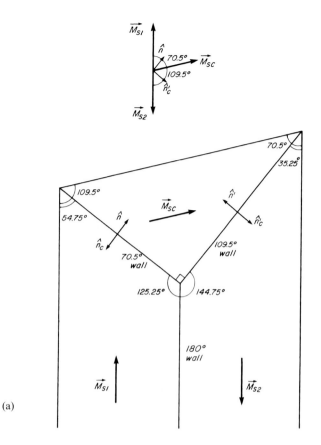

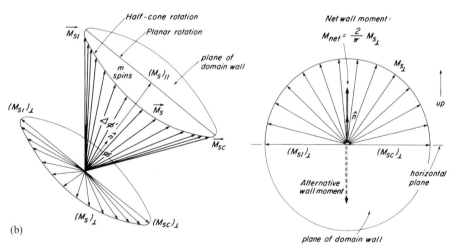

Figure 5.9 (a) Sketch of a single closure domain like the ones in Fig. 5.8(b) and adjacent parts of the body domains, showing angles between M_s vectors and between adjacent walls, and outward normals to the boundaries of the closure and body domains. The viewing plane is {110}. (b) Left: half-cone rotation of spins across a $\phi = 71°$, 90° or 109° Bloch wall. The angle $\Delta\phi'$ between adjacent spins is larger than for a planar rotation. Right: the in-plane component of M_s rotates through 180°, producing a net wall moment directed either up or down.

{110} viewing plane, such as that in Fig. 5.9(a). This means that the outward normals $\hat{n}$, $\hat{n}'$, $\hat{n}_c$ and $\hat{n}'_c$ to the boundaries of the body and closure domains in Fig. 5.9(a) lie in the plane of view. To eliminate surface poles on the 70.5° wall between body domain 1 and the closure domain, we require

$$\boldsymbol{M}_{s1} \cdot \hat{n} + \boldsymbol{M}_{sc} \cdot \hat{n}_c = \boldsymbol{M}_{s1} \cdot \hat{n} - \boldsymbol{M}_{sc} \cdot \hat{n} = 0, \qquad (5.13)$$

so that $\hat{n}$ must bisect the angle between $\boldsymbol{M}_{s1}$ and $\boldsymbol{M}_{sc}$. Similarly $\hat{n}'_c$ for the 109.5° wall bisects the angle between $\boldsymbol{M}_{sc}$ and $\boldsymbol{M}_{s2}$.

The structure within the wall is shown in Fig. 5.9(b). Spins rotate through a half-cone of angle 70.5° across the 70.5° wall. By this means, the component of $\boldsymbol{M}_s$ parallel to the wall normal, $M_{s\parallel} = \boldsymbol{M}_s \cdot \hat{n} = M_s \cos(\frac{1}{2} \times 70.5°)$ is kept constant, so that $\nabla \cdot \boldsymbol{M} = \partial M_{s\parallel}/\partial n = 0$ and no poles appear within the wall. Simultaneously, the component of $\boldsymbol{M}_s$ in the plane of the wall (i.e., perpendicular to $\hat{n}$), $\boldsymbol{M}_s \times \hat{n}$, with magnitude $M_{s\perp} = M_s \sin(\frac{1}{2} \times 70.5°)$, rotates through 180° in equal increments (Fig. 5.9b, right).

Notice that there is a net moment $\pm (2/\pi) M_{s\perp}$, either up or down ($\perp \{110\}$), as a result of the in-plane components $\boldsymbol{M}_s \times \hat{n}$ of $\boldsymbol{M}_s$. These moments, and their counterparts in 180° walls, behave like single-domain moments, independent of the surrounding domains. They will be the subject of §5.8. The poles due to these moments appear on crystal boundaries a large distance above and below the plane of view (labelled 'horizontal plane' in Fig. 5.9b) and, in large crystals, do not generate a magnetostatic energy E_d for the wall large enough to affect our calculations.

The main modification to our previous calculation of wall thickness and energy is that the angle between neighbouring spins in a 70.5° wall containing lines of $m + 1$ spins across the wall is not $\Delta\phi = (1/m) \times 70.5°$ as it would be for a planar rotation of spins, but $\Delta\phi' = (1/m)\pi \sin(\frac{1}{2} \times 70.5°) = (1/m) \times 103.9°$ (cf. Fig. 5.9b, left). The corresponding result for a 109.5° wall is $\Delta\phi' = (1/m) \times 147.0°$. This has the effect of increasing the exchange energy slightly. Its effect on the crystalline anisotropy energy is hard to gauge because the half-cone rotation generates a curved path on the unit sphere of Fig. 5.7(b). Since the path is more constrained than the easiest plane for a 180° wall, $\langle E_K \rangle$ and K almost certainly increase. These increases are made necessary to avoid an even larger increase in magnetostatic energy. Magnetostrictive energy makes an important contribution to the total energy of the closure domains (§5.4.2), but is unimportant compared to magnetocrystalline energy in the walls.

The results for domain wall thickness and energy are

$$\delta_w = \pi \sin\left(\frac{\phi}{2}\right)\left(\frac{A}{K}\right)^{1/2} \qquad (5.14)$$

and

$$\gamma_{\rm w} = 2\pi \sin\left(\frac{\phi}{2}\right)(AK)^{1/2}, \tag{5.15}$$

where ϕ is the angle between $M_{\rm s}$ vectors on either side of the wall: either 70.5° or 109.5°. The same formulas apply for $\langle 100 \rangle$ easy axes, e.g., in TM60 or iron. In this case, $\phi = 90°$. The thickness and energy per unit area of the walls bounding closure domains will be comparable to the values calculated in §5.3.1 for 180° walls. Without a precise calculation of K for each type of wall, we cannot be more exact.

5.4 Width and energy of domains

5.4.1 Equilibrium number and width of lamellar domains

The number n and width d of lamellar domains (Fig. 5.2a–c) in a grain is determined by the balance between demagnetizing energy and the energy of the $n - 1$ domain walls. From eqns. (4.9), (5.3) and (5.11) or (5.15),

$$E_{\rm tot} \approx \frac{1}{2}\mu_0 N_{\rm SD} M_{\rm s}^2 LWD\frac{1}{n} + \gamma_{\rm w}LW(n-1). \tag{5.16}$$

At equilibrium, setting $dE_{\rm tot}/dn = 0$,

$$n_{\rm eq} = \left(\frac{\mu_0 N_{\rm SD} M_{\rm s}^2}{2\gamma_{\rm w}}\right)^{1/2} D^{1/2} \tag{5.17}$$

and

$$d_{\rm eq} = \left(\frac{2\gamma_{\rm w}}{\mu_0 N_{\rm SD} M_{\rm s}^2}\right)^{1/2} D^{1/2}. \tag{5.18}$$

Both the number and average width of lamellar domains are expected to increase with increasing grain size D, in proportion to $\sqrt{D}$. Titanomagnetite follows this predicted relation quite well, pyrrhotite not so well (Fig. 5.10).

According to (5.17) and (5.18), if $M_{\rm s}$ is large, grains should be subdivided into many narrow domains, while weakly magnetic minerals should have a few broad domains. Generally speaking this is what is observed. A 200-μm grain of pyrrhotite ($M_{\rm s} \approx 80\,{\rm kA/m}$) or TM55 ($M_{\rm s} \approx 125\,{\rm kA/m}$), might contain 20 or 30 domains (Fig. 5.10), whereas the same-sized grain of hematite ($M_{\rm s} \approx 2.5\,{\rm kA/m}$) contains about 2 domains (Halgedahl, 1995).

By the same reasoning, a 200-μm grain of magnetite ($M_{\rm s} = 480\,{\rm kA/m}$) might be expected to have four or five times as many domains as a 200 μm pyrrhotite or TM55 grain. In reality, an $\approx$ 200-μm subgrain of magnetite contains 7 domains (Özdemir and Dunlop, 1993a; Fig. 1.2). Similarly, a 3 mm magnetite crystal is predicted from (5.18) to have a domain width $d_{\rm eq} \approx 8\,\mu{\rm m}$ but the widths observed (Fig. 5.8a,b) are 25–65 μm. The explanation lies in closure domains, which form

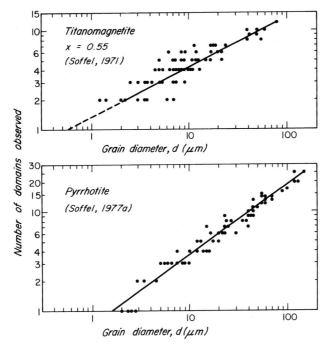

Figure 5.10   Observed number of domains as a function of grain size in TM55 and pyrrhotite. The straight lines have expected slopes of $\frac{1}{2}$ (eqns. (5.17), (5.23)). Their intercepts give approximate values for the critical SD size d_0. The dispersion of points reflects, in part, alternative states available to a grain of given size. [After Soffel (1971) and (1977a), with the permission of the author and the publisher, Deutsche Geophysikalische Gesellschaft, Köln, Germany.]

readily in magnetite (Özdemir *et al.*, 1995) but not in pyrrhotite or titanomagnetite. By reducing the demagnetizing energy, closure domains reduce n_{eq} and increase d_{eq}, as we shall show in the next section.

It is interesting to calculate d_{eq} for a much smaller magnetite grain, 1 µm in size. Equation (5.18) predicts 0.14 µm, which is less than the observed domain wall width (cf. §5.3.1). Even allowing for the inadequacies of (5.18), we anticipate that the structure of 0.2 µm and smaller magnetite grains may resemble a wall more than a classical domain. We shall return to the question of micromagnetic structures in magnetite in Chapter 7.

5.4.2   The effect of closure domains

Closure domains greatly reduce E_d, as described in §5.1. In order to further reduce E_d by minimizing surface poles, M_{sc} rotates into near parallelism with the crystal surface. The evidence for this is shown in Fig. 5.11. Closure domains observed at a {111} face of magnetite, which is perpendicular to the ⟨111⟩ easy axes in the body domains, are bounded by $\approx 90°$ walls rather than 70.5° and

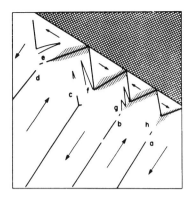

75μm

Figure 5.11 Closure domains bounded by $\approx 90°$ walls at a $\{111\}$ crystal surface in magnetite ($\{110\}$ viewing plane). M_{sc} directions in the closure domains are believed to be almost parallel to the crystal surface. [After Özdemir *et al.* (1995) © American Geophysical Union, with the permission of the authors and the publisher.]

109.5° walls, and take the form of Fig. 5.1(a). Because M_{sc} is rotated away from $\langle 111 \rangle$, there is an increase in E_K, but the main energies to be considered are the magnetoelastic energy E_σ due to magnetostrictive strain of the closure domains, and the wall energy E_w, which in large crystals is due mainly to the 180° walls of the body domains. Our calculation follows that of Özdemir *et al.* (1995). For an alternative approach, see Shcherbakov and Tarashchan (1990).

The total volume V_c of closure domains is easily shown to be $\frac{1}{2}WDd = \frac{1}{2}V(d/L)$, where L, W, D are the crystal dimensions (see Fig. 5.2), crystal volume $V = LWD$, and d is the width of either body or closure domains. E_σ arises from the incompatible magnetostrictions of the closure and body domains. Assuming that the resulting elastic deformation is entirely accommodated by the closure domains, we can show from eqns. (2.60) and (2.66) that

$$E_\sigma = \lambda \sigma V_c = E_{\sigma 0} - \tfrac{3}{2}\lambda_{100} V_c(\sigma_{11}\alpha_1^2 + \sigma_{22}\alpha_2^2 + \sigma_{33}\alpha_3^2)$$
$$- 3\lambda_{111} V_c(\sigma_{12}\alpha_1\alpha_2 + \sigma_{23}\alpha_2\alpha_3 + \sigma_{31}\alpha_3\alpha_1), \qquad (5.19)$$

where $E_{\sigma 0}$ is the magnetoelastic energy when there are no closure domains. In (5.19), the α_i refer to M_{sc} in the closure domains, while the stress tensor σ_{ij} results from the magnetostriction of the body domains. With the body domains magnetized along $\langle 111 \rangle$, $\sigma_{11} = \sigma_{22} = \sigma_{33}$ and $\sigma_{12} = \sigma_{23} = \sigma_{31} = c_{44}\lambda_{111}$ (Kittel, 1949; cf. eqn. (2.62)). The λ_{100} term in (5.19) contributes only a constant, since $\alpha_1^2 + \alpha_2^2 + \alpha_3^2 = 1$. Rewriting (5.19) without the constant terms and substituting $V_c = \frac{1}{2}V(d/L)$, we obtain

$$E_\sigma = \frac{3}{2}\lambda_{111}^2 c_{44} V \frac{d}{L}(1 - \alpha_1\alpha_2 - \alpha_2\alpha_3 - \alpha_3\alpha_1)$$
$$= \frac{9}{4}\lambda_{111}^2 c_{44} V \frac{d}{L}, \qquad (5.20)$$

setting $\alpha_1 = \alpha_2 = 1/\sqrt{6}$, $\alpha_3 = -2/\sqrt{6}$ (M_{sc} is along a $\langle 112 \rangle$ axis, $\perp \langle 111 \rangle$).

The energy of the $180°$ walls bounding body domains, each of area approximately LW (here we ignore the existence of the closure domains, which are a fine structure in large grains), is

$$E_w = \gamma_w LW\left(\frac{D}{d} - 1\right) = \gamma_w V \frac{1}{d} - \text{const.} \tag{5.21}$$

since a crystal with $n = D/d$ body domains has $n - 1$ $180°$ walls.

The total energy is $E_{tot} = E_\sigma + E_w$ (both E_d and E_K are much smaller terms after M_{sc} deflects into near-parallelism with the crystal surface). Setting $dE_{tot}/dd = 0$ for minimum energy, we find

$$d_{eq} = \left(\frac{4\gamma_w}{9\lambda_{111}^2 c_{44}}\right)^{1/2} L^{1/2} \tag{5.22}$$

and

$$n_{eq} = \left(\frac{9\lambda_{111}^2 c_{44}}{4\gamma_w}\right)^{1/2} \frac{1}{q} L^{1/2}, \tag{5.23}$$

where $q = L/D$ is particle elongation. As in (5.17) and (5.18), the number and width of body domains are predicted to increase with grain size as the square root of particle dimension. However, the actual values of d_{eq} and n_{eq} are rather different from those given by (5.17) and (5.18).

For magnetite, $c_{44} = 9.55 \times 10^{10} \, \text{N/m}^2$ (Doraiswami, 1947) and $\lambda_{111} = +72.6 \times 10^{-6}$ (Table 3.1). Then (5.22) predicts for a crystal with $L = 3 \, \text{mm}$ that $d_{eq} \approx 50 \, \mu\text{m}$. This is about the domain size observed in Figs. 5.8 and 5.11. For a $200 \, \mu\text{m}$ sub-grain with $q \approx 1$, like that in Fig. 1.2, (5.23) gives $n_{eq} \approx 15$ domains, an order of magnitude less than the 80–120 estimated in §5.4.1 when closure domains were not considered, and of the same order as the number observed, $n = 7$.

Equations (5.22) and (5.23) provide a means of calculating γ_w provided c_{44} is known (which is the case for magnetite, but not for other magnetic minerals). In Fig. 1.2, the sub-grain tapers from bottom to top, so that the domain length L varies. The domain width d can be seen to increase, on average, as L increases. These and other similar observations are summarized in Fig. 5.12. Then using eqn. (5.22), one finds $\gamma_w = 0.91 \, \text{erg/cm}^2$ or $0.91 \times 10^{-3} \, \text{J/m}^2$ for magnetite (Özdemir and Dunlop, 1993a; §5.3.1).

5.4.3   Wall energy in magnetite estimated from Néel spikes

Spike domains have M_s vectors reversed to the direction of M_s in the body domains they penetrate. They are typical of materials with strong uniaxial crystalline anisotropy, e.g., TM60 with dominant magnetoelastic anisotropy due to internal stress. Small spike domains, often associated with closure domains, are

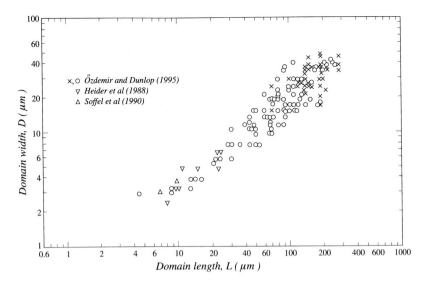

Figure 5.12 Observed domain width as a function of domain length in magnetite. The data lead to an estimate of 0.91 erg/cm^2 for the wall energy γ_w. [After Özdemir and Dunlop (1993a) © American Geophysical Union, with the permission of the authors and the publisher.]

common in magnetite as well (see Figs. 5.8a,b, 5.11). Spikes also form around cavities or non-magnetic inclusions (Fig. 5.13a: Özdemir and Dunlop, 1996b).

Néel (1944b) pointed out that a void or inclusion within a domain would have quite a large demagnetizing energy E_d because of the $+$ and $-$ poles that would form on its faces (Fig. 5.13b). He predicted that a pair of spike domains would form in order to spread the poles over a larger surface area and increase the distance between $+$ and $-$ poles (Fig. 5.13c), thereby reducing E_d. (In addition, two sets of small closure domains, not shown in Fig. 5.13c, are required to eliminate poles on the cavity walls.)

Following Néel, the spikes flanking a cavity of diameter d can be approximated by an elongated ellipsoid of revolution with diameter $(2/\pi)^{1/2}d$ (giving the same cross-sectional area $\frac{1}{2}d^2$ as the square cavity) and length kd (Fig. 5.13d). The spheroid has volume $kd^3/3$ and surface area $(\pi^3/8)^{1/2}kd^2$. Its demagnetizing factor is approximately $N_{cgs} = (8/k^2)\{\ln[(2/\pi)^{1/2}k] - 1\}$ or $N_{SI} = N_{cgs}/4\pi$; this formula is valid for an axially magnetized cylinder or ellipsoid of large k. The demagnetizing energy, following eqn. (4.9), is $E_d = \frac{1}{2}\mu_0 VN(2M_s)^2$, since the spikes are equivalent to an ellipsoid of magnetization $-2M_s$ superimposed on a body domain of uniform magnetization M_s. Thus

$$E_d = 2\mu_0 VNM_s^2 = \frac{\mu_0}{4\pi}\frac{16d^3 M_s^2}{3k}\left[\ln\left(\sqrt{\frac{2}{\pi}}k\right) - 1\right], \tag{5.24}$$

the factor $(\mu_0/4\pi)$ being omitted in cgs.

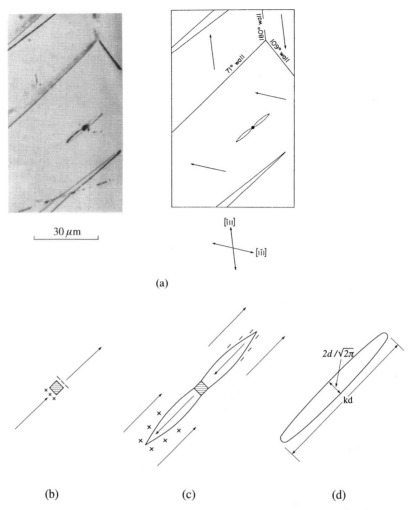

Figure 5.13 (a) Néel spikes flanking a cavity in a magnetite crystal ({110} viewing plane). (b), (c) Sketches of a cavity of square cross-section and the resulting distribution of poles, without and with Néel spikes. (d) Approximation of the pair of Néel spikes by an ellipsoidal region of reverse magnetization. [After Özdemir and Dunlop (1996b) © American Geophysical Union, with the permission of the authors and the publisher.]

The added wall energy due to the ellipsoidal spike region is

$$E_{\text{w}} = \gamma_{\text{w}} \sqrt{\pi^3/8}\, kd^2. \tag{5.25}$$

Minimizing the total energy $E_{\text{tot}} = E_{\text{d}} + E_{\text{w}}$ by setting $\mathrm{d}E_{\text{tot}}/\mathrm{d}k = 0$, one finds

$$\gamma_{\text{w}} = \frac{\mu_0}{4\pi}\sqrt{\frac{2}{\pi^3}}\,\frac{32dM_{\text{s}}^2}{3k^2}\left[\ln\left(\sqrt{\frac{2}{\pi}}k\right) - 2\right], \tag{5.26}$$

the $(\mu_0/4\pi)$ again being omitted in cgs.

The pair of spike domains in Fig. 5.13(a) are not really suitable for Néel's calculation because they are not axially magnetized (they are bounded by 71° rather than 180° walls). Nevertheless, substituting their length to diameter ratio, $k = 21\,\mu m/1.7\,\mu m = 12.35$, and using $M_s = 480\,emu/cm^3$ for magnetite, one obtains $\gamma_w \approx 1\,erg/cm^2$, which is of the correct order of magnitude.

5.4.4   Non-equilibrium domain structures

In calculating equilibrium values n_{eq} and d_{eq} as in §5.4.1 and 5.4.2, we are making the assumption that enough energy is available for a grain to overcome energy barriers between alternative states like those shown in Fig. 5.2 and reach the equilibrium state. This is not always the case. Grains of similar size and composition often have different numbers of domains. This is one reason for the dispersion of n values for a particular grain size D (or of D values for a fixed n) in Fig. 5.10. Likewise, a single grain may have different numbers of domains on different occasions, depending on its magnetic history. An example is shown in Fig. 5.14.

Each alternative state of a magnetic grain represents a **local energy minimum (LEM)**. However only the state with the lowest energy – the **global energy minimum (GEM)** state – is an equilibrium state. All other LEM states are metastable, although they can be very long-lived. In Chapter 7, we will consider what conditions of time, temperature and field can produce **transdomain transitions** between LEM states.

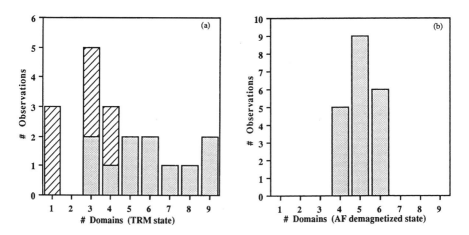

Figure 5.14   Histograms of the number of domains observed in the same titanomagnetite grain (a) after replicate TRM experiments, and (b) following AF demagnetization. [After Halgedahl (1991) © American Geophysical Union, with the permission of the author and the publisher.]

5.5 The single-domain range

5.5.1 Critical single-domain size

The single-domain (SD) state has great practical interest because of its high remanence and unparalleled stability (§5.7). The critical grain size d_0 below which the SD structure has a lower energy than a two-domain (2D) or other structure is the single most important parameter in rock magnetism. Grains with $D > d_0$ may be trapped in metastable SD states because of the energy required to nucleate a domain wall (§9.9, 12.8). However, grains smaller than d_0 are unlikely to retain a distinct 2D structure because the wall will fill most of the particle.

Néel (1947) used this latter idea to estimate d_0: when $\delta_w \to D$, the grain is at or near the critical SD size. In §5.3.1, δ_w (for thin walls with negligible demagnetizing factors) was estimated to be $\approx 0.28\,\mu\text{m}$ in magnetite. In grains near critical SD size, the wall is not thin compared to its lateral dimensions. It fills much of the grain volume and has a substantial demagnetizing energy E_d. To offset this effect, the wall must lower its demagnetizing factor N by becoming thinner than it would be if only exchange and anisotropy energies were involved. For magnetite, $\delta_w \approx 0.1\text{–}0.2\,\mu\text{m}$ as $D \to d_0$ (Enkin and Dunlop, 1987).

Following Kittel (1949) and Butler and Banerjee (1975a,b), we can also estimate d_0 by finding the critical value of D at which the energies of SD and 2D structures (Fig. 5.2a,b) are equal:

$$(E_d)_{SD} = (E_{d0})_{2D} + \gamma_w L W. \tag{5.27}$$

Using eqns. (4.9) for $(E_d)_{SD}$, (5.3) for $(E_{d0})_{2D}$, and (5.11) for γ_w, we arrive at

$$d_0 = D_{crit} = \frac{4\gamma_w}{\mu_0 N_{SD} M_s^2}. \tag{5.28}$$

Substituting $\gamma_w = 10^{-3}\,\text{J/m}^2$ and $M_s = 480\,\text{kA/m}$ for magnetite, d_0 is estimated to be about $0.04\,\mu\text{m}$. This is only 50 lattice spacings. It is also considerably less than the wall width. Thus the thin-wall model of Fig. 5.2(b) is inappropriate, and the 2D energy must be revised upward to account for E_d of the wall, thereby increasing d_0.

Although (5.28) underestimates d_0, it does give insight into the factors that control the upper size limit for equilibrium SD behaviour. Elongated grains with smaller N_{SD} (§4.2.4) will support SD structure to much larger sizes than equidimensional grains. Since $d_0 \propto 1/M_s^2$, and M_s decreases with rising temperature, grains which are 2D at ordinary temperatures may transform to SD at high temperature, particularly near the Curie point. These effects are illustrated in Fig. 5.15.

The effect of M_s on d_0 is also clear when comparing different minerals. Weakly magnetic hematite and goethite have critical SD sizes orders of magnitude

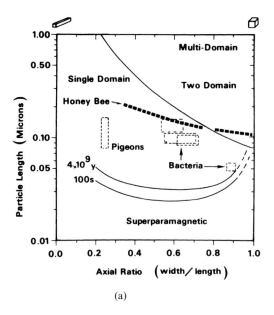

(a)

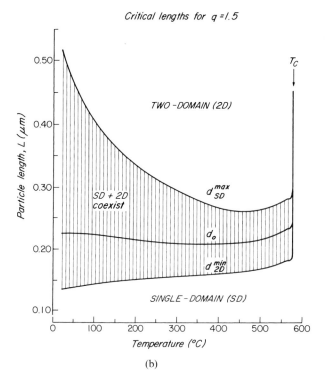

(b)

Figure 5.15　Calculated critical single-domain size d_0 for magnetite as a function of (a) grain elongation and (b) temperature. [After Butler and Banerjee (1975b) © American Geophysical Union, Kirschvink (1983) © Plenum Publishing, New York, and Dunlop *et al.* (1994) © American Geophysical Union, with the permission of the authors and the publishers.]

greater than those of strongly magnetic magnetite or iron. Table 5.1 lists theoretical and experimental room-temperature values of d_0 for a few of the commoner magnetic minerals. Experimental determinations of d_0 are not easy. They usually involve extrapolating from observations of the number of domains in much larger grains (e.g., Fig. 5.10) with the aid of equations like (5.17) or (5.23), or interpreting remanence and coercive force data of samples whose (often broad) grain size distributions span the SD threshold.

Micromagnetic calculations (Chapter 7) predict that SD structures can evolve to a variety of LEM states in larger grains without undergoing a sharp transformation. The meaning of d_0 is then problematic. However, in all such

Table 5.1 Upper and lower size limits, d_0 and d_s, for single-domain behaviour in equidimensional grains at 20 °C

Mineral	Superparamagnetic size, d_s (μm)	Critical single-domain size, d_0 (μm)
Iron	0.008[1]	0.023[1]
	0.026[2]	0.017[2]
Magnetite	0.025–0.030[3–5]	0.05–0.06[4,6]
		0.079–0.084[7,8]
Maghemite		0.06[9]
Titanomagnetite	0.08[10]	0.2[10]
($x = 0.55$–0.6)		≈ 0.6[11,12]
Titanomaghemite		
($x = 0.6, z = 0.4$)	0.05[12]	0.75[12]
($x = 0.6, z = 0.7$)	0.09[12]	2.4[12]
Hematite	0.025–0.030[13,14]	15[14,15]
Pyrrhotite		1.6[16]

[1]Kneller and Luborsky (1963) – experimental; [2]Butler and Banerjee (1975a) – theoretical; [3]McNab *et al.* (1968) – experimental; [4]Dunlop (1973a) – experimental; [5]Dunlop and Bina (1977) – experimental; [6]Argyle and Dunlop (1984) – theoretical; [7] Enkin and Dunlop (1987) – theoretical; [8]Enkin and Williams (1994) – theoretical; [9]Morrish and Yu (1955) – theoretical; [10]Butler and Banerjee (1975b) – theoretical; [11]Soffel (1971) – experimental (see Fig. 5.10); [12]Moskowitz (1980) – theoretical; [13]Bando *et al.* (1965) – experimental; [14]Banerjee (1971) – experimental; [15]Chevallier and Mathieu (1943) – experimental; [16]Soffel (1977a) – experimental (see Fig. 5.10).

evolutions, the remanence is predicted to drop catastrophically as grain size increases, providing a practical definition of d_0 for a particular family of evolved SD states.

In magnetic recording, where the switching of SD states is of paramount importance, the transition from coherent (whole-particle) parallel rotation of atomic moments to incoherent modes, in which a non-uniform structure exists during reversal (§7.9), is sometimes taken as marking the SD threshold. Many incoherent modes resemble stable structures found in slightly larger particles. They can be thought of as transitory LEM states.

5.5.2 Superparamagnetic threshold

Under ordinary conditions, spontaneous reversals of an SD grain are unlikely because the energy barrier ΔE due to crystalline, magnetoelastic or shape anisotropy is much larger than the available thermal energy. However E_K, E_σ and E_d are all proportional to particle volume V. In small enough grains, the barrier becomes comparable to $25kT$, where k is Boltzmann's constant and T is temperature. The cumulative effect of thermal excitations once every 10^{-9} s or so then makes it probable that the SD moment will reverse once every few minutes.

This thermally activated condition is called **superparamagnetism** (Bean and Livingston, 1959). Thermal activation theory will be developed fully in Chapters 8, 9 and 10. For the moment, we will simply remark that the superparamagnetic (SP) threshold forms an effective lower limit to the stable SD range.

To estimate the SP threshold size d_s, we equate the energy barrier VK to the thermal energy $25kT$ available on a time scale of seconds to minutes. For a cubic grain,

$$d_s = (25kT/K)^{1/3}. \tag{5.29}$$

K as used here refers to a rotation of spins during some interval of time. However, we can estimate K from the magnetocrystalline value $K = 1.64 \times 10^3 \, \text{J/m}^3$ (eqn. (5.8)), which refers to a spatial rotation of spins. (The two are not quite equivalent because during thermal activation spins are not confined to a plane, as they were in Bloch wall calculations, and may find an easier route in Fig. 5.7b.) The room-temperature SP size for magnetite is then calculated to be $0.04 \, \mu\text{m}$. Estimates based on shape anisotropy, which generally produces a large barrier, are somewhat smaller and agree well with experimental values of 0.025–$0.03 \, \mu\text{m}$ (Table 5.1).

The room-temperature stable SD range is very narrow for magnetite and iron. Grains of these minerals in equilibrium SD states must be quite rare in nature. The SD range is broader for titanomagnetite and pyrrhotite. Equilibrium SD states occur in grains as large as $1 \, \mu\text{m}$, and metastable SD states in even larger grains. Hematite has an SP threshold size similar to magnetite's but an estimated critical SD size almost three orders of magnitude larger. Throughout much of its naturally occurring grain-size range, hematite should be in a stable SD state.

The SP threshold size d_s is only weakly dependent on mineral properties such as anisotropy constant and grain shape. It is also time dependent, but again only weakly. The coefficient 25 in (5.29) approximately doubles for a 1 Ga time scale, leading to a mere 30% increase in d_s (Fig. 5.15a). On the other hand, d_s is strongly dependent on temperature T. K due to crystalline, magnetoelastic, or shape anisotropy (cf. eqns. (5.34), (5.35)) decreases rapidly to zero near the Curie point (Figs. 2.11, 3.13). Even the largest grains then become SP. This ability to pass readily between stable and thermally unstable conditions at high temperatures is at the heart of thermoremanent magnetization (Chapter 8).

5.6 Magnetic hysteresis of multidomain grains

5.6.1 Wall displacement, nucleation, and domain rotation

So far we have considered domain configurations mainly in the absence of an external field. A multidomain (MD) grain responds to an applied field H_0 in one of three ways. The domains whose M_s vectors are closest to the direction of H_0 are energetically favoured (eqn. (4.8)) and will enlarge at the expense of other domains, limited largely by the resulting increase in E_d (eqn. (5.1)). This process amounts to **domain wall displacement**. The coupled spins that form the wall rotate in a reversible fashion, thereby changing the wall's position. In a perfect crystal, γ_w would be the same for any location of the wall and very little energy would be expended in wall displacement.

In real crystals in rocks, imperfections like impurity atoms, inclusions of a second phase, voids, and lattice defects such as dislocations locally change γ_w, result in the appearance of magnetic poles as in Fig. 5.13 (thereby changing E_d), or create stress fields. In all cases, the net effect is to create a potential well pinning the wall at or near the defect. Energy must be supplied, by increasing H_0, before the wall will move, and when H_0 is subsequently decreased, the wall will not return exactly to its former position. **Wall pinning** thus gives rise to magnetic hysteresis and explains both remanence and coercivity (the fields necessary to move walls and change remanence).

An applied field may also cause the **nucleation** of new domains. Nucleation requires local rotation of spins against an anisotropy barrier like the one calculated in eqn. (5.8). The fields required are generally much higher than wall pinning fields, but demagnetizing fields near sharp corners (e.g., Figs. 4.10, 5.3) or local reductions in K around defects may aid in nucleation. Note that a high concentration of defects aids in nucleation but impedes wall displacement. Thus a newly created domain nucleus and its bounding wall may remain pinned and only later, in higher fields, move away from the edge of the grain or propagate transversely as a wedge or spike splitting an existing domain in two. This process is referred to as secondary nucleation.

Eventually, in high enough fields, reverse domains will shrink and disappear ('denucleation'). The only further response possible is for domains that are unfavourably oriented (perpendicular to H_0 or nearly so) to rotate their M_s vectors towards H_0. **Domain rotation** is usually a higher-energy process than wall displacement or nucleation because it is opposed by both shape and crystalline anisotropies. Domain rotation is the only possible response of an SD grain. For this reason, a much higher field is generally needed to change the magnetization of an SD grain than to change the magnetization of an MD grain. SD grains are said to be magnetically 'hard' whereas MD grains tend to be magnetically 'soft'.

5.6.2   Susceptibility, remanence and coercivity of MD grains

An idealized hysteresis (M versus H) loop of an MD grain is depicted in Fig. 5.16. Although the magnetization does not return exactly to its original state after a cycle of increasing and decreasing H_0, the irreversibility is fairly small. The linearity of the loop results from the balance between the applied and demagnetizing fields:

$$H_0 - NM = 0 \quad \text{or} \quad M = H_0/N. \tag{5.30}$$

The initial **susceptibility** $\chi = \mathrm{d}M/\mathrm{d}H_0 = 1/N$ and the saturation field is NM_s.

The hysteresis or width of the loop results from wall pinning or nucleation or both. A direct measure of the strength of these effects is the **coercive force** H_c, the reverse field necessary to reduce M to zero in the descending loop. Figure

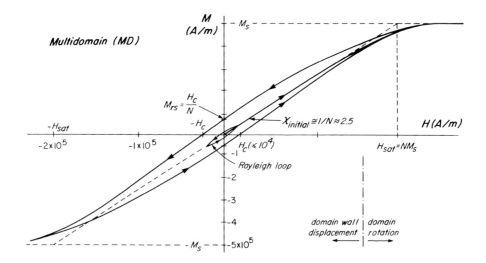

Figure 5.16   Idealized hysteresis loop for a large multidomain grain of magnetite.

5.17 is an example in which coercivity arises from pinning of a 71° wall in a magnetite crystal by an inclusion and its associated Néel spikes. In Fig. 5.17(a) and (b), the 71° wall truncates the tip of the right-hand spike. If H_0 is applied parallel to M_s in the body domain containing the inclusion, the wall will move to the right and, at some critical value of H_0, will break loose and jump to the right (Fig. 5.17c,d). The field at which the wall snaps loose is called the **(micro)coercivity**, h_c, and the irreversible wall motion is called a **Barkhausen jump**. Néel (1944b) gives for the critical field

$$h_c = \frac{5}{4}\sqrt{\frac{\pi^3}{8}}\,\frac{\gamma_w}{\mu_0 M_s r},$$

(5.31)

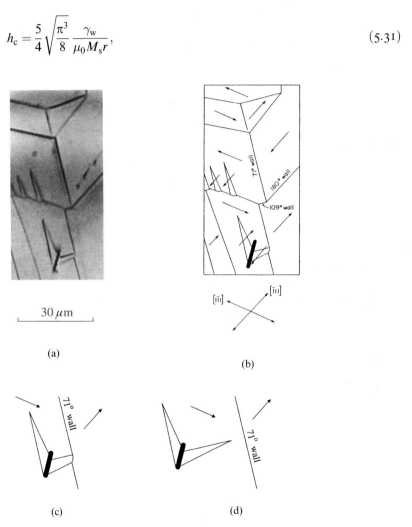

Figure 5.17   (a), (b), (c) A 71° wall intersecting, and pinned by, a Néel spike anchored on an inclusion (black) in a magnetite crystal ({110} viewing plane). (d) In a sufficiently high applied field, the wall snaps loose from its pin and jumps to the right. [After Özdemir and Dunlop (1996b) © American Geophysical Union, with the permission of the authors and the publisher.]

where r is the radius of the inclusion. Substituting $\gamma_w = 1 \text{ erg/cm}^2$ and approximating r by the length $l = 7.5 \,\mu\text{m}$, we find $h_c \approx 7 \,\text{Oe}$. The bulk coercive force H_c, an average value of h_c for all pinned walls, was measured to be about $5 \,\text{Oe}$ for this crystal.

It is easy to see from the geometry of the hysteresis loop (Fig. 5.16) that the **saturation remanence** M_{rs} is related to H_c by

$$M_{rs} = H_c/N. \tag{5.32}$$

In the course of tracing out a saturation hysteresis loop, H_0 drives walls past many pinning sites in a sequence of Barkhausen jumps. If nucleation is the dominant process, many domains nucleate and propagate in the course of approaching saturation. The magnetization process is discontinuous, occurring by many discrete steps, but the steps are so small that they are seldom detectable in macroscopic magnetization measurements. Similarly the overall remanence is the sum of many individual domain-wall displacements, which are usually not resolvable macroscopically.

If only small fields are applied, a Rayleigh loop is traced. The Rayleigh magnetization law is

$$M = AH_0 + BH_0^2, \tag{5.33}$$

starting from a demagnetized state. In small fields, some walls move only short distances, staying within the potential well of their pinning centre. These account for the first Rayleigh coefficient A, which is equal to the reversible susceptibility χ. Other walls move past one or more pins. The second coefficient B describes such irreversible processes, including wall jumps or transverse domain propagation. Figure 5.18 illustrates observations of wall pinning by defects and reversible wall movements in an applied field.

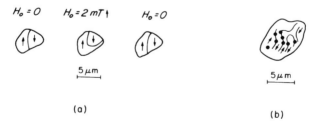

(a)   (b)

Figure 5.18   (a) Reversible bending of a 180° domain wall in TM55. (b) Bent domain walls in TM55 pinned by dislocations (dots). [After Soffel (1971), with the permission of the author and the publisher, Deutsche Geophysikalische Gesellschaft, Köln, Germany.]

5.7   Magnetic hysteresis of single-domain grains

5.7.1   Susceptibility, remanence and coercivity of SD grains

An idealized hysteresis curve for an assemblage of uniaxial SD grains with randomly oriented easy axes is shown in Fig. 5.19. (The curve will be derived in §8.3.) The magnetization process is domain rotation, opposed by crystalline, magnetoelastic or shape anisotropy (§2.8.1, 2.8.4, 4.2.4), shape anisotropy being dominant for strongly magnetic minerals like magnetite and iron. We will first consider uniaxial anisotropy, which is described by an energy $E_K = K_u V \sin^2 \theta$, where θ is the angle between $\boldsymbol{M}_s$ and the easy direction $\theta = 0$ (eqn. (2.48), to second order). K_u may be magnetocrystalline in origin, but more commonly

$$K_u = \tfrac{3}{2}\lambda_s \sigma \qquad \text{(magnetoelastic)} \tag{5.34}$$

or

$$K_u = \tfrac{1}{2}\mu_0(N_b - N_a)M_s^2 \qquad \text{(shape-controlled)} \tag{5.35}$$

from eqns. (2.66) or (4.11).

The total energy E_{tot} is the sum of E_K and E_H from eqn. (4.8). In the special case where $\boldsymbol{H}_0$ is applied parallel to the easy axis, $\theta = 0$ or π, we have

$$E_{tot} = K_u V \sin^2 \theta - \mu_0 V M_s H_0 \cos \theta. \tag{5.36}$$

When H_0 reaches a critical value $(H_0)_{crit}$, called the SD microcoercivity or anisotropy field H_K, $\boldsymbol{M}_s$ will rotate irreversibly from one easy direction to the other. We find this field by setting $dE_{tot}/d\theta = 0$ (minimum energy for a given H_0) and $d^2E_{tot}/d\theta^2 = 0$ (condition for instability of the minimum-energy state), which yields

$$\begin{aligned} H_K = (H_0)_{crit} &= \frac{2K_u}{\mu_0 M_s} \qquad \text{(general)} \\ &= \frac{3\lambda_s \sigma}{\mu_0 M_s} \qquad \text{(magnetoelastic)} \\ &= (N_b - N_a)M_s. \qquad \text{(shape)} \end{aligned} \tag{5.37}$$

(A more detailed treatment appears in §8.3.) When $\boldsymbol{H}_0$ is applied at an angle to the easy axis, H_K is reduced. The bulk coercive force H_c in Fig. 5.19, which is the average $\langle H_K \rangle$ over randomly oriented grains, is about one-half the values given by (5.37).

For multiaxial anisotropy, the calculation of critical fields is more intricate but similar in principle. Suppose $\boldsymbol{M}_s$ is close to [111] and has direction cosines α_i with respect to $\langle 100 \rangle$ principal axes, while $\boldsymbol{H}_0$ is applied along $[\bar{1}\bar{1}\bar{1}]$, with direction cosines $(-1/\sqrt{3}, -1/\sqrt{3}, -1/\sqrt{3})$. Then, using (2.47) (4th-order terms only) and (4.8), we have

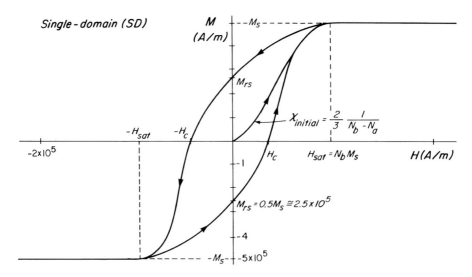

Figure 5.19 Idealized hysteresis loop for randomly oriented uniaxial single-domain grains. Numerical values are for magnetite.

$$E_{\text{tot}} = K_1 V(\alpha_1^2\alpha_2^2 + \alpha_2^2\alpha_3^2 + \alpha_3^2\alpha_1^2) + (1/\sqrt{3})\mu_0 V M_s H_0(\alpha_1 + \alpha_2 + \alpha_3).$$
$$(5.38)$$

Setting $\partial E_{\text{tot}}/\partial\alpha_i = 0$ for each of the α_i (minimum-energy condition) and $\partial^2 E_{\text{tot}}/\partial\alpha_i^2 = 0$ (instability condition), one obtains

$$H_K = (H_0)_{\text{crit}} = \frac{4}{3}\frac{|K_1|}{\mu_0 M_s} \qquad (5.39)$$

as the critical field at which M_s flips from [111] to [$\bar{1}\bar{1}\bar{1}$]. This field is approximately half the critical field for uniaxial anisotropy. Considerably lower fields result if H_0 is applied at an angle to M_s rather than antiparallel to M_s. Notice that the problem of a 'Barkhausen jump' in the orientation of M_s when H_0 is increased to a threshold of instability is a different problem from either rotation of spins against K in a Bloch wall (§5.3.1) or thermal activation of M_s between states of opposite orientation (§5.5.2). In each case, a different route will be taken in Fig. 5.7(b), and the energy barrier and the effective anisotropy will be different.

We can now calculate some typical coercivities for SD magnetite, using $|K_1| = 1.35 \times 10^4 \, \text{J/m}^3$ and $M_s = 480 \, \text{kA/m}$. Cubic crystalline anisotropy, according to (5.39), gives a microscopic coercive force $\mu_0 H_K$ of 37.5 mT ($H_K = 375$ Oe), but a more realistic figure for randomly oriented grains is about half this, or 20 mT. This is a minimum or threshold field, below which H_0 cannot cause any irreversible rotation of M_s in an equidimensional grain of magnetite. Such a threshold field is a hallmark of SD behaviour. The magnetocrystalline value of H_K is a substantial field, 1–2 orders of magnitude larger than typical

MD coercivities in magnetite (cf. §5.6.2). It ensures complete stability of SD remanence against changing geomagnetic fields (which are ≈ 1 Oe or 0.1 mT) at ordinary temperatures, making SD grains the ideal paleomagnetic recorder.

Even higher stability results if grains are elongated. A 10% difference between the longest and shortest particle axes results in a coercivity $H_K = (N_b - N_a)M_s \approx 200$ Oe (16 kA/m or 20 mT). Magnetite grains in rocks tend to be more elongated than this, on average about 1.5:1. Their shape-controlled coercivities are typically 30–50 mT (300–500 Oe), considerably above the magnetocrystalline threshold value.

Induced magnetization results from reversible rotations of M_s vectors in grains whose easy axes are oriented approximately perpendicular to H_0. The rotations are complete when $H_0 = H_K$ as given by (5.37) (the exact theory will be developed in Chapter 8). Since grains with easy axes (i.e., long axes for shape anisotropy) perpendicular to H_0 are twice as frequent as those parallel to H_0 in a random assemblage,

$$\chi = \frac{2}{3}\frac{M_s}{H_K} = \frac{2}{3}\frac{1}{N_b - N_a}, \tag{5.40}$$

or an equivalent equation for crystalline or magnetoelastic anisotropy. An exact calculation gives the same result.

Although it is often stated that SD grains have lower susceptibility than MD grains, this is not necessarily the case. From (5.30), $\chi \approx 1/N \approx 3$ (0.25 in cgs) for MD magnetite or other strongly magnetic minerals. For SD magnetite, χ has a lower limit according to (5.40) of $(\frac{2}{3})(\frac{1}{2})^{-1} = 1.33$ (0.11 in cgs) for needle-like grains (§4.2.5). However, an upper limit is set by the magnetocrystalline barrier for equidimensional grains, and this gives χ as high as about 20 (1.6 in cgs). Although wall motion usually requires lower fields than domain rotation, it does not necessarily lead to higher susceptibility because wall displacements are effectively limited by the demagnetizing field.

In the remanent state ($H_0 = 0$) following saturation, the M_s vectors of all grains are in the easy direction nearest H_0. For uniaxial anisotropy, they are isotropically distributed over a half-sphere and so

$$M_{rs} = \int_0^{\pi/2} M_s \cos\phi \sin\phi \, d\phi = 0.5 M_s, \tag{5.41}$$

where ϕ is the angle between M_s and H_0 as $H_0 \to 0$. If magnetocrystalline anisotropy is dominant,

$$M_{rs} = 0.832 M_s \qquad (\langle 100 \rangle \text{ easy axes}) \tag{5.42}$$

or

$$M_{rs} = 0.866 M_s \qquad (\langle 111 \rangle \text{ easy axes}). \tag{5.43}$$

It is these very high remanence levels, together with substantial coercivities preventing changes in the remanence, that make SD grains ideal for magnetic recording, both in technology and in nature.

5.7.2   Superparamagnetic magnetization

In its thermally activated or SP condition, an SD grain preserves no remanence. The minimum-energy states defined by shape or crystalline easy axes still exist but energy barriers between them are low. Thermal energy rapidly establishes an equilibrium partition among the states following any change in H_0 or T. In particular, if $H_0 = 0$, $M = 0$.

Under these conditions, each SD particle acts like a Langevin paramagnetic moment (§2.4), except that its moment $\mu = VM_s$ is giant compared to that of a single paramagnetic atom. Hence the term **superparamagnetism**.

The reversible magnetization can be calculated from eqns. (2.15) or (2.16). Uniaxial grains aligned with H_0 would have

$$M(H_0, T) = M_s \tanh(\alpha), \quad \alpha \equiv \frac{\mu_0 V M_s H_0}{kT} \tag{5.44}$$

but a more realistic approximation for a randomly oriented assemblage is the Langevin function $L(\alpha)$ for a continuum of possible alignments,

$$M(H_0, T) = M_s L(\alpha) = M_s \left[\coth(\alpha) - \frac{1}{\alpha} \right]. \tag{5.45}$$

Notice that according to either expression, magnetization curves measured at different temperatures should superimpose if plotted as a function of H_0/T. This property of H_0/T superposition can be used as a test of SP behaviour.

The SP magnetization curve represented by (5.45) is plotted in Fig. 5.20. The curve rises steeply and saturates at low fields compared to stable SD or MD magnetization curves. The initial susceptibility $\chi = dM/dH_0 = \mu_0 V M_s^2 / 3kT$ for $0.03\,\mu m$ cubes of magnetite at room temperature is about 650 (52 in cgs), two orders of magnitude higher than typical SD or MD susceptibilities. Even a small fraction of SP material in a sample will tend to dominate the induced magnetization.

The iron-rich mica biotite hosts magnetite inclusions with a wide range of grain sizes. Magnetization curves for three biotite crystals from the same drill core are compared in Fig. 5.21. They are very close to the ideal MD, SD and SP curves depicted in Figs. 5.16, 5.19 and 5.20. More complex magnetization curves, if due wholly to magnetite, can be deconvolved with the aid of MD, SD and SP type curves. This is a useful procedure for estimating the volume fraction of each in a rock.

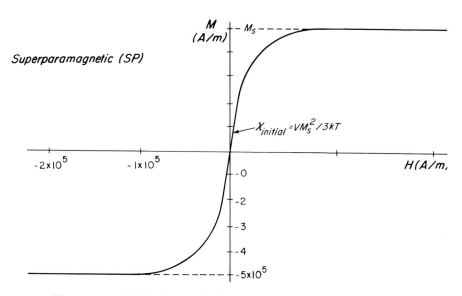

Figure 5.20 Idealized magnetization curve for superparamagnetic magnetite particles.

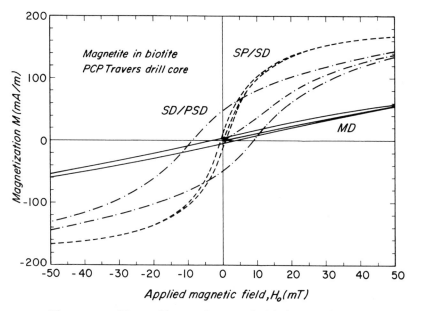

Figure 5.21 Observed hysteresis curves for biotite crystals containing magnetite inclusions. The inclusions range in size from superparamagnetic (SP) through single-domain (SD) to pseudo-single-domain (PSD; Chapter 12) and multidomain (MD).

5.8   Domain wall magnetization

In crossing a 180° Bloch wall, the spins rotate in the plane of the wall through 180°. In §5.3.2 and Fig. 5.9(b), we saw that the spins in a 70.5° or 109.5° wall rotate on the surface of a cone, rather than in a plane. The same is true of 90° walls. There are two possible senses of rotation, tracing out two half-cones (or in the limit of a 180° wall, two half-circles). The components of M_s vectors parallel to the wall normal, $M_{s\parallel}$, are identical for either sense of rotation. However, the components $M_{s\perp}$ in the plane of the wall sum to oppositely directed moments, each with a net magnetization

$$M_{\text{net}} = \frac{2}{\pi} M_{s\perp} = \frac{2}{\pi} M_s \sin\left(\frac{\phi}{2}\right), \qquad (5.46)$$

where ϕ is the angle (70.5°, 90°, 109.5° or 180°) between M_s vectors in adjacent domains.

These domain-wall moments lie in the plane of the wall, midway between the in-wall projections of M_s vectors of the neighbouring domains (Fig. 5.22). When the wall displaces, the wall moments are unaffected. Similarly, the wall moment can reverse without altering the position of the wall or the domain magnetizations. Wall moments thus behave like SD moments imbedded in, and independent of, their MD host. They potentially could explain pseudo-single-domain (PSD) behaviour (Chapter 12) and therefore have been dubbed 'psarks' (Dunlop, 1977).

A domain wall may itself be subdivided into two wall domains with opposite magnetizations M_{net}. The situation is depicted in Fig. 5.22 for a 180° Bloch wall. Across the Bloch line between the wall domains, spins rotate in half-cones of varying angle. If a field H_0 is applied parallel or nearly parallel to one of the wall moments, the Bloch line will move to enlarge the favourably oriented wall domain.

Bloch line displacement is analogous to wall displacement in MD magnetization. It provides a low-energy alternative to reversal of wall moments. Walls will only have SD-like properties if they contain no Bloch line, i.e., if they consist of only a single wall domain. Calculations of the critical SD wall size (see §12.7) indicate that walls should be 'SD' in magnetite grains up to a few µm in size, but subdivided or 'MD' in iron (Dunlop, 1977). Observations on magnetite (Chapter 6) support this view.

A grain containing psarks or undivided walls should in principle have a magnetization curve that is a combination of MD and SD types. However, in large MD grains the walls are narrow and the moments of alternate walls should be oppositely directed in order to reduce E_d overall. These two facts imply that psarks are intrinsically small and mutually cancelling in large grains. It is only in small grains, where the walls occupy a substantial fraction of the grain volume,

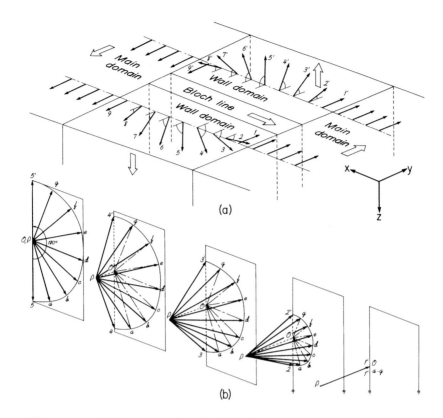

Figure 5.22 (a) A cutaway view of a 180° Bloch wall in a two-domain particle. Large open arrows indicate net magnetization directions in the main or body domains, the wall domains and the Bloch line. (b) Half-cone rotation of spins 1–5 across the Bloch line. [Reprinted from Dunlop (1981), with kind permission of the author and Elsevier Science – NL, Sara Burgerhartstraat 25, 1055 KV Amsterdam, The Netherlands.]

that psark moments are likely to be resolvable in hysteresis curves or significant contributors to remanence processes. 2D grains are particularly important hosts because there is only one wall and it likely occupies 30% or more of the particle volume (see Chapter 7).

Chapter 6

Domain observations

6.1 Introduction

In the last fifteen years, there has been a resurgence of interest in observing domain patterns on naturally occurring minerals, as well as a revolution in ideas about how the observed structures control magnetic properties. The strained surface layer that results when sections are mechanically polished for domain observations by the Bitter method can now be efficiently removed by final polishing with a suspension of amorphous silica microspheres (Hoffmann *et al.*, 1987). The previously used method, ionic polishing, required many hours' exposure to an ion beam and still failed to remove the stressed layer completely. Surface domains can also be observed using the scanning electron microscope (SEM), whose greater depth of field eliminates the need for precision polishing. The interior of domains and crystals can be imaged directly with the transmission electron microscope (TEM) and the magneto-optical Kerr effect (MOKE). The magnetic force microscope (MFM) provides the ultimate in spatial resolution, allowing us to view fine structures, e.g., the structure of domain walls.

Traditional Bitter-pattern observations use a colloidal suspension of ultra-fine magnetite particles to make visible the walls between domains, where flux leakage from the surface is maximum. These observations will be discussed in §6.2–6.6. SEM, TEM, MOKE and MFM observations are described in §6.7–6.9.

The observation of domains in minerals of paleomagnetic interest, in the $\approx 1\mu m$ grain sizes that probably carry the strongest and most reliable NRM in rocks, coupled with direct magnetization measurements on the same domains (e.g., Metcalf and Fuller, 1987a), should soon establish a direct link between the magnetic response of a rock sample and the nature of the domains responsible.

Domain observations in magnetite and other minerals of rock magnetic concern were scarcely mentioned in Nagata's (1961) classic, *Rock Magnetism*. At that time, rock magnetists had to rely on observations made on materials of industrial interest, mostly coarse polycrystalline grains at temperatures much below T_C. We now know that magnetite and TM60, the minerals responsible for most of the NRM of continental and oceanic rocks respectively, have quite different domain structures and responses to field and temperature changes. Magnetite in ≥ 30 μm grains has lamellar domains which reconfigure with mild heating (§6.6.2), while <10 μm magnetites have persistent simple structures that could carry significant TRM. TM60, on the other hand, has both lamellar and irregular, stress-dominated domains, neither of which changes configuration greatly with heating (§6.4, 6.6.3).

A few words of caution are in order. Domain patterns observed on polished surfaces of crystals can never be truly representative of the remanence or hysteresis behaviour of the same crystals in a rock. Removing part of the crystal during polishing changes the grain size and shape. Inevitably the domains must suffer some rearrangement. The stress environment of the crystal is changed with removal of part of the confining matrix and the magnetostatic environment may change as well. Particularly important in stress-sensitive minerals like TM60 are the surface stresses introduced during polishing, which can never be entirely removed. Crystal defects aid in nucleating domains but impede subsequent wall motion (§5.6). Thus MD coercivities and susceptibilities may be quite different after polishing compared to those of undisturbed material.

6.2   Bitter-pattern observations on pyrrhotite

Pyrrhotite has been the subject of many domain structure investigations. Its magnetocrystalline anisotropy is high (11.8×10^4 J/m^3 according to Hunt *et al.*, 1995a), about ten times K_1 for magnetite, but its magnetostriction is low ($\lambda < 7 \times 10^{-6}$, about five times less than λ_s for magnetite), resulting in simple lamellar domains, unobscured by maze patterns due to residual polishing stresses. In massive polycrystalline pyrrhotite, Halgedahl and Fuller (1981) observed classic Kittel structures in the NRM and saturation isothermal remanence (SIRM; i.e. M_{rs}) states or after alternating field (AF) demagnetization in fields decreasing slowly from an initial peak amplitude of 100 mT (80 kA/m, 1 kOe). Undulating walls that moved easily in small applied fields were seen after thermal demagnetization (i.e., zero-field cooling) or production of TRM by cooling in a 50-μT field from above the Curie temperature of 320 °C. Straight walls could be made wavy by heating to 315 °C, presumably because γ_w decreases so much at high temperatures that the decrease in E_d more than offsets the increase in E_w.

Halgedahl and Fuller (1983) described the nucleation and growth of domains in <30 μm primary pyrrhotite grains from a diabase. Figure 6.1 illustrates some of their results. The ≈10-μm 2D grain shown had a substantial domain-wall displacement in its NRM state. Domain imbalance preserved by cooling from high temperature in the weak geomagnetic field can therefore lead to a surprisingly intense primary TRM – 20–30% of M_s – in individual grains. Field cycling from 5 mT demagnetized some, but not all, of this initial TRM. Following saturation in a 200 mT field, reverse domains nucleated but remained pinned at the surface as $H_0 \rightarrow 0$. The net magnetization was therefore close to saturation in the SIRM state. In a back field of −9 mT, the nuclei expanded to form a large reverse domain, which enlarged by three Barkhausen jumps of the domain wall before negative saturation.

A positive domain nucleated as the field was reduced in strength, but at a different site than in the first field half-cycle. Indeed the site is exactly where the original positive domain disappeared from view in approaching negative saturation. Thus true denucleation and renucleation of domains did not occur, despite effective saturation of the grain's magnetic moment. In the second half-cycle, the wall jumps occurred at similar fields and were similar in size to those in the first half-cycle, but they occurred at different sites. This asymmetric hysteresis would scarcely be detectable in magnetic measurements, but it is a fundamental distinction between true nucleation of reverse domains and secondary nucleation due to surface pinning of previously existing nuclei.

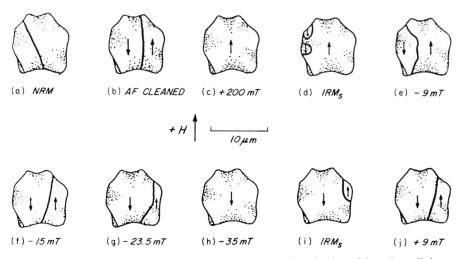

(a) NRM   (b) AF CLEANED   (c) +200 mT   (d) IRM$_s$ (e) − 9 mT

+H   10 μm

(f) − 15 mT   (g) − 23.5 mT   (h) − 35 mT   (i) IRM$_s$ (j) + 9 mT

Figure 6.1 The nucleation, displacement, and denucleation of domain walls in a 10 μm pyrrhotite grain with changing fields. The asymmetry between + and − field cycles shows that the nucleation is secondary, i.e., domain nuclei are trapped at the surface, not destroyed. [After Halgedahl and Fuller (1983) © American Geophysical Union, with the permission of the authors and the publisher.]

Halgedahl and Fuller (1983) also observed many grains which initially contained two or more domains, but failed to nucleate any walls following saturation. Usually these grains were <10 μm in size, but in one case a 50 μm grain, initially containing seven domains, preserved a metastable SD state in saturation remanence. The condition for nucleation of a reverse domain is

$$|H_d + H_0| > \frac{2K_u}{\mu_0 M_s} \qquad (6.1)$$

(Brown, 1963a), H_d being the demagnetizing field, whose average value is $-NM$ (§4.3.4), and $2K_u/\mu_0 M_s$ being the critical field for reversal of a uniaxial SD grain (§5.7.1). If nucleation occurs while H_0 is still positive (i.e., in the same direction as M), the grain will contain at least one wall in its SIRM state and $M_{rs} \ll M_s$. If, however, $H_d < 2K_u/\mu_0 M_s$, Brown's condition can only be fulfilled in negative applied fields (backfield nucleation). The particle remains in a metastable SD state at SIRM, i.e., $M_{rs} = M_s$. For an equant pyrrhotite grain, $N \approx \frac{1}{3}$ in the saturated or SD state, so that H_d is about $80/3 = 27$ kA/m ($\mu_0 H_d = 34$ mT). The anisotropy field $2K_u/\mu_0 M_s$ is comparable to the AF coercivity, which is typically ≈100 mT for SD pyrrhotite (e.g., Clark, 1984). Thus H_d alone will not be sufficient to nucleate a reverse domain in a saturated perfect crystal of pyrrhotite. Additional reverse domains will be even more difficult to nucleate because M, and therefore H_d, decrease with the appearance of the first wall.

By this argument, *all* pyrrhotite grains, once saturated, should remain saturated in their remanent state. Since even a weak field H_0 is sufficient to saturate a particle at temperatures near the Curie point, TRM as well as SIRM should be carried entirely by metastable SD particles. This is not at all the case. Except in the smallest grains, metastable SD remanence is the exception rather than the rule. The resolution of this conundrum – Brown's paradox – lies in the imperfection of real grains. Crystal defects, particularly at the surface, may locally lower K_u, while sharp edges or corners lead to a locally enhanced H_d (§4.3.4), allowing Brown's condition to be satisfied at specific nucleation sites. The proability that a grain contains at least one such site decreases as the grain size decreases (see §12.8.2).

Backfield nucleation may be followed by a wall jump to a temporarily stable position in the grain, or the wall may sweep across the particle, reversing M. In the latter case, the coercive force H_c equals the nucleation field H_n, while in the former case, $H_c > H_n$ and is governed by conventional wall pinning. Experimental nucleation fields in pyrrhotite range from about 50 mT for 7 μm grains to about 20 mT for 40 μm grains (Halgedahl and Fuller, 1983), with a wide variation at any particular size. Coercive forces vary from 82.5 to 22 mT over the same size range (Clark, 1984). Cycling a grain through positive and negative H_0 is an effective way of adding further walls.

Both Soffel (1977a) and Halgedahl and Fuller (1983) measured the average domain spacing d in pyrrhotite as a function of grain length L. The data in Fig. 6.2 are for grains in their NRM states, but the results are qualitatively similar for the AF cleaned and SIRM states. The average trend of domain width values has approximately the $\sqrt{L}$ dependence predicted by eqns. (5.18) and (5.22). However, grains of a particular size do not all have the same number n of domains. There may be as many as six different multiplicities for a single grain size. It is true that grains of the same size but different shapes can have different equilibrium domain structures, but many of the values in Fig. 6.2 undoubtedly represent metastable LEM states. An extrapolation of the average trend to $D = L$ (i.e., $n = 1$) gives an estimate of $\approx 1.5\,\mu$m for the critical SD size d_0 in pyrrhotite. Soffel (1977a) obtained a similar value, 1.6 μm (Fig. 5.10, Table 5.1).

6.3 Bitter-pattern observations on magnetite

Magnetite is the most frequent NRM carrier in nature, but apart from a few early studies (Hanss, 1964; Soffel, 1965, 1966; Bogdanov and Vlasov, 1965, 1966a,b), it has been comparatively neglected in domain observations. One reason is the need for precise orientation of the viewing plane. Magnetite's magnetostriction is moderate ($\lambda_s = 36 \times 10^{-6}$), so that maze patterns due to polishing stress are

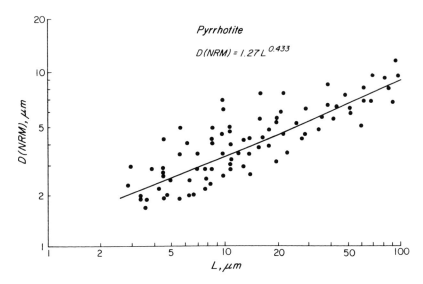

Figure 6.2 Observed domain widths and lengths in pyrrhotite grains in their NRM states. [After Halgedahl and Fuller (1983) © American Geophysical Union, with the permission of the authors and the publisher.]

not unduly difficult to remove. However, demagnetizing spikes and other extraneous patterns, generated by surface poles with density $M_s \cdot \hat{n}$, are a severe problem in magnetite compared to pyrrhotite or TM60 because of magnetite's high M_s value. Kneller (1962, Fig. 324) shows that 'fir-tree' patterns obscure the intrinsic domain structures in silicon-iron if the viewing surface deviates $\approx 1°$ from a plane containing the magnetocrystalline easy axes. Similarly, in magnetite, clear domain images are only obtained on a plane carefully oriented during polishing to within a few degrees of $\{110\}$.

Özdemir and Dunlop (1993a) and Özdemir *et al.* (1995) have studied domain patterns on a well-oriented $\{110\}$ surface of a natural single crystal of magnetite. The large M_s and relatively high anisotropy ($K_1 = -1.35 \times 10^4$ J/m^3) of magnetite ensure narrow walls which image sharply. Walls are generally straight (i.e., planar) because any curvature in a high-M_s material generates magnetic poles, thus increasing E_d. Surface closure domains, which have seldom been observed in previous studies, were common (e.g., Figs. 1.2, 5.8b, 5.11). Domain structures away from crystal boundaries are illustrated in Fig. 6.3. The lamellar body domains are magnetized along two sets of $\langle 111 \rangle$ easy axes and are separated by 70.5°, 109.5° and 180° walls. Because $\{110\}$ contains two sets of $\langle 111 \rangle$ axes, M_s vectors can follow magnetocrystalline easy directions without intersecting the viewing surface. Thus no poles or demagnetizing spikes are generated on the plane of view, and the domain patterns observed are representative of those in the crystal interior before sectioning.

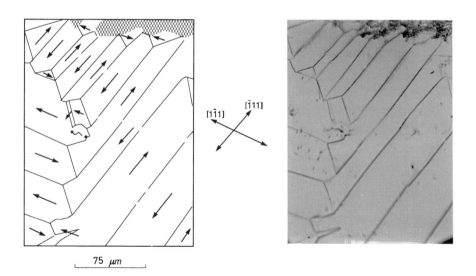

75 μm

Figure 6.3 Interior domain structure observed on a $\{110\}$ polished surface of a natural 3-mm magnetite crystal. Closure domains form at the boundary of a chemically altered (oxidized) region at top. [After Özdemir *et al.* (1995) © American Geophysical Union, with the permission of the authors and the publisher.

An interesting feature in Fig. 6.3 is the set of closure domains formed at the margin of a chemically altered region (top), which acts as an internal magnetic boundary. Such marginal closure structures may play a role in partial remagnetization accompanying chemical alteration of magnetic minerals (§16.2).

Much more complex structures appear on viewing surfaces that do not contain $\langle 111 \rangle$ easy axes, for example the $\{111\}$ octahedral face viewed in Fig. 6.4. The three sets of dark straight lines at 120° angles are replicas of underlying simple domain walls. They follow $\langle 112 \rangle$ axes, which are the surface projections of interior $\langle 111 \rangle$ easy axes. The zig-zag and 'coathanger' patterns between the main walls are demagnetizing effects, extraneous to the interior domain structure. They mark the boundaries of Néel spikes of reverse magnetization, which offset the effect of surface poles and reduce E_d (cf. §5.4.3). They are a direct demonstration that M_s cuts the $\{111\}$ surface and generates magnetic poles.

Özdemir *et al.* (1995) also observed that domain structures in magnetite are very sensitive to crystal defects, such as voids (Fig. 5.13), inclusions (Fig. 5.17), dislocations, cracks and other grain boundaries, and chemically altered regions (Fig. 6.3). All of these can generate closure structures (e.g., Figs. 1.2, 5.8b, 6.3) or reverse spikes (e.g., Figs. 5.8b, 5.13, 5.17), and thus can serve as nucleating sites for reverse domains. Evidence of the effect of dislocations are 'colloid gaps' in 180° walls in magnetite. These tend to line up, e.g., line abd in Fig. 5.8(b) and lines abcd, efgh in Fig. 5.11, and are probably due to deflection of spins in the walls where they cross a dislocation line.

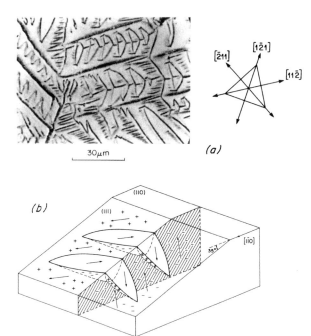

(a)

(b)

Figure 6.4 Complex domain patterns observed on a $\{111\}$ surface of the magnetite crystal of Fig. 6.3. Underlying domains with M_s along $\langle 111 \rangle$ axes produce the heavy black wall images. Because the $\langle 111 \rangle$ easy axes cut the $\{111\}$ surface, generating magnetic poles, demagnetizing spikes appear between the main walls to dilute the poles and reduce E_d. [After Özdemir *et al.* (1995) © American Geophysical Union, with the permission of the authors and the publisher.]

Sharp bends and corners can also be regarded as defects. They too serve as nucleation sites for reverse domains. Figure 6.5 shows a reverse domain that has formed in response to H_d at a bend in a deep crack dividing the crystal internally. The reverse domain is on the opposite side of the crack from the domain that generates the large demagnetizing field.

Soffel and Appel (1982) synthesized hysteresis loops by measuring the difference in total area between positive and negative domains at different stages of field cycling. Examples are given in Fig. 6.6. The hysteresis loops vary with particle size in the expected way. A moderately large grain (30 μm) exhibits distinctively MD hysteresis: a narrow loop with constant slope determined by self-demagnetization, saturating around $H_0 = NM_s$. (No published example is available for a low-x titanomagnetite, whose saturation field NM_s is expected to be 3 or 4 times larger.) A small (5 × 10 μm) grain, near magnetite in composition, has a much broader (larger M_{rs}) and harder (larger H_c) loop; only a minor loop was traced out in the fields available. The samples were sintered polycrystalline aggregates. Small grains were comparatively isolated within pores and voids, but they likely interact significantly with the surrounding crystallites. Nevertheless, the results demonstrate in a very direct way that classic wall displacements do lead to individual particle hysteresis loops that are similar to macroscopic loops measured on bulk material (e.g., Fig. 5.16).

The most reliable determinations of the number of domains, n, in magnetite, determined by a number of different techniques, are plotted in Fig. 6.7(a) (Özdemir and Dunlop, 1996b). By extrapolation, the critical SD size d_0 is

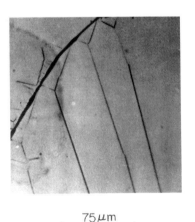
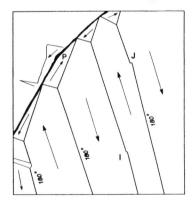

75 μm

Figure 6.5 Closure domains observed along a crack in the magnetite crystal of Figs. 6.3 and 6.4 ({110} viewing surface). At point P, where the crack bends, a reverse domain forms on the opposite side of the crack in response to the demagnetizing field at the bend. [After Özdemir *et al.* (1995) © American Geophysical Union, with the permission of the authors and the publisher.]

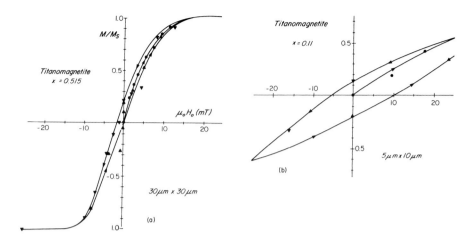

Figure 6.6 (a) MD-type and (b) SD-type hysteresis loops synthesized by measuring the areas of + and − domains viewed under the microscope at different field strengths. [Reprinted from Soffel and Appel (1982), with kind permission of the authors and Elsevier Science − NL, Sara Burgerhartstraat 25, 1055 KV Amsterdam, The Netherlands.]

estimated to be ≈ 0.06 μm, which is quite reasonable (cf. Table 5.1). However, the slope of the graph is about $\frac{1}{3}$, not the $\frac{1}{2}$ predicted by eqn. (5.23). Another data set (Geiss *et al.*, 1996; Fig. 6.7b) gives a wide variety of n for grains of a given size (or equivalently, a wide range of grain size for a particular domain multiplicity n). It extrapolates to a larger value of d_0, about 0.2 μm. We can draw three conclusions.

1. The predicted $\sqrt{L}$ dependence of n does not seem to be followed in magnetite. In grains approaching SD size, this may perhaps indicate structures more complex than those modelled in §5.4.

2. The critical single-domain size d_0 predicted by extrapolation (Fig. 6.7a) is reasonably consistent with direct measurements and calculations (Table 5.1). Notice that there are no observations of metastable SD magnetite grains in Figs. 6.7(a) or (b).

3. However, magnetite grains in the 1–10 μm size range do exhibit a variety of domain states. Some of these must be metastable LEM states, in which a wall or walls have failed to nucleate, rather than equilibrium structures.

Nucleation should be easy in magnetite. The average value of H_d at saturation is about $480/3 = 160 \, \text{kA/m}$ $(\mu_0 H_d = 200 \, \text{mT})$, about four times the anisotropy field $2K_1/M_s = 2.7 \times 10^4/480 = 55 \, \text{mT}$. No special nucleation sites should be needed. On the other hand, Boyd *et al.* (1984) reported examples of natural 10–20 μm magnetite grains which remain metastably SD in the SIRM state. They pointed out that since these grains are more than two orders of magnitude larger

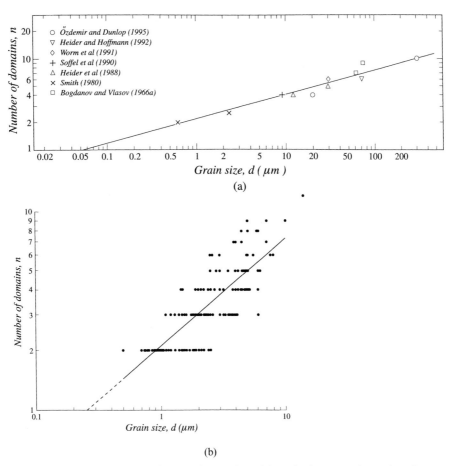

Figure 6.7 (a) Selected data on the number of domains in magnetite grains of widely varying sizes, giving an extrapolated value $d_0 \approx 0.06 \, \mu m$ for the critical single-domain size. (b) Similar observations on unoriented magnetite grains grown in a glass matrix give a wide variation in the number of domains in a grain of fixed size and an extrapolated value $d_0 \approx 0.2 \, \mu m$. [After Özdemir and Dunlop (1996b) © American Geophysical Union and Geiss *et al.* (1996), with the permission of the authors and the publishers, The American Geophysical Union, Washington, DC and Blackwell Science, Oxford, UK.]

than d_0, they possess giant moments that overshadow those of equilibrium SD structures. Indeed these individual moments equal those of typical one-inch cores of weakly magnetized rocks used in paleomagnetism, and they should be measurable with modern cryogenic magnetometers. Boyd *et al.*, following Becker (1973), suggested that these non-equilibrium SD structures may not reflect nucleation failure as such, but rather very strong surface pinning of reverse nuclei (presumably smaller than those of Fig. 6.1 and not resolvable optically).

Becker's model of secondary nucleation was proposed for high-anisotropy cobalt–rare-earth magnets and does not seem a very natural one for a cubic mineral like magnetite with only moderate anisotropy. Nevertheless, there is no question that at least some magnetite grains which initially possess domains fail to renucleate them after the walls are driven out of the grain. Boyd *et al.* reported no field cycling results, but they did discuss the effects of stress cycles to 30 MPa (0.3 kb) and temperature cycles through the isotropic point $T_I \approx 130$ K, at which K_1 changes sign and temporarily vanishes (Fig. 3.7a). Either type of cycling was sufficient to nucleate walls, some of which disappeared when the stress was released.

6.4   Bitter-pattern observations on titanomagnetite

Domains have been viewed throughout the range TM10–TM75 (Soffel *et al.*, 1982). However, most observations in recent years have been made on titano-magnetites near TM60 in composition, in the form of synthetic polycrystalline aggregates or natural grains in submarine basalts. Although TM60 has been favoured for observations above room temperature (e.g., Soffel and Appel, 1982) because of its low Curie point ($\approx 150°C$), it has a number of drawbacks. It is always surface oxidized to some degree to titanomaghemite, producing an inhomogeneous variation in composition, Curie temperature, and internal stress due to changing lattice parameter. Crystalline anisotropy is low; $K_1 \approx 0.2 \times 10^4$ J/m^3 for TM61 and ≈ 0 for TM55. The walls are therefore broad and give more diffuse Bitter patterns than in magnetite. Magnetostriction is high ($\lambda_s \approx 114 \times 10^{-6}$), resulting in persistent stress-controlled patterns that obscure lamellar structures even after prolonged ionic polishing.

For TM60, Appel and Soffel (1984, 1985) calculated that the magnetoelastic contribution $1.5\lambda_s\sigma$ to E'_K (per unit volume) equals the magnetocrystalline contribution $K_1/3$ when $\sigma = 10$ Mpa (0.1 kb), and that a mere 1% differential Δz in oxidation parameter within the particle generates an internal stress $\sigma_i = 30$ MPa. Appel (1987) reported that average σ_i rises sharply from about 40 to 60 MPa as the overall z value increases from 0.05 to 0.15, then drops in more highly (and uniformly) oxidized titanomaghemites. σ_i will tend to be strongest and most inhomogeneous at the particle surface, where Bitter patterns are observed. Lamellar domains, which reflect either low stress (Appel and Soffel, 1984) or higher but uniform stress (Appel, 1987; Halgedahl, 1987), are therefore a rarity in high-Ti titanomagnetites. Closure domains are lacking, evidence that cubic crystalline anisotropy is outweighed by uniaxial magnetoelastic anisotropy.

Focusing on grains with simple patterns, Soffel (1971) and Soffel and Appel (1982) measured the domain spacing and found an average $\sqrt{L}$ dependence over the broad size range $L = 1$–80 µm (cf. Fig. 5.10). For TM55 and TM62, d_0 values of 0.6 and 0.54 µm respectively were predicted by extrapolation. Equilibrium domain calculations predict lower values of d_0 (0.2 μm, Table 5.1) and narrower domains throughout this entire size range (Halgedahl, 1987, Fig. 10). Metastable states with fewer than the equilibrium number of domains seem to be the explanation. Nucleation is expected to be localized, based on the low average value of demagnetizing field, $\langle H_d \rangle = 125/3 \approx 42\,\text{kA/m}$ or 52 mT, compared to the anisotropy field, which must be comparable to typical SD coercivities (100–200 mT for stoichiometric and slightly oxidized TM60: Özdemir and O'Reilly, 1978, 1982a).

Actual nucleation fields and coercivities resulting from wall pinning in MD TM60 grains are much lower than SD coercivities, however. Figure 6.8 illustrates forward-field nucleation at a surface site. The nucleation site continues to pin the newly formed reverse domain well off centre in the SIRM state, producing a strong remanence. Halgedahl and Fuller (1983) point out that even if the new domain were centred, a large M_{rs} would persist because the number of domains is odd. Similar but smaller domain imbalance moments are predicted by theory (§12.6.3).

The walls move readily in backfields as small as 3 mT and an additional domain nucleates after field cycling. Notice that nucleation and transverse propagation of a lamellar structure create *two* new walls. The 3-domain particle becomes a 5-domain particle, and the domain imbalance moment remains considerable.

Nucleation failure in a nominally 2-domain grain was first described by Halgedahl and Fuller (1980) for the small TM60 particle shown in Fig. 6.9. Following saturation in 150 mT, a reverse domain nucleated at the edge in a very small backfield (-1.2 mT) and made successive Barkhausen jumps at fields of -3, -4 and -4.8 mT. During the succeeding half-cycle from negative saturation, a positive domain nucleated at the same site and the wall jumped to the same locations in the identical sequence of positive backfields. The same symmetry was recorded for the hysteresis cycle of Fig. 6.8, and is characteristic of true saturation followed by primary nucleation.

The grain in Fig. 6.9 is of special interest because at and following nucleation, the wall can be seen to extend down the side of the particle. The surface domain

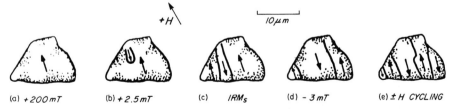

Figure 6.8 True nucleation and wall propagation in a 10 μm TM60 grain and alternative structures in the remanent state (c,e). [After Halgedahl and Fuller (1983) © American Geophysical Union, with the permission of the authors and the publisher.]

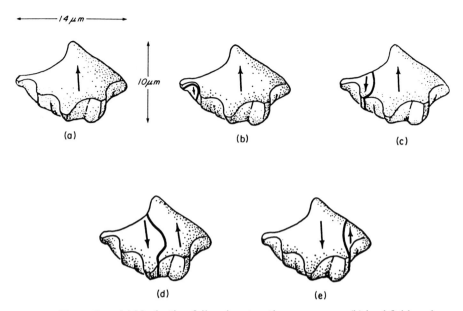

Figure 6.9   (a) Nucleation failure in saturation remanence, (b) backfield nucleation of a wall in a −1.2 mT field, and (c, d, e) successive Barkhausen jumps of the
wall in increasing negative fields for a 14 μm TM60 grain. Notice that the wall
extends through the body of the grain. [After Halgedahl and Fuller (1983) ©
American Geophysical Union, with the permission of the authors and the publisher.]

structure penetrates the grain. It is also noteworthy that fields applied at an angle
to the domain magnetizations did not noticeably rotate the structure in any of
Halgedahl and Fuller's observations. In low fields at least, nucleation and wall
displacement are the key processes, not domain rotation.

The role of complex domain patterns in TM60 is controversial. Closely spaced
walls, undulating to tightly arcuate, sometimes chaotic, typically occupy
patches, separated by islands of simpler or no structure. Appel and Soffel (1984,
1985) and Appel (1987) believe these are characteristic structures that reflect the
rapid spatial variation of internal stress in TM60. Their model envisages hard
SD-like remanence in small islands of uniform stress, separated by regions of
very inhomogeneous stress. They concede that magnetostatic and especially
exchange coupling of the regions compromise their proposed SD-like behaviour.
Halgedahl (1987) argues that regions of wavy walls are surficial only and are
underlain by lamellar domains. Metcalf and Fuller (1987a,b) remark that superficial complex patterns can sometimes be seen to be traversed by underlying simple body domains, rather in the manner of Fig. 6.4. They agree that the
contorted structures are unresponsive to fields, even at high temperatures, but
doubt that the amount of remanence they represent is significant compared to
that of simpler structures, especially non-equilibrium SD states.

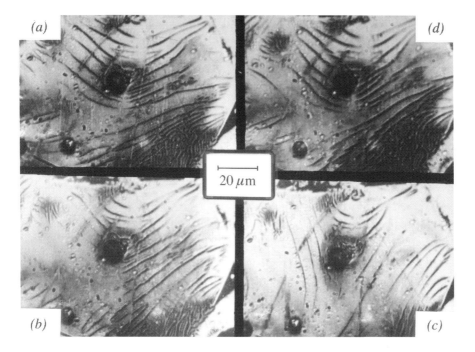

Figure 6.10 The response of domain structures in polycrystalline Al-doped TM62 to a horizontally directed compressive stress σ of (a) 0, (b) 15.5 MPa, (c) 25.5 MPa, (d) 0. [After Appel and Soffel (1985), with the permission of the authors and the publisher, Deutsche Geophysikalische Gesellschaft, Köln, Germany.]

External directed stress σ_0 simplifies and eventually eliminates complex patterns (Fig. 6.10). The patch of intricate branching walls at the lower right shrank under 15.5 MPa compression and disappeared at 25.5 MPa. Lamellar domains respond to stress in several ways. The unfavourably oriented domains (subparallel to the compressive axis) on the left of the picture were driven out by 15.5 MPa and reformed almost normal to the stress by 25.5 MPa. The domains on the right simply rotated, an uncommon response to applied fields. Nucleation of new domains was not observed, although existing favourably oriented small domains grew. The stress response seems fundamentally different from that observed for magnetite over the same stress range by Boyd *et al.* (1984), in which domain nucleation was dominant. The fact that TM60 is three times more magnetostrictive than magnetite may explain some of the differences.

6.5 Bitter-pattern observations on hematite

Hematite has a low spontaneous magnetization ($M_s \approx 2.5$ kA/m), which should result in low stray fields and negligible colloid concentrations above domain walls. In fact, however, walls in hematite are visible by both Bitter and electron-beam techniques (Gustard, 1967; Gallon, 1968; Eaton and Morrish, 1969). The critical SD size of hematite is large ($d_0 \approx 15$ μm, Table 5.1) and so even 100 μm size grains contain only a few domains. M_s is confined to the basal plane and coercivities are high (hundreds of mT) in the SD state at least. To measure hysteresis and observe the changes in domain structure that cause the magnetization changes require strong and carefully oriented fields H_0.

Halgedahl (1995) measured hysteresis for about 50 oriented single-crystal platelets of hematite and observed Bitter patterns during field cycling for many of the platelets. Figure 6.11(a) illustrates the first, abrupt appearance of a domain wall, approximately bisecting the crystal, in a backfield of ≈ 3 mT (30 Oe). The platelet was previously saturated, and this wall nucleation/propagation event served to essentially demagnetize the remanence. The hysteresis loop (Fig. 6.11b) exhibits this same large Barkhausen jump around 3 mT, together with a large number of much smaller jumps. The hysteresis loop is symmetrical, indicating that nucleation is primary and occurs at the same location and in the same field following positive or negative saturation. Direct domain observations during field cycling bore this out.

The hysteresis loop has a very single-domain-like form (cf. Fig. 5.19), except that following nucleation, the wall does not sweep across the entire grain, reversing its moment, but jumps to a position that approximately demagnetizes the grain and then must be driven in small steps away from the demagnetized state. This latter behaviour is characteristic of multidomain hysteresis opposed by H_d.

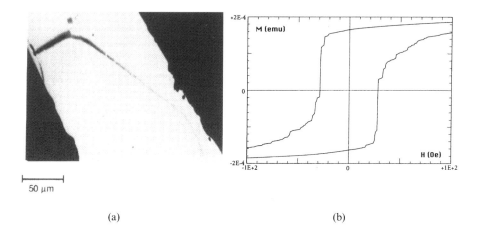

50 μm

(a)   (b)

Figure 6.11   (a) First appearance of a major wall in a previously saturated hematite platelet after application of a backfield $H_0 = -3$ mT. (b) Hysteresis loop for the same platelet, showing symmetrical Barkhausen jumps at ± 3 mT, corresponding to the wall nucleation observed in (a). [After Halgedahl (1995) © American Geophysical Union, with the permission of the author and the publisher.]

These hematites have a combination of SD-like remanence ($M_{rs}/M_s \approx 0.7$) and MD-like field response following nucleation, which occurs at fields much less than typical SD coercivities.

Halgedahl's observations directly link domain-wall movements to measured hysteresis in grains just above SD size. As the grain size increases, nucleation becomes easier. Halgedahl found $H_c = H_n \propto d_{eff}^{-0.64}$, where the effective grain size d_{eff} was determined from particle magnetic moment in the saturated state. Similar grain-size dependences are known for other magnetic minerals and are hallmarks of pseudo-single-domain (PSD) behaviour (Chapter 12).

6.6   High-temperature domain observations

6.6.1   Pyrrhotite

In polycrystalline pyrrhotite, Halgedahl and Fuller (1981) noted that undulating walls that moved readily in small fields were preserved by cooling from the Curie point, with or without a weak field present. They speculated that the greater wall area permitted at high temperature to reduce E_d is preserved metastably by cooling. Field cycling at room temperature eliminated the waviness (which was also not seen in the initial NRM state), and the resulting straight walls remained unchanged until after heating to within $10\,°C$ of the Curie point. Although no observations were made *at* high temperature, the results imply that the highest-blocking-temperature partial TRM, which accounts for most of the TRM intensity, has a metastable fraction that is soft to AF demagnetization and could also decay viscously at room temperature.

6.6.2   Magnetite

Using nearly perfect euhedral crytals of magnetite, embedded in an epoxy matrix, Heider *et al.* (1988) studied domain patterns at temperatures up to $350\,°C$. The walls faded to obscurity at higher temperatures, presumably because of wall broadening and the decrease in M_s of both the section being observed and the magnetite colloid marking the wall positions. An example of the results appears in Fig. 6.12. This $\approx 30\,\mu m$ grain was heated by absorption of energy from the transmitted beam of the microscope, which was focused in the viewing plane, while domain patterns were observed with the reflected beam. The transparent matrix and magnetite crystals outside the field of view were negligibly heated. Contact with the immersion oil of the microscope objective conducted away some heat and was allowed for in calibrating the temperature, which increased linearly with the light-source voltage up to $400\,°C$. Similar domain patterns were observed at the lower temperatures using a conventional copper

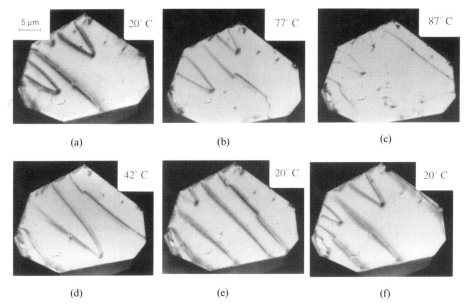

Figure 6.12 Variation of domain structure in a 30 μm magnetite crystal (a)–(e) during a heating–cooling cycle, and (f) after a second heating. [After Heider *et al.* (1988) © American Geophysical Union, with the permission of the authors and the publisher.]

block heating stage. The focused-beam method has the advantages of quick thermal response and much higher temperature operating range.

The initial domain pattern after AF cleaning consisted of two block domains, two large spikes and a small spike. By 77 °C, the minor spike had been driven out, the neighbouring large spike had shrunk, and the main domain wall traversing the grain jumped to a central position that greatly reduced the net moment. By 87 °C, the wall had crosssed most of the grain. By 122 °C, the simple domain structure was replaced by irregular faint walls meandering across the grain. There were few changes in this complex pattern up to 200 °C.

The domain structure that appears on cooling resembles but does not duplicate the original structure. It differs from one heating–colling cycle to another, and may contain from two to five domains.

None of the changes observed can be dictated by magnetostatic energy since M_s decreases by only about 10% over this temperature range. The initial patterns up to 87 °C, are classic domain structures following magnetocrystalline easy axes. However K_1 decreases very rapidly with temperature in magnetite (Fig. 3.7b) and stress-controlled patterns take over at higher temperatures. The magnetostriction constants decrease approximately as $M_s^{2.3}(T)$ in magnetite (Fig. 3.8b), and at higher temperatures the stress-controlled patterns fade in their turn.

Almost all large (> 30 μm) synthetic grains behaved in this way. So also did ≈ 70 μm grains of natural magnetite (Heider, 1990). At room temperature, they contained a few simple domains, but this structure changed enormously with mild heating. Such domains could not possibly carry stable TRM blocked near the Curie point. Fields of 0.8 kA/m or 1 mT applied in various orientations had little effect on the thermal pattern changes, which seem to be controlled by the intrinsic anisotropies. Large MD grains of magnetite are thus dubious carriers of primary NRM, in agreement with conventional wisdom.

Among rather small magnetite grains (< 10 μm) Heider *et al.* (1988) observed that 10–20% had simple body plus closure domain structures which did not change appreciably up to 350 °C. The walls did respond to applied fields but at higher threshold fields than in the temperature dependent structures. These small grains are more promising candidates for carriers of stable, high-blocking-temperature NRM.

Ambatiello *et al.* (1995) have observed domains on a {110} polished surface of a 6 mm magnetite single crystal using the magneto-optical Kerr effect (MOKE) up to 550 °C. In contrast to Heider *et al.*'s results, they observed simple lamellar body and closure domains with only minor and reversible thermal changes in domain width up to 400 °C. At higher temperatures, domain width increased upon heating and decreased on cooling, but there were no dramatic changes in the style of the domain structure.

6.6.3 Titanomagnetite

Domain structures in Ti-rich titanomagnetites behave in an entirely different way near the Curie temperature. Soffel (1977b) studied a natural TM75 grain in which walls gradually faded in the interior and eventually also at the surface between −40 and +70 °C, without moving appreciably. His interpretation was that differential oxidation between the interior and surface produced a wide range of Curie temperatures. Soffel and Appel (1982) produced 1.5 mT TRM's in synthetic TM72 ($T_C = 60$ °C) and also noted that the domains did not change much in number or pattern, although displaced walls led to a large TRM ($M_{tr} = 0.16 M_s$).

Halgedahl (1987) likewise noted gradual fading of the interior walls during heating of natural TM60 grains with $T_C = 150$ °C, without significant shifts in any of the walls. Wavy walls did not change their pattern with heating either. Halgedahl showed that the number n of walls theoretically changes with T as $M_s K^{-1/4} A^{-1/4}$, where A is the exchange constant. In magnetite, $A(T) \propto M_s^{1.7}$ (Heider and Williams, 1988). Assuming a similar dependence in TM60, a magnetoelastic rather than a crystalline origin of K is the likely explanation of a temperature-invariant number of walls. Although the walls had become faint by 105 °C, Halgedahl deduced that n had not changed because the addition of a new wall would have changed H_d, thereby shifting the other walls.

When similar natural TM60 grains were heated in $H_0 = 100$ µT, the number of domains decreased with rising temperature (Metcalf and Fuller, 1987b). The grains were almost saturated by 100 °C and the last small wall vanished at 120 °C. Although surprising at first sight, these observations are reasonable. The magnetostatic energy $E_H = -\mu_0 V \boldsymbol{M} \cdot \boldsymbol{H}_0$ due to the external field decreases less rapidly than E_d, E_{ex} or E_K, and at high enough temperatures will dominate the total energy. Under these conditions, even a weak field $\boldsymbol{H}_0$ will suffice to saturate the particle. In confirmation of these ideas, the field required to drive out all walls was observed to decrease as T increased. Even rather weak fields are effective in driving walls at high temperatures. Microscopes themselves are sufficiently magnetic that it is practically impossible to eliminate all fields during domain observations. It is possible that in experiments to date the intrinsic temperature dependence of domain structure near the Curie point has not been isolated from the field response of the domains.

Metcalf and Fuller (1987b, 1988) found that quite large TM60 grains were in a metastable SD state after acquiring TRM by cooling in $H_0 = 3$ mT and an almost saturated state after acquiring TRM in a 0.1 mT field. From their observations of the fraction of metastable SD grains as a function of the field producing TRM, they were able to synthesize a $M_{tr}(H_0)$ curve that resembles measured curves quite closely. We will return to this subject in §9.9.

Halgedahl (1991) observed a variety of LEM states, ranging from one to nine domains, when synthetic 25–50 µm TM60 grains containing 10 mole% Al and Mg ($T_C = 75$ °C) acquired TRM in $H_0 = 42$ µT in replicate experiments. Metastable SD states were quite common and did not follow the same distribution as the other LEM states (Fig. 5.14). A much smaller range of states (usually 4, 5 or 6 domains) was observed following AF demagnetization.

The mechanism for this variety of TRM states seemed to be denucleation of one or more domains during cooling. An example is shown in Fig. 6.13. As in the experiments of Heider *et al.* (1988) on magnetite and Metcalf and Fuller (1987b) on TM60, denucleation consisted of the constriction of a lamellar domain into a spike and its eventual collapse transversely across the grain. This process of transverse nucleation or denucleation changes the number of domains by two. The changes occurred over a temperature range of at most 4 °C, and sometimes much less than this.

6.7 Electron microscope observations

6.7.1 SEM observations

To take advantage of the higher magnification and resolution available with the scanning electron microscope (SEM), Goto and Sakurai (1977) introduced the

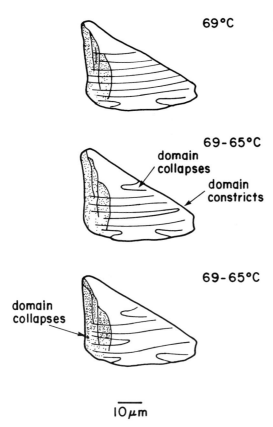

69°C

69-65°C
domain
collapses
domain
constricts

69-65°C
domain
collapses

$\overline{10\,\mu m}$

Figure 6.13 Domain denucleation observed in a 30-μm TM60 grain during cooling from the Curie temperature. [After Halgedahl (1991) © American Geophysical Union, with the permission of the author and the publisher.]

dried-colloid technique of wall observation. Moskowitz *et al.* (1988b) have used the method to reveal domain structures in synthetic polycrystalline TM60 with Al and Mg impurities, with a crystallite size ≈ 100 μm and a Curie temperature of 80 °C. A hydrocarbon-based magnetite colloid was evaporated on polished or unpolished surfaces and the dried Bitter patterns were observed with the SEM after coating the surface with carbon or gold-palladium. The large depth of field of the SEM permits domains to be imaged on quite rough surfaces, eliminating the need for mechanical and ionic polishing. Furthermore, with high magnifications (20 000–30 000×), fine (< 1 μm) structures can be resolved and the widths of walls measured with an uncertainty of ±0.05 μm.

On unpolished surfaces, long straight walls, wavy walls, and nested arrays of spikes and wavy walls were observed, as with the optical Bitter method on ionically polished surfaces. However, consistent patterns covered much broader areas than in polished sections. Assuming that the uniaxial style of the domain patterns is stress-controlled, σ_i is now more uniform in direction but is still sufficient that magnetoelastic anistropy outweighs crystalline anisotropy.

A major advance in Moskowitz *et al.*'s study was the first attempt to measure the wall width δ_w. The quantitative relationship between Bitter patterns and the widths of the underlying walls is contentious (Hartmann, 1987; Williams *et al.*, 1992a). Nevertheless, Moskowitz *et al.* estimated δ_w to be 0.17 μm for straight walls bounding lamellar domains, 0.24 μm for wavy walls, and 0.40 μm for nested spikes. In one case, straight walls on one crystal face were seen to connect with wider wavy walls on another face. Since $\delta_w \propto K^{-1/2}$ (eqns. (5.10), (5.14)), the regular change in δ_w suggests that increasingly complex surface patterns in TM60 correspond to a decrease in $K_\sigma = 1.5\lambda\sigma$, i.e. to a *decrease* in overall stress level.

From the amplitude and wavelength of wavy walls, the specific wall energy $\gamma_w = 2\pi(AK)^{1/2}$ (eqn. (5.11)) can be found by using the theory of Szymczak (1968), reviewed by Halgedahl (1987). Equations (5.10) and (5.11) then yield A and K_σ. For their Al- and Mg-doped TM60, Moskowitz *et al.* found $A \approx 0.55 \times 10^{-11}$ J/m, about one-half that of magnetite. Their calculated K_σ values implied average σ_i levels of about 20, 10 and 4 MPa (0.2, 0.1 and 0.04 kb) for crystallites displaying lamellar domains, wavy walls, and nested structures respectively. These stresses are about an order of magnitude less than those calculated by Appel (1987) for slightly oxidized natural TM60's with Al and Mg impurities (see §6.4).

A recently introduced technique uses spin-polarized SEM observations to directly measure surface magnetizations (Haag and Allenspach, 1993). Secondary scattered electrons have polarized spins which 'remember' the spin direction of the surface atoms from which they were scattered. In principle, the surface magnetization can be mapped out along three orthogonal axes, giving complete vector information about M_s at the sample surface. The spin-polarization signal is independent of the topography, so that no corrections need to be applied, unlike the situation in magnetic force microscopy (§6.9).

6.7.2   TEM observations

The transmission electron microscope (TEM) offers high magnifications and the possibility of viewing interior domain structure, rather than just surface patterns. The major drawback is that specimens must be thinned by ion bombardment to a <0.5 μm foil. As a result, virgin crystal surfaces cannot be examined and, more seriously, the domain structures observed are not necessarily those of bulk three-dimensional material. Electron beams were used as long ago as the late 1950's to detect the surface leakage fields associated with domains in hematite and pyrrhotite at various temperatures (Blackman *et al.* 1957, 1959; Gustard, 1967).

More recently, P. P. K. Smith has been the leader in TEM observations relevant to rock magnetism (Smith, 1979, 1980; Morgan and Smith, 1981). He has employed Lorentz microscopy, which makes use of the deflection of the electron beam as it passes through magnetic material. Oppositely magnetized regions cause opposite deflections of the beam. The magnetic field of the objective lens is so strong that it will saturate the specimen. To image domain structure in a reasonably low-field state, the objective must be switched off and reduced magnifications result. The image is purposely defocused until the magnetically deflected beam either converges or diverges at the domain walls, which then show up as narrow intense bright lines or broader dark lines.

Smith (1980) and Morgan and Smith (1981) show Fresnel defocus images of walls in ≈ 50 μm magnetite regions within natural magnetite–ilmenite high-temperature oxyexsolution intergrowths (§3.1, 14.4). Energy dispersive analysis of the characteristic X-ray emission confirmed that the spinel examined contained $< 1\%$ Ti. The main walls are long, straight and parallel (Fig. 6.14). They image as alternating bright and dark lines, indicating that the wall moments alternate in polarity, as expected from minimum-energy considerations (see §5.8). Closure domains form near the edge of the magnetite. The spacing of the main domains is ≈ 1 μm, in reasonable accord with theory (eqn. (5.22)).

Two- and three-domain structures have been imaged by Smith (1980) in 0.5–1 μm natural magnetite inclusions in a garnet matrix. Examples appear in Fig. 6.15. Since the grain dimensions are similar to the foil thickness, the structures observed are probably representative of those in the original grains. The walls were straight and parallel to the long axis of each grain, thus minimizing E_d. The number of domains is again about the theoretically expected equilibrium number. Furthermore, when the domain pattern was saturated with the objective lens field and then allowed to reform, the same structure always reappeared. There were no nucleation difficulties and no alternatives to the equilibrium domain structure.

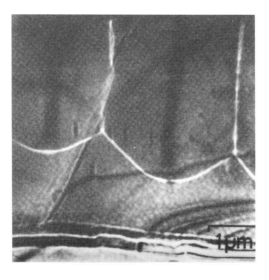

Figure 6.14 Fresnel TEM image of body and closure domains near the edge of a ≈ 50 μm magnetite sub-grain in an oxyexsolution intergrowth. Walls of opposite polarization image as dark and light lines. The domain width is about 1 μm. [After Smith (1980), with the permission of the publisher, Institute of Physics Publishing, Bristol, UK.]

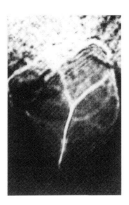

Figure 6.15 Fresnel TEM images of two-domain structures, with and without closure domains, in ≈ 1 μm magnetite grains in garnet. [After Smith (1980), with the permission of the publisher, Institute of Physics Publishing, Bristol, UK.]

6.8   MOKE observations

The magneto-optical Kerr effect (MOKE) is the rotation of the plane of polarization of light by a magnetic medium. By this technique, one can examine variations in the magnetization within domains and not simply the location of 'discontinuities', i.e., domain walls. Disadvantages are the extremely smooth surface required and the small Kerr rotation in magnetite and TM60. By polishing surfaces with a suspension of amorphous silica spheres and digitally processing and enhancing the Kerr images, Hoffmann *et al.* (1987) produced the first Kerr-effect domain pattern photographs for natural and synthetic TM60's. All the features seen using Bitter patterns – patches of lamellar domains, separated by regions of undulating walls and spikes, or featureless areas – were also seen in the Kerr images.

Heider and Hoffmann (1992) observed simple lamellar domains in about 30% of the synthetic magnetite crystals they examined. Since the crystals were not specially oriented for polishing, this fraction presumably represents those crystals whose ⟨111⟩ easy axes happened to be sufficiently close to the plane of observation for simple patterns to form. A 70 μm crystal was observed in the course of field cycling to saturation (Fig. 6.16a). At most six body domains formed, considerably less than the equilibrium number predicted. All reverse domains vanished by 100 mT and in the SIRM state there was apparently still only a single domain. A reverse domain nucleated and grew to fill more than half the grain in a back-field of −30 mT. By −40 mT, the normal domain had almost disappeared and the crystal was nearing negative saturation.

The SD-like hysteresis implied by the domain observations is at odds with measured hysteresis for similar crystals, however (Fig. 6.16b). A bulk sample containing dispersed 80 μm crystals gives a classic MD linear ramp, saturating around 200 mT or 160 kA/m (cf. Fig. 5.16). Heider and Hoffmann propose that the domains are being viewed broadside on, rather than edge-on. Thus the domains appear to be less numerous and much wider than they are in reality.

Figure 6.16
(a)i

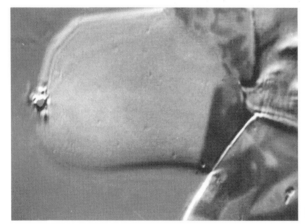

(a)ii

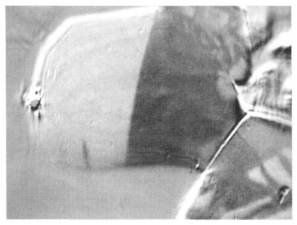

(a)iii

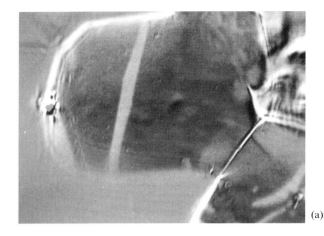

(a)iv

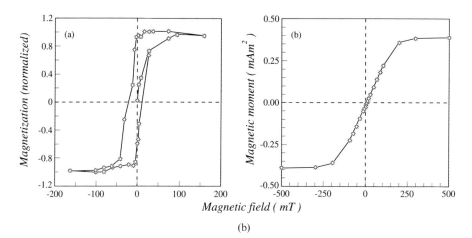

(b)

Figure 6.16 (a) MOKE images of body domains in a ≈ 70 μm magnetite grain for different applied fields H_0. At most six domains are observed (i), and the grain is metastably single-domain in the SIRM state (ii). Wall nucleation followed by two large Barkhausen jumps bring the grain to a demagnetized and then a reversely saturated state (iii,iv). (b) Simulated SD-like hysteresis loop, based on the domain observations in (a), compared with the MD-like hysteresis curve measured for grains of similar size. [Reprinted from Heider and Hoffmann (1992), with kind permission of the authors and Elsevier Science – NL, Sara Burgerhartstraat 25, 1055 KV Amsterdam, The Netherlands.]

This is a potential problem for any method of domain observation and points up the need for careful orientation of crystals before polishing. Only a {110} plane contains two sets of ⟨111⟩ easy axes and is in addition perpendicular to the wall plane (§5.3.1). Thus only if {110} is the plane of observation does one have much chance of imaging interior domain structures in a favourable (i.e., edge-on) orientation (cf. Figs. 1.2, 5.8, 6.3).

6.9 Magnetic force microscopy

Magnetic force microscopy is a recently developed technique that allows the magnetic stray field above the surface of a grain to be mapped in great detail (Hartmann, 1994; Dahlberg and Zhu, 1995; Proksch *et al.*, 1995). The magnetic force microscope (MFM) consists of a magnetized needle-like tip on the end of a small cantilever which scans over the grain surface, ideally at constant height (typically about 0.1 μm). The principle is analogous to flying an aeromagnetic survey with constant terrain clearance and flight-line spacing, but on a micro-scale (in fact, a nanoscale, since a resolution of 10 nm or 0.01 μm is achievable). To remove the effects of surface topography, a similar image of Coulomb forces measured with a non-magnetic tip (an atomic force microscope or AFM) is subtracted from the magnetic image.

The first MFM images of domain walls in magnetite were reported by Williams *et al.* (1992b) on a polished {110} surface of a 5 mm natural single crystal. A symmetrical profile with a single force maximum was obtained across the wall between lamellar domains (Fig. 6.17). This is the expected profile for a 180° Bloch wall (see §7.5.4). Asymmetric Bloch walls or Néel 'caps' (near-surface portions of walls), if present, could not be resolved.

One striking observation was that the force peak reversed sign partway along the domain wall in one 20 μm-square section. The force reversal implies reversal of the sense of rotation of spins in the Bloch wall and of the associated domain wall moment (§5.8). Thus walls are not invariably of a single polarity in magnetite. Measurements of the typical length of a wall domain should soon give us an experimental estimate of the critical size for SD walls or 'psarks'.

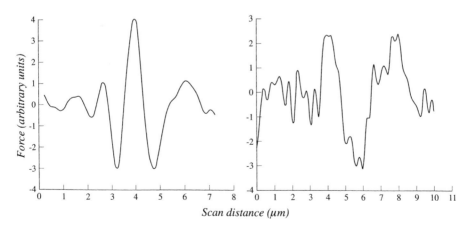

Scan distance (μm)

Figure 6.17 MFM force profiles across the same 180° Bloch wall, viewed on a {110} surface of a natural magnetite crystal, in different parts of a 20 μm-square section. The wall has reversed polarization and magnetic moment between the two profiles. [After Williams *et al.* (1992b), with the permission of the authors and the publisher, Blackwell Science, Oxford, UK.

Proksch *et al.* (1994a) also reported MFM images of domain-wall structures in the {110} plane of a single crystal of magnetite. An image of a 180° wall, parallel to the [111] easy axis, in a 5 µm × 5 µm region and the corresponding stray-field profile over the wall are shown in Fig. 6.18. The half-width of the profile is 210 ± 40 nm, which is quite consistent with our rough calculation of δ_w in §5.3.1, but somewhat larger than more refined estimates of 95–160 nm for δ_w. A major problem in MFM studies is deconvolving the overall response to remove the effect of the magnetized tip, whose width and direction of magnetization are often only approximately known. The tip response may be responsible for broadening the wall profile in Fig. 6.18(b).

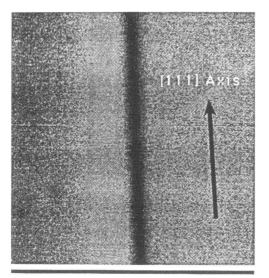

(a)

5µm

Figure 6.18 (a) MFM image of a 180° Bloch wall in a single crystal of magnetite ({110} viewing plane). (b) MFM response profile across the wall compared to a model of the wall structure. The tip geometry and the scan height have been adjusted for best fit. The half-width of the profile agrees reasonably well with predictions of the Bloch-wall width δ_w in magnetite. [After Proksch *et al.* (1994a) © 1994 IEEE, with the permission of the authors and the publisher.]

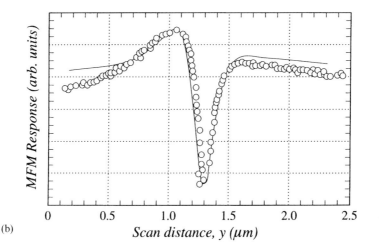

(b)

Chapter 7

Micromagnetic calculations

7.1 Introduction

The theory of domain structures in magnetite and titanomagnetite has made great strides in the last decade. Most earlier calculations used Kittel or Amar models like those of §5.4. These were deterministic; no interesting and novel structures could emerge. Block-like walls and domains without internal structure were imposed in advance. Only their numbers and widths were adjustable. We will review some of these calculations in §7.2.

About the mid-1980's, less constrained **micromagnetic calculations** (Brown, 1963a) were introduced in rock magnetism. In micromagnetism, model crystals are divided into cells and the magnetization direction of each cell is varied independently until an overall structure emerges that minimizes the total energy. Most such structures are **local energy minimum (LEM)** states rather than the **global energy minimum (GEM)** state. That is, the equilibrium number of domains has usually not been found. However, at least domain structure can be expected to emerge in a natural way instead of being imposed at the outset. Fine structures within domains and domain walls also appear without forcing.

In model magnetite crystals larger than $\approx 0.5\,\mu m$, the calculations do converge on structures resembling classic domains. Closure domains and finer structures such as vortices feature in the solutions. In smaller grains, the fine structures fill much of the particle. Domains as such are barely recognizable until the SD range ($< 0.1\,\mu m$) is reached. These structures are described in §7.3–7.5. Changes in particular LEM states or families of states with changes in applied field H_0, grain volume V, and temperature T are the subject of §7.6–7.8.

Transformations between LEM states, particularly at high T, are considered in §7.9 and 7.10.

The major limitation on micromagnetic calculations is computation time. For reasonable resolution of structures, the shortest dimension of cells (which are cubes, rods or sheets, respectively, for three-, two- or one-dimensional calculations) must be ≈ 10–$20\,\mathrm{nm}$. The number of pair interactions among cubes required to evaluate E_d at each step in a three-dimensional variational calculation grows rapidly as the crystal size increases, making calculations prohibitively long and expensive, even using supercomputers. The practical upper size limit for such calculations is presently about $1\,\mu\mathrm{m}$, and for two-dimensional calculations (which are no longer completely unconstrained) about $5\,\mu\mathrm{m}$. Almost all domain observations have been made on $> 5\,\mu\mathrm{m}$ size grains. Direct comparison of theory and experiment therefore remains difficult.

Recent theoretical results indicate some possible problems in observing domains. The larger the number n of domains in a crystal of a given size, the broader the walls are predicted to be (§7.3). Broad walls do not attract magnetic colloid particles as effectively as narrow walls and may not image clearly. It is conceivable that metastable SD states of grains that are theoretically much too large to support such a structure are observational artefacts. Direct measurements of magnetic moments of individual particles would conclusively distinguish between SD and large-n LEM states.

Micromagnetic calculations have pointed out what in hindsight is obvious. Near-surface spins deflect to minimize surface poles, with the result that interior Bloch walls evolve into Néel-like walls at the crystal surface (§7.5.4). The flux leakage from these domain-wall terminations (called Néel caps) is less than conventional estimates based on constrained structures with Bloch structure throughout, and they may image poorly.

Closure domains of major size are a feature of two-dimensional and some three-dimensional solutions for magnetite (§7.5). They have been omitted from most previous theorizing, although it is well known that they raise equilibrium domain widths substantially (§5.4.2). Only for certain favourable orientations, namely a $\{110\}$ viewing plane perpendicular to the plane of $71°$, $109°$ and $180°$ walls, are closure domains well imaged by the colloid method.

The major theoretical advance of the last decade has been the recognition that domain structures other than the ground state – non-equilibrium LEM states – not only occur frequently, but could conceivably dominate the remanence and impart quite unexpected grain-size dependences (§7.4). Since metastable SD states were observed as early as 1980, the theoretical advance may seem rather belated. Nevertheless, the idea of transformation from one equilibrium domain structure to another at a set of fixed particle sizes was so firmly rooted that the notion of alternative states and 'transdomain' transitions between them in a crystal of fixed size was revolutionary (§7.9.2). The implications for theories of TRM acquisition (§7.10) have not been worked out fully and will be a source of excitement in the years to come.

7.2   Constrained calculations

To find a stable magnetization structure, the sum of the exchange, anisotropy, demagnetizing and field energies (E_{ex}, E_K, E_d and E_H, respectively) must be minimized. E_{ex}, E_K and E_H are local energies which are readily evaluated for any model structure. E_d, however, is a long-range pair-interaction energy whose computation has been a major stumbling block. Simple formulations of E_d like eqns. (5.1), (5.3) and (5.4) are exact only for large numbers of domains, and moreover ignore the structure of the walls. Most modern formulations use Rhodes and Rowlands' (1954) analytic result for the demagnetizing energy of arrays of rectangular prisms, which can represent 180° body domains of variable width in constrained calculations or small cells of fixed size in micromagnetic calculations. The RR formulation of E_d and its implementation were described in §4.4.

Amar (1958) extended constrained calculations to include E_d of the walls, which he modelled as rectangular blocks uniformly magnetized at ±90° to the domains. The Rhodes–Rowlands–Amar (RRA) expressions for E_d (e.g., eqns. (4.34), (4.35)) are complicated, but have the great advantage that the RRA coefficients need only be calculated once for a given domain and wall geometry.

Shcherbakov (1978) and Moskowitz and Banerjee (1979) used RRA formulations to calculate energies and equilibrium size ranges for 2D, 3D and 4D structures in magnetite cubes and rectangular prisms. Argyle and Dunlop (1984) improved the RRA formulation slightly by subdividing the domain walls into two equal parts, each with an effective magnetization $2/\pi$ times that of the domains (Fig. 4.11). They calculated sizes of 0.06 μm for the transition from SD to 2D equilibrium structure (i.e., d_0), 0.13 μm for the 2D–3D transition, and 0.30 μm for the 3D–4D transition (Fig. 7.1). Moskowitz and Banerjee (1979) reported 0.08, 0.15 and 0.31 μm, respectively, for these critical sizes.

Argyle and Dunlop also pointed out that the internal structure of 2D and 3D grains is quite variable. Near the lower limit of the 2D or 3D equilibrium size range, structures with broad domain walls that fill much of the grain volume have lower energy than conventional structures with narrow walls and broad domains (Fig. 7.1). Near the upper limit of the equilibrium size range, walls narrow considerably.

The fractional volume occupied by domain walls, which for crystals of rectangular form is simply δ_w/D, determines the net moment of the walls relative to the net moment of the domains. For 180° domains and Bloch walls, these moments are orthogonal and independent (§5.8). If the wall fills most of a 2D

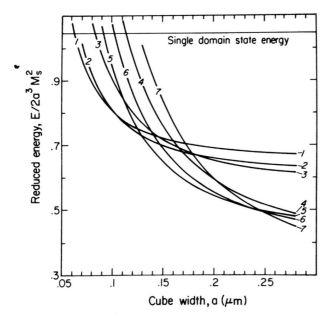

Figure 7.1 Theoretical estimates of the total magnetic energy of magnetite cubes as a function of size for different 1-dimensional domain structures: SD, 2D (curves 1, 2, 3: domain-wall widths of 85%, 66% and 40%, respectively), 3D (curves 4, 5, 6: wall widths of 30%, 49% and 37.5%, respectively), and 4D (7: wall width 31%). In the size range shown, all are permissible alternative LEM states. [After Argyle and Dunlop (1984) © American Geophysical Union, with the permission of the authors and the publisher.]

grain just above d_0, its moment will be almost as intense and stable as that of an SD grain. In 3D grains, however, it is energetically favourable for the spins to rotate in opposite senses in the two walls (Argyle and Dunlop, 1984; Enkin and Dunlop, 1987), so that the wall moments cancel. In 4D and larger grains, the wall moments alternate in polarity (cf. Fig. 6.14). If the number of domains is even, there is always one uncompensated wall, but M_{net} due to wall moments becomes relatively minor.

If the number of domains is odd, on the other hand, the wall moments balance but the domains have a net moment, even in zero applied field (Craik and McIntyre, 1969; Shcherbakov, 1978; Dunlop, 1983a). The volume of the two end domains and other domains of the same polarity is greater than the volume of the central domain and its family of domains. This spontaneous remanence due to domain imbalance cannot be demagnetized. Just as in the SD state, the grain is in a minimum-energy condition. However the domain-imbalance remanence M_r is much weaker than SD remanence. It amounts to about 12% of the SD magnetization M_s for 3D cubes and 2–4% of M_s for 5D and 7D cubes (Dunlop, 1983a).

7.3   One-dimensional micromagnetic calculations

An important advance was made by Moon and Merrill (1984), who used a one-dimensional variational method to determine the micromagnetic structure of a model cube. The analogous problem for a sphere was solved much earlier by Stapper (1969) but had little influence in rock magnetism. Moon and Merrill divided their model crystal into lamellae within which M_s was confined to the plane of the lamella, just as in 180° domains and Bloch walls. Within a particular lamella M_s was at a specified angle θ_i to the easy axis of anisotropy, and all values of θ_i were permitted, not just 0, 90° and 180° as in conventional RRA calculations. The set of θ_i were varied until the total energy converged on a minimum. In this approach, magnetic charges appear on the four outer faces of each lamella but not on the two interior faces. The calculation is micromagnetic, i.e., variational, but not completely unconstrained.

Most of the volume of a cube of equilibrium 2D size was found to be non-uniformly magnetized. The 'wall' filled the particle, as had been concluded by Veitch (1983) and Argyle and Dunlop (1984) on the basis of more restrictive calculations. A new feature was 'overshoot' of the θ_i beyond 0 or 180° at either edge of the wall, forming 'skirts' with moments opposed to the wall moment (Figs. 5.6b, 7.2).

In agreement with Argyle and Dunlop's simple RRA calculation, Moon and Merrill, and later Enkin and Dunlop (1987), found that the volume fraction occupied by walls decreases in grains much larger than the critical minimum size for a particular equilibrium domain state. Figure 7.2 illustrates this effect

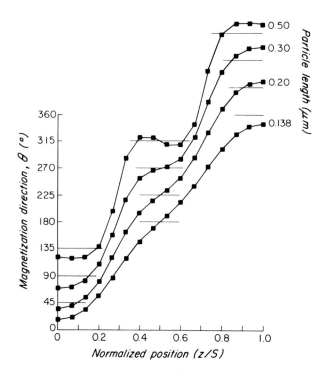

Figure 7.2 Minimum-energy 1-dimensional micromagnetic structures (spin vector orientation as a function of position z, normalized to cube edge S) in magnetite cubes of different sizes. The curves have been offset by 45° for clarity. Notice the evolution from broad to narrow walls as particle size increases, which was also evident in Fig. 7.1. [After Enkin and Dunlop (1987) © American Geophysical Union, with the permission of the authors and the publisher.]

for 3D magnetite cubes ranging from 0.14 μm to 0.50 μm in size. As well as being narrower, the walls in the larger particles develop larger skirts which effectively compensate or 'screen' the wall moment. This effect imparts an extreme grain-size dependence to wall-moment magnetization and effectively rules out the 'psark' model for SD-like remanence except in the smaller 2D grains.

7.4   Non-equilibrium or LEM states

Moon and Merrill (1984) noted that, in addition to observations of metastable SD structures (§6.2–6.6), there were good reasons for believing that particles are not always in their lowest energy states. Consider the 0.20 μm magnetite grain of Fig. 7.2, whose theoretical equilibrium structure is 3D. Because the grain size is not much above the critical 2D–3D size, the walls are broad and would image poorly with the Bitter technique. On the other hand, 0.20 μm is well above the SD–2D transition size d_0 and if the grain is in a (metastable) 2D state, the wall will be narrow and will image well. The Bitter method cannot in practice be used with such small grains, but the same argument applies to optically visible magnetites containing larger numbers of walls. The states that yield well-defined narrow walls and are most readily observed under the microscope tend to be LEM states containing fewer than the equilibrium number of walls.

The realization that higher-energy branches like the ones in Fig. 7.1 are not merely mathematical constructs but represent frequently occupied, perhaps even typical, physical states is a breakthrough in understanding ferromagnetic particles. For example, although the number and total moment of domain walls increases with increasing grain size if particles are in their equilibrium states, the moment decreases sharply if grains of all sizes remain in a 2D state, because of the narrowing of the walls.

There are limits to how large or how small a grain can support a given metastable domain structure. Too large or too small a grain will spontaneously nucleate or denucleate a domain wall. Moon and Merrill (1985) approached the nucleation problem by stepping a 180° wall of prescribed structure into a particle, starting from one edge. When such an edge perturbation spontaneously propagated into the particle, the upper stability limit had been reached. This is a constrained rather than a micromagnetic calculation and suffers from the obvious limitation that only stability to a particular form of perturbation is tested. However, the results (Fig. 7.3) agree quite well with micromagnetic calculations of stability of metastable SD states by Shcherbakov and Lamash (1988) and of SD, 2D and 3D LEM states by Enkin and Dunlop (1987). The latter calculations are described next.

At each step of their energy minimization, Enkin and Dunlop calculated not only the gradient of the total energy E_{tot} with respect to each of the θ_i but also the Hessian matrix of second derivatives of E_{tot}. When all gradients approach zero and all second derivatives are positive, a LEM state which is stable against all possible perturbations of the θ_i has been found. Enkin and Dunlop also computed complete energy surfaces (Fig. 7.4). The θ_i values were fixed at opposite

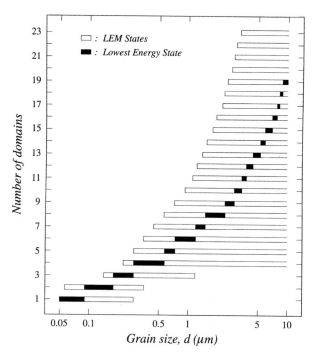

Figure 7.3 Theoretical stable size range for various 1-dimensional LEM states in magnetite. The stability ranges greatly exceed the equilibrium size ranges (in black) for most structures. [Reprinted from Moon and Merrill (1985) with kind permission of the authors and Elsevier Science – NL, Sara Burgerhartstraat 25, 1055 KV Amsterdam, The Netherlands.]

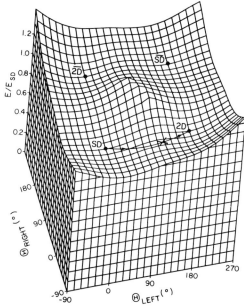

Figure 7.4 Theoretical total energy surface for 1-dimensional micromagnetic structures (specified by the spin orientations at opposite edges of the particle) in a 0.08 μm magnetite cube. Notice that there are two SD LEM states, $(0, 0)$ and $(180°, 180°)$, with opposite moments and two 2D LEM states, with opposite wall polarities and moments. A typical energy barrier and transition path between states are shown. [After Enkin and Dunlop (1987) © American Geophysical Union, with the permission of the authors and the publisher.]

edges of the grain and a minimum-energy structure subject to this constraint was calculated. The total energy as a function of the constrained angles, which were changed independently in 15° increments, maps out an energy surface on which the locations of stable minimum-energy structures are clear.

In the example shown, a 0.08 μm magnetite cube, there are two SD and two 2D energy minima, each pair having oppositely directed moments (in the 2D case, these are wall moments). In zero applied field, the two SD states have equal energies and are equivalent equilibrium structures. Of only slightly higher energy than the SD ground states are the two degenerate 2D structures, which represent excited states. However, a sizeable energy barrier separates the LEM and GEM structures.

In grains smaller than 0.08 μm, the 2D energy increases until the energy barrier vanishes and the domain wall denucleates. In larger grains, the SD energy increases and a 3D state eventually appears at approximately the same location on the energy surface. The calculated minimum and maximum sizes for stability of SD, 2D and 3D LEM states agree quite well with Moon and Merrill's values.

It is evident from Fig. 7.3 that in >1 μm magnetite grains, a variety of LEM states are possible. A 1 μm grain, for example, is predicted to have a 6D GEM state but can host as many as 10 domains or as few as 3 domains. Excess domain walls have not been observed experimentally; nucleation of walls seems to be the problem in the real world, not denucleation. A serious discrepancy is that metastable SD and 2D remanent states theoretically should not exist in >1 μm magnetites, whereas in fact they are well documented (Boyd et al., 1984). The imperfection of real crystals should if anything aid in nucleating walls, and so there is no immediate explanation to this paradox.

7.5 Two- and three-dimensional micromagnetic calculations

7.5.1 Methodology

Two-dimensional and especially three-dimensional micromagnetic calculations with reasonably large arrays of cells can only be implemented on a supercomputer or a parallel-processor computer (see, however, Shcherbakov et al., 1990). Wright et al. (1996), using a three-dimensional fast-Fourier-transform (FFT) algorithm without restrictive boundary conditions to calculate E_d, were able to determine minimum-energy structures with a resolution of $64 \times 64 \times 64$ cells in about 16 hours of CPU time on a 16 000-processor parallel computer. With a cell edge of 10–20 nm, model grains as large as 1 μm can be studied. By way of comparison, earlier calculations (Williams and Dunlop, 1989) with $12 \times 12 \times 12$ resolution (maximum grain size of 0.2 μm) required up to 24 hours on a supercomputer. Efficient coding is the key to calculating structures in larger grains.

Apart from the recent introduction of FFT methods for transforming E_d into a form suitable for parallel processing (each processor calculating a small number of pair interactions), the formulation of the various energy terms is fairly standard (Schabes and Aharoni, 1987; Newell et al., 1993a,b; Fabian et al., 1996; Wright et al., 1996). However, the algorithms used to find and test energy minima vary in their efficiency and reliability. The conjugate-gradient technique has been most widely used. At each step, it adjusts the direction of search to the steepest gradient of descent. The modified Newton method calculates second derivatives at each step, thereby avoiding false minima, but the memory required to store the Hessian matrix of second derivatives may be prohibitive. Monte Carlo and simulated-annealing methods are random walks and therefore very time-consuming. They do allow the local environment of minima to be explored, sometimes resulting in escape from a shallow minimum to a deeper nearby one, with significant reductions in LEM energies (Fukuma and Dunlop, 1995). Simulated annealing (e.g., Thomson et al., 1994) imitates thermal agitation and permits uphill as well as downhill walks. It provides a means of exploring possible transition paths between LEM states, but the low-resolution grids that can be used and limitations on the number of trials at present limit its use to excitations of SD or quasi-SD grains under nearly SP conditions.

7.5.2   Three-dimensional LEM states

Williams and Dunlop (1989, 1990) calculated unconstrained three-dimensional micromagnetic states for magnetite cubes and elongated prisms ranging from $0.05\,\mu m$ to $0.50\,\mu m$ in size. E_d was calculated using modified RRA expressions. M_s vectors were not constrained to lie in planes and θ_i for each of the 12^3 or 22^3 cells was varied independently. In this case, magnetic charges or poles appear on all six faces of each cell, and interactions between orthogonal as well as parallel pole sheets must be computed.

LEM states of five different types were found for $0.20\,\mu m$ cubes (Fig. 7.5). Some are recognizable variants of conventional domain structures which incorporate features suppressed in one-dimensional modelling. For example, the 2D structure incorporates large closure domains. In the quasi-SD structure, the magnetization vectors curl into or out of the cube corners to form external flux loops. These deflections could have been predicted from the H_d vectors at the corners of the SD cube of Fig. 4.9. Schabes and Bertram (1988) call this the flower state.

Other states are entirely non-lamellar. One such state is a 5D structure with a large core domain and four oppositely magnetized domains along lateral edges of the crystal. Another (not illustrated) is superficially similar but has internal

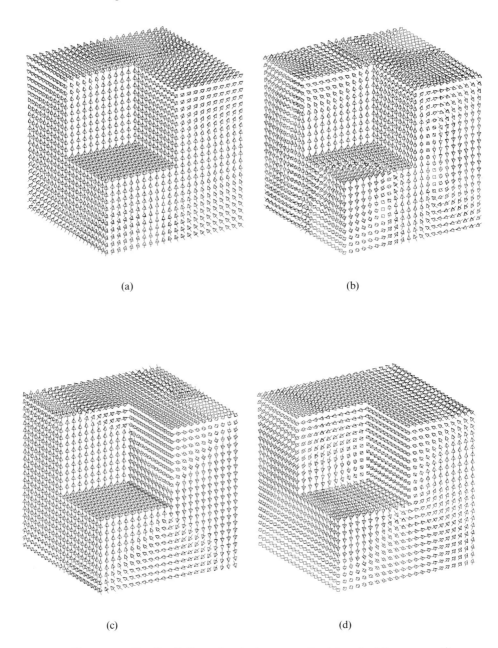

(a)

(b)

(c)

(d)

Figure 7.5 Possible 3-dimensional micromagnetic states of a 0.2 μm magnetite cube. (a) Modified SD (1A, or flower) state. (b) A related (1B) state, in which corner spins reverse and form reverse domains along four edges. (c) 2D state with closure domains. (d) Vortex state, with spins curling about a central axis. Vortex and 2D structures are similar in providing complete internal flux closure. [After Williams and Dunlop (1989), reprinted with permission from *Nature*. Copyright © 1989 Macmillan Magazines Ltd.]

spin curling instead of conventional domain walls. The third has no recognizable domains but consists of a single vortex of spins curling about the particle axis (Schabes and Bertram's vortex state).

Fabian *et al.* (1996), using an FFT method and grids up to $23 \times 23 \times 63$, found flower, vortex and double-vortex states in model magnetite grains similar in size to those in Fig. 7.5. (Double-vortex states will be illustrated in Fig. 7.10.) A variety of double-vortex configurations were possible in elongated grains.

Because three-dimensional models allow spins to deflect so as to reduce or eliminate surface poles, as they no doubt do in real crystals, there is less flux leakage from the surface and complex structures might not be imaged as such by colloid patterns. In other words, there is a possible problem with *observations* using the Bitter technique, and internal structures in real grains could in some cases be more complicated, instead of simpler, than observed surface patterns.

In larger magnetite grains, particularly elongated ones (maximum size: $0.5\,\mu\mathrm{m} \times 0.83\,\mu\mathrm{m}$), structures containing two or three approximately lamellar domains form spontaneously (Williams and Dunlop, 1990). They are terminated and magnetically linked by closure domains. (These states evolve from vortex and double-vortex states in smaller grains, as we will show in §7.7.) If the initial trial structure is an SD state parallel to the short axis, the domain magnetizations are also parallel to the short axis (a so-called head-on state). The energy would be lower for M_s vectors parallel to the long axis, but the transformation is prevented by configurational anisotropy, the barrier represented by the still higher energy of a diagonally magnetized state. Luo and Zhu (1994) have confirmed experimentally, using the MFM, that a head-on 2D state exists in elongated iron particles, although it is unstable and will form only if a saturating field is applied exactly along the short axis of the grain.

Wright (1995), with an FFT routine to compute E_d and high-resolution grids up to 64^3, has traced the development of lamellar multidomain structure as grain size increases from the SD range through PSD sizes (Chapter 12). He also modelled octahedral grains, resembling real magnetite crystals, and grains of irregular shape like those frequently found in the confining mineral matrix of a rock.

Three-dimensional models are at an early stage of development, but they offer tantalizing glimpses of what transitional structures in real crystals may look like. They have already demonstrated that there are distinct quasi-SD (i.e., flower) states and curled or vortex states, resembling a 2D structure with closure domains (Fig. 5.2d), over some ranges of grain size, rather than a single blended state that evolves from SD-like to 2D-like with increasing grain size (e.g., Veitch, 1983). The possibility that distinct LEM states are an artefact of the constraints built into one-dimensional models can now be ruled out.

7.5.3   Two-dimensional LEM states

Since two-dimensional lamellar structures develop spontaneously in three-dimensional modelling of 0.5–1 μm magnetite grains, it is reasonable to use two-dimensional modelling as a means of exploring structures in grains larger than 1 μm. Rod-shaped cells become the modelling elements and the number of pair interactions for a particular grain size is reduced enormously. For example, a three-dimensional calculation with 50^3 cells requires $\frac{1}{2} \times 50^6$ pair interactions to be evaluated at each iteration. A two-dimensional calculation with 50^2 cells requires $\frac{1}{2} \times 50^4$ such evaluations, a factor 2500 less. The computational problem is eased, so that more modest computers can be used. Alternatively, with the same computing power, one can model, at the same spatial resolution, grains ≈ 7 times larger (i.e., 7×7 times larger arrays of two-dimensional cells).

At present, 5 μm magnetite grains are the largest that have been modelled. Figure 7.6 shows two of the many possible LEM states. The first is a checkerboard pattern that develops if the initial trial structure is an upwardly magnetized SD state. The SD state is not stable in grains this large. Instead pairs of closure domains form in the four corners, providing a set of closed flux paths for the two diamond-shaped central domains, which retain the initial M_s orientation and are responsible for the remanence.

The second state develops from an initial lamellar 3D structure. Four large closure domains evolve, closing the flux internally. However, at the corners there are small vortices of the opposite sense, which are reminiscent of the flower state in very much smaller grains. Notice that this 3D state also has a remanence because the outer body domains have a larger volume than the central domain. This domain-imbalance moment is considerably less than the net moment of the checkerboard state, however.

Among the lamellar LEM states for a 5 μm grain, the 4D structure has a minimum energy. Thus the equilibrium domain width is about 1 μm, in agreement with the observations in Fig. 6.14.

Figure 7.7 compares the total energies of the vortex or curled state and various lamellar states as determined by one-dimensional and two-dimensional calculations in a 1 μm magnetite cube. The vortex structure, which has no one-dimensional equivalent, has the lowest energy. Closure domains greatly reduce the energies of two-dimensional lamellar states compared to their one-dimensional counterparts. The predicted equilibrium domain width is ≈ 0.5 μm (2D being the GEM state), in agreement with experiment (Fig. 6.15).

The distribution of two-dimensional LEM states is narrower than the distribution of one-dimensional states, and it is shifted to lower numbers of domains. The reason is again the saving in demagnetizing energy caused by the appearance of closure domains. The one-dimensional GEM state is 6D but the two-dimensional GEM state is 2D, which is not even a permitted one-dimensional LEM

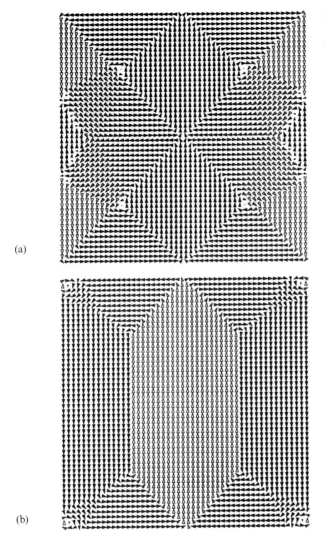

(a)

(b)

Figure 7.6 High-symmetry 2-dimensional micromagnetic states of a 5 μm magnetite cube. (a) Checkerboard pattern developed from an initial SD state with upward magnetization. Only the two diamond-shaped central domains preserve this original M_s orientation. Four flux-closure loops (each like Fig. 7.5c) occupy the four quadrants of the grain. (b) 3D structure with closure domains, developed from an initial lamellar 3D state. The pattern consists of two flux-closure loops in the left and right halves of the grain. [After Xu *et al.* (1994) © American Geophysical Union, with the permission of the authors and the publisher.]

state. Less-constrained calculations can thus achieve quite spectacular reductions in energy and changes in structures, including the range of permitted LEM states. Notice that in 1 μm magnetite grains metastable SD grains are prohibited as one-dimensional LEM states but are permitted as two-dimensional states. In reality they are only marginally stable and tend to degenerate into checkerboard states according to recent Monte Carlo calculations (Fukuma and Dunlop, 1995).

7.5.4   Domain-wall structure

The usual cubes or rods used as elements in micromagnetic modelling are 10–20 nm in their short dimension. Each side encompasses 12–25 unit cells of magnetite. In order to examine the internal structure of domain walls, a higher-

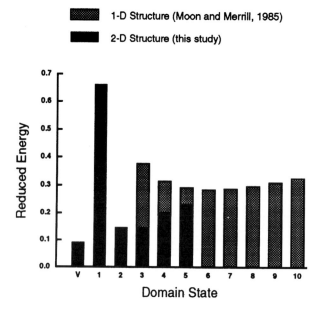

Figure 7.7   Comparison of energies of 1-dimensional and 2-dimensional micromagnetic states in a 1 μm model magnetite cube. Both the LEM-state energies and the number of domains in the GEM state are much lower when the 1-dimensional constraint is removed. [After Xu *et al.* (1994) © American Geophysical Union, with the permission of the authors and the publisher.]

resolution grid is needed in the wall region. The most common approach is to impose uniform domains beyond a distance four or five times that of the expected wall width (e.g., Scheinfein *et al.*, 1991), so that micromagnetic calculations are confined to the wall region and its immediate surroundings.

In the resulting structures (Fig. 7.8), the 70.5°, 109.5° or 180° Bloch wall (cf. Fig. 5.9b) in the interior of a magnetite crystal gradually changes to a Néel wall near the crystal surface. In this **Néel cap**, the spins rotate in the plane of the surface and generate no surface poles. The Néel cap is offset with respect to the underlying Bloch wall, which has a width $\delta_w \approx 0.15\,\mu m$ (Fig. 7.8a,b), and is wider than the Bloch wall. Predicted leakage-field profiles over these composite walls are basically symmetrical, showing the predominant influence of the underlying Bloch wall (Fig. 7.8c). The offset Néel cap results in some minor asymmetry. The predicted profiles reproduce most features of measured MFM profiles (Xu and Dunlop, 1996; Proksch *et al.*, 1994a; Figs. 6.17, 6.18).

7.6   Modelling hysteresis

All the structures described thus far are remanent ($H_0 = 0$) states. Williams and Dunlop (1995) have simulated hysteresis in 0.1–0.7 μm magnetite grains by carrying out successive three-dimensional micromagnetic minimizations when the field is stepped in 10 mT increments from 0 to +150 mT, and then, in 2 mT steps, reduced to −150 mT and increased back to +150 mT. The initial guess for each structure was the final structure from the preceding step. This procedure

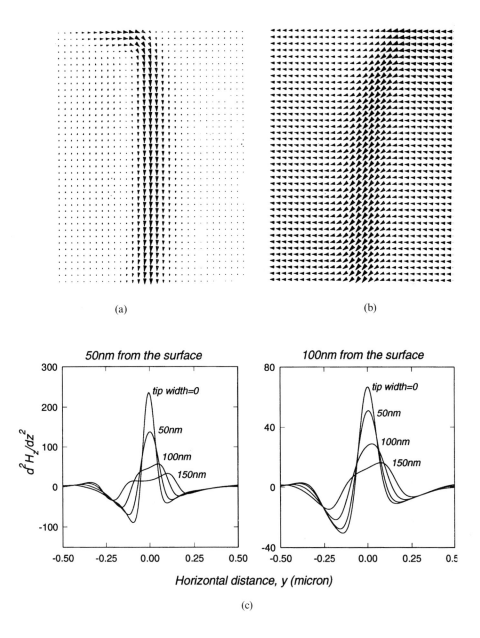

(a)   (b)

50nm from the surface   **100nm from the surface**

tip width=0   *tip width=0*

50nm   *50nm*

100nm   *100nm*

150nm   *150nm*

Horizontal distance, y (micron)

(c)

Figure 7.8 Néel caps developed at the crystal surface (top) for (a) a $180°$ and (b) a $109.5°$ Bloch wall in magnetite. Both the Néel and Bloch wall structures were predicted by 2-dimensional micromagnetic modelling. The area viewed is $0.6\,\mu m$ wide and extends into the crystal a depth of $0.8\,\mu m$. Uniform domains were assumed at large distances from the walls, but well beyond the area shown. (c) Profiles of d^2H_z/dz^2, which determines MFM response, for scanning heights of 50 and 100 nm and various MFM tip widths. The profiles show only a small amount of asymmetry due to the Néel cap. The profile half-width for small tip widths indicates a Bloch-wall width $\delta_w \approx 0.15\,\mu m$. [After Xu and Dunlop (1996) © American Geophysical Union, with the permission of the authors and the publisher.]

generates a quasi-static succession of metastable states rather than the true dynamic response (Schabes, 1991), but provided the damping of gyroscopic effects resulting from field changes is rapid compared to the rate at which the field changes, experimental hysteresis loops should be well simulated. Typical energy barriers are several orders of magnitude larger than $25kT$ even for the smallest grains modelled, so that thermal activation has a negligible effect compared to field-driven changes in structure.

A 0.1 µm magnetite grain initially in the quasi-SD flower state reverts irreversibly to the curling or vortex state during the descent from positive saturation (Fig. 7.9). In successive cycles between $+$ and $-$ saturation, M_s vectors of cells in the outer part of the grain rotate reversibly. The core reverses discontinuously at a higher field. The overall hysteresis cycle thus has both irreversible SD and reversible MD features.

In larger grains, the vortex state remains the ultimate GEM state reached during field cycling and the outer shell and core of the grain continue to react to H_0 almost independently of each other. However, irreversible jumps become smaller, reversals occurring in a series of steps rather than in SD style as a single 180° rotation. Halgedahl (1995) observed a similar change in measured hysteresis loops of hematite grains just above SD size as the grain size increased (see §6.5). Edge and core regions both undergo reversible and irreversible changes, core changes occurring before edge changes as grain size increases. In 0.7 µm grains, a cylindrical domain wall separates the edge and core, and its motion, continuous and discontinuous, produces magnetization reversal. The overall hysteresis curve for this grain size is quite narrow and ramp-like, with the expected slope $1/N$ (cf. Fig. 5.16).

The theoretical values of coercive force, H_c, and saturation remanence ratio, M_{rs}/M_s, change continuously with grain size, not in step-like fashion as micromagnetic remanence models without field cycling predict. The match to experimental measurements of H_c and M_{rs}/M_s is quite good (see Fig. 12.22). The key to understanding the grain-size dependence of hysteresis parameters is therefore in directly modelling spontaneous changes in LEM states due to changing H_0. Conclusions based on comparing a single remanent LEM state in different sizes of grains do not allow for field-induced transitions between states and may be misleading.

7.7   Modelling grain size changes

In some situations, for example the acquisition of chemical or crystallization remanent magnetization (CRM) (Chapter 13) by the growth of grains in a small H_0 like the geomagnetic field, changes in LEM state aided by the field are unlikely. The growing grain remains in a single family of LEM states whose pro-

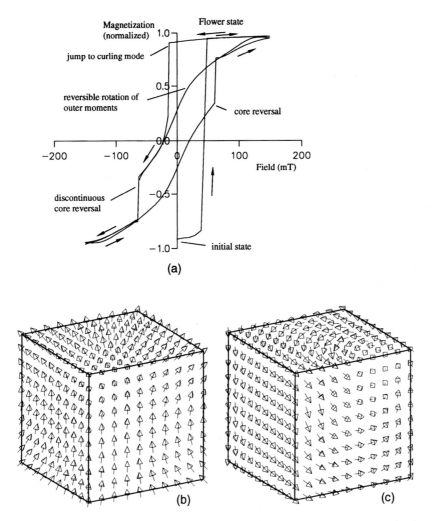

Figure 7.9 (a) Simulated hysteresis curve of a 0.1 μm cube of magnetite, based on a succession of 3-dimensional micromagnetic states for different applied fields H_0. (b) The initial flower state, and (c) the curling or vortex state to which the initial state transforms during the second hysteresis cycle. [After Williams and Dunlop (1995) © American Geophysical Union, with the permission of the authors and the publisher.]

genitor was the LEM state when the particle nucleated and began to grow (Newell *et al.*, 1993b). One such succession of states evolves from the diagonal quasi-SD state in small magnetite grains (Fig. 7.10). The $\langle 111 \rangle$ magnetocrystalline easy axis lies along a body diagonal but a $\langle 110 \rangle$ face-diagonal orientation of M_s vectors reduces magnetostatic energy E_d (Fig. 7.10a). This simple structure in 0.1 μm grains develops into a 2-vortex pattern in 0.2 μm and 0.3 μm grains (Fig. 7.10b) and a 4-vortex state in 0.4 μm grains (Fig. 7.10c). Notice that pairs

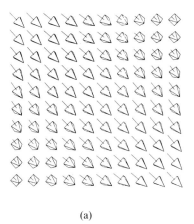

(a)

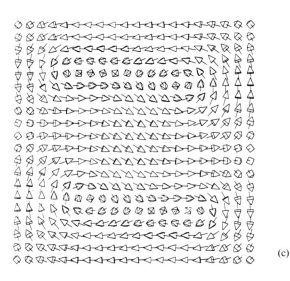

(b)

(c)

Figure 7.10 A family of related 2-dimensional diagonal micromagnetic states, developed by 'growing' a very small magnetite cube with a ⟨110⟩ diagonal SD state, using small steps in grain size. The initial structure for each grain size is the final structure for the preceding smaller size; this procedure simulates grain-growth CRM. The grain sizes illustrated are (a) 0.1 μm, (b) 0.2 μm, (c) 0.4 μm, (d) 1.0 μm. (e) A map of surface field H^2 over the 1 μm grain. The concentration of colloid in the Bitter technique is proportional to H^2; thus this map simulates the Bitter patterns that would be observed. Only the two vortex lines image clearly; most fine structure, including closure-domain boundaries, is lost. [After Newell *et al.* (1993b) © American Geophysical Union, with the permission of the authors and the publisher.]

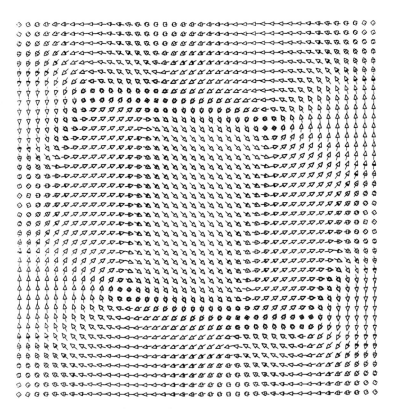

(d)

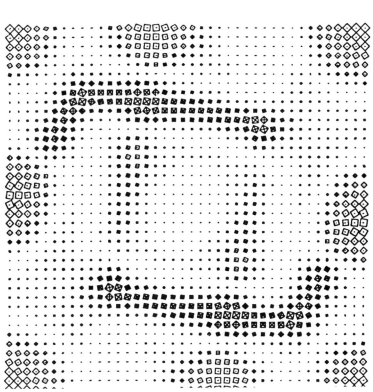

(e)

of vortices are linked by what amounts to a Bloch wall. The overall structure in Fig. 7.10(c) resembles a classic 3D grain with closure domains, except that the spins in the central body domain tend to be along $\langle 110 \rangle$ rather than $\langle 100 \rangle$ and the moments of the two 'walls' are parallel, not antiparallel. In these calculations, as in the hysteresis simulations, the initial guess for any step is the final minimum-energy structure for the previous step, in this case for a smaller grain size. By taking sufficiently small steps in grain size, continuity in the structures can be shown.

Some families of LEM states, for example the axial or $\langle 100 \rangle$ state of Newell *et al.* (1993b), destabilize and transform to another family above some critical size. The diagonal family of states, however, is continuous up to 1 μm at least. The lineal descendant of the simple diagonal structure of Fig. 7.10(a) in a 1 μm grain is illustrated in Fig. 7.10(d). Although complex, it is recognizably kin to the 4-vortex structure in a 0.4 μm grain. The vortex pairs are somewhat decoupled, but the 'vortex line' linking them still has basically Bloch character. The central body domain has separated into three domains, each along one of the two $\langle 110 \rangle$ axes in the plane of observation.

A surface-field map of the 1 μm grain (Fig. 7.10e) shows that only the four vortices with their linking vortex lines would gather significant amounts of colloid in the Bitter technique of observing domains. This structure would image as a 3D grain with two bent walls that fade toward the grain boundaries, without recognizable closure domains. The real structure, on the other hand, contains three central $\langle 110 \rangle$ body domains, two $\langle 100 \rangle$ body domains at the top and bottom, and four large wedge-shaped closure domains along the left and right edges.

Bitter patterns may fail to image much of the detail in micromagnetic structures and could actually be misleading. On the other hand, the structure in Fig. 7.10(d) can probably only be generated as a CRM. Such a complex structure never appears spontaneously in hysteresis simulations on only slightly smaller grains. Experimentally, Lorentz microscopy detects simple 2D and 3D states, with straight walls and sometimes closure domains, in ≈1 μm magnetite grains (Fig. 6.15). These grains have been exposed to strong fields in the electron microscope and are therefore in field-driven states.

7.8   Modelling temperature changes

Theoretical study of high-temperature domain structures has been hampered by uncertainties about the temperature variation of material properties like the crystalline anisotropy constant K_1, the magnetostriction λ, and especially the exchange constant A, which control domain-wall width and energy (§5.3). $A(T)$ is now reasonably well determined for magnetite (Heider and Williams, 1988), but the major theoretical study of equilibrium numbers and sizes of domains by

Moskowitz and Halgedahl (1987) is for TM60, whose material constants are less well known. Because of the latitude in permissible temperature dependences of the exchange and anisotropy constants, and uncertainty about whether magnetocrystalline or magnetostrictive effects are dominant at high temperatures. Moskowitz and Halgedahl were obliged to consider six models in which $A(T) \propto M_s(T)$ or $M_s^3(T)$ and $\lambda(T) \propto M_s^2(T)$, $M_s^4(T)$ or $M_s^6(T)$ respectively. $K_1(T) \propto M_s^8(T)$, as for magnetite (Fig. 3.7b), was assumed in all models. (Recent measurements favour $\lambda(T) \propto M_s^2(T)$ and $K_1'(T) \propto M_s^6(T)$ (Sahu and Moskowitz, 1995; Fig. 3.13).) Both Kittel and RRA formulations were used and gave quite similar results, except near the Curie temperature.

A typical result is shown in Fig. 7.11. Here a homogeneous internal stress of 100 MPa (about twice the typical σ_i suggested in §6.4 for TM60 and about five times σ_i calculated in §6.7.1) in a 2 μm grain leads to a general increase in the equilibrium number of domains as the temperature rises, no matter what model is assumed. Magnetocrystalline anisotropy (including magnetostriction, but with $\sigma_i = 0$) resulted in quite similar patterns, as did reducing the grain size to 0.2 μm or increasing it to 25 μm. Observations on Ti-rich titanomagnetites near the Curie temperature (§6.6.3) revealed either no change in domain patterns (Soffel, 1977b; Soffel and Appel, 1982; Halgedahl, 1987) or a decrease in the number of domains (Metcalf and Fuller, 1987b). As suggested earlier, the destabilization of domains in 'large' fields (near the Curie point, even a small H_0 may saturate a grain) may be masking the intrinsic zero-field response of the domains to heating.

Since wall width δ_w varies as $(A/K)^{1/2}$ (eqns. (5.10), (5.14)) and K of either crystalline or magnetoelastic origin usually has a more rapid thermal decrease than A, walls tend to enlarge with temperature. In some of Moskowitz and Halgedahl's models, they expanded to fill a large fraction of the particle volume.

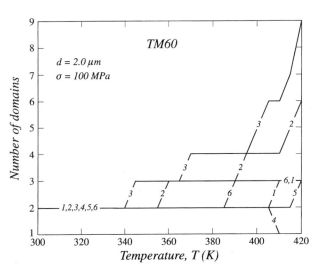

Figure 7.11 Theoretical equilibrium domain structures at high temperature in 2 μm TM60 cubes with 100 MPa homogeneous stress (magnetoelastic anisotropy dominant). Models 1–6 are explained in the text. [After Moskowitz and Halgedahl (1987) © American Geophysical Union, with the permission of the authors and the publisher.]

The resulting increase in wall energy limits the number of walls and domains at high temperature in RRA modelling.

One-dimensional micromagnetic modelling for magnetite at high temperatures has been attempted only for small numbers of domains (Newell *et al.*, 1990; Dunlop *et al.*, 1994). Xu and Merrill (1990a) have used an RRA model with subdivided walls (Fig. 4.11) to extend calculations up to ≈ 30 domains. With $A(T) \propto M_s^{1.7}(T)$ (Heider and Williams, 1988), walls are expected to expand when 0.2–0.5 μm grains are heated, but 2D wall moments should *decrease* in the larger grains because skirting becomes more effective at high temperature. The stability ranges for SD, 2D and 3D LEM states all contract at high temperature, with the result that only over rather restricted grain-size ranges does a particle cooling from the Curie point have a choice of domain structures as it passes through its blocking temperature.

7.9   Domain structure changes

With changing temperature T, stress σ or field H_0, a particular domain structure may become unstable and transform into another. One familiar example is the transformation to SD structure by driving out domain walls in a saturating field, either at ordinary temperatures or as a result of heating in a weak field to near the Curie point. Another is spontaneous nucleation of a reverse domain when a backfield is applied to a saturated particle (Figs. 6.1, 6.9). In theoretical simulations, a transformation from the flower to the vortex state occurred in Fig. 7.9. If the before and after states are not too different in energy, a spontaneous thermally excited transformation may occur. In each case, the questions to be answered are:

1. What is the form of the transformation mode?
2. At what critical value of the field or energy barrier does the transformation become overwhelmingly probable?

7.9.1   Single-domain reversals

Incoherent reversals of SD particles have been much studied in the permanent-magnet and magnetic-recording industries. Classic incoherent modes like curling, buckling (Frei *et al.*, 1957) and fanning (Jacobs and Bean, 1955) are dominant in elongated particles of strongly magnetic materials (Fig. 7.12). They lead to considerably reduced critical fields for reversal compared to coherent rotation, particularly at small angles between M_s and H_0 (see Fig. 8.7) and larger grain sizes (Fig. 7.13a). Like coherent rotation, curling, buckling or fanning bring

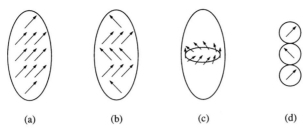

Figure 7.12   Sketches of (a) coherent and (b)–(d) incoherent reversal modes in SD ellipsoids and chains of spheres. Crystalline anisotropy is ignored compared to shape anisotropy. The patterns shown occur during the transition between upward and downward SD states caused by an applied field directed downward (not shown). The modes are (a) coherent rotation (or rotation in unison), (b) buckling, (c) curling, (d) fanning.

into play all spins simultaneously. They represent a dynamic instability: a complete change of state is effected very rapidly at a critical value of H_0.

In the roughly equant crystals of magnetite and titanomagnetite that typify most rocks, other modes may be important. Moon and Merrill (1988) calculated that domain wall nucleation and propagation across the short dimension of a grain should dominate in all SD magnetites except quite elongated particles, in which longitudinal modes (somewhat analogous to buckling or fanning) would be preferred. Coherent rotation would occur only in a restricted size range below 0.05 μm.

The advent of micromagnetic calculations has allowed refinement of these ideas. Enkin and Williams (1994) have made three-dimensional calculations of transitional states triggered by coherent and anti-coherent (spins rotated in opposite senses at opposite ends of a grain) perturbations. Reversal proceeds by vortex nucleation and propagation in magnetite grains larger than ≈0.06 μm in size (Fig. 7.13b), even though the SD state is energetically favoured over the vortex state up to 0.08 μm. The vortex reversal pattern is not unlike that of curling, except that spins have more freedom of orientation (Fig. 7.14). Instead of being a simultaneous mode of all spins, the curled region propagates transversely and longitudinally in the grain. In this respect, it resembles wall nucleation and propagation, but the actual pattern of spin deviations is unlike that in a domain wall.

In §5.8, we pointed out that domain walls have two possible polarities or senses of spin rotation, with oppositely directed moments. If the wall is SD, its moment may reverse at a critical value of H_0, whereas if the wall is subdivided into wall domains of alternating polarity, magnetization can change gradually by Bloch-line displacement (Fig. 5.22). Incoherent modes of SD wall reversal are quite possible. One such mode would be Bloch-line nucleation and propagation, although experience with whole-particle reversals suggests that a less constrained mode would be energetically favoured.

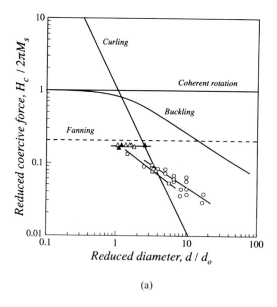

(a)

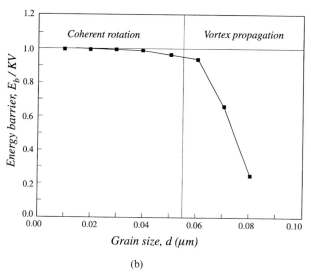

(b)

Figure 7.13 (a) The grain size dependence of the critical field for SD magnetization reversal by coherent rotation, buckling and curling in an infinite cylinder and fanning in a chain of spheres. Data on elongated SD iron and iron-cobalt particles have the $d^{-0.6}$ size dependence predicted for buckling, but their magnitudes of coercivity are more consistent with curling reversals. [After Luborsky (1961), with the permission of the publisher, The American Institute of Physics, Woodbury, NY.] (b) Theoretical energy barrier to magnetization reversal in magnetite cubes as a function of grain size for coherent rotation and vortex propagation modes. The vortex state has a higher energy than the SD flower state up to 0.08 μm, but between 0.06 and 0.08 μm, the SD state reverses by the transitory passage of a vortex across the grain (see Fig. 7.14). [After Enkin and Williams (1994) © American Geophysical Union, with the permission of the authors and the publisher.]

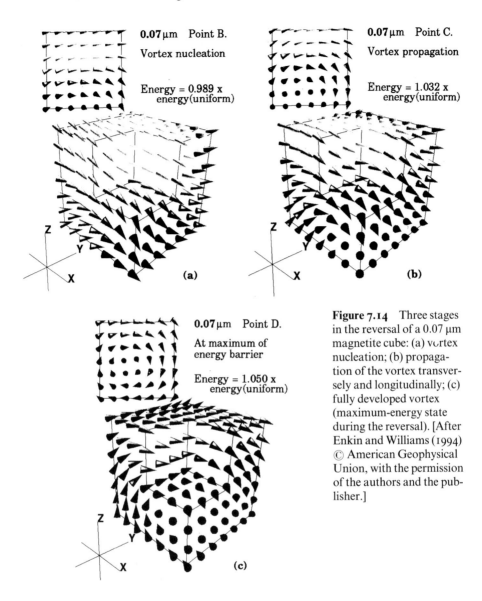

0.07 μm   Point B.

Vortex nucleation

Energy = 0.989 x
    energy(uniform)

0.07 μm   Point C.

Vortex propagation

Energy = 1.032 x
    energy(uniform)

(a)

(b)

0.07 μm   Point D.

At maximum of
energy barrier

Energy = 1.050 x
    energy(uniform)

(c)

Figure 7.14   Three stages
in the reversal of a 0.07 μm
magnetite cube: (a) vortex
nucleation; (b) propaga-
tion of the vortex transversely and longitudinally; (c)
fully developed vortex
(maximum-energy state
during the reversal). [After
Enkin and Williams (1994)
© American Geophysical
Union, with the permission
of the authors and the publisher.]

7.9.2   Transdomain transitions

In grains large enough to host both SD and other LEM states, SD reversals can
occur in two stages via an intermediate non-SD resting state. Such a transition
between one-dimensional micromagnetic states, was illustrated in Fig. 7.4. The
path between oppositely oriented SD structures was via a 2D state and involved
two separate transformations, SD → 2D and 2D → SD. Transformations of
this type, in which one LEM state is reconfigured into another LEM state of
entirely different structure, are called **transdomain transitions**. One-dimensional
models are appealing for transdomain calculations because each LEM structure

can then be specified by only two parameters, the spin angles at either edge of the grain, and the transition path can be followed on a two-dimensional energy surface. The topography of the transition is easy to imagine and the energy barrier between states is readily calculable.

If one follows the minimum-energy path over the saddlepoint between LEM states (e.g., the SD → 2D path in Fig. 7.4), the demagnetizing energy decreases but the anisotropy and especially the exchange energies increase to peaks (Dunlop *et al.*, 1994). In the example of Fig. 7.15, the total energy peaks when the

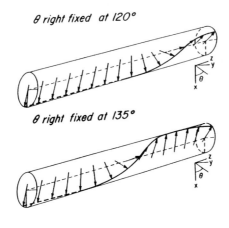

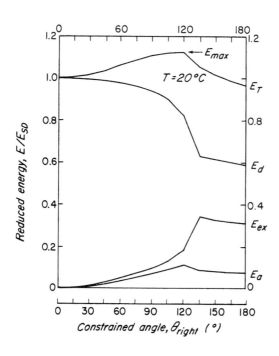

Figure 7.15 Changes in anisotropy, exchange, demagnetizing and total energies accompanying edge nucleation of a domain wall in 1-dimensional modelling of a $L = 0.255\,\mu\text{m}$ magnetite grain with elongation $q = 1.5$. A domain wall nucleates and propagates into the grain between edge spin angles of 120° and 135°, the same range in which the energy terms change dramatically. [After Dunlop *et al.* (1994) © American Geophysical Union, with the permission of the authors and the publisher.]

spin angle at one edge of the grain has increased from 0 to 120°, but the most rapid changes in the component energy terms occur between 120° and 135°. The micromagnetic structures reveal that a domain wall nucleates at the right when $\theta = 120°$ and propagates into the grain by $\theta = 135°$. Thus we have calculated the energy required for **edge nucleation** of a domain wall, as well as the structure of the wall that forms.

In an earlier model of edge nucleation, Moon and Merrill (1985) stepped a wall of specified form inwards from the edge of a cube of magnetite until it propagated of its own accord. The energy barrier for the transdomain transformation is overestimated in such a calculation. First, the structure of the wall is prescribed instead of resulting from a micromagnetic minimization as in Fig. 7.15. Secondly, other modes such as **transverse nucleation**, in which a spike nucleates and propagates transversely, splitting an existing domain (Figs. 6.12, 6.13), cannot be accommodated in one-dimensional models. Nevertheless, an order-of-magnitude estimate does result. For the easiest transition, from 5D to the 6D GEM state in a 1 µm magnetite grain, the energy barrier was in excess of several thousand kT at room temperature. Since only $\approx 25kT$ of thermal energy is available on laboratory timescales and $\approx 60kT$ on geological timescales (§5.5.2), such transitions cannot be thermally excited at ordinary temperatures.

7.10   Transdomain TRM

At temperatures close to the Curie point, T_C, the exchange and demagnetizing energies decrease to such an extent that energy barriers between LEM states drop below $25kT$ (or $60kT$). Thermal fluctuations can then excite changes in domain state and transdomain thermoremanent magnetization (TRM) becomes possible on laboratory (or geological) timescales. Dunlop $et\ al.$ (1994) have explored the size ranges over which SD, 2D and 3D states are stable in 0.1–0.5 µm magnetite grains at high temperatures and energy barriers E_b between these states. Calculated energy surfaces, with stable LEM states and transition paths just below $T_C = 580\,°C$, are shown in Fig. 7.16. A weak field, $H_0 = 100\,\mu T$ (similar to the earth's field), saturates the grain at T_C, but only 1–2 °C below T_c, E_d begins to outweigh E_H and a 2D LEM state of lower M develops.

A series of total-energy profiles along the least-energy transition path between SD and 2D states, for a 0.21 µm magnetite grain at temperatures of 574 °C, 560 °C and 500 °C, appears in Fig. 7.17. Above 560 °C, E_b for SD $\rightarrow$ 2D transitions, $\Delta E_{12} = E_{max} - E_{SD}$, and E_b for 2D $\rightarrow$ SD transitions, $\Delta E_{21} = E_{max} - E_{2D}$, are both $< 60kT$. Transitions in either direction are easy, on geological timescales, and grains are in a thermally unblocked or superparamagnetic condition. Around 560 °C, ΔE_{12} becomes $> 60kT$ and SD $\rightarrow$ 2D transitions

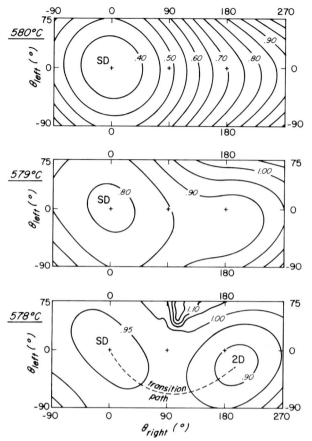

Figure 7.16 Total-energy surfaces (applied field $H_0 = 100\,\mu\text{T}$) for the $L = 0.255\,\mu\text{m}$ magnetite grain of Fig. 7.15 at and just below the Curie temperature. The grain is saturated at T_C and has only an SD state available. With only 2 °C of cooling, a 2D state develops and becomes the GEM state. [After Dunlop *et al.* (1994) © American Geophysical Union, with the permission of the authors and the publisher.]

become *blocked*. Even on a geological timescale, they will not occur with the aid of thermal energy alone. Around 500 °C, E_{2D} becomes lower than E_{SD}. 2D rather than SD is now the GEM state. However, the grains remain trapped in metastable SD states because the barriers E_b for transitions between states are now about $160kT$. Transdomain TRM has been acquired.

Notice that the occupation probability of the favoured SD state at the **blocking temperature**, T_B, of 560 °C was calculated using Boltzmann statistics to be 100%. There was no choice between SD and 2D LEM states at T_B. This turns out to be the case whenever the energies of two states differ by $4kT$ or more. The GEM state at T_B is overwhelmingly populated, and transdomain TRM occurs only if the GEM state changes during cooling, as it does in Fig. 7.17.

In an attempt to counteract this determinancy, Ye and Merrill (1995) have proposed that a great variety of 'predomain' states with only short-range order exist within $\approx 1\,°\text{C}$ of T_C and give rise to different LEM states with cooling. However, just below the predomain range, probably within a few °C at most of T_C and well above the range of typical T_B's, long-range ordering occurs and LEM

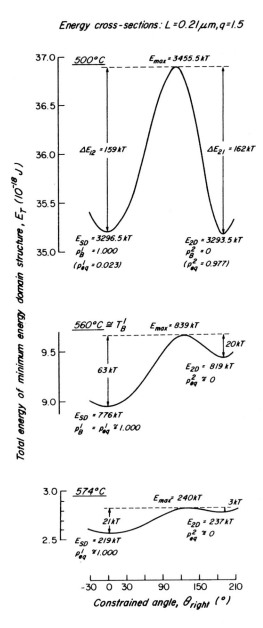

Energy cross-sections: L =0.21 μm, q=1.5

Figure 7.17 Total energy profiled along transition paths like the one in Fig. 7.16 (bottom) between SD and 2D 1-dimensional LEM states ($\theta_{\text{right}} = 0$ and $\approx 180°$, respectively). Energy barriers are low at 574 °C for transitions in either direction. At $T_{\text{B}} = 560$ °C, the barrier for SD → 2D transitions becomes larger than $60kT$ and the grain is trapped in the SD state, which has a Boltzmann occupation probability $p_{\text{eq}} \approx 1$ at this temperature. At 500 °C, 2D becomes the GEM state, but the grain remains trapped in a metastable SD state, acquiring transdomain TRM. [After Dunlop *et al.* (1994) © American Geophysical Union, with the permission of the authors and the publisher.]

states form. Transitions between LEM states are easy at these temperatures, as Fig. 7.16 shows, and LEM states of different predomain parentage will rapidly drop down into the prevailing GEM or ground state. It seems impossible for inherited predomain structures to survive unchanged until T_{B}.

Another problem pointed out by Moon (1991) and Dunlop *et al.* (1994) is that the size ranges over which various LEM states are stable become so narrow near T_{C} that there is very little overlap of states. Only for particular grain sizes, such as the 0.21 μm grains of Fig. 7.17, is there a choice of more than one type of LEM state at T_B. For larger grains which, at room temperature anyway, have a wider choice of stable LEM states (Fig. 7.3), transdomain TRM is more likely. Moskowitz and Halgedahl (1987) have pointed out yet another problem. In their larger TM60 grains, they predict that substantial numbers of domains should be added or lost (typically lost) during cooling (cf. Fig. 7.11). They speculate that blocking cannot occur until all such transdomain changes have ceased. The energy barriers opposing these proposed transdomain changes have not been calculated, however.

Moskowitz and Halgedahl's calculations show that the equilibrium number n_{eq} of domains in TM60 is typically larger near T_C than at T_0 (Fig. 7.11). One might then wonder how it is that grains, when they cool, come to be trapped in states with *fewer* than n_{eq} domains. In particular, how do metastable SD states of TRM (Halgedahl, 1991; Halgedahl and Fuller, 1983; Fig. 5.14a), come to exist? The flaw in the reasoning is that Moskowitz and Halgedahl's calculations are for zero applied field, and are therefore relevant to thermal demagnetization, i.e., unblocking rather than blocking. Above T_B, even a field as weak as the earth's field may suffice to drive out most or all of the walls (Fig. 7.16, top). It is presumably this SD state existing within a degree or so of T_C that is somehow preserved during cooling.

Chapter 8

Single-domain thermoremanent magnetization

8.1 Introduction

When magnetic minerals cool in a weak field H_0 from above their Curie temperatures, they acquire **thermoremanent magnetization** (TRM) in the direction of H_0 (or rarely, in the opposite direction) with an intensity proportional to H_0. The fidelity of TRM in recording directions and intensities of ancient geomagnetic fields is the justification for paleomagnetism. TRM is the primary NRM of igneous rocks and some high-grade metamorphic rocks. The NRM of individual detrital grains in sediments and sedimentary rocks is frequently also a TRM inherited from the eroded source rocks.

TRM is much more intense than isothermal remanence (IRM) acquired in the same weak field H_0 at room temperature. TRM is also very stable over long periods of time against changes in field (e.g., polarity reversals) or reheating. The reason for this high intensity and stability compared to room-temperature remanence is that TRM is acquired at high temperatures, where energy barriers and coercivities are low, and stabilized by cooling to ordinary temperatures, where barriers and coercivities are high.

TRM is a frozen-in high-temperature equilibrium distribution achieved by thermally excited transitions among different magnetic states (cf. §7.10). Transitions cease below the **blocking temperature**, T_B, because in the course of cooling, the energy barriers E_b between different magnetization states grow larger than the available thermal energy ($\approx 25kT$ for experimental times of a few minutes; $\approx 60kT$ for long geological times). In the last chapter, we discussed transdomain TRM, a partition between different domain structures governed by nucleation or denucleation of domains. In the present chapter and the one that follows, we

201

will show how other changes in magnetic state, such as thermally activated reversals of single-domain (SD) grains and thermally excited Barkhausen jumps of domain walls, also provide the ingredients for stable TRM.

Néel (1949, 1955) established a theoretical basis for understanding TRM which endures today. Néel's SD theory (§8.4–8.6) gives the following results. At time t in an ensemble of identical non-interacting SD grains, coherent reversals between states in which moments are either parallel or antiparallel to a weak field H_0 control the evolution of the net magnetization $\boldsymbol{M}$ of the ensemble towards equilibrium:

$$M(t) = M_0 e^{-t/\tau} + M_{eq}(1 - e^{-t/\tau}) \tag{8.1}$$

where $M_0 = M(0)$. The thermal equilibrium magnetization is

$$M_{eq} = M(\infty) = M_s \tanh\left(\frac{\mu_0 V M_s H_0}{kT}\right) \tag{8.2}$$

and the relaxation time τ is

$$\frac{1}{\tau} = \frac{1}{\tau_0} e^{-E_b/kT} = \frac{1}{\tau_0} \exp\left[-\frac{VK}{kT}\left(1 - \frac{H_0}{H_K}\right)^2\right], \tag{8.3}$$

in which $\tau_0 \approx 10^{-9}$ s is the atomic reorganization time or interval between successive thermal excitations. Notice that τ changes extremely rapidly, exponentially in fact, with changes in temperature T, grain volume V, or microscopic coercive force H_K, the critical field for rotations in the absence of thermal energy. $H_K = 2K/\mu_0 M_s$ for uniaxial shape anisotropy (SD magnetite) or magnetoelastic anisotropy (SD hematite) (see eqns. (5.34)–(5.37)).

According to (8.1), M will be close to its thermal-equilibrium value M_{eq} if we wait a time $t \approx \tau$. Small changes in T have an enormous effect on how long we have to wait. Close to T_C, where E_b is small, equilibrium is attained almost instantaneously (an unblocked or superparamagnetic condition). At ordinary temperatures, where E_b is large, equilibrium is attained sluggishly or not at all (a blocked or frozen condition). At the blocking temperature T_B, equilibrium is just achieved in a typical wait time t. Under laboratory conditions, this might be ≈ 60 s, whereas under geological conditions, t might be as long as 1 Ga (10^9 years). Solving (8.3), we obtain

$$E_b = \ln(t/\tau_0)kT_B \approx 25kT_B \quad \text{(laboratory)} \tag{8.4}$$
$$\approx 60kT_B \quad \text{(geological)}$$

Equations (8.4) justify our earlier estimates (§5.5.2, 7.10) of the barriers that can just be crossed with the aid of thermal energy. Notice that the available thermal energy is $\gg kT$; this is because the characteristic time τ_0 for a single thermal excitation is 10 orders of magnitude less than ordinary experimental times. The collective effect of $\geq 10^{10}$ thermal impulses helps the magnetic system over the barrier between states.

When a grain cools below T_B, M_{eq} is frozen-in as TRM. According to (8.2), weak-field TRM should be proportional to H_0, since $\tanh(\alpha) \approx \alpha$ for small α. TRM should be relatively intense because $\tanh(\alpha)$ saturates rapidly as α increases and M_{eq} then approaches the SD saturation magnetization M_s. (Physically this means that all SD moments are parallel to H_0.) Two of the properties of TRM are thus explained. The high thermal stability of TRM is also accounted for, because M_{eq} can only be unfrozen and reset to zero by reheating (in $H_0 = 0$) to the unblocking temperature, T_{UB}. According to (8.3), if H_0 is $\ll H_K$, $T_{UB} \approx T_B$. TRM can only be demagnetized by reheating to its original high blocking temperature.

Néel's theory has been criticized as simplistic, but more rigorous treatments (e.g., Brown, 1959, 1963b) give essentially identical results, while experiments on SD maghemite (γFe_2O_3) particles in a frozen ferrofluid (Bacri *et al.*, 1988) confirm Néel's predicted T, V and H_0 dependences. A more serious criticism is that coherent rotation is the reversal mode over only part of the stable and metastable SD range in magnetite (§7.9.1). Equation (8.3) can readily be adapted for numerical calculations by using theoretical energy barriers for incoherent reversals. One can go further and reformulate eqn. (8.2) for a general Boltzmann partition among competing SD and non-SD LEM states. Relaxation times must then be computed for all possible transitions between transdomain or other states (e.g., states with a constant number of walls but varying wall displacements). However, in view of its attractive simplicity and the ease with which it can be generalized, we shall revert to Néel's analytic SD formulation for the remainder of this chapter.

8.2   Some experimental properties of TRM

The intensity of a weak-field TRM, and the intensities of all the partial TRM's (fractions with different T_B; see §1.1.3) that compose it, are proportional to the applied field strength H_0. The TRM data of Fig. 8.1 (Özdemir and O'Reilly, 1982b) are for oxidized synthetic SD grains of TM60 with Al replacing 20 mole % of the Fe (ATM60/20). The data are interesting for two reasons. First, the TRM-carrying capacity falls by almost a factor 4 as z increases from 0.16 to 0.59, much more than the corresponding decrease in M_s. The same maghemitization process degrades primary titanomagnetites in lavas erupted on the ocean floor, damping the TRM signal that is the principal direct evidence for seafloor spreading and plate motions (Chapter 14). Secondly, Özdemir and O'Reilly (1982c) have compared similar data for unoxidized TM60 grains with the predictions of eqn. (8.2). Their theoretically predicted values of V are compatible with the $\approx 0.04 \, \mu m$ sizes observed in electron micrographs.

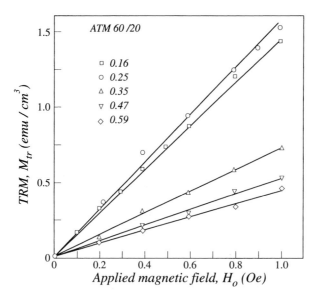

Figure 8.1 The acquisition of TRM in weak fields for SD-size titanomaghemites with oxidation parameters z ranging from 0.16 to 0.59. [After Özdemir and O'Reilly (1982b), with the permission of the authors and the publisher, Journal of Geomagnetism and Geoelectricity, Tokyo, Japan.]

A second important property of single-domain TRM is the equality of the blocking temperature T_B during field cooling and the unblocking temperature T_{UB} during zero-field reheating. The data of Fig. 8.2 illustrate the thermal demagnetization of partial TRM's acquired over a narrow range of blocking temperatures by a natural sample containing SD maghemite. Upon reheating,

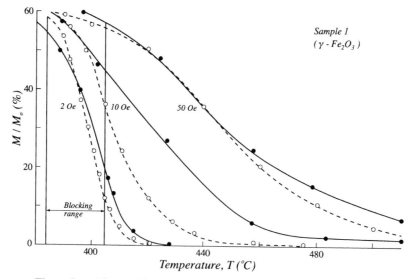

Figure 8.2 Thermal demagnetization of partial TRM's acquired over a narrow blocking-temperature range in fields of 2, 10 and 50 Oe by a natural SD maghemite. The dashed curves are predicted by Néel SD theory (eqn. (8.19)). For weak fields, the pTRM unblocks over its original blocking range: $T_{UB} = T_B$. [After Dunlop and West (1969) © American Geophysical Union, with the permission of the authors and the publisher.]

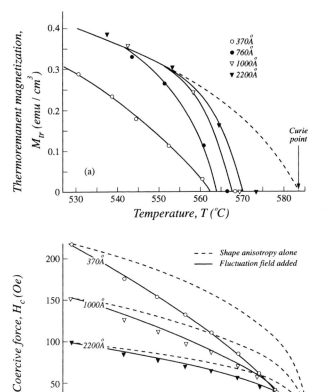

Figure 8.3 Dependence of unblocking temperatures, determined from (a) thermal demagnetization of TRM and (b) from coercive force data, on grain size for four SD or nearly SD magnetites. The smaller the grain size, the lower is T_{UB} or T_B. [After Dunlop and Bina (1977), with the permission of the authors and the publisher, Blackwell Science, Oxford, UK.]

the partial TRM acquired in $H_0 = 2$ Oe (0.2 mT) disappears over almost exactly the original blocking range. Larger acquisition fields make T_{UB} and T_B increasingly different, but the data are quite well described by Néel theory (dashed curves). The equality of T_B and T_{UB} for weak fields and the proportionality of TRM to field H_0 form the basis for determining paleofield intensity by the Thellier and Thellier (1959) method (§8.9).

Blocking and unblocking temperatures vary with grain volume V and microcoercivity H_K according to (8.3). If they did not, partial TRM's would not exist. The experimental dependence of T_{UB} on V is demonstrated in Fig. 8.3. T_{UB} is indicated directly in Fig. 8.3(a) by TRM unblocking and indirectly in Fig. 8.3(b) as the temperature at which the bulk coercive force $H_c \rightarrow 0$ in high-temperature hysteresis measurements. As predicted, small SD grains have lower unblocking (and blocking) temperatures than large SD grains.

TRM intensity varies strongly with grain size, as the data in Fig. 8.4 illustrate. Equation (8.2) predicts an increase in M_{eq} as V increases: larger SD grains should align their moments more efficiently in the direction of H_0 than small grains.

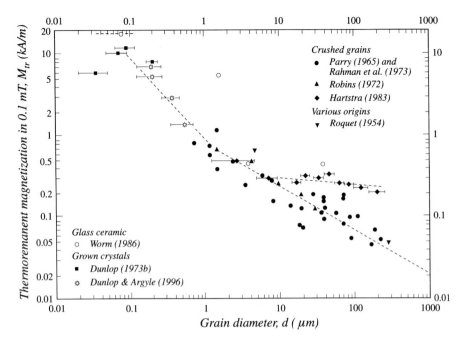

Figure 8.4   The grain-size dependence of the intensity of weak-field TRM in magnetite. [After Dunlop (1990), with the permission of the author and the publisher, Institute of Physics Publishing, Bristol, UK.]

This dependence is observed in the equilibrium SD range, i.e. for $d \leq d_0$. (In this and succeeding chapters, we shall use d for grain size, rather than D or L.) Above the critical SD size, TRM decreases with increasing grain size, as strongly as d^{-1} for <1 μm magnetites grown from solution (Dunlop, 1973b; Dunlop and Argyle, 1996) but less strongly for >1 μm magnetites (Fig. 8.4). The data for titanomagnetites are contradictory. TRM intensities for sintered and ground TM40 and TM60 grains have been reported to vary as weakly as $d^{-0.2}$ over the size range <0.1 μm to 30 μm (O'Donovan et al., 1986) and as strongly as $d^{-0.7}$ (Day, 1977).

   Many other properties of TRM and partial TRM's are well established, but these four – the dependences of TRM intensity on grain size and applied field, the equivalence of blocking and unblocking temperatures in heating and cooling, and the size dependence of the blocking/unblocking temperature – are the decisive constraints on SD TRM theory.

8.3   Coherent rotation without thermal activation

Before embarking on TRM theory proper, we need to know the energy barrier opposing reversal of SD magnetization. As a by-product we will also derive ele-

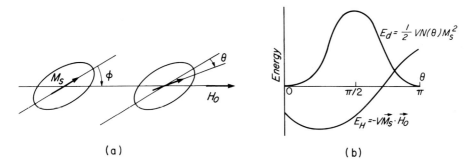

Figure 8.5 Simple model of coherent rotation in a spheroidal SD grain. (a) Orientation of M_s without and with a field H_0 applied at angle ϕ to the easy axis, $\theta = 0$, of shape anisotropy. (b) Contributions to the total energy.

mentary SD hysteresis loops in the absence of thermal fluctuations. Following Stoner and Wohlfarth (1948), we restrict our attention to spheroidal SD grains (which have uniform internal fields and simple demagnetizing tensors; §4.2.4) whose reversals are coherent. Because all spins rotate in unison, we can deal with them collectively via the grain moment VM_s, which is pinned by uniaxial shape anisotropy. For the moment we will ignore multiaxial magnetocrystalline anisotropy, whose barrier to rotation is usually minor compared to shape anisotropy in magnetite, maghemite or iron.

The model spheroidal SD grain shown in Fig. 8.5(a) has its easy (long) axis at angle ϕ to an applied field H_0. H_0 has the effect of rotating M_s through an angle θ away from the easy axis. The total energy, ignoring E_K, and also E_{ex} which does not change as a result of coherent rotation, is

$$E(\theta, \phi) = E_H + E_d = -\mu_0 VM_s \cdot H_0 + \tfrac{1}{2}\mu_0 VN(\theta)M_s^2$$
$$= -\mu_0 VM_s H_0 \cos(\phi - \theta) + \tfrac{1}{2}\mu_0 V(N_b - N_a)M_s^2 \sin^2 \theta, \quad (8.5)$$

using (4.8), (4.9) and (4.11). N_a and N_b are the demagnetizing factors when M_s is directed parallel or perpendicular, respectively, to the long axis. In magnetite, if N_a and N_b differ by more than 10%, which requires $< 10\%$ elongation (Fig. 4.5), shape anisotropy outweighs crystalline anisotropy.

A plot of E_d vs. θ appears in Fig. 8.5(b). E_d provides an energy barrier separating the (zero-field) energy minima or easy directions $\theta = 0$ and π. The demagnetizing field H_d is minimized for these two orientations. Unlike the situation in multidomain grains, there is no demagnetized state with $M = 0$ towards which H_d can drive the magnetization. (An *ensemble* of SD grains, on the other hand, does have a demagnetized state, as we shall see in §8.4.)

For equilibrium, $dE/d\theta = 0$ and

$$H_0 \sin(\phi - \theta) = (N_b - N_a)M_s \sin \theta \cos \theta = \tfrac{1}{2}(N_b - N_a)M_s \sin(2\theta).$$
$$(8.6)$$

Usually more than one value of θ satisfies (8.6) for a particular value of H_0. These θ values represent energy minima, inflection points or maxima according as

$$\frac{\mathrm{d}^2 E}{\mathrm{d}\theta^2} = \mu_0 V M_{\mathrm{s}}[H_0 \cos(\phi - \theta) + (N_{\mathrm{b}} - N_{\mathrm{a}})M_{\mathrm{s}} \cos(2\theta)] \begin{matrix} > 0 \\ = 0 \\ < 0 \end{matrix} \qquad (8.7)$$

Detailed numerical solutions of (8.6) and (8.7) are given by Stoner and Wohlfarth (1948). Analytical solutions are possible if H_0 is either parallel or perpendicular to the easy axis ($\phi = 0$ and $\pi/2$, respectively). When $\phi = 0$, $\theta = 0$ and π (M_{s} parallel or antiparallel to H_0) are both energy minima, provided $|H_0| < (N_{\mathrm{b}} - N_{\mathrm{a}})M_{\mathrm{s}} = H_{\mathrm{K}}$. ($H_{\mathrm{K}}$ may also be due to magnetoelastic or magnetocrystalline anisotropy: see eqns. (5.37), (5.39).) H_{K} is the **microscopic coercive force** (also called the microcoercivity or the anisotropy field). It is a critical field above which one of the energy minima becomes unstable and the grain moment undergoes an irreversible rotation to the remaining stable orientation. The resulting hysteresis loop is shown in Fig. 8.6(a). If H_0 is applied perpendicular to the easy axis (i.e., $\phi = \pi/2$), the magnetization rotates reversibly at all fields, producing a constant susceptibility $M_{\mathrm{s}}/H_{\mathrm{K}}$ (Fig. 8.6b).

If H_0 is applied at some other angle ϕ, the M vs. H_0 curve exhibits a combination of reversible and irreversible behaviour. In each case, the magnetization makes a single jump at a critical field H_{K} that is less than H_{K} for $\phi = 0$ (Fig. 8.6c). Mathematically, the condition $H_0 = H_{\mathrm{K}}$ corresponds to an inflection point in $E(\theta)$. Setting $\mathrm{d}^2 E/\mathrm{d}\theta^2 = 0$ in (8.7) and combining with (8.6) leads to the result $\tan^3 \theta_{\mathrm{c}} = -\tan \phi$ for the critical angle θ_{c}, at which irreversible rotation occurs. $H_{\mathrm{K}}(\phi)$ is obtained by substituting θ_{c} in (8.6).

Figure 8.7 compares the very dissimilar dependences of microcoercivity on field angle ϕ for coherent and incoherent SD rotation and MD domain-wall displacement. Incoherent modes like curling and fanning (§7.9.1) give lower critical fields than coherent rotation if $\phi < 45°$. The MD function is $\sec \phi$ and follows

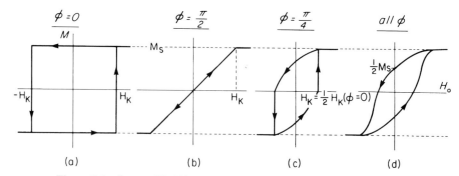

Figure 8.6 Stoner–Wohlfarth theory: low-temperature isothermal magnetization curves of spheroidal SD grains with easy axes at various angles ϕ to H_0.

from eqn. (4.8) (see discussion in Chapter 9). A field applied nearly perpendicular to the domain magnetizations provides little driving force for 180° wall motion (although 71° and 109° walls in magnetite or 90° walls in TM60 and iron will move readily). Rock specimens are often 'tumbled' during alternating-field (AF) demagnetization to present domains in their most favourable orientation relative to the field.

The average hysteresis curve of a randomly oriented array of SD grains, widely enough separated that their mutual magnetostatic interactions can be neglected, is shown in Fig. 8.6(d). It bears an obvious resemblance to experimental curves (Figs. 5.19, 5.21). The average coercive force is approximately $\frac{1}{2} H_K(\phi = 0)$:

$$H_c \approx \frac{1}{2} \frac{2K_u}{\mu_0 M_s} = \frac{1}{2}(N_b - N_a)M_s, \tag{8.8}$$

or the equivalent for magnetoelastic anisotropy (eqn. (5.37)).

Following saturation, all the M_s vectors are in the easy direction nearest H_0, and so

$$M_{rs} = \int_0^{\pi/2} M_s \cos \phi \sin \phi \, d\phi = 0.5 M_s. \tag{8.9}$$

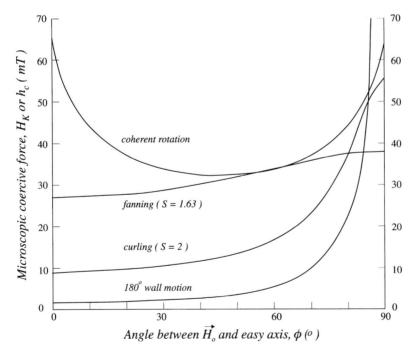

Figure 8.7 Critical anisotropy field or microscopic coercive force for SD reversal or multidomain wall displacement as a function of the angle ϕ between the domain magnetization(s) M_s and the applied field H_0.

Saturation remanence is much stronger than in the MD case (eqn. (5.32)). M_{rs}/M_s is even larger for multiaxial anisotropy (eqns. (5.42), (5.43), (11.16)).

The initial susceptibility, χ_i, can be calculated approximately by replacing the randomly oriented array by one having one-third of the grains parallel to H_0 with zero low-field susceptibility, and the other two-thirds perpendicular to H_0, with susceptibility M_s/H_K (Fig. 8.6b). Thus

$$\chi_i = \frac{2}{3}\frac{M_s}{H_K} = \frac{2}{3}\frac{1}{N_b - N_a}. \tag{8.10}$$

8.4   Magnetization relaxation due to thermal fluctuations

8.4.1   A simple model of single-domain TRM

As T approaches T_C, M_s drops rapidly to zero (see Figs. 3.5, 3.12, 3.13, 3.20, 3.27 for magnetite, titanomagnetites, hematite and pyrrhotite). Since $H_K \propto M_s$ for shape anisotropy, $\propto \lambda/M_s$ for magnetoelastic anisotropy, or $\propto K/M_s$ for crystalline anisotropy, with $K(T)$ or $\lambda(T) \propto M_s^n(T)$ (e.g., Figs. 3.7b, 3.8b), hysteresis loops of individual SD grains will decrease in width as well as height as the temperature rises (Fig. 8.8a, left). Eventually, very close to T_C, H_K will become less than a small applied field H_0, and the grain moment will flip to align M_s with H_0. During cooling, the grain moments will remain aligned with H_0, and when $H_0 \to 0$ at room temperature T_0, the grain will be left with an intense TRM. The TRM will moreover be very resistant to changes in field, including the fields used in AF demagnetization, because $H_K(T_0)$ is $\gg H_0$.

This theory of TRM is simple and straightforward, but it has a number of failings. First, it predicts a TRM intensity equal to the saturation remanence, no matter how small the applied field, H_0. Second, the blocking temperature T_B is determined by the condition $H_K(T_B) = H_0$, independent of grain volume V. Experimentally, however, blocking and unblocking temperatures are grain-size dependent, small SD grains having lower T_B and T_{UB} than larger grains (Fig. 8.3). Finally, the theory would not account for time-dependent or *viscous magnetization* changes below the blocking temperature. Viscous magnetic changes are a familiar phenomenon in rocks at room temperature (see Chapter 10).

In reality, as we shall see, thermal excitations as well as decreasing anisotropy help grain moments surmount energy barriers at high temperatures. Then H_c, the effective coercive force, is $< H_K$ and near T_C may be $\ll H_K$. The blocking temperature due to thermal relaxation is determined by the condition $H_c(T_B) = H_0$, which gives much lower values of T_B than the anisotropy blocking

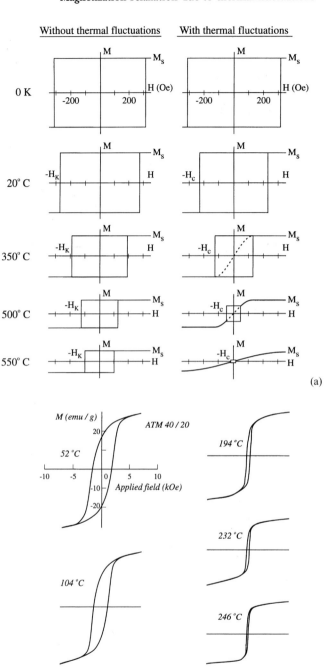

Without thermal fluctuations   With thermal fluctuations

Figure 8.8 (a) Hysteresis loops, calculated without and with thermal fluctuations, for one individual SD particle aligned with $\boldsymbol{H}_0$ at various temperatures from 0 K to near T_C. [Unpublished calculations by D. J. Dunlop.] (b) Measured hysteresis loops at various temperatures up to T_C for synthetic SD TM40 containing 20 mol% Al. The rapid drop in H_c and the SP curves at high temperature are evidence of thermal fluctuations. [Reprinted from Özdemir and O'Reilly (1981), with kind permission of the authors and Elsevier Science – NL, Sara Burgerhartstraat 25, 1055 KV Amsterdam, The Netherlands.]

(a)

(b)

temperature. Because H_c is grain-size dependent (§8.7), so too are blocking temperatures. High-temperature hysteresis loops when thermal excitation is taken into account are shown in Fig. 8.8(a, right). Actual measured loops at various temperatures (Fig. 8.8b; Özdemir and O'Reilly, 1981) show clear indications of thermal activation, for example in the rapid decrease in H_c with rising T and in the superparamagnetic (SP) curves that appear well below T_C (cf. Fig. 5.20).

8.4.2   Thermal relaxation theory

In thermodynamic terms, although TRM may be very long-lived, it is never truly stable because when $H_0 = 0$, the ultimate equilibrium state is $M = 0$. An individual SD grain has no $M = 0$ state available; it is always magnetized to saturation. An *ensemble* of SD grains, however does possess a state of zero or near-zero magnetization, in which the M_s vectors of various grains are oriented at random. At any temperature, random thermal energy will aid the assembly in relaxing towards this equilibrium demagnetized state. Although an unusually large thermal excitation at any temperature could activate a reversal in an individual particle, the likelihood of all particles being activated only becomes significant at temperatures where the available thermal energy ($25kT$–$60kT$, depending on the time scale) is comparable to the energy barrier to be scaled.

Because thermal excitations have a statistical distribution of magnitudes, the magnetization does not change abruptly (as in field-driven changes) by reversal of all grains but relaxes gradually as individual grains reverse at random. The crucial question for paleomagnetism is the kinetic one: How long does the relaxation take to reach equilibrium? We shall find that SD grains only slightly above SP size have room-temperature relaxation times greater than the age of the earth, but relaxation times of only a few minutes at blocking temperatures of a few hundred °C. TRM can therefore be 'frozen-in' when a rock is cooled in an applied field through the thermal relaxation blocking temperature T_B (which is always lower than the blocking temperature due to the vanishing of anisotropy, as we argued above). The TRM will remain stable at room temperature in the absence of the field, or in the presence of subsequently applied fields, for times of geological length. This answer is at first sight surprising but, from our point of view, fortunate. Were it not so, magnetic memory for most igneous rocks would not exist.

Let us follow the classic theory of Néel (1949) and examine the behaviour of an assembly of spheroidal SD grains aligned with H_0, i.e. with $\phi = 0$. It is, of course, not necessary to restrict our attention in this way, but the behaviour of an aligned assembly is representative of that of a random assembly while being much easier to deal with mathematically. Each grain has minimum-energy states $\phi_1 = 0$ with $M_1 = +M_s$ (grain moment aligned with H_0) and $\theta_2 = \pi$ with $M_2 = -M_s$ (moment antiparallel to H_0). The thermal equilibrium magnetization M_{eq} of the assembly, if the number N of grains is statistically large, is a Boltzmann partition between the energy states:

$$M_{eq} = \frac{\sum M_i e^{-E_i/kT}}{\sum e^{-E_i/kT}} = M_s \tanh\left(\frac{\mu_0 V M_s H_0}{kT}\right). \tag{8.11}$$

The energies E_1, E_2 are given by eqn. (8.5). $M_{eq} = 0$ for $H_0 = 0$ is a special case of eqn. (8.11).

At any time, n grains are in state 1 $(\theta_1 = 0)$ and $N - n$ are in state 2 $(\theta_2 = \pi)$. To rotate from state 2 to state 1, a grain moment $V M_s$ must gain thermal energy ΔE_{21} equal to the energy barrier $E_{max} - E_2$ (Fig. 8.9). Rotation from state 1 to state 2 requires energy $\Delta E_{12} = E_{max} - E_1$. From eqns. (8.6) and (8.7), it is easy to show that $E = E_{max}$ when $\theta = \cos^{-1}(-H_0/H_K)$, where $H_K = (N_b - N_a)M_s$ for shape anisotropy, $2K_u/\mu_0 M_s$ for uniaxial anisotropy generally (eqn. (5.37)), or $(4/3)(|K_1|/\mu_0 M_s)$ for cubic magnetocrystalline anisotropy with $\langle 111 \rangle$ easy axes (eqn. (5.39)). Hence, substituting from (8.5),

$$\begin{aligned} \Delta E_{12} &= E_{max} - E(0) = \tfrac{1}{2}\mu_0 V M_s H_K (1 + H_0/H_K)^2, \\ \Delta E_{21} &= E_{max} - E(\pi) = \tfrac{1}{2}\mu_0 V M_s H_K (1 - H_0/H_K)^2. \end{aligned} \tag{8.12}$$

If $H_0 > 0$, state 1 $(\theta = 0)$ is energetically favoured and the number n of grains in this state will grow with time until equilibrium in the ensemble is reached.

From the kinetic point of view, because $\Delta E_{21} < \Delta E_{12}$, rotations from $\pi \to 0$ are easier than those from $0 \to \pi$ and the same conclusion holds. The kinetic equation is

$$\frac{dn}{dt} = K_{21}(N - n) - K_{12}n = \frac{N - n}{\tau_{21}} - \frac{n}{\tau_{12}}. \tag{8.13}$$

Equation (8.13) is the precise analog of the radioactive decay equation for an intermediate member in a multistage decay scheme. K_{12}, K_{21} are probabilities or rate constants for the $0 \to \pi$ and $\pi \to 0$ transitions, respectively, and τ_{12}, τ_{21}

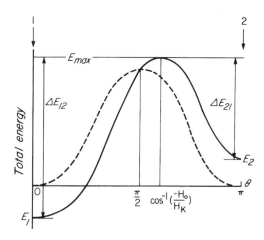

Figure 8.9 Variation of the total energy, $E_d + E_H$ (Fig. 8.5b), with deflection θ of M_s when H_0 is applied along the easy axis $\theta = 0$. When $H_0 = 0$, $E_{tot}(\theta) = E_d(\theta)$ (dashed curve) and the energy barriers between states 1 and 2 are equal. When H_0 is applied along $\theta = 0$, state 1 is favoured, $E_1 < E_2$, and the energy barriers become unequal $(\Delta E_{12} > \Delta E_{21})$, promoting transitions to state 1.

are the corresponding **relaxation times**, analogous to mean lifetimes of radio-active elements. Since $M = [(2n - N)/N]M_s$ for the ensemble, (8.13) can be rewritten in terms of M and solved to give

$$M(t) = M_{eq} + (M_0 - M_{eq})e^{-t/\tau} = M_0 e^{-t/\tau} + M_{eq}(1 - e^{-t/\tau}), \qquad (8.14)$$

M_0 being $M(0)$ and M_{eq}, which is given by eqn. (8.11), being $M(\infty)$. Equations (8.14) and (8.11) are equivalent to eqns. (8.1) and (8.2) of the Introduction. The relaxation time τ is related to the individual relaxation times τ_{12}, τ_{21} by

$$\frac{1}{\tau} = \frac{1}{\tau_{12}} + \frac{1}{\tau_{21}}. \qquad (8.15)$$

8.5   Relaxation times for identical SD grains

How long does it take to achieve the thermal equilibrium magnetization M_{eq}? It takes a short time if ΔE_{12}, ΔE_{21} are comparable to the cumulative thermal impulse energy, 25–$60kT$, but a very long time otherwise. In fact, τ_{12} and τ_{21} obey an Arrhenius thermal activation equation with activation energies ΔE_{12}, ΔE_{21}:

$$\frac{1}{\tau_{12}} = \frac{1}{\tau_0}e^{-\Delta E_{12}/kT} = \frac{1}{\tau_0}\exp\left[-\frac{\mu_0 V M_s H_K}{2kT}\left(1 + \frac{H_0}{H_K}\right)^2\right],$$

$$(8.16)$$

$$\frac{1}{\tau_{21}} = \frac{1}{\tau_0}e^{-\Delta E_{21}/kT} = \frac{1}{\tau_0}\exp\left[-\frac{\mu_0 V M_s H_K}{2kT}\left(1 - \frac{H_0}{H_K}\right)^2\right],$$

substituting ΔE_{12}, ΔE_{21} from eqn. (8.12). Lederman *et al.* (1994) have measured the thermal switching of individual SD particles on 0.1–30 s time scales using an MFM as a magnetometer. They propose that dynamic reversal occurs via a complex path and not by thermal activation over a single barrier. However, the single-barrier model of (8.16) leads to predictions about the time, temperature, and grain-size dependence of magnetization which can be verified experimentally on time scales from ≈ 60 s to millions of years (see §8.6, 8.7, 8.12, 10.6, 16.3). It seems a fairly sound basis for thermal activation theory.

Unlike the situation in radioactive decay, the relaxation times τ_{12} and τ_{21} in (8.16) (or the rate 'constants' K_{12}, K_{21}) are not constants. They depend very strongly on temperature T and applied field H_0, as well as on the size (V) and shape (H_K) of the grains. The atomic reorganization time, $\tau_0 \approx 10^{-9}$ s (McNab *et al.*, 1968), is the typical time between thermal impulses. Strictly speaking, τ_0 is not constant. It depends on T, H_0 and H_K, but the exponential factors have an overwhelming influence on τ_{12} and τ_{21}.

If $|H_0| \geq 4kT/\mu_0 V M_s$ (about 0.3 mT or 3 Oe for 500 Å or 50 nm magnetite cubes at room temperature), τ_{12} is negligible compared to τ_{21} for $H_0 > 0$ and $\theta = 0$ (state 1, M_s parallel to H_0) is strongly favoured. If $H_0 < 0$, $\tau_{21} \ll \tau_{12}$ and $\theta = \pi$ (state 2, M_s again parallel to H_0) is favoured. (In physical terms, one of the energy barriers in Fig. 8.9 becomes so much lower than the other that transitions towards the state favoured by H_0 drain the less-favoured state and $M_{eq} \rightarrow M_s$.) In either case, using (8.15),

$$\frac{1}{\tau} = \frac{1}{\tau_0} \exp\left[-\frac{\mu_0 V M_s H_K}{2kT}\left(1 - \frac{|H_0|}{H_K}\right)^2\right]. \tag{8.17}$$

On the other hand, if $H_0 = 0$, the two barriers are of equal height (Fig. 8.9, dashed curve) and $\tau_{12} = \tau_{21}$. Thus, from (8.15) and (8.16),

$$\frac{1}{\tau} = \frac{2}{\tau_0} \exp\left(-\frac{\mu_0 V M_s H_K}{2kT}\right). \tag{8.18}$$

For intermediate fields, the expression for τ is complicated and awkward to use. Equations (8.11), (8.17) and (8.18) are the central results of the Néel (1949) theory. Equation (8.17) is the counterpart of eqn. (8.3) of the Introduction.

Figure 8.10 illustrates the strong dependence of τ on grain size and temperature. SD magnetite grains about 280 Å or 28 nm in size with moderate shape anisotropy ($H_{K0} = H_K(T_0) = 20$ mT) behave superparamagnetically at T_0, since $\tau < 60$ s, a typical experimental time. In §5.5.2 and Table 5.1, the estimated and measured room-temperature SP–SD threshold size d_s was 25–30 nm. An assemblage of 320 Å grains, on the other hand, has magnetic stability for times of the order of 1 month, and an assemblage of 370 Å grains is stable for ≈ 1 Ga or, practically speaking, forever.

The dependence of τ on temperature is equally spectacular. Notice in Fig. 8.10(b) that grains with $\tau \approx 1$ Ga at 60 °C have $\tau \approx 1$ s (SP or unblocked condition) at 275 °C. Heating by ≈ 200 °C has destabilized (or unblocked) the remanence of these grains. Conversely if the same grains were cooled in a field H_0 from 275 °C to T_0, they would pass from a SP condition, with $\tau = 1$ s and $M = M_{eq}$ as given by (8.11), to a stable or blocked condition, with $\tau > 1$ Ga. At T_0, H_0 can be changed or removed without affecting M, which remains frozen at the value it had at the *blocking temperature*, T_B, when τ became $\approx t$ (an observational time controlled by the cooling rate dT/dt). For this grain ensemble, T_B would be about 250 °C for $t \approx 60$ s (rapid cooling) or about 170 °C for $t \approx 1$ year (e.g., slowly cooling interior of a thick lava flow).

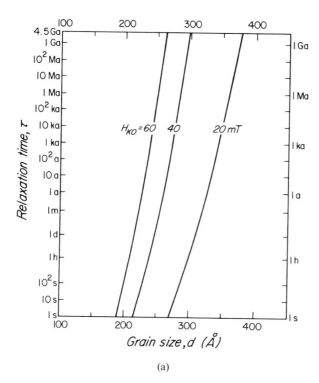

(a)

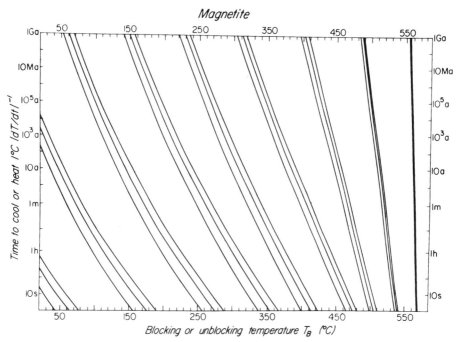

(b)

Figure 8.10 Theoretical relaxation times of SD magnetite grains as a function of (a) grain size ($T = T_0$, $H_0 = 0$) and (b) temperature. Part (b) is equivalently regarded as a graph of the time dependence of blocking temperature T_B. The three curves in each set denote 5%, 50% and 95% blocking or unblocking of TRM for ensembles with a particular value of VH_{K0}. [Reprinted from Dunlop (1981), with kind permission of the author and Elsevier Science – NL, Sara Burgerhartstraat 25, 1055 KV Amsterdam, The Netherlands.]

8.6 Blocking temperature and TRM of identical SD grains

We have just sketched the TRM blocking process in an ensemble of identical SD grains. At high temperature, τ is $\ll t$, a typical time characteristic of the cooling and the net magnetization M of the ensemble relaxes, effectively instantaneously, to M_{eq}. Just above T_{B}, τ remains $< t$ and, according to eqn. (8.14), equilibrium is still attained in time t. However, at T_{B}, magnetization relaxation becomes sluggish, and just below T_{B}, since τ is now $> t$, M does not change appreciably in time t from M_0 (eqn. (8.14)). M_0 in this case is the equilibrium magnetization achieved just a few °C higher, namely $M_{\mathrm{eq}}(T_{\mathrm{B}})$. This same magnetization, apart from the reversible increase in M_{s} between T_{B} and T_0, remains frozen as TRM when $H_0 \rightarrow 0$ at T_0.

What must be emphasized is the *sharpness* of the blocking temperature, that is, the rapidity with which magnetization relaxation changes from a near-instantaneous to a sluggish to a frozen process when a rock is cooled. Most blocking temperatures in a rock tend to be within 100 °C or so of T_{C} (580 °C for magnetite). In this region, the τ–T contours in Fig. 8.10(b) are nearly vertical and the 5% and 95% blocking contours are closely spaced. τ changes from a few seconds to a few years over a temperature interval of one or two degrees. Small SD grains just above SP threshold size, for example the ones we described in the last section with $T_{\mathrm{B}} \approx 250\,°\mathrm{C}$, have TRM blocking ranges of 20–30 °C (Fig. 8.10b). The statistical nature of thermal relaxation is more evident in these low-T_{B} ensembles, but TRM blocking still occurs comparatively rapidly during cooling.

Substituting $\tau = t$ in eqn. (8.17) gives an explicit formula for calculating T_{B}:

$$\frac{T_{\mathrm{B}}}{\beta^2(T_{\mathrm{B}})} = \left(\frac{\mu_0 V M_{s0} H_{\mathrm{K}0}}{2k \, \ln(t/\tau_0)}\right) \left[1 - \frac{|H_0|}{H_{\mathrm{K}0}\beta(T_{\mathrm{B}})}\right]^2, \tag{8.19}$$

in which $M_{s0} \equiv M_{\mathrm{s}}(T_0)$, $H_{\mathrm{K}0} \equiv H_{\mathrm{K}}(T_0)$ and $\beta(T) \equiv M_{\mathrm{s}}(T)/M_{s0}$. T_{B} is a strong function of grain size (i.e., V) and shape (since $H_{\mathrm{K}0} = (N_{\mathrm{b}} - N_{\mathrm{a}})M_{s0}$) but a weak function of t.

T_{B} depends quadratically on H_0, as confirmed experimentally (Fig. 8.11; Sugiura, 1980; Clauter and Schmidt, 1981). The percentage change in T_{B} for H_0 ranging from 0 to 1 mT is negligible. This is the justification for assuming T_{UB} in zero-field heating is identical to T_{B} during in-field cooling, for example in paleointensity determination (§8.9). The predicted dependence of T_{B} on V is followed quite closely by the experimental data in Fig. 8.3(a). A computer simulation of the TRM blocking process, using three-dimensional micromagnetic modelling (Thomson *et al.*, 1994), also confirmed (8.19) for magnetite grains within the SD range, provided they reversed coherently ($d \leq 0.06$ μm; cf. Fig. 7.13b).

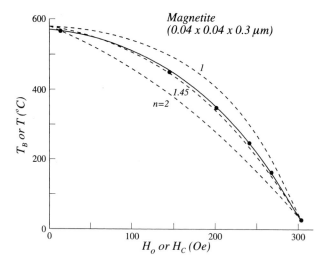

Figure 8.11 The dependence of T_B on H_0 (eqn. (8.19)) tested by plotting $H_c(T)$ data (equivalent eqn. (8.24)). Single-domain theory (solid line) gives a better fit than multidomain theory (dashed curves). T_B is almost independent of H_0 for weak H_0 (≤ 10 Oe). [After Dunlop (1982) © American Geophysical Union, with the permission of the author and the publisher.]

The thermal-equilibrium partition between energy states, measurable as M/M_s of the assembly, is frozen in at T_B. Therefore, M/M_s at T_0, either before or after H_0 is removed, equals M_{eq}/M_s at T_B, as given by (8.11):

$$\frac{M_{tr}(T_0)}{M_{s0}} = \frac{M(T_0)}{M_{s0}} = \frac{M_{eq}(T_B)}{M_{sB}} = \tanh\left(\frac{\mu_0 V M_{sB} H_0}{k T_B}\right), \qquad (8.20)$$

in which $M_{sB} \equiv M_s(T_B)$. Because we restricted our attention to grains aligned with H_0, the TRM $M_{tr}(T_0)$ described by (8.20) approaches M_{s0} in large applied fields. For a randomly oriented ensemble of uniaxial grains, however, the saturation value of TRM should be $M_{rs}(T_0) = 0.5 M_{s0}$ (eqn. (8.9) or (5.41)). Thus we should write

$$M_{tr}(T) = M_{rs}(T)\tanh\left(\frac{\mu_0 V M_{sB} H_0}{k T_B}\right), \qquad T < T_B. \qquad (8.21)$$

Equation (8.21) predicts the intensity of TRM at any temperature below T_B, including T_0. It is immaterial to the particular ensemble with blocking temperature T_B whether H_0 is applied during cooling from T_B to T_0 or removed immediately below T_B, although ensembles with smaller (V, H_{K0}), and hence lower blocking temperatures, *will* be affected (see §8.8).

The TRM intensity predicted by (8.21) reaches saturation when $|H_0| \geq 4kT/\mu_0 V M_s$, as explained in connection with eqn. (8.17). For 500 Å magnetite cubes at $T_B = 500\,°C$, the saturation field is predicted to be ≈ 1.5 mT or 15 Oe. Experimentally TRM usually saturates in much larger fields, on the order of 10 mT (Fig. 8.12). As far as predicting TRM intensity is concerned, Néel's (1949) theory is not much more successful than the simplistic model of §8.4.1 that ignored thermal agitation and predicted saturation of the TRM in arbitrarily small fields.

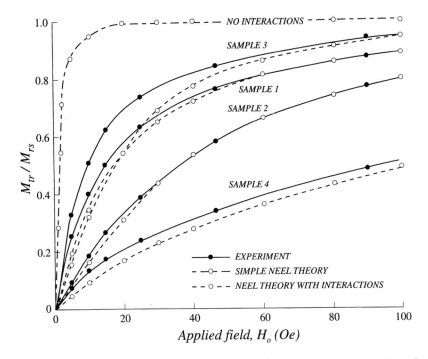

Figure 8.12 Measured dependence of TRM intensity on field strength H_0 for four samples containing SD magnetite or maghemite grains, compared with predictions of Néel theory, with and without grain interactions. Interaction field distributions were estimated from experimental Preisach diagrams (Chapter 11). [After Dunlop and West (1969) © American Geophysical Union, with the permission of the authors and the publisher.]

In multidomain grains, TRM cannot saturate in small fields because H_0 is opposed by the internal demagnetizing field H_d, which can be thought of as the resultant interaction field of all the domains. In SD assemblies, magnetostatic interactions among the grains in principle play an analogous role. Shcherbakov *et al.* (1995) have carried out a Monte-Carlo simulation of TRM acquisition in the presence of interactions. We will take a simpler approach, along the same lines as Dunlop and West (1969), which gives a feeling for the magnitude of interaction fields required to explain TRM intensities.

Each grain has a TRM given by (8.21), with a local interaction field H_{int}, varying in direction and intensity from grain to grain, added to H_0. The TRM of the assembly is reduced to observed levels if grain-to-grain interaction fields having strengths up to 3–5 mT (30–50 Oe) are present (Fig. 8.12; Dunlop and West, 1969). These are not large fields compared to H_d within an SD grain, but they do require the magnetic grains to be clustered. A roughly equidimensional rock sample containing 1% by volume of uniformly spaced, randomly oriented SD grains would have an average value $\langle H_d \rangle$ at any point in the non-magnetic matrix of

$$\langle H_{\rm d} \rangle \leq 0.01 N_{\rm SD} M_{\rm rs} = 0.01 \times \tfrac{1}{3} \times 0.5 \times 480 \times 10^3$$
$$= 800 \text{ A/m } (\approx 10 \text{ Oe or 1 mT}).$$
$$(8.22)$$

A more potent source of $H_{\rm int}$ is found in dipole–dipole interaction between *non-uniformly spaced* quasi-spherical grains, which according to eqn. (2.2) and §4.2.3 gives rise to

$$H_{\rm int} \approx \frac{\mu}{4\pi r^3} = \frac{1}{24} M_{\rm s} \left(\frac{d}{r} \right)^3 \tag{8.23}$$

for grains of diameter d and centre-to-centre separation r. For magnetite, if $r \approx 2d$, $H_{\rm int} \approx 2.5 \text{ kA/m or} \approx 3 \text{ mT (30 Oe)}$.

There is indirect evidence from other magnetic measurements (Dunlop and West, 1969) of interaction fields this large in single-domain clusters, but it is not likely that the magnetic grains in rocks are normally so inhomogeneously distributed as to interact this strongly. In fact, if they did, the law of additivity of partial TRM's, which is fundamental to determining paleomagnetic field intensity (see §8.8), probably would not hold.

Fortunately, the Néel theory predicts directional stability of TRM much more successfully than TRM intensity. The blocking temperature at which TRM is frozen in upon cooling and the unblocking temperature for thermal demagnetization upon zero-field heating predicted by eqn. (8.19) are quite close to experimental vaues for magnetite grains of accurately determined sizes (Fig. 8.3; Dunlop, 1973b).

8.7   Effect of thermal fluctuations on observed coercivity

Magnetic hardness or coercivity of a rock is usually determined by measuring resistance to alternating-field (AF) demagnetization (§8.11). The unblocking or destructive field, $H_{\rm c}$, for a particular SD ensemble is the field H_0 that makes $\tau = t$, allowing the magnetization to relax. From eqn. (8.17),

$$H_{\rm c}(V, H_{\rm K}, T) = H_{\rm K} - \left(\frac{2H_{\rm K} kT \ln (t/\tau_0)}{\mu_0 V M_{\rm s}} \right)^{1/2}$$
$$= H_{\rm K} - H_{\rm q}(V, H_{\rm K}, T). \tag{8.24}$$

The temperature dependence of $H_{\rm c}$ mimics the field dependence of $T_{\rm B}$, since both are simply rearrangements of eqn. (8.17). We took advantage of this fact in plotting the data of Fig. 8.11.

Equation (8.24) indicates that the destructive field or effective microcoercivity $H_{\rm c}$ of SD grains is less than $H_{\rm K}$ by an amount equal to $H_{\rm q}$, the 'fluctuation field'

(Néel, 1955). H_q is by no means negligible: for 500 Å magnetite cubes, with $H_K = 30\,\text{mT}\,(300\,\text{Oe})$, it amounts to about $10\,\text{mT}\,(100\,\text{Oe})$ at room temperature.

$H_q(T)$, representing the effect of thermal excitations, increases with rising T, whereas $H_K(T)$, representing the pinning effect of anisotropy, decreases with heating. When the two become equal, the barrier opposing rotation vanishes and unblocking occurs. Thus an alternative statement of the blocking/unblocking condition is

$$H_c(T_B) = H_K(T_B) - H_q(T_B) = 0. \tag{8.25}$$

The thermal-fluctuation blocking temperature defined by (8.25) is always lower, and sometimes much lower, than the anisotropy blocking temperature of the simple TRM model of §8.4.1, which is defined by the condition $H_K(T_B) = 0$.

The predicted dependences of H_c and H_q on V, T and H_K have been tested experimentally (Kneller and Luborsky, 1963; Kneller and Wohlfarth, 1966; Dunlop, 1976; Dunlop and Bina, 1977). To appreciate the tests better, it is useful to first transform (8.24) to a simpler form. From (8.18), a grain passes from a thermally stable to an SP condition, or vice versa, when its volume is just equal to the critical blocking volume (in zero field).

$$V_B = \frac{2kT \ln (2t/\tau_0)}{\mu_0 M_s H_K}. \tag{8.26}$$

The critical SP threshold size for a cubic grain at a given temperature T is then

$$d_s = V_B^{1/3} \approx (25kT/K)^{1/3}, \tag{8.27}$$

as in eqn. (5.29), using the fact that $H_K = 2K/\mu_0 M_s$ for uniaxial anisotropy and setting $t = 60$ s. Returning to eqn. (8.26), we can now substitute in (8.24) and show that

$$H_q(V, H_K, T) = H_K(V_B/V)^{1/2}$$

and $\qquad\qquad\qquad\qquad\qquad\qquad\qquad\qquad\qquad\qquad\qquad\qquad$ (8.28)

$$H_c(V, H_K, T) = H_K[1 - (V_B/V)^{1/2}].$$

In Fig. 8.13(a), $H_q(T_0)/H_K(T_0)$ is plotted against $V^{-1/2}$ for a series of SD and quasi-SD magnetites (Dunlop and Bina, 1977). The plot is a straight line, as (8.28) predicts, and its slope gives an experimental estimate of 250 Å (25 nm) for d_s in magnetite at room temperature. This d_s value is in excellent agreement with other independent estimates (Table 5.1).

Figure 8.13(b) presents a test of the temperature dependence of H_c predicted by (8.24). If we assume shape anisotropy to be dominant, so that $H_K(T) \propto M_s(T) = M_{s0}\beta(T)$, we can linearize (8.24):

$$\frac{H_c(T)}{\beta(T)} = H_{K0} - \left[\frac{2kH_{K0} \ln (t/\tau_0)}{\mu_0 V M_{s0}}\right]^{1/2} \frac{T^{1/2}}{\beta(T)}. \tag{8.29}$$

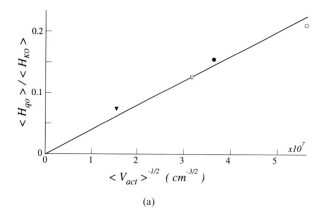

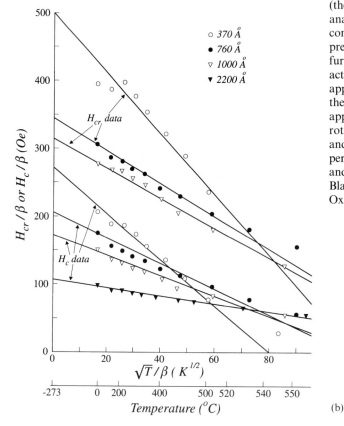

Figure 8.13 Tests of the predicted dependences of fluctuation field H_q, coercive force H_c and remanent coercive force H_{cr} (§11.3.2) on (a) grain size (eqns. (8.28)) and (b) temperature (eqn. (8.29)). The experimental materials are the four SD or nearly SD magnetites of Fig. 8.3. The graph in (a) verifies Néel theory and yields $d_s = 25$ nm for the SP–SD threshold size. The graphs in (b) (thermal fluctuation analysis or TFA) also confirm Néel's predictions and furthermore predict activation volumes approximately equal to the particle volume, as appropriate for coherent rotation. [After Dunlop and Bina (1977), with the permission of the authors and the publisher, Blackwell Science, Oxford, UK.]

Plots of H_c/β vs. $\sqrt{T}/\beta$ for the same four magnetites tested in Fig. 8.13(a) are indeed linear, verifying thermal relaxation theory. From their slopes, one can derive an estimate of the thermally activated volume V_{act}. This method of granulometry or size determination is called **thermal fluctuation analysis** (TFA) (Dunlop, 1976). The three smallest magnetites in Fig. 8.13(b) yielded V_{act}

$= V$, the particle volume. These are truly SD grains, whose entire volumes are thermally activated. In the largest magnetite, only a fraction of each grain was activated, possibly the domain wall in a 2D structure (§5.8).

8.8 Grain distributions and laws governing partial TRM's

In previous sections, we have dealt with idealized ensembles of identical single-domain grains. A real rock contains many such ensembles, with a wide variety of grain sizes and shapes, described by a **grain distribution**, $f(V, H_{K0})$. Indeed the notion of an isolated ensemble of identical grains is something of a mathematical abstraction, since grains of a particular (V, H_{K0}) are intermingled with grains of other (V, H_{K0}) and physically dispersed through the rock. Since T_B and H_c both depend on V and on H_K (eqns. (8.19), (8.24), (8.28)), single-domain TRM exhibits both a **spectrum of blocking temperatures** and a **spectrum of microcoercivities** or destructive fields. For the moment, we shall focus attention on the blocking-temperature spectrum.

The total TRM of a rock can be thought of as comprising a **spectrum of partial TRM's**, each carried by grain assemblies with similar (V, H_{K0}). Experimentally, if a rock is cooled from T_C to T_1 in zero applied field, from T_1 to T_2 in a *weak* field H_0, and from T_2 to T_0 in zero field, only those SD ensembles with $T_2 \leq T_B \leq T_1$ acquire TRM. The remanent magnetization of this blocking-temperature fraction is a **partial TRM**, $M_{ptr}(T_1, T_2, H_0)$.

Thellier (1938) demonstrated that partial TRM's obey three experimental laws (cf. §1.1.3), which are readily explicable in the light of Néel's theory:

1. The law of **reciprocity**: $M_{ptr}(T_1, T_2, H_0)$, acquired between T_1 and T_2 during cooling in H_0, is thermally demagnetized over precisely the interval (T_2, T_1) when heated in zero field. That is, the blocking and unblocking temperatures are identical, a result of the fact that T_B is almost independent of H_0 for weak fields (eqn. (8.19), Fig. 8.11).

2. The law of **independence**: $M_{ptr}(T_1, T_2, H_0)$ is independent, in direction and intensity, of any other partial TRM produced over a temperature interval that does not overlap (T_1, T_2), since the grains carrying the two partial TRM's represent different parts of the blocking-temperature spectrum.

3. The law of **additivity**: Partial TRM's produced by the same H_0 have intensities that are additive:

$$M_{ptr}(T_C, T_1, H_0) + M_{ptr}(T_1, T_2, H_0)$$
$$+ \ldots + M_{ptr}(T_n, T_0, H_0) = M_{tr}(T_C, T_0, H_0), \qquad (8.30)$$

since the blocking-temperature spectrum can be decomposed into non-overlapping fractions, each associated with one of the partial TRM's.

The sharpness of the blocking temperature is responsible for the simplicity of these laws.

None of the Thellier laws can be expected to hold if grain-to-grain interaction fields are $\gg H_0$. For one thing, $T_B(H_0 + H_{\text{int}})$ will not usually equal $T_B(0)$ for a particular ensemble. Furthermore, since SD ensembles with different blocking temperatures are physically intermingled in the rock, partial TRM's with non-overlapping blocking temperature ranges will interact magnetostatically. They will not be independent or additive. The apparent generality of the Thellier laws implies that interaction fields do not play a central role in single-domain TRM.

8.9   Theoretical basis of paleofield intensity determination

Experimentally, TRM intensity, M_{tr}, is proportional to the strength of the laboratory field H_{lab} in which it was acquired, for fields ≤ 0.1 mT or 1 Oe (Fig. 8.1). If the NRM of a rock was acquired as a TRM in an ancient geomagnetic field $H_a \leq 0.1$ mT, NRM intensity M_{nr} should likewise be proportional to H_a. The intensity of the paleofield is therefore determined by the formula

$$H_a = \frac{M_{\text{nr}}}{M_{\text{tr}}} H_{\text{lab}}. \tag{8.31}$$

The proportionality factors relating M_{tr} to H_{lab} and M_{nr} to H_a are inadequately predicted by the Néel theory, but luckily they are identical and so do not appear in (8.31). This simple method of paleointensity determination has the drawback that part of the NRM may have been viscously remagnetized (Chapter 10) or the rock may alter chemically either in nature or during laboratory heating, producing either too-high or too-low an estimate of H_a.

Chemical alteration in the laboratory occurs at high temperature, while viscous remagnetization affects only the lowest blocking-temperature fraction (§10.3). The Thellier and Thellier (1959) method of paleointensity determination takes advantage of these facts. A *partial* TRM and a *partial* NRM with the same range of blocking temperatures satisfy eqn. (8.31); otherwise partial TRM's would not be independent and additive. A series of non-overlapping partial TRM's and the corresponding partial NRM's therefore give independent replicate determinations of H_a (Thellier, 1941). Spurious values at the low-temperature and high-temperature ends of the spectrum caused by viscous or chemical changes are readily recognized and rejected.

In the Coe (1967) version of the Thellier method now in general use, the sample is twice heated to a temperature T_a (say $100\,^\circ\mathrm{C}$) and cooled to T_0, first in zero field, then in a known field H_{lab}. The decrease in M_{nr} after the first heating gives $M_{\mathrm{pnr}}(T_a, T_0, H_a)$ and the change in M_r between the first and second heatings gives $M_{\mathrm{ptr}}(T_a, T_0, H_{\mathrm{lab}})$. The double heatings are repeated in steps (T_b, T_0), etc. to higher and higher temperatures. The first heating (in zero field) to T_b erases both $M_{\mathrm{ptr}}(T_a, T_0, H_{\mathrm{lab}})$ and $M_{\mathrm{pnr}}(T_b, T_0, H_a)$. The second heating and cooling (in H_{lab}) adds in $M_{\mathrm{ptr}}(T_b, T_0, H_{\mathrm{lab}})$. Of course, all the M_{pnr} must be in the same direction or else M_{nr} is not a single component TRM.

The results are usually plotted as a graph of (scalar) NRM remaining after the first of the two heating–cooling steps to temperature T_i, namely $M_{\mathrm{nr}}(T_i) = M_{\mathrm{nr}}(T_0) - M_{\mathrm{pnr}}(T_i, T_0)$, versus partial TRM gained in the second step to T_i, $M_{\mathrm{ptr}}(T_i, T_0)$. Ideally the graph, often called an Arai plot (Nagata, Arai and Momose, 1963), is linear with a slope $-H_a/H_{\mathrm{lab}}$. Non-linearity indicates non-ideal behaviour. Some typical paleointensity plots are shown in Fig. 8.14, with the causes of non-ideal behaviour diagnosed.

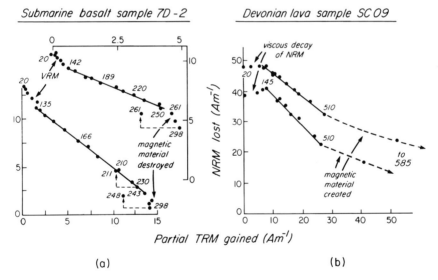

(a)   (b)

Figure 8.14   Typical semi-ideal Thellier paleointensity plots for (a) a submarine basalt containing TM60 with $T_c \approx 250\,^\circ\mathrm{C}$ and (b) a subaerial basalt containing magnetite ($T_c = 580\,^\circ\mathrm{C}$). Numbers along the curves are temperatures of double heatings in $^\circ\mathrm{C}$. Non-ideal behaviour results from viscous remagnetization or decay of low-T_B NRM (see Chapter 10) and irreversible chemical changes in high-temperature steps, e.g., the destruction of magnetic material in (a) evidenced by the non-repeatability of the 248 $^\circ\mathrm{C}$ and 261$^\circ\mathrm{C}$ pTRM's. [After Grommé *et al.* (1979) © American Geophysical Union and Kono (1979), with the permission of the authors and the publishers, The American Geophysical Union, Washington, DC and Blackwell Science, Oxford, UK.]

The original Thellier method does not require a zero-field environment. Both coolings from T_i take place in a field H_{lab} but the sample is rotated 180° with respect to H_{lab} between heatings. M_{pnr} and M_{ptr} are obtained vectorially. An important requirement in either version of the Thellier method is that the temperature T_i be precisely repeated in the second of each pair of heating–cooling cycles. Nowadays reliable field-free spaces that use magnetic shielding or feedback coil systems are widely available and most laboratories use the modified Thellier method.

A complete Thellier-type paleointensity determination is time-consuming. Of course, this disadvantage can be offset by treating a large number of samples simultaneously, since measurement time is a fraction of the duration of a heating–cooling cycle. There have been many proposals of alternative methods that eliminate repeated heatings and coolings. One widely used method compares the AF demagnetization curves of NRM and a laboratory TRM (Van Zijl *et al.*, 1962; McElhinny and Evans, 1968; Shaw, 1974). This process amounts to making replicate NRM/TRM determinations for different fractions of the coercivity spectrum. In the Thellier technique, TRM acquisition and thermal demagnetization are inverse processes, and the TRM can be added gradually, beginning with low-temperature heatings, in the course of demagnetizing the NRM. In AF methods, the TRM is induced in a single high-temperature heating. If the rock alters in the process, all paleointensity information is lost, while low-temperature steps in the Thellier method might have yielded a reliable paleointensity value. Supplementary reliability criteria have been suggested to accompany AF and other alternatives to the Thellier method (e.g., Shaw, 1974). None provides a built-in check that operates *during* the paleointensity determination, as in the Thellier method. They require separate experiments, on the basis of which one may reject results or, at best, apply a correction factor, whose reliability is itself difficult to assess. The reliability of AF paleointensity methods has been tested by Kono (1978).

8.10   Néel diagrams and the blocking-temperature spectrum

Partial TRM's carried by different parts of the grain distribution, as well as other remanent magnetizations of SD assemblies, can be modelled graphically in a manner suggested by Néel (1949). Equations (8.17)–(8.19) show that all grains with the same value of VH_{K0} have the same relaxation time τ under given experimental conditions (T, H_0) or, equivalently, the same blocking temperature for fixed (t, H_0). Suppose we represent each SD grain ensemble in a sample by a point (V, H_{K0}) on a graph of V vs. H_{K0} (the Néel diagram). The density of points is $f(V, H_{K0})$, the **grain distribution**. The curves $VH_{K0} = $ const. represent relaxation-time contours $\tau(T, H_0) = $ const. or blocking-temperature contours $T_B(t, H_0) = $ const.

The contour of central importance is the blocking contour, defined by substituting $\tau = t$ in eqns. (8.17) and (8.18) or $T_B = T$ in eqn. (8.19). Assuming shape anisotropy as the source of H_K, the equation of the blocking contour is

$$VH_{K0} = \frac{2kT \ln(t/\tau_0)}{\mu_0 M_{s0}\beta^2(T)}\left[1 - \frac{|H_0|}{H_{K0}\beta(T)}\right]^{-2} = C(T, t, H_0) \tag{8.32}$$

or

$$VH_{K0} = \frac{2kT \ln(2t/\tau_0)}{\mu_0 M_{s0}\beta^2(T)} = C(T, t, 0). \tag{8.33}$$

Equations (8.32) and (8.33) describe the **blocking curve** joining the representative points of all grains whose magnetizations are in the process of relaxing (unblocking) or being blocked under conditions (T, t, H_0) or $(T, t, 0)$. Since the blocking process is relatively sharp, the blocking curve marks a well-defined boundary between *unblocked* or relaxed grains, with $VH_{K0} < C(\text{or } \tau \ll t)$ and $M = M_{eq}$, and *blocked* grains, with $VH_{K0} > C$ (or $\tau \gg t$) and $M = M_0$. By changing T, t or H_0, the blocking curve can be made to sweep through the entire grain distribution. Thermal, viscous (time-dependent), and field-induced magnetization changes can be modelled in this fashion.

Figure 8.15 illustrates the modelling of TRM acquisition and thermal demagnetization. Curve 1 is the room-temperature, zero-field blocking curve

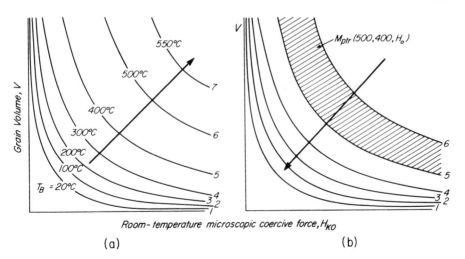

(a)   (b)

Room-temperature microscopic coercive force, H_{K0}

Figure 8.15 (a) Modelling thermal demagnetization of remanence carried by a distribution of single-domain grains using blocking curves ($T_B = $ const) on the Néel diagram. (b) Modelling partial TRM acquisition with the same set of contours.

$VH_{K0} = C(T_0, t, 0)$, with $t \approx 60$ s as a typical experimental time. $\tau(T_0, 0) = 60$ s or $T_B(t, 0) = 20\,°C$ are other ways of writing the equation of curve 1. Curves 2, 3, etc. are contours $T_B(t, 0) = 100\,°C$, $200\,°C$, etc. Grains below and to the left of curve 1 are unblocked or SP even at room temperature. Grains above and to the right of curve 1 can carry a stable remanent magnetization at room temperature.

Stepwise thermal demagnetization can be modelled in the following way. If the sample is heated to $100\,°C$, the blocking curve is $VH_{K0} = C(100\,°C, t, 0)$, coincident with contour 2. In the course of heating, the curve (here an **unblocking curve**) has swept through all the representative points between contours 1 and 2, thermally demagnetizing the pre-existing remanent magnetizations (whatever their origin) of the corresponding grain ensembles. Stepwise thermal demagnetization to $100\,°C$, $200\,°C$, $300\,°C$, etc. is represented by the sweeping out of the areas between contours 1 and 2, 2 and 3, 3 and 4, etc. by the unblocking curve.

Equal increments of C (eqn. (8.33)) represent equal areas swept out by the blocking curve, and if $f(V, H_{K0})$ is uniform, equal increments of magnetization. $\beta(T)$ is, however, a highly nonlinear function, especially near the Curie point (e.g., Fig. 3.5), so that equal temperature increments produce increasing increments of C, or accelerated motion of the blocking/unblocking curve, near T_C. The effect is clearly shown in Fig. 8.15(a). For this reason, the blocking-temperature spectrum is usually skewed toward high T_B's. If only large grains (or elongated grains) are present, falling above the $500\,°C$ contour (curve 6 in Fig. 8.15a) for example, the T_B spectrum will consist of a single peak just below T_c. Such samples possess **discrete blocking temperatures** (Irving and Opdyke, 1965). If only small and/or near-equidimensional grains are present, falling below the $500\,°C$ contour in Fig. 8.15(a) for example, the T_B spectrum is broad and fairly uniform. Such samples possess **distributed blocking temperatures**.

In practice, it is the spectrum of *unblocking* temperatures T_{UB} of NRM that is obtained experimentally, via stepwise thermal demagnetization. In the ideal case where the NRM is a weak-field TRM, $T_{UB} = T_B$, but if H_0 or t are very different in the laboratory setting from what they were in initial cooling or if the grains have altered in size, shape or composition since they formed, the original blocking temperature spectrum will be difficult or impossible to estimate. Thellier paleointensity determination will be unsuccessful in these circumstances.

Because of skewing of the T_B spectrum, much of the information in Thellier plots comes from a comparatively narrow temperature range within $100\,°C$ or so below T_C (e.g., Fig. 8.14a). High-resolution paleointensity determination or thermal demagnetization demands closely-spaced temperature steps near the Curie point. This is particularly the case if the NRM is carried by hematite, because the $M_s(T)$ curve for hematite (Fig. 3.20) is almost a step-function.

When a sample is cooled from T_C in a weak field H_0, acquisition of TRM can be modelled as in Fig. 8.15(b). The blocking contours are nearly identical to the zero-field unblocking contours of Fig. 8.15(a), but TRM is now produced by the applied field as the blocking curve sweeps through the representative point (V, H_{K0}) of a given grain ensemble. Since the blocking curve sweeps out the entire diagram, except for the area below $VH_{K0} = C(T_0, t, H_0)$, i.e., curve 1, *all* grains (except ones that are SP at T_0), no matter how large their microscopic coercive forces, acquire TRM in a weak field. It is easy to see why weak-field TRM is so much more intense and resistant to demagnetization than weak-field IRM.

If H_0 is applied during part of the cooling only, from 500 to 400 °C say, only grain ensembles in the area between contours 6 and 5 have a non-zero magnetization at the instant they are blocked. The partial TRM is in effect confined to the area between these contours. The laws of independence and additivity of partial TRM's are easy to understand from this viewpoint of areas on the Néel diagram.

8.11 AF demagnetization and the coercivity spectrum

Figure 8.16 illustrates a series of blocking contours, for steady or alternating fields. These contours are obtained by setting $H_c = H_0$ or $H_0 + \tilde{H}$ (where $\tilde{H}$ is AF) in eqns. (8.17) or (8.24). The contours shown are for $H_c = 0$, 10 mT, 20 mT, (0, 100, 200 Oe), etc. Note the important effect of H_q on the coercive force of small grains. For example, the grain ensemble indicated by the dot has $H_K \approx 27$ mT, but its magnetization can be changed by H_0 or $\tilde{H} = 10$ mT. The difference, $H_q \approx 17$ mT, is supplied by thermal agitation.

If an AF of peak value 10 mT is applied to a sample and the amplitude of the AF is smoothly decreased to zero, grain ensembles between the contours $H_c = 10$ mT and $H_c = 0$ are activated and their magnetizations, of whatever ori-

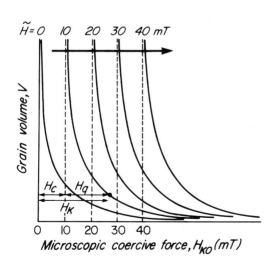

Figure 8.16 The effect of steady or alternating fields on SD blocking curves, e.g., during AF demagnetization. Note the strong effect of the thermal fluctuation field H_q on the coercive force of small grains. The grain(s) indicated by the dot have $H_{K0} = 27$ mT, but the particle moment(s) switch in a field of only 10 mT.

gin, are randomized. Subsequently applied AF's of peak values 20 mT, 30 mT, etc. erase the successive areas bounded by $H_c = 20$ mT, $H_c = 30$ mT, etc. in Fig. 8.16. This is the process of stepwise **alternating-field** (AF) **demagnetization**, univerally used in paleomagnetic studies to erase NRM components of low coercivity and consequent low reliability.

Just as incremental thermal demagnetization defines the blocking- or unblocking-temperature spectrum (or, for a TRM, the spectrum of partial TRM's), incremental AF demagnetization defines the **(micro)coercivity spectrum** or distribution of destructive fields of TRM and other remanences. The median value, $\tilde{H}_{1/2}$, of this spectrum (the value of H_c for which one-half the initial remanence is randomized) is the **median destructive field** (MDF) frequently cited as a measure of NRM hardness in paleomagnetic studies. TRM, anhysteretic remanence (Chapter 11) and saturation IRM are carried by all grain ensembles, whatever their coercive forces. These magnetizations, to a first approximation at least, have the same normalized coercivity spectra and AF demagnetization curves. This intrinsic spectrum is determined for single-domain grains by the relative numbers of grain ensembles, $f(V, H_{K0})$, in the areas between various unblocking contours in Fig. 8.16.

Other remanences, e.g., partial TRM's, magnetize only part of the Néel diagram. It is apparent that the contours bounding $M_{ptr}(500\,°C, 400\,°C, H_0)$ in Fig. 8.15(b) have a different shape than the room-temperature contours associated with AF demagnetization in Fig. 8.16. Partial TRM's will be AF demagnetized gradually, not sharply as in thermal demagnetization. Furthermore, higher-temperature partial TRM's, which lie farther to the right on the Néel diagram, will require higher AF's to demagnetize. Experimental confirmation of these predictions is seen in Fig. 8.17(a). Each partial TRM has a broad spectrum of coercivities but the median destructive field of each partial TRM increases (approximately linearly) with its mean blocking temperature.

The intrinsic or total TRM demagnetization curve, described in the previous paragraph, resembles the AF curve of the $T_B = 580\,°C$ partial TRM because most of the total TRM (for weak fields) has blocking temperatures near T_C. The shape of this AF curve is typical of SD remanence. There is an initial plateau of no demagnetization, extending to 10 mT (100 Oe) or so, dictated by magnetocrystalline anisotropy, which sets a lower limit for coercivity in equidimensional grains. Most of the demagnetization occurs in a range of intermediate fields, controlled by shape anisotropy, producing a sigmoid-shaped curve.

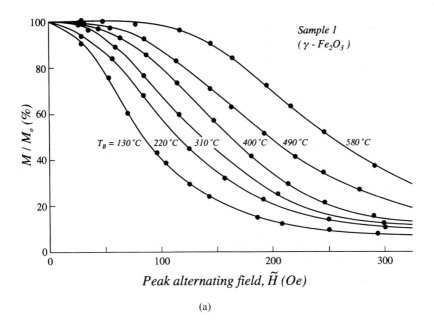

(a)

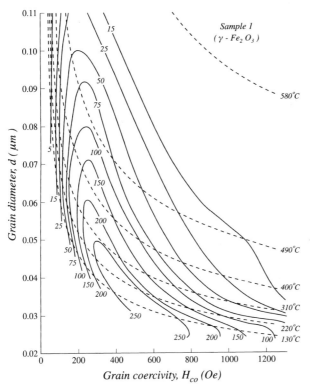

(b)

Figure 8.17
(a) Experimental AF demagnetization curves of partial TRM's acquired by the natural SD maghemite sample of Fig. 8.2 in six narrow temperature intervals centred on the T_B's indicated. Each pTRM has a wide spectrum of AF destructive fields but the MDF or average coercivity increases in proportion to T_B.
(b) Grain distribution $f(V, H_{K0})$ determined from the data of (a). [After Dunlop and West (1969) © American Geophysical Union, with the permission of the authors and the publisher.]

8.12   **Measuring and using the grain distribution,** $f(V, H_{K0})$

Sets of AF demagnetization curves like those of Fig. 8.17(a) are useful in working out the experimental grain distribution $f(V, H_{K0})$ of a particular sample. A set of partial TRM's acquired in narrow temperature intervals magnetizes a set of narrow ribbons on the Néel diagram, and these are demagnetized in small increments by the AF. Thus we can determine the magnetization associated with small areas $\Delta V \Delta H_{K0}$ on the Néel diagram. The grain distribution $f(V, H_{K0})$ for the natural SD maghemite sample of Fig. 8.17(a) is sketched in Fig. 8.17(b).

Once the grain distribution has been determined, we can use the predictive power of Néel's theory to generate synthetic results for other experiments, quite different from those used to determine $f(V, H_{K0})$ in the first place. An example early in this Chapter is Fig. 8.2, in which the thermal demagnetization curves of partial TRM's produced by different field strengths H_0 (for the same maghemite sample of Fig. 8.17) were correctly predicted by theory.

Other successful predictions have been made of the spectrum of unblocking temperatures T_{UB} of *total* TRM's produced in fields of different strengths (Dunlop and West, 1969). An example of such thermal demagnetization data, for TRM's acquired by SD titanomagnetite in fields of 1–700 Oe, appears in Fig. 8.18 (Özdemir and O'Reilly, 1982c). The T_{UB} spectrum, which is discrete and

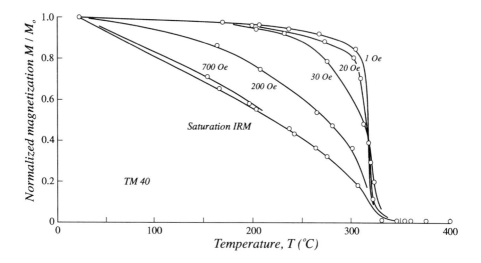

Figure 8.18   Stepwise thermal demagnetization of TRM's in unoxidized SD TM40. Unblocking occurs at lower temperatures as the field in which TRM was acquired increases from 1 Oe to saturation. [After Özdemir and O'Reilly (1982c), with the permission of the authors and the publisher, Blackwell Science, Oxford, UK.]

concentrated just below T_C for weak-field TRM becomes broadly distributed for strong-field TRM or saturation IRM. The AF spectrum of these remanences is also dependent on inducing field H_0 (the Lowrie–Fuller test; Chapter 11) but in a less spectacular fashion.

Chapter 9

Multidomain thermoremanent magnetization

9.1   Introduction

Multidomain (MD) grains have three likely sources of TRM:

(1) pinning of domain walls;

(2) domain nucleation or denucleation (transdomain TRM);

(3) nucleation failure (metastable SD grains).

Pinned domain walls have received the most attention and will be the subject of §9.2–9.7. Transdomain TRM was described from a theoretical point of view in §7.10; the experimental evidence and its interpretation will be dealt with in §9.8. Metastable single-domain TRM is the subject of §9.9.

Pinned domain walls can be thermally activated just as SD moments can. Wall displacements in response to a weak field applied at high temperatures, like reversals of SD grains, can be frozen in by cooling to room temperature, resulting in TRM. Analog theories of TRM due to pinned walls (Néel, 1955; Everitt, 1962a; Schmidt, 1973) have therefore been constructed along the lines of the Néel (1949) SD theory outlined in Chapter 8. In these theories, V becomes the volume of one Barkhausen jump of a wall or wall segment and H_K becomes the critical field for such a Barkhausen jump.

We should recognize, however, certain fundamental differences between SD and MD TRM. No individual wall can be treated independently of its neighbours in an MD grain. The internal demagnetizing field H_d favours a set of wall displacements yielding minimum net moment for the grain. In order to achieve this minimum-internal-field condition, walls may undergo a series of jumps during

234

heating or cooling, even if the field remains constant. Thus there is no single blocking or unblocking temperature for a given domain wall, let alone for a grain containing many walls in which a jump by one wall alters the internal field at the location of every other wall. Self-demagnetization also leads to non-linear dependence of TRM intensity on the applied field, but as with SD grains, TRM is approximately proportional to field strength in the weak-field (≤ 0.1 mT) region of interest in paleointensity determination.

9.2   Wall displacement and hysteresis in a 2-domain grain

9.2.1   Wall pinning and total energy

To avoid the complexities of wall interactions, much of our treatment will model 2-domain (2D) grains containing only a single wall. Any real crystal is imperfect. It contains lattice defects such as inclusions, vacancies, and dislocations. All such defects tend to anchor or pin domain walls, vacancies and inclusions because they reduce the volume of a wall and create magnetic poles, affecting both γ_w and E_d (see discussion in §5.6 and Figs. 5.13, 5.17), and dislocations by virtue of the magnetoelastic interaction between their stress fields and the spins in the wall, which are rotated out of easy directions (see discussion in §6.3 and Figs. 5.8b, 5.11, 5.18). Domain walls have minimum energy if they incorporate as many pinning centres as possible. The pinning effect is sufficiently strong that domain walls in titanomagnetite will bend rather than breaking free of their pins, in spite of the resultant energy of the poles that appear on the bent wall. Figure 5.18 illustrated reversible bending of a pinned wall when a 2 mT (20 Oe) field was applied and removed.

We can represent pinning by a fluctuating wall energy $E_w = \gamma_w A$, A being the area of the domain wall. Figure 9.1(a) shows how E_w in a 2D grain might depend on displacement x of the wall from the central (demagnetized) position. Minima in E_w correspond to locations where the wall (here assumed *not* to bend) is anchored by defects. To move the wall away from one of the minima requires energy. For simplicity, we will assume that H_0 is applied in the direction of one of the domain magnetizations M_s. The net magnetization of the grain is $M = (x/a)M_s = rM_s$ (see Fig. 9.1b). Physically, the quantity r represents either the fractional wall displacement x/a or the fractional magnetization M/M_s.

The total energy of the grain is

$$E_{tot}(x) = E_w + E_H + E_d = E_w - 2\mu_0 A M_s H_0 x + \mu_0 N A M_s^2 x^2 / a. \quad (9.1)$$

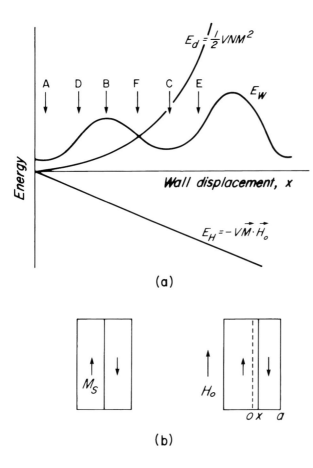

Figure 9.1 Simple model of domain-wall displacement in a two-domain grain. (a) Contributions to the total energy. A, B, C mark minima and maxima in E_w: D, E, F are points of steepest slope $|dE_w/dx|_{max}$. (b) Domain configurations without and with a field H_0 applied parallel to one of the domains. [After Dunlop and Xu (1994) © American Geophysical Union, with the permission of the authors and the publisher.]

Equivalent cgs formulas can be obtained by setting $\mu_0 = 1$. Equations (4.8) and (4.9) have been used for E_H and E_d, and $N = N_{2D}$ (§5.2.2). At equilibrium, E_{tot} is a minimum if $dE_{tot}/dx = 0$. Thus

$$dE_w/dx = 2\mu_0 A M_s(H_0 - NM) = 2\mu_0 A M_s H_i, \tag{9.2}$$

H_i being the total internal field $H_0 - NM$ (eqn. (4.5)).

9.2.2   Susceptibility and microscopic coercive force

The **initial reversible susceptibility**, χ_i, intrinsic to the ferromagnetic material, usually measured in a field $H_0 \leq 0.1$ mT (1 Oe), is

$$\chi_i = \frac{\partial M}{\partial H_i} = \frac{2\mu_0 A M_s^2 x}{a}\left(\frac{dE_w}{dx}\right)^{-1}. \tag{9.3}$$

χ_i is constant only if E_w is a quadratic function of x for small displacements from equilibrium. As well as χ_i due to wall displacements in domains parallel to H_0, there is a contribution to intrinsic susceptibility from the rotation of domains oriented perpendicular to H_0. Stacey and Banerjee (1974) call these susceptibilities $\chi_{\parallel}$ and $\chi_{\perp}$, respectively. χ_i due to both sources is about 12 (1 in cgs) for magnetite and several times this for iron. There is a difference of 4π between susceptibilities in SI and cgs units, just as there was for demagnetizing factors (§4.2.3), arising from the different definitions of B in SI and cgs (eqn. (2.1)).

In practice, χ_i is difficult to determine precisely. What is normally measured is the **apparent susceptibility** $\chi_0 = \partial M/\partial H_0$. Since $M = \chi_i H_i = \chi_i(H_0 - NM)$, we see that

$$\chi_0 = \frac{\partial M}{\partial H_0} = \frac{\chi_i}{1 + N\chi_i} \to \frac{1}{N} \quad \text{if } \chi_i \gg \frac{1}{N}. \tag{9.4}$$

The limiting value $1/N$, reflecting the shape of grains and not χ_i, was quoted in §5.6.2. This limit is approached quite closely in the case of iron. For equidimensional grains of magnetite, $\chi_0 \approx 0.75/N$.

Larger fields $(H_0 \gg 0.1\,\mathrm{mT}\ (1\,\mathrm{Oe}))$ may displace the domain wall past an energy maximum, like B in Fig. 9.1(a), to the next energy minimum, C. The internal field required is the **microscopic coercive force** or **microcoercivity**, h_c. An SD grain has a single microscopic coercive force H_K because the particle moment VM_s rotates against anisotropy K in a single critical jump. Each wall in a 2D or MD grain, on the other hand, makes a series of jumps past many different energy barriers. Multidomain grains thus have a spectrum of microcoercivities. The average of the **microcoercivity distribution** $f(h_c)$ is the bulk coercive force H_c of §5.6.2.

From eqn. (9.2), the microcoercivity or critical value of $|H_i|$ due to a particular barrier depends on the maximum slope $|dE_w/dx|_{max}$ of the barrier:

$$h_c = \frac{1}{2\mu_0 A M_s}\left|\frac{dE_w}{dx}\right|_{max}. \tag{9.5}$$

However, h_c, being an internal field, cannot easily be measured. Externally, when a critical value of field H_0,

$$(H_0)_{crit} = h_c + NM, \quad \text{i.e., } (H_i)_{crit} = (H_0)_{crit} - NM = h_c, \tag{9.6}$$

is applied to the grain, the wall is unpinned and will move spontaneously from D to a point E of slightly steeper slope in the neighbouring E_w minimum, with no increase in H_0. The critical external field $(H_0)_{crit}$ is greater than h_c by an amount $NM = |H_d|$, M being the grain magnetization when the wall is at point D where $|dE_w/dx|$ is maximum.

Such a spontaneous movement of the wall is called a **Barkhausen jump**, and is irreversible. The wall will not return to the first E_w minimum unless H_0 is applied in the opposite direction. The *negative* critical field required to move the wall back from C to A is

$$(H_0')_{crit} = -h_c + NM, \quad \text{i.e., } (H_i)_{crit} = (H_0)_{crit} - NM = -h_c, \tag{9.7}$$

in which h_c and M at the point of steepest negative slope F are to be substituted. Notice that $|(H'_0)_{\text{crit}}|$ is now numerically *less* than h_c, since M is positive at F. Thus $\boldsymbol{H}_d$ acts as a restoring field assisting the wall towards a state of lower magnetization, while $\boldsymbol{H}_0$ acts as a driving field and h_c as a resistive field impeding displacements in either direction.

9.3   A simple model of multidomain TRM

As $T \to T_C$, $M_s \to 0$ and the restoring force exerted on domain walls by $\boldsymbol{H}_d = -N\boldsymbol{M}$ is reduced and eventually lost. At the same time, resistance to wall motion disappears because $h_c \to 0$, as M_s, λ and $K_1 \to 0$. A weak field $\boldsymbol{H}_0$ applied near the Curie temperature consequently produces a large wall displacement. During cooling, both the restoring force H_d and the resistive force h_c increase, but at different rates. The high-temperature wall displacement, or a substantial fraction of it, is frozen in at a **blocking temperature** T_B determined by the balance between H_d and h_c. Néel (1955) and Stacey (1958) based theories of multidomain TRM on this picture of the blocking process. We consider Stacey's theory first and Néel's theory in §9.4.

Equation (9.4) tells us that if $\chi_i \gg 1/N$, which is certainly the case at high temperature, $\chi_0 \approx 1/N$ and so $M = H_0/N$. The same result follows from the principle of minimum internal field: $H_i = H_0 - NM \to 0$. At the blocking temperature T_B, the wall displacement x is described by

$$r(T_B) = \frac{x(T_B)}{a} = \frac{M(T_B)}{M_s(T_B)} = \frac{H_0}{NM_s(T_B)}. \tag{9.8}$$

During cooling below T_B, the wall's displacement is frozen, so that M changes only by virtue of the reversible increase in $M_s(T)$. Thus the TRM at room temperature T_0 is given by

$$M_{tr} = r(T_B)M_s(T_0) = \frac{M_{s0}}{M_{sB}}\frac{H_0}{N} = \frac{H_0}{N\beta(T_B)}, \tag{9.9}$$

where $M_{s0} \equiv M_s(T_0)$, $M_{sB} \equiv M_s(T_B)$, and $\beta(T) \equiv M_s(T)/M_{s0}$.

Equation (9.9) appears to predict the linear dependence of TRM on applied field observed experimentally for weak fields. In reality, as Néel (1955) and Schmidt (1976) have shown, T_B is itself dependent on H_0. TRM as given by (9.9) actually varies approximately as $H_0^{1/2}$, a dependence observed experimentally only when $H_0 > 1\,\text{mT}\,(10\,\text{Oe})$.

However, we can make quite an accurate estimate of the **Koenigsberger ratio**, $Q_t = M_{tr}/\chi_0 H_0$, using (9.9). The companion ratio $Q_n = M_{nr}/\chi_0 H_0$, where H_0 is the present earth's field ($\approx 50\,\mu\text{T}$ or $0.5\,\text{Oe}$) and M_{nr} is NRM, is frequently quoted in paleomagnetic studies. The TRM predicted by eqn. (9.9) gives rise to an internal demagnetizing field H_d that can move loosely pinned walls. The soft magnetization, M_{in} induced in this way partially shields or cancels M_{tr}. Since $M_{in} = \chi_i H_d = \chi_i[-N(M_{tr} + M_{in})]$, the resultant magnetization or observed TRM M'_{tr} is

$$M'_{tr} = M_{tr} + M_{in} = M_{tr} - N \frac{\chi_i}{1 + N\chi_i} M_{tr} = \frac{1}{1 + N\chi_i} M_{tr}. \qquad (9.10)$$

For equidimensional magnetite grains, the screening or shielding factor $(1 + N\chi_i)^{-1}$ is about $\frac{1}{5}$ and $\beta(T_B) \equiv M_s(T_B)/M_s(T_0)$ is about $\frac{1}{3}$ (Stacey, 1963). Therefore

$$Q_t = \frac{M'_{tr}}{\chi_0 H_0} \approx \left(\frac{3}{5}\frac{H_0}{N}\right) \bigg/ \left(\frac{H_0}{N}\right) = 0.6. \qquad (9.11)$$

Experimental Q_t values for MD magnetite are close to this figure (Parry, 1965; Stacey, 1967). Values for SD magnetite are much higher, usually > 10.

9.4   The Néel theory of TRM in 2D grains

9.4.1   TRM in the absence of thermal fluctuations

The defining equation for TRM in Stacey's theory (§9.3) was $H_0 = NM$ or $H_i = H_0 - NM = 0$. Comparing this to eqns. (9.6) and (9.7), we see that $\pm h_c$ has been set to zero. But wall pinning, as manifested in h_c, is responsible for TRM blocking. Therefore the blocking temperature T_B cannot be obtained from Stacey's theory. T_B can be obtained from the theories of Néel (1955), Everitt (1962a) and Schmidt (1973), which are described next. We will consider only 2D grains in which the wall encounters a series of identical E_w barriers, each with the same microcoercivity h_c for positive or negative jumps. A more general treatment incorporating distributions of h_c (Xu and Dunlop, 1994) gives generally compatible results.

Figure 9.2 graphs the total energy, $E_w + E_d + E_H$, in a 2D grain with many identical E_w barriers. These energies were plotted separately in Fig. 9.1. The cyclic variations due to E_w are minor compared to the broad parabolic shape imparted by E_d, but the small potential wells can nevertheless trap walls in a variety of local energy minimum (LEM) states. As H_0 increases, an increasing gradient tilts the energy function down to the right, until the energy barrier holding the wall in its original well disappears and the wall jumps to a new LEM state. The equivalent condition in terms of E_w alone is displacing the wall to a position where dE_w/dx is a maximum. Repeated jumps of the wall generate the ascending branch of the hysteresis loop, which has slope $1/N$ (apart from the sawtooth pattern due to the discrete jumps). In decreasing fields, the wall undergoes a set of jumps to the left and generates a descending loop offset from the ascending loop, but also with slope $1/N$.

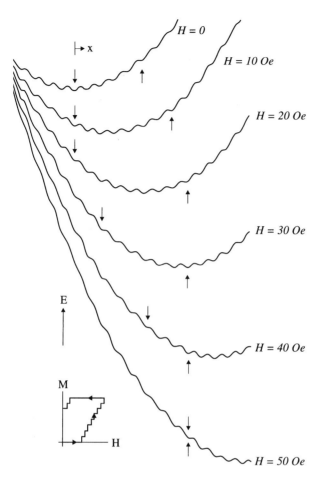

Figure 9.2 $E_{tot} = E_w + E_d + E_H$ as a function of domain-wall displacement x from the demagnetized state. Downward arrows mark successive local energy minima in which the wall is trapped as H_0 increases from 0 to 50 Oe. Upward arrows indicate minima where the wall is pinned as H_0 subsequently decreases from 50 Oe to 0. The wall occupies different minima in increasing and decreasing fields of the same strength, giving rise to magnetic hysteresis and remanence (inset). Analogous wall jumps and hysteresis occur when H_0 is constant but the temperature changes (see text). [Reprinted from Schmidt (1973), with kind permission of Elsevier Science – NL, Sara Burgerhartstraat 25, 1055 KV Amsterdam, The Netherlands.]

Quite strong fields were used to generate the curves of Fig. 9.2. However, the same effect can be achieved with a weak field of fixed strength by increasing the temperature, which has the effect of decreasing h_c and making the potential wells shallower and shallower. Near T_C, the wells are so shallow that even a weak field will push the wall right to the boundary of the grain. The grain becomes SD and $M \rightarrow M_s$. During cooling, the parabolic envelope due to E_d deepens as M_s grows, tending to spill the wall to the left. On the other hand, the LEM wells also deepen, resisting spillage, because h_c is also growing. At high temperatures, the restoring force to the left due to E_d is the stronger effect and the wall is pushed back, but eventually the deepening of individual wells becomes more important and the wall begins to be pushed to the right. However, the wall cannot now escape from its local energy well (unless H_0 increases) and is trapped in this LEM state in further cooling. It has reached its blocking temperature.

The total energy equation can be used to find T_B, following this scenario. This was the approach taken by Everitt (1962a) and Schmidt (1973). We will follow instead the mathematically equivalent approach of Néel (1955), in which $E_w + E_H$ and E_d are treated separately. Equations (9.2) and (9.5) indicate that if all E_w barriers are identical, with the same $(dE_w/dx)_{max}$ and h_c, the internal field hysteresis loop, $M(H_i)$, will be rectangular with height $\pm M_s$ and width or bulk coercive force $\pm H_c = \pm h_c$. A series of such loops at different temperatures is sketched in Fig. 9.3(a). However to obtain the externally measured $M(H_0)$ loops, we must remember that $H_i = H_0 - NM$ or

$$M = (1/N)(H_0 - H_i). \qquad (9.12)$$

Equation (9.12) generates a temperature-independent line of slope $-1/N$ and intercept H_0, the demagnetizing line, which is also plotted in Fig. 9.3(a). An intersection between the demagnetizing line and the rectangular loop for a particular T gives $M(H_0, T)$.

It is clear graphically that at high enough temperatures (e.g., T_1, T_2 in Fig. 9.3a), $M = M_s$. However, at lower temperatures (T_3, T_4), the demagnetizing line intersects the descending loop and $M < M_s$. The wall has been driven back from the grain boundary. At temperature T_5, the intersection point reascends the descending loop. This is not permitted physically because when the wall reverses its motion, it cannot surmount the right-hand barrier in its local well and remains trapped, as explained above. We can determine T_B, at which the wall becomes trapped, by solving for $dr/dT = 0$, the point at which the wall reverses its motion ($r \equiv x/a$ or M/M_s).

Mathematically, we first solve the equation for the vertical arms of the rectangular loop, $H_i = \pm H_c$, with eqn. (9.12) to obtain the external-field hysteresis loop,

$$M = (1/N)(H_0 \mp H_c). \qquad (9.13)$$

The measured loop is said to be sheared: the ascending and descending branches have slopes $1/N$, like measured MD hysteresis curves (Fig. 5.16). Two such sheared loops are shown in Fig. 9.3(b). For TRM blocking we choose the descending branch of (9.13) and obtain

$$M(T) = (1/N)[H_0 + H_c(T)]. \qquad (9.14)$$

Domain-wall pinning is temperature-dependent by virtue of the T dependences of K_1 and magnetostriction λ. Both of these material properties vary approximately as a power of $M_s(T)$ (see §2.8 and Figs. 3.7b, 3.8b). Thus we can assume $H_c(T) = H_{c0}\beta^n(T)$, where $H_{c0} \equiv H_c(T_0)$ and $\beta(T) \equiv M_s(T)/M_{s0}$. Making this substitution and solving for $r(T) = M(T)/M_s(T)$, we find

$$r(T) = \frac{H_0 + H_{c0}\beta^n(T)}{NM_{s0}\beta(T)}. \qquad (9.15)$$

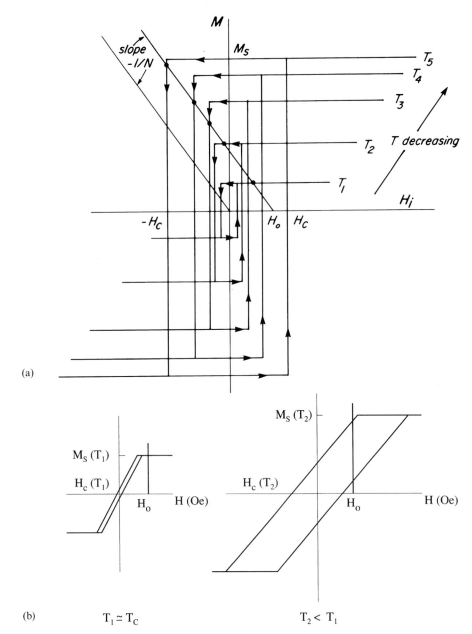

Figure 9.3 (a) Illustration of the process of field-blocking of TRM. The intersections (dots) of the demagnetizing line, of slope $-1/N$ and intercept H_0, and rectangular M–H_i hysteresis loops for various temperatures $T_1, T_2 \ldots$ give $M(H_0, T)$. M/M_s cannot reascend the descending loop because x/a cannot increase during cooling in a constant field H. When the wall attempts to reverse its motion, between T_4 and T_5, it remains trapped in its local potential well, i.e., blocked. (b) M–H_0 hysteresis loops, obtained from (a) by making demagnetizing lines vertical, thereby 'shearing' the rectangular M–H_i loops. [After Dunlop and Xu (1994) © American Geophysical Union and Day (1977), with the permission of the authors and the publishers, The American Geophysical Union, Washington, DC and Journal of Geomagnetism and Geoelectricity, Tokyo, Japan.]

Finally, since blocking occurs when $dr/dT = 0$, the blocking temperature is given by

$$H_c(T_B) = \frac{H_0}{n-1}$$

or   (9.16)

$$\beta(T_B) = \left[\frac{H_0}{(n-1)H_{c0}}\right]^{1/n}.$$

This situation, in which TRM is blocked by the growth of energy barriers to wall motion and thermal fluctuations are ignored, is referred to as **field-blocked TRM** (Dunlop and Xu, 1994). The blocking temperature predicted by eqn. (9.16) for magnetite, with $n = 2$ and various assumed values of H_{c0}, is plotted as a function of H_0 in Fig. 9.4. It is generally similar to the field dependence of T_B

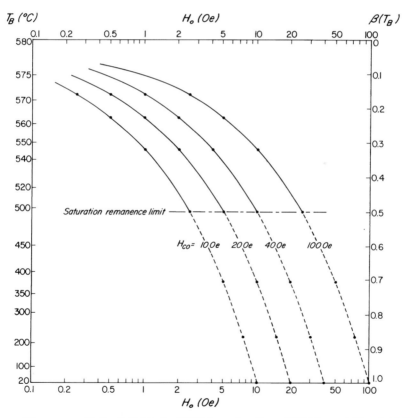

Figure 9.4 T_B for field-blocked TRM as a function of TRM acquisition field H_0 for various microcoercivities H_{c0} when $n = 2$. The $T_B(H_0)$ curves are approximately quadratic, as for SD TRM (Fig. 8.11), but they are truncated at the saturation limit, when TRM becomes equal to the room-temperature saturation IRM. [After Dunlop and Xu (1994) © American Geophysical Union, with the permission of the authors and the publisher.]

in SD grains (Fig. 8.11), in that T_B decreases approximately quadratically as H_0 increases. However, there is an effective lower limit to T_B (the saturation remanence limit: 500 °C for the parameters chosen in Fig. 9.4), when H_0 has pushed the wall so far from equilibrium that 'blocking' during in-field cooling is only temporary. When $H_0 \to 0$ at T_0, H_d causes the wall to unpin and spring back to a position corresponding to the saturation remanence $M_{rs}(T_0)$. This process is called **wall re-equilibration**.

Provided T_B is above the saturation remanence limit, walls remain blocked even when $H_0 \to 0$ at T_0. Then the TRM intensity is equal to $M(T_B)/\beta(T_B)$, which from (9.14) and (9.16) gives

$$M_{tr} = \frac{n}{(n-1)^{1-\frac{1}{n}}} \frac{H_{c0}^{1/n}}{N} H_0^{1-\frac{1}{n}}. \tag{9.17}$$

With the field dependence of T_B taken into account, TRM intensity is no longer proportional to field strength H_0. If $n = 2$, as Néel (1955) assumed, $M_{tr} \propto H_0^{1/2}$.

Some TRM acquisition data are given in Fig. 9.5. They lend general support to the field dependence predicted by eqn. (9.17), but only in the range $1 \, \text{mT} < H_0 < 10 \, \text{mT}$. For higher fields, the TRM saturates at a value equal to $M_{rs}(T_0)$ (see eqn. (9.26)), through wall re-equilibration. For lower fields, thermal activation governs TRM acquisition, as we shall show in the next section, and TRM is proportional to H_0, as required for paleointensity determination.

Notice that theoretical TRM's for MD grains in Fig. 9.5 are much weaker than TRM's for SD grains, even when the microscopic coercive forces are comparable. Experimentally, this is the pattern observed (Fig. 8.4). The internal demagnetizing field is responsible for the contrast between MD and SD intensities of TRM. By opposing displacements of walls away from the demagnetized state, H_d severely limits the intensity of magnetization M in an MD grain. In an SD grain, however, M always has magnitude M_s and H_d can only influence the *orientation* of M. The SD remanence-carrying potential is therefore much greater. Note that eqn. (9.17) strictly applies only to 2D grains. Shielding by soft walls in MD grains could be taken into account in the manner of §9.3 and would reduce predicted TRM intensities by about another factor 5.

9.4.2   TRM in the presence of thermal fluctuations

Néel (1955) pointed out that field blocking of TRM occurs very close to T_C if H_0 is small. The blocking condition (9.16), for $n = 2$ as assumed by Néel, is $H_c(T_B) = H_0$ and only within ≈ 10 °C of T_C do microcoercivities fall to 1 Oe or less (Fig. 9.4). At such high temperatures, thermal fluctuations can unpin a domain wall below its nominal blocking temperature. A parallel situation in SD grains was discussed in §8.7.

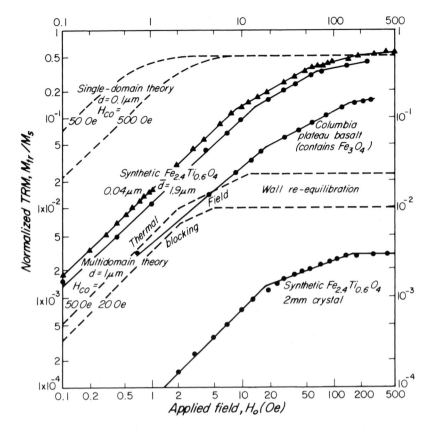

Figure 9.5   The dependence of TRM intensity on applied field H_0. Data for multidomain samples are from Dunlop and Waddington (1975) (Columbia plateau basalt) and Tucker and O'Reilly (1980a) (synthetic TM60, $d = 1.9\,\mu m$ and 2 mm, respectively). Data for single-domain TM60 ($d = 0.04\,\mu m$) are from Özdemir and O'Reilly (1982c). Single-domain and multidomain theoretical curves (dashed) follow eqn. (8.21) and eqns. (9.23), (9.17) and (9.26), respectively. In the thermal blocking (low-field) range, a grain size of 1 μm and Barkhausen jumps of 30 nm were assumed. [After Dunlop and Xu (1994) © American Geophysical Union, with the permission of the authors and the publisher.]

As a simple example of thermally activated wall motion, consider a single Barkhausen jump of a domain wall over a barrier of height $\Delta E_w = (E_w)_{max} - (E_w)_{min}$. If the jump occurs as a result of an applied field H_0, the wall moves from D to E (Fig. 9.1). If the jump is thermally activated, the wall moves from A to C without the assistance of a field. In either case, the wall jumps one wave-length, λ, of E_w and the magnetic moment of the grain changes by

$$\Delta m = 2A\lambda M_s = 2V_{Bark}M_s, \tag{9.18}$$

where $V_{Bark} = A\lambda$ is the 'Barkhausen volume' swept out by the moving domain wall. We shall assume that Δm is too small to appreciably affect H_d, although this is seldom the case in practice. Then, for a field-induced jump,

$$H_c = \frac{\Delta E_w}{2\mu_0 A M_s c(\lambda/2)} = \frac{\Delta E_w}{c\mu_0 V_{Bark} M_s} \tag{9.19}$$

from eqn. (9.5), c being a number of order unity relating the maximum slope of a barrier to its height. For example, $c = 1$ if $E_w(x)$ is a sawtooth function and $c = 2/\pi$ if $E_w(x)$ is sinusoidal. For a thermally activated jump,

$$\frac{1}{\tau} = \frac{1}{\tau_0} \exp^{-\Delta E_w/kT} = \frac{1}{\tau_0} \exp\left(-\frac{c\mu_0 V_{Bark} M_s H_c}{kT}\right). \tag{9.20}$$

τ_0 here is slightly different from τ_0 in eqns. (8.16)–(8.18), since the mechanisms by which thermal energy is imparted to domain walls and to SD moments are different (Gaunt, 1977), but the distinction is not important for our purposes. The significant point is that if we consider the thermal activation of a domain wall to be confined to the volume V_{Bark} affected by the wall's displacement, eqn. (9.20) is formally identical (except for minor numerical factors) to eqn. (8.18) for SD grains under zero-field conditions. The microscopic coercive force H_c (or h_c for non-identical barriers) due to wall pinning corresponds to the microscopic coercive force H_K for domain rotation.

As a condition for blocking in the presence of thermal fluctuations (**thermal blocking**), Néel suggested (by analogy with eqn. (8.25) for SD grains) replacing the field blocking condition $H_c(T_B) = H_0$ (for $n = 2$) by

$$H_c(T_{Bf}) = H_f(T_{Bf}), \tag{9.21}$$

where T_{Bf} is the thermal fluctuation blocking temperature, which is always lower than the field blocking temperature T_B. The fluctuation field H_f is given by

$$H_f(T_{Bf}) = \frac{kT \ln(t/\tau_0)}{c\mu_0 V_{Bark} M_s(T_{Bf})}, \tag{9.22}$$

using eqn. (9.20) and setting $\tau = t$ for blocking. Notice that H_f for 2D (or MD) grains has a different dependence on T and V than H_q for SD grains (eqn. (8.24)). Then, following the same reasoning as in §9.4.1, we find for the TRM intensity

$$M_{tr} = \frac{H_{c0}^{1/n}}{N H_f^{1/n}(T_{Bf})} H_0. \tag{9.23}$$

Thermally blocked TRM should be proportional to H_0, as observed experimentally for small fields (Fig. 9.5).

In practice, eqn. (9.23) is not very useful for predicting actual magnitudes of TRM because H_f depends on the Barkhausen volume swept out by a domain wall. V_{Bark} could be estimated by direct domain observations of jumping walls or wall segments, but to date there are few published data. However, from Fig. 9.5, the change from linear to quadratic dependence of M_{tr} on H_0 occurs between 1 and a few Oe. $H_f(T_{Bf}) \approx 1$ Oe ($100\,\mu$T) is therefore an order-of-magnitude value, allowing T_{Bf} and M_{tr} to be estimated approximately. When $H_0 \leq 1\,Oe$, M_{tr} in the presence of thermal fluctuations is even weaker than M_{tr} estimated from (9.17) for field blocking.

9.5   **Thermal demagnetization of total and partial TRM**

The detailed theory of thermal demagnetization is given by Dunlop and Xu (1994) and Xu and Dunlop (1994). We shall give only a brief synopsis. As a preliminary, consider a *partial* TRM acquired by cooling from T_C to T_B (the field-blocking temperature for total TRM) in field H_0 and in zero field from T_B to T_0. Combining eqns. (9.14) and (9.16), we see that the magnetization $M(T_B)$ in the presence of H_0 is

$$M(T_B) = \frac{n}{n-1}\frac{H_0}{N}, \tag{9.24}$$

but when $H_0 \to 0$ at T_B, the same equations give for the remanence $M_r(T_B)$

$$M_r(T_B) = \frac{H_c(T_B)}{N} = \frac{1}{n-1}\frac{H_0}{N}. \tag{9.25}$$

Thus the partial TRM will be only $1/n$ as intense as the total TRM.

The mechanism is pushing back, or re-equilibration, of the wall by the internal demagnetizing field $\boldsymbol{H}_d$ when $H_0 \to 0$. The same process may occur even for a total TRM, when $H_0 \to 0$ at T_0, since TRM cannot exceed the saturation remanence H_{c0}/N. This is the saturation limit referred to in Fig. 9.4. Saturation, accompanied by room-temperature re-equilibration of wall positions, can occur for surprisingly small fields: $H_0 \geq 5\,Oe$ for $H_{c0} = 20\,Oe$ and $n = 2$ (Figs. 9.4, 9.5). The wall re-equilibration process is illustrated in Fig. 9.2; here a much larger H_0 (initially 50 Oe) has been used to make the re-equilibration easy to see.

Let us now turn our attention to reheating a TRM, with $H_0 = 0$, from T_0 to T_B. When we reach T_B, we know the TRM intensity must be reduced to a fraction $1/n$ of the intensity it had when cooled in H_0 through T_B. This does not occur abruptly, but in a series of small Barkhausen jumps. In fact, the process is exactly analogous to that illustrated in Fig. 9.2, except that $H_0 = 0$ (the uppermost curve) and the amplitude of the energy variations falls with heating. As the energy barriers decrease, the wall can successively jump to the left, towards the demagnetized state, one barrier at a time. Since H_c, which represents the wall pinning, only vanishes at T_C, this process continues to very high temperatures, essentially to T_C.

The thermal demagnetization process for 2D (and MD) grains is very different from that of SD grains, in which each grain had a single T_{UB} which was equal, for weak H_0, to T_B during cooling. Because of the active influence of $\boldsymbol{H}_d$ on magnetization intensity, i.e., wall position, a wall which is field-blocked during cooling at temperature T_B has a spectrum of T_{UB}'s extending from well below T_B to well above T_B, in fact virtually to T_C. This thermal demagnetization 'tail' is a distinctive characteristic of MD grains. It is also of considerable practical importance, because it means that thermal 'overprints' (secondary partial TRM's acquired in nature) cannot be easily removed by thermal demagnetization if they are carried by MD grains.

Some theoretical thermal demagnetization curves for total TRM's acquired in different fields and for partial TRM's acquired between $T_2 = 550\,°C$ and various T_1 in a fixed field appear in Fig. 9.6. All the 'curves' actually decrease linearly with respect to $\beta(T) \equiv M_s(T)/M_{s0}$, a dependence we shall explain momentarily. Total TRM's at the saturation limit (for $H_0 \geq 5\,Oe$ in this example) have been re-equilibrated at T_0 and undergo further re-equilibration as soon as they are heated above T_0. TRM's acquired in weaker fields did not have their walls pushed so far from equilibrium that they re-equilibrated when $H_0 \to 0$ at T_0 during the original cooling. They can be heated substantially above T_0 before H_d exerts enough force to begin pushing the walls back toward the equilibrium (demagnetized) state. For example, when $H_0 = 0.5\,Oe$ ($50\,\mu T$), a typical geomagnetic field strength, thermal demagnetization does not begin until $\approx 540\,°C$. However, in all cases, thermal demagnetization is only complete at T_C.

Partial TRM's (Fig. 9.6b), even in a field of 5 Oe which would saturate a total TRM, are relatively weak, and they demagnetize like weak-field total TRM's. For the conditions chosen, re-equilibration occurs when $H_0 \to 0$ at T_1 during original cooling if $T_1 \leq 482\,°C$. For the complete set of partial TRM's thermal demagnetization begins at 470–545 °C and ends at T_C.

In order to explain the linear dependence of thermal demagnetization on $\beta(T)$, we consider first a saturated total TRM. Its intensity at T_0 is given by eqn. (9.14), for $H_0 = 0$, as

$$M_{rs}(T_0) = \frac{H_{c0}}{N},\tag{9.26}$$

which is the justification for eqn. (5.32). When heated to temperature T in zero field, eqn. (9.14) predicts

$$M_r(T) = \frac{H_c(T)}{N} = \frac{H_{c0}\beta^n(T)}{N}.\tag{9.27}$$

Thus M_{rs} must be reduced to this level, by wall re-equilibration, at T. If the measurement of M_r is made at T (continuous thermal demagnetization), then

$$M_r(T)/M_r(T_0) = \beta^n(T).\tag{9.28}$$

More commonly in paleomagnetic and paleointensity studies, M_r is measured after cooling in zero field to T_0 (stepwise thermal demagnetization). M_{r0} measured at T_0 is larger than M_r measured at T by a factor

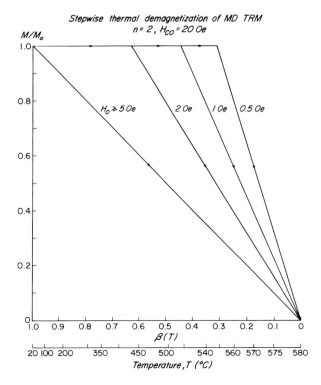

(a)

Figure 9.6 (a) Theoretical stepwise thermal demagnetization of total TRM ($H_{c0} = 20$ Oe, $n = 2$) for various TRM acquisition fields. Weak-field TRM has higher unblocking temperatures than stronger-field TRM. (b) Theoretical stepwise thermal demagnetization of high-temperature partial TRM's ($H_0 = 5$ Oe, $H_{c0} = 20$ Oe, $n = 2$). Unblocking begins at or above the lower T_B of each pTRM but continues above the upper T_B of 550 °C until the Curie point T_C. These unblocking temperatures are much higher than those of total TRM for the same H_0 in (a). [After Dunlop and Xu (1994) © American Geophysical Union, with the permission of the authors and the publisher.]

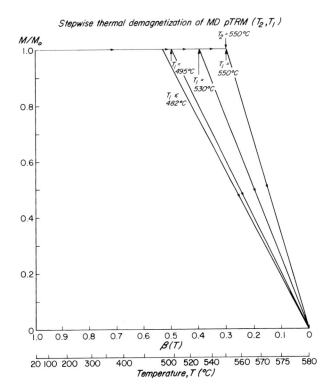

(b)

$M_s(T_0)/M_s(T) = \beta^{-1}(T)$, because M_s increases reversibly in cooling from T to T_0. In this case, the thermal demagnetization is described by

$$M_{r0}(T)/M_r(T_0) = \beta^{n-1}(T). \tag{9.29}$$

For $n = 2$, as assumed in Fig. 9.6, the remanence decreases in proportion to $\beta(T)$.

Equations (9.28) and (9.29) are quite general descriptions of continuous or stepwise thermal demagnetization due to wall re-equilibration driven by H_d. However, when the initial TRM or partial TRM is below its saturation limit at T_0, the onset of re-equilibration during heating does not come immediately above T_0, but at some higher temperature, as explained above. Once walls begin to be driven back, subsequent demagnetization is described by (9.28) or (9.29).

All the above theory is predicated on field-blocking as the process by which TRM or partial TRM is acquired. In §9.4, we concluded that thermal fluctuations actually govern TRM acquisition for fields of geomagnetic strength. However, the re-equilibration model is actually independent of how the walls were originally blocked, until we heat to T_{Bf}, at which point any remaining wall displacements should disappear. Since T_{Bf} is itself a high temperature, thermal demagnetization tails are still a prominent feature of MD remanence, although the tail should be truncated somewhat below T_C. In practice, thermal demagnetization data tend to become noisy at high temperatures and the truncation is not usually obvious.

Some experimental data are shown in Fig. 9.7. Thermal demagnetization curves for total TRM's of small MD TM60 grains (average grain size: 6.4 μm) as a function of acquisition field H_0 (Fig. 9.7a) resemble theoretical curves (Fig. 9.6a), although n has to be increased to unrealistically high values to explain the 1 Oe data. The likely explanation is that these grains are too small to be 'truly' MD in behaviour. As well as wall pinning, SD or SD-like processes likely play some role in weak-field remanence (see Chapter 12). (Notice the resemblance between the weak-field curves in Fig. 9.7(a) and the corresponding curves for SD TM40 in Fig. 8.18.) Strong-field remanence is predominantly due to pinned displaced walls and follows MD theory quite well, assuming $n \approx 3$ (Fig. 9.7a).

Continuous thermal demagnetization of partial TRM's acquired by 3 μm magnetite grains over the range (400 °C, 350 °C) is independent of H_0, at least for the field range examined here (Fig. 9.7b). This is another prediction of the more detailed theory (Xu and Dunlop, 1994). A theoretical fit using (9.28) with $n = 4$ (solid curve) is unconvincing, however. The data actually show a combination of SD and MD character. About two-thirds of the partial TRM demagnetizes in SD-fashion when heated to just above $(T_B)_{max}$ $= 400\,°C$: $T_{UB} \approx T_B$. The remaining one-third displays a MD-style thermal

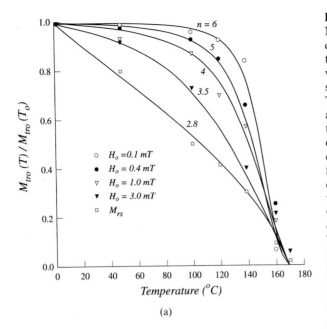

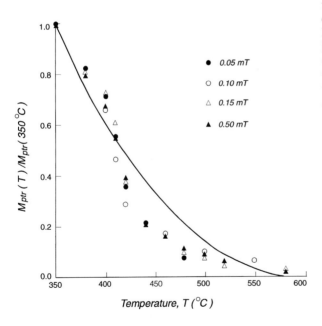

Figure 9.7 (a) Normalized thermal demagnetization curves of total TRM's acquired in various fields H_0 and of saturation IRM for 6.4 μm TM60 grains. Data are after Day (1977) and theoretical curves were calculated from the equivalent to eqn. (9.29) for an exponential distribution $f(h_c)$ and the n values shown. (b) Continuous thermal demagnetization of partial TRM's acquired in the range 400–350 °C in the field indicated; data from Worm *et al.* (1988). A sharp drop in remanence around 400 °C is interpreted to be due to unblocking of a SD-like component of pTRM. The long tail extending to T_C is a multidomain effect. MD theory (solid curve) can explain the tail, but not the lower-temperature data. [After Xu and Dunlop (1994) © American Geophysical Union, with the permission of the authors and the publisher.]

demagnetization tail extending to $T_C = 580\,°C$. Thus in 3 μm magnetite grains, which are much larger than the theoretical micromagnetic limit for metastable SD behaviour (§7.5.3), SD-like carriers are responsible for more than half the weak-field partial TRM. (The possible origin of these SD-like carriers will be the subject of Chapter 12.) Less than half the pTRM resides in pinned domain walls.

9.6. Acquisition of partial TRM and paleointensity determination

The Thellier double-heating procedure for determining paleofield intensity was described in §8.9. In the first heating to temperature T, which in the modified Thellier experiment now in general use is in zero field, a fraction of the NRM (which must be a TRM in order for the experiment to work) is thermally demagnetized. In the second heating, the sample is cooled in a laboratory field H_{lab} and acquires a partial TRM. In the case of SD grains, because $T_B = T_{UB}$ for grains of a particular size and shape, i.e., (V, H_{K0}), the partial TRM gained exactly replenishes the NRM lost if H_{lab} matches the paleofield H_a in direction and strength. If $H_{lab} \neq H_a$, which is the usual situation, the partial TRM gained can be calculated as the vector difference between the total remanences after the second and first steps, and the ratio ΔM_{nr} (the loss in NRM) to M_{ptr} gives H_a/H_{lab}, provided $M_{tr} \propto H_0$. The results are usually presented as a graph of NRM remaining versus partial TRM acquired, whose slope is $-H_a/H_{lab}$ (e.g., Fig. 8.14).

The success of the Thellier procedure is a result of the experimental laws of reciprocity, independence and additivity of partial TRM's acquired in non-overlapping temperature intervals by SD grains (§1.1.3, 8.8). MD grains do not obey these laws and will respond differently in a Thellier experiment. Reciprocity is not a feature of MD partial TRM: as we have seen, a partial (or total) TRM with blocking temperature T_B has not one but a spectrum of T_{UB}'s, spanning a broad temperature interval from below T_B to near the Curie point. Clearly different partial TRM's cannot be independent, since their unblocking ranges overlap. More fundamentally, the lack of independence is rooted in the fact that different partial TRM's are carried by the same wall, or set of walls, displaced by different amounts, whereas in SD grains, different partial pTRM's are carried by physically distinct grains. Finally, the law of additivity of partial TRM's, if it holds at all (e.g., Levi, 1979; McClelland and Sugiura, 1987), must be obeyed only in a statistical average sense, since the various partial TRM's are not independent of one another. Theoretically (Dunlop and Xu, 1994, Table 2), the sum of intensities of non-overlapping partial TRM's can vary from $\ll$ total TRM to $\gg$ TRM.

In the last section, we considered the simple case of a partial TRM acquired between T_C and T_B. In all other cases, the theory of partial TRM acquisition is mathematically intricate (Dunlop and Xu, 1994; Xu and Dunlop, 1994). However, much insight can be gained from our previous analysis (§9.4, 9.5). Partial TRM's acquired in high-temperature intervals are generally the most intense, because the energy barriers grow more during cooling and walls remain firmly pinned. Note that below T_B, the increase in wall pinning or h_c with cooling outweighs the increase in H_d, so that the walls, within their local wells, tend to be pushed slightly *away* from the demagnetized state. Partial TRM's acquired in lower-temperature intervals are less well stabilized by barrier growth and tend to re-equilibrate when $H_0 \to 0$ at T_0, with consequent loss of intensity. Thus most partial TRM is acquired at high temperatures. On the other hand, thermal demagnetization of a *total* TRM, which is the NRM in a Thellier experiment, may begin soon above room temperature. Comparing Figs. 9.6(a) and (b), it is clear that for the same field (here, 5 Oe), total TRM demagnetizes at lower average temperatures than the various partial TRM's.

Experimentally, NRM–pTRM graphs of MD magnetites vary from mildly to strongly convex downwards (Fig. 9.8a). Similar results, but over a more limited size range, were found by Levi (1977). Even 1 μm grains fall distinctly below the ideal single-domain line, indicating that more NRM is lost in low-temperature heating intervals than is replenished by partial TRM gained in the same intervals. The largest magnetites (135 μm average size) give results falling quite close to MD field-blocking theory. A plot of the NRM unblocking spectrum and the pTRM acquisition spectrum on the same graph (Fig. 9.8b; cf. Fig. 1.6 for SD grains) demonstrates that thermal demagnetization is biased toward lower temperatures and partial TRM acquisition toward higher temperatures.

Since the final partial TRM, cooled from T_C, reproduces the total TRM, the final point for all grain sizes in Fig. 9.8(a) returns to the ideal or SD line. Joining the first and last points will therefore give the correct paleointensity. However, irreversible chemical alteration of grains during high-temperature heatings also commonly leads to a sagging or convex-down shape of the Thellier graph (e.g., Fig. 8.14b). The usual procedure is to fit a straight line to the low-temperature points, below the assumed onset of alteration. Such a procedure would be disastrous with the data of Fig. 9.8(a). A best-fit line to the 11 lowest points for the 135 μm sample gives a paleointensity that is more than twice the true value.

How can one diagnose reversible MD behaviour as opposed to irreversible behaviour due to alteration of the grains? A widely used practical procedure is partial TRM checks, illustrated in Fig. 9.9. Following a particular double heating, one performs a third, zero-field heating to the same temperature, with the intent of erasing the partial TRM produced in the second heating. Then the partial TRM intensity at a lower temperature is checked by reheating to that temperature and cooling in field. If the partial TRM's check, no irreversible

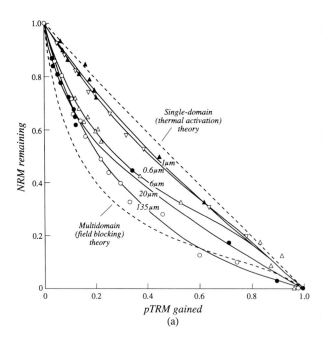

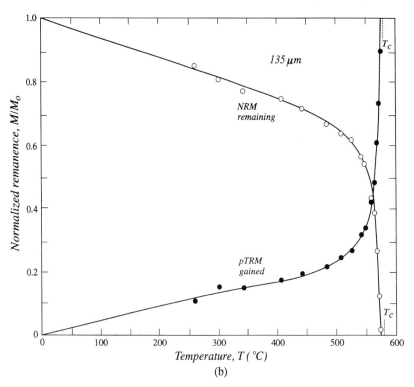

Figure 9.8 (a) Simulated Thellier paleointensity determinations for magnetites of various grain sizes. All the data sag below the ideal SD line. Data for the largest magnetites are close to the predictions of MD field-blocking theory. (b) Comparison of experimental pTRM acquisition and NRM unblocking spectra of 135 μm MD grains, showing that NRM unblocking is biased toward low temperatures and pTRM blocking toward high temperatures. For SD grains (Fig. 1.6), the two curves are exactly reciprocal. [After Xu and Dunlop (1995a) © American Geophysical Union, with the permission of the authors and the publisher.]

alteration has occurred up to the highest temperature reached. We might first point out that there is no compelling reason to believe that partial TRM checks should work for MD grains. The third heating will not completely erase the second-step partial TRM: it must be heated close to T_C to be removed completely. An indication of this is seen in Fig. 9.9 in the 558 °C–484 °C and 508 °C–409 °C checks: following the third (zero-field) heatings to 558 °C or 508 °C, about 10% of the pTRM remains undemagnetized. Nevertheless, the lower-temperature pTRM's reproduce reasonably well in every case.

NRM demagnetization and pTRM acquisition tend to begin at higher temperatures in MD grains than in SD grains. More than two-thirds of the demagnetization/acquisition in Fig. 9.9 occurs between 500 °C and 580 °C and more than half occurs above 560 °C. However, this is no more than a qualitative indi-

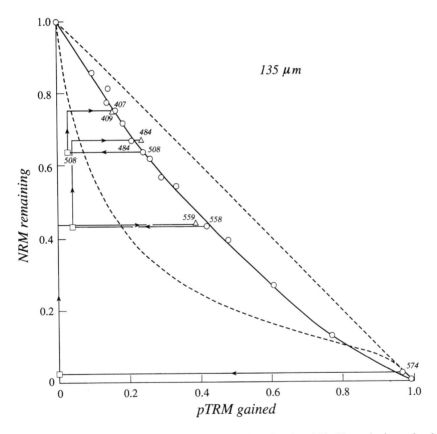

Figure 9.9 Successful pTRM checks in a simulated Thellier paleointensity determination on 135 µm MD grains of magnetite. The dashed theoretical curves are the same as in Fig. 9.8(a), but the data lie closer to the ideal SD line because pTRM's were produced perpendicular to NRM, rather than parallel to NRM. [After Xu and Dunlop (1995a) © American Geophysical Union, with the permission of the authors and the publisher.]

cator. Single-domain grains can also have concentrations of very high blocking/unblocking temperatures. Furthermore, chemical alteration tends to occur at the highest temperatures and often produces large amounts of new magnetic material.

9.7 AF demagnetization and the microcoercivity spectrum

Alternating-field (AF) demagnetization and the microcoercivity spectrum of single-domain TRM were described in §8.11. There is no satisfactory equivalent theory for multidomain TRM, although an adequate theory does exist for saturation remanence (§11.5). For MD grains, the microcoercivity spectrum describes the distribution of barriers to wall motion. The MD spectrum is much softer than the SD spectrum because domain walls are generally much less strongly pinned than are single-domain moments. In addition, H_d in MD grains assists in the demagnetizing process.

The internal demagnetizing field is largely responsible for the characteristic exponential form of the AF demagnetization curve of MD materials. Because magnetocrystalline anisotropy provides a minimum rotational barrier, the single-domain coercivity spectrum has a lower cut-off field (about 10 mT or 100 Oe for magnetite), resulting in an initial plateau in the AF demagnetization curve. As a consequence, AF cleaning in quite modest fields eliminates much of the NRM carried by multidomain grains while scarcely affecting single-domain NRM. In view of the persistent thermal demagnetization tail in MD grains, AF rather than thermal cleaning is the sensible procedure for randomizing MD remanence.

9.8 Domain nucleation and TRM

Bol'shakov and Shcherbakova (1979) were among the first to measure thermal demagnetization tails extending to T_C for MD magnetites. They interpreted their observations as indicating a series of unblocking temperatures at which trapped domain walls became unpinned and reorganized their configurations in response to the internal field, as we have done in previous sections. However, there is growing evidence that the reorganization of walls may be in part a response to another process, the nucleation or denucleation of domains during heating or cooling.

McClelland and Sugiura (1987) made continuous measurements, during cooling and reheating, of intensity changes in high-temperature partial TRM's carried by MD magnetites ranging in size from 10–15 μm to 100–150 μm. In zero-field cooling below the lower pTRM acquisition temperature the pTRM intensity increased less rapidly than $M_s(T)$ and, in the case of the 100–150 μm magnetite, actually decreased. The normalized magnetization $r(T) = M(T)/M_s(T)$, which measures the net relative displacement x/a of all the walls, *decreased* with cooling, particularly in the larger grains, rather than remaining constant or blocked (Fig. 9.10). The domain walls must be adjusting towards a demagnetized state below the minimum 'blocking' temperature of the pTRM. The decrease is not simple viscous decay (Chapter 10): no change was seen if the sample was held at a constant temperature. Similar observations were reported by Sugiura (1981) and Markov *et al.* (1983).

In reciprocal experiments, McClelland and Sugiura traced the growth of total or partial TRM's during in-field cooling to T_0 (Fig. 9.10). The increase in moment, measuring the response of domains and walls to H_0, continued down to ≈400 °C, showing that walls can still be driven in the face of an internal demagnetizing field $H_d \gg H_0$ that will force the same walls back if H_0 is switched off. There is a remarkable symmetry between the in-field cooling curves and the zero-field cooling curves (dashed) of pTRM's below their minimum T_B's. For example, the 580–560 °C and 560 °C–T_0 cooling curves are nearly mirror images,

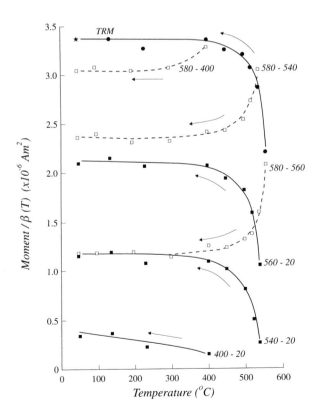

Figure 9.10 Continuous in-field (solid) and zero-field (dashed) cooling curves of TRM and of partial TRM's acquired between the temperature limits shown. Measurements were made continuously at high temperature but the data have been reduced to equivalent room-temperature moment by removing the reversible $M_s(T)$ variation. The curves therefore represent real gains or losses of magnetization. [Reprinted from McClelland and Sugiura (1987), with kind permission of the authors and Elsevier Science – NL, Sara Burgerhartstraat 25, 1055 KV Amsterdam, The Netherlands.]

as are the 580–540 °C, 540 °C–T_0 and 580–400 °C, 400 °C–T_0 pairs of curves. This reciprocity would be very much in the spirit of the single-domain Thellier laws, except that here remanence is being erased in zero-field *cooling*, not heating.

Domain observations made during heating and cooling by Metcalf and Fuller (1987b), Heider *et al.* (1988), and Halgedahl (1991) (Figs. 6.12 and 6.13), give some clues to the underlying mechanisms. In ≥ 30 μm magnetite crystals heated in the small residual field of the microscope (≈ 20 μT or 0.2 Oe) to only 120 °C, walls unpinned and moved substantially in response to the change in dominant anisotropy from crystalline to magnetoelastic (Fig. 6.12) . In addition, walls denucleated during heating and nucleated during cooling. A spike domain nucleated at a corner on the surface of one of the body domains and spread across the crystal through opposite displacements of its bounding walls until its tip reached the other edge of the crystal. The bounding walls then separated and straightened, splitting the existing domain in two and adding a central domain of reverse polarity. This transverse nucleation process of domain splitting or sandwiching increases the number of domains and walls by two not one. During denucleation, a lamellar domain transformed into a large spike when its planar bounding walls bent and joined at one edge of the crystal. Small wall displacements drove this spike transversely across the crystal, destroying the domain. Although a single lamellar domain could in principle nucleate or denucleate at one side of the crystal (edge or longitudinal nucleation: a domain stacking instead of a splitting), the addition or subtraction of a pair of walls by a spike traversing the crystal seemed to be typical. Most of the observed domain changes in TM60 followed the same pattern (Fig. 6.13), which can be likened to a 'magnetic dislocation' spreading across the grain.

A further point to note is that when a new domain appeared, splitting an existing domain in two, all the walls moved so as to maintain roughly equal spacing. The reason for this adjustment was made clear in Dunlop *et al.*'s (1990a) calculations of internal field distributions. Domains of unequal size had average demagnetizing fields that were different by a factor of 2 or more.

The observations can be summarized as follows.

1. Walls move readily, even in weak fields at low temperatures.
2. Domain-wall nucleation/denucleation by transverse spike motion seems to be easy in > 30 μm grains of magnetite and TM60.
3. Wall nucleation/denucleation can occur far below T_C.
4. Following nucleation/denucleation, all the walls adjust their positions in response to the altered internal demagnetizing field.

McClelland and Sugiura (1987) based the interpretation of their TRM and pTRM data on two main observations. First, remanences seem to change in exactly the same way when cooled below a given temperature T, whether a field H_0 is switched on or off at T, except that the moment increases in the first case and decreases in the second (Fig. 9.10). In the simplest approximation, there are just two equilibrium states, magnetized and demagnetized. As a result of the change in conditions at T (switching the field on or off), the system is left in an out-of-equilibrum state from which it relaxes, given sufficient energy to overcome the barrier between states. Second, a total TRM, which experiences no change in H_0 at any temperature, is almost constant in cooling from $\approx 500\,°C$ to T_0. Partial TRM's, on the other hand, increase or decrease considerably over this range, and the lower the T at which H_0 is switched on or off, the greater the percentage change (Fig. 9.10). Total TRM is interpreted to maintain an equilibrium magnetized state, achieved very close to T_C, throughout its cooling. Switching H_0 on or off at lower and lower temperatures leaves the system farther and farther from its equilibrium state, requiring greater changes in moment.

It is clear in Fig. 9.10 that the bulk of the increase or decrease in moment occurs at or just below the temperature at which field conditions changed. Presumably this reflects the movement of soft walls in response to the changed internal field. The real puzzle is why the system continues to evolve towards equilibrium with further cooling, since barriers are growing and less and less thermal energy is available to overcome them. The answer must be that domain nucleation during cooling is responsible for altering the internal fields and allowing the walls to re-equilibrate. The larger the grain, the more the equilibrium number of domains is predicted to change during cooling (Moskowitz and Halgedahl, 1987; see Fig. 7.11). Thus larger grains have more opportunities to reach the equilibrium magnetic state (TRM or demagnetized) and their moments change the most with cooling.

Although the concepts in this theory are not yet fully quantified, it does offer hope of explaining the mechanism of TRM in a system where all the walls interact and no wall can be blocked independently of the others. It provides an explanation for the strongly skewed distribution of pTRM intensities towards high blocking temperatures, since changes towards the fully magnetized TRM state are thermally activated and occur most readily at high temperature.

9.9   Nucleation failure and TRM

A contrary view of nucleation during cooling is presented by Halgedahl and Fuller (1983) and Halgedahl (1991). Halgedahl found that titanomagnetite grains repeatedly given TRM's under identical conditions achieved a wide variety of domain multiplicities whereas the same grains had only a few possible domain states after AF demagnetization. In the example shown in Fig. 5.14, the

grain usually contained three to nine domains after being given TRM but had
four to six domains after AF demagnetization. Domain nucleation/denucleation
was easy in AF demagnetization, resulting in a narrow distribution of states
about the equilibrium or GEM state. But judging by the wide variety of LEM
states preserved in TRM acquisition, the same grain was often unable to nucleate
or denucleate enough walls during TRM acquisition to achieve the low-energy
AF state and frequently remained trapped far from the GEM state. Domain
nucleation must have been relatively *difficult during cooling*. That is, *nucleation
failure*, not *easy nucleation*, is implied, contrary to the ideas of McClelland and
Sugiura (1987) and Shcherbakov *et al.* (1993).

Sometimes a grain fails to nucleate any walls at all as it cools from its saturated
state near T_C. Such metastable single-domain grains were described in Chapters
6 and 7 and will be discussed further in Chapter 12. Halgedahl's observations
imply that these states are special and not part of the overall LEM state distribution. Notice in Fig. 5.14 that no 2D states were observed. There is a gap in the distribution between SD and 3D or higher states. This gap is a general feature of
other grains as well.

Metcalf and Fuller (1988) synthesized the field dependence of TRM intensity
for 2–50 μm TM60 grains in an oceanic basalt on the assumption that the
moments of metastable SD grains outweigh all other sources of remanence (Fig.
9.11). The numbers of metastable SD grains observed under the microscope
after successive field coolings increased as the field H_0 increased. By estimating

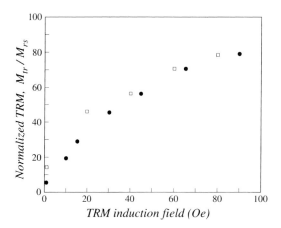

Figure 9.11   TRM acquisition by 2–50 μm TM60 grains as a function of field
strength H_0 (squares), synthesized by dividing the total volume of metastable SD
grains observed under the microscope at each H_0 by the corresponding metastable
SD volume in the saturation IRM state. Dark points are TRM intensities for a
single-domain magnetite sample measured by Dunlop and West (1969). [After
Metcalf and Fuller (1988) © American Geophysical Union, with the permission of
the authors and the publisher.]

the total volume of such grains relative to their total volume in the saturation remanence state, they generated a reasonable fit to $M_{\mathrm{tr}}(H_0)$ data of Dunlop and West (1969) (which were actually measured for SD magnetite). Thus even in quite large grains, TRM may have an SD source.

Chapter 10

Viscous and thermoviscous magnetization

10.1 **Introduction**

Viscous magnetization, or magnetic viscosity, is the gradual change of magnetization with time at constant temperature. Useful reviews are given by Dunlop (1973c) and Moskowitz (1985). Viscous magnetization is a thermally activated phenomenon and is governed by the equations derived in Chapters 8 and 9:

$$M(t) = M_{eq} + (M_0 - M_{eq})e^{-t/\tau} = M_0 e^{-t/\tau} + M_{eq}(1 - e^{-t/\tau}), \qquad (10.1)$$

where $M_0 = M(0)$ and $M_{eq} = M(\infty)$. The relaxation time τ for SD grains is

$$\frac{1}{\tau} = \frac{1}{\tau_0} \exp\left[-\frac{\mu_0 V M_s H_K}{2kT} \left(1 - \frac{|H_0|}{H_K} \right)^2 \right], \qquad (10.2)$$

while

$$\frac{1}{\tau} = \frac{1}{\tau_0} \exp\left(-\frac{c\mu_0 V_{Bark} M_s H_c}{kT} \right) \qquad (10.3)$$

for thermally activated wall jumps in MD grains. Following a change in H_0, a TRM or any other remanent magnetization will, according to eqn. (10.1), spontaneously relax from its initial value M_0 to its new equilibrium value M_{eq} within a time t comparable to τ.

In most rocks, the bulk of the SD grains have $VH_{K0} \gg C(T_0, t, H_0)$ (see eqns. (8.32), (8.33)), so that $\tau \gg t$ at ordinary temperatures even for times of geological length. If this were not so, paleomagnetic remanences could not survive changes in the geomagnetic field after their formation. However, there is always a fraction of grains just above superparamagnetic size that have VH_{K0}

$\approx C(T_0, t, H_0)$ or $\tau \approx t$. The initial remanent magnetization of these grains is gradually replaced by a **viscous remanent magnetization** (VRM) in the direction of H_0.

It is commonly observed that magnetization does not change exponentially with time, as (10.1) predicts, but more or less logarithmically with time (e.g., Street and Woolley, 1949; Le Borgne, 1960a). The reason for this is the broad range of grain sizes and shapes in any real sample, as described by the grain distribution $f(V, H_{K0})$ (§8.8, 8.10). At any instant t after a field change, the initial magnetization of ensembles with $\tau \ll t$ relaxes immediately, while M_0 of ensembles with $\tau \gg t$ remains almost perfectly blocked. Only ensembles with $\tau \approx t$ display a sluggish or viscous change in magnetization in the short interval of observation centred on t. Now as t increases, the 'unblocking window' moves gradually through the relaxation time spectrum determined by $f(V, H_{K0})$. Since linear changes in V or H_{K0} result in logarithmic changes in τ (eqn. (10.2)), and V and H_{K0} are fairly uniformly distributed over small intervals, to a first approximation magnetization changes are proportional to log t. Figure 10.1 illustrates how a distribution of grains, each with exponential viscosity, collectively exhibit a log t time dependence. Experimentally, log t behaviour is so common that viscous changes in M are often described by a viscosity coefficient $S = \partial M / \partial \log t$.

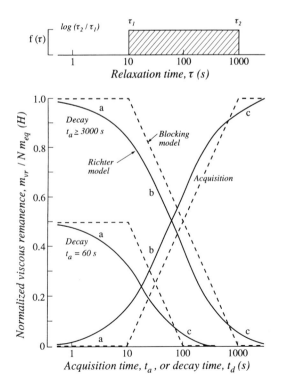

Figure 10.1 Acquisition and decay of viscous magnetization for a narrow but uniform distribution $f(\log \tau)$. If $t_a \gg \tau_{max}$, the superposition of exponential relaxations of grains with different τ produces log t acquisition and decay curves, with identical slopes or viscosity coefficients S, for both the exact theory (Richter, 1937; eqn. (10.4)) and the blocking approximation of eqn. (10.8), except at very short and very long times. If $t_a \ll \tau_{max}$, the underlying exponential nature of decay is evident even for intermediate times. [After Dunlop (1973c) © American Geophysical Union, with the permission of the author and the publisher.]

In nature, rocks are exposed to a field of fairly constant direction and variable intensity, but the polarity of the field reverses a few times every million years (Figs. 1.4, 1.5), resetting the magnetic system. Kok and Tauxe (1996) have pointed out that following a field reversal, a fraction of the pre-reversal remanence of opposite polarity will decay viscously, giving rise to a 'sawtooth' or ramp-like pattern in field intensity records. Figure 1.5 does show such ramp-like descents following the two reversals bounding the Jaramillo event (time t runs from right to left). It is presently uncertain how much of this pattern is real geomagnetic field behaviour and how much is a viscous artefact.

The NRM of virtually all rocks contains a low-temperature VRM acquired during the present Brunhes normal epoch, the product of **viscous remagnetization** by the geomagnetic field over the last 0.78 Ma and carried by grain ensembles with τ less than ≈ 1 Ma. **Viscous decay** of remanence under zero-field conditions, although frequently investigated in the laboratory, is never found in nature. Brunhes-epoch VRM obscures older NRM's. Its removal is usually the main objective of laboratory cleaning methods like AF or thermal demagnetization. Such VRM is easily recognized because it roughly parallels the present geomagnetic field. It is usually 'softer' or more easily cleaned than older NRM's of TRM, DRM (depositional remanence; Chapter 15) or CRM (crystallization or chemical remanence; Chapter 13) origin. However, we shall find that VRM in magnetically hard minerals like hematite, pyrrhotite and iron is not easily AF demagnetized.

Slowly cooled plutons and rocks reheated orogenically or by burial in sedimentary or volcanic piles acquire viscous partial thermoremanent magnetization (VpTRM) (see Chapter 16). VpTRM blurs the distinction between purely thermal and purely viscous magnetization or remagnetization. Cooling times of millions or tens of millions of years encompass several polarity reversals. VpTRM is thus potentially a bipolar remanence of reduced net intensity. Through careful thermal demagnetization of very slowly cooled rocks, it may be possible to recover, from different blocking-temperature fractions of VpTRM, a geomagnetic polarity record (e.g., Rochette *et al.*, 1992a), or even to detect apparent polar wander due to plate motion. Unless acquired since the late Mesozoic, VpTRM does not approximately parallel the present geomagnetic field, nor is it easily AF or thermally cleaned. On the plus side, it constitutes a paleomagnetic signal rather than noise, although dating the time of remagnetization is a difficult problem.

10.2   Experimental properties of viscous magnetization

10.2.1   Size effects in magnetic viscosity

Room-temperature viscosity coefficients S_a for the growth of viscous induced magnetization (VIM) and S_d for the zero-field decay of VRM in magnetite are shown in fig. 10.2. The data cover four decades of grain diameter or 12 decades of grain volume. There are only two size ranges in which magnetic viscosity rises above background levels: in 20–40 nm SD grains just above room-temperature SP size and perhaps in ≈ 5–10 μm grains, presumably as a result of thermally excited displacements of small wall segments (see also Zhilyaeva and Minibaev, 1965).

The narrowness of the windows in τ or V that permit viscous relaxation at a particular $(T,\ t)$ poses a severe experimental problem. Only a minor fraction of even the most carefully sized synthetic grains is viscous at T_0 and ΔM is invariably a small fraction of the TRM that can be blocked thermally by the same sample. An even smaller fraction of the broad size spectrum of a typical rock is viscous on a laboratory time scale. Some of the scatter in Fig. 10.2 may result from contamination of samples by viscous ultrafine grains, since the > 1μm grains were obtained by crushing large crystals. Viscosity measurements are needed on well-sized crystals grown by modern techniques, such as hydrothermal recrystallization (Heider *et al.*, 1987), precipitation in a glass-ceramic matrix (Worm and Markert, 1987), and flux growth (Smith, 1988).

The viscous decay process is reciprocal to the acquisition process but the initial state is more complicated. In an acquisition experiment starting from $M_0 = 0$ and lasting a time t_a, ensembles with $\tau \approx t_a$ will be only partially

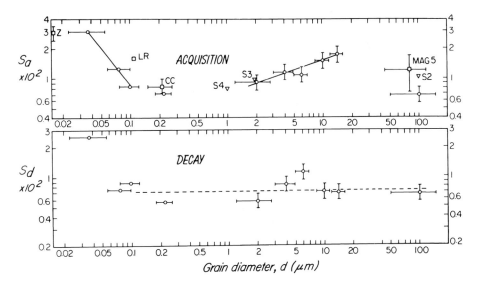

Figure 10.2 Viscosity coefficients for acquisition (S_a) and decay (S_d) of viscous magnetization in magnetite, in units of kA/m or emu/cm^3, as a function of grain size. S2, S3, S4 are data from Shimizu (1960); Z, LR, CC, MAG5 are from Tivey and Johnson (1981, 1984); all other data are from Dunlop (1983b). [After Dunlop (1990), with the permission of the author and the publisher, Institute of Physics Publishing, Bristol, UK.]

activated. Their initial state for decay will be $M(t_a) = pM_{eq}$, where $p = 1 - \exp(-t_a/\tau)$ from eqn. (10.1). A fraction p of this initial M will decay in time $t_d = t_a$, so that $S_d = -pS_a$ for such ensembles. Overall we expect S_d to be slightly less than S_a, the exact difference depending on $f(V, H_{K0})$, and this is seen in the data (Fig. 10.2).

Magnetic viscosity measurements can be extended to very short times by measuring the dependence of initial susceptibility χ on the frequency f of an applied alternating field. A comparison of susceptibilities at two fixed frequencies is often used in studies of soils and sediments to detect superparamagnetic (in reality, nearly superparamagnetic or viscous) grains. For soils containing SD magnetite and maghemite, $\partial\chi/\partial\log f = \text{const.}$ (Mullins and Tite, 1973). No frequency dependence was detectable for MD magnetite.

10.2.2   Temperature dependence of magnetic viscosity and susceptibility

Magnetic viscosity, like any thermal activation phenomenon, increases as the temperature rises (Fig. 10.3). Viscous magnetization rises to a dramatic peak 50–100 °C below T_C (Pozzi and Dubuisson, 1992). The cause is rapid decreases in M_s and H_K or H_c, and for MD grains also in V_{Bark} (Bina and Prévot, 1994), which cause τ to plummet (eqns. (10.2), (10.3)). For the same reason, this is also the principal range of TRM blocking/unblocking temperatures and of enhancement of susceptibility or short-term induced magnetization (the Hopkinson effect; Fig. 10.4).

The expected increase in viscous acquisition and decay with rising temperature is observed for magnetite and TM60 (Fig. 10.3; Kelso and Banerjee, 1994) and also for pyrrhotite (Dubuisson *et al.* 1991). In Fig. 10.3, viscous changes level off above 500 °C (80 °C below T_C) in magnetite and decrease at 100 and 130 °C (just below T_C) in TM60. These decreases in viscosity indicate that most values of τ have dropped to such an extent that they are shorter than the times used in the laboratory experiments, i.e., most grains are above their blocking temperatures and behave superparamagnetically. These SP grains will have high induced magnetization or susceptibility (Fig. 5.20).

As expected, experimental Hopkinson peaks in magnetic susceptibility χ occur just above the peaks in magnetic viscosity (Dubuisson *et al.*, 1991) and in the pTRM blocking-temperature spectrum (Fig. 10.4). The Hopkinson peak can be large: the SP susceptibility of SD grains above their unblocking temperatures is about two orders of magnitude higher than the susceptibilities of the same grains when thermally blocked (§5.7.2).

The high-temperature enhancement of both 'non-viscous' and viscous magnetizations (which in reality are the short-time and long-time ends of a continuous spectrum, rather than physically distinct phenomena) is potentially important in explaining regional magnetic anomalies with deep crustal sources near the

Synthetic magnetite (d = 0.076 μm)

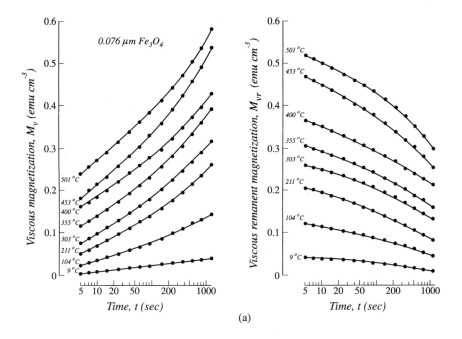

(a)

Submarine basalt (contains Fe$_{2.4}$ Ti$_{0.6}$ O$_4$)

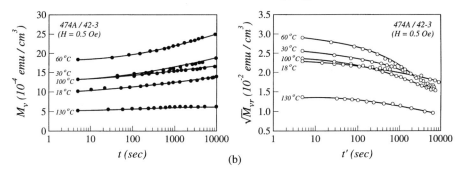

(b)

Figure 10.3 Acquisition and decay of viscous magnetization at various temperatures for synthetic fine-grained (SD) magnetite and a submarine basalt containing coarse-grained (MD) TM60. Both acquisition and decay accelerate compared to a log t function. Viscosity coefficients increase with rising temperature until just below T_C, when most grains become SP and viscosity drops. [After Dunlop (1983b) and Bina and Prévot (1994), with the permission of the authors and the publisher, Blackwell Science, Oxford, UK.]

Curie-point isotherm (see §1.2.2). However, MD grains generally have a much more muted Hopkinson peak than SD grains (Clark and Schmidt, 1982; Soffel *et al.*, 1982, Shive and Fountain, 1988; Bina and Henry, 1990), because the internal demagnetizing field limits χ to values $\leq 1/N$ at all temperatures (§5.6.2,

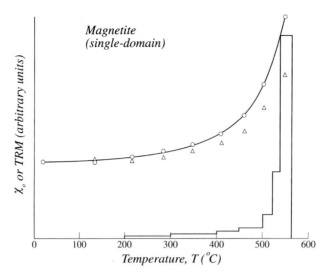

Figure 10.4 The Hopkinson effect or high-temperature peak in measured initial susceptibility of a single-domain magnetite. The theoretical SD susceptibility, $\chi = 2M_s/3H_K$ (eqn. (8.10)), (shown as triangles, calculated from measured M_s and remanent coercive force data), and the spectrum of partial TRM's both peak at similar temperatures. All are manifestations of remanence unblocking. [After Dunlop (1974), with the permission of the author and the publisher, Deutsche Geophysikalische Gesellschaft, Köln, Germany.]

9.2.2). Since MD material is likely to dominate the induced magnetization of most deep crustal rocks, Hopkinson-type enhancement is probably a partial explanation at best of the large regional and Magsat anomalies observed (Dubuisson *et al.*, 1991; Pozzi and Dubuisson, 1992).

At intermediate temperatures, viscous changes accelerate compared to a log *t* relationship, so that acquisition and decay curves take on a convex-down or convex-up aspect when plotted on a logarithmic time scale (Fig. 10.3). The convexity is more pronounced when many decades of *t* are covered in the experiment (Smith, 1984; Tivey and Johnson, 1984; Moskowitz, 1985) or for inherently more viscous materials (Özdemir and Banerjee, 1981; Tivey and Johnson, 1981; Dunlop, 1983b).

10.2.3   Magnetic viscosity of TM60 and oceanic basalts

There has been considerable debate about the source of strong VRM's in oceanic basalts containing TM60 as their main magnetic constituent (e.g., Lowrie and Kent, 1978). Both oxidized fine grains and unoxidized coarse grains have unusually large viscosity. Dunlop and Hale (1977), Smith (1984) and Hall *et al.* (1995) showed that viscous growth and decay coefficients increase with time on a time scale of a few seconds to a few hours. They increase even more with heating to $\approx 100\,^\circ$C, which is just below the TM60 Curie point and is similar to temperatures reached at shallow (a few km) depths in the oceanic crust. The practical importance of these observations is that if the high viscosity coefficients persisted, virtually all primary TRM in the upper oceanic crust would be viscously overprinted within ≈ 1 Ma after any change in field, about the typical duration of one polarity epoch. Lowrie and Kent, however, found in much longer experiments that viscosity coefficients levelled off and began to decline after a few days.

In fine ($< 0.1\,\mu$m) SD-size synthetic TM60 particles, oxidized to varying degrees to titanomaghemites, Özdemir and Banerjee (1981) showed that high VRM is restricted to high oxidation states (Fig. 10.5a). When the oxidation parameter z increased from 0.4 to 0.9, the VRM growth coefficient increased tenfold (Fig. 10.5b), while coercive force decreased by a similar factor. The explanation seems to be an increase in the superparamagnetic or critical blocking volume V_B (eqn. (8.26)) and an accompanying decrease in effective coercivity H_c (eqn. (8.28)) as M_s decreases with oxidation. No grain cracking was detected.

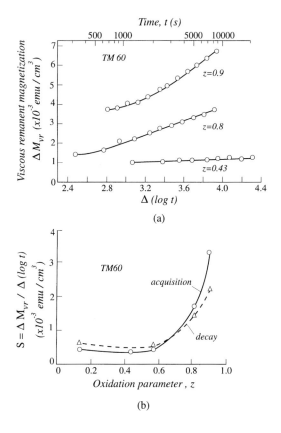

(a)

(b)

Figure 10.5
(a) Acquisition of room-temperature VRM by three fine-grained titanomaghemite samples. (b) Viscous acquisition and decay coefficients, S_a and S_d, as a function of oxidation parameter z for titanomaghemites with $x = 0.6$. [After Özdemir and Banerjee (1981) © American Geophysical Union, with the permission of the authors and the publisher.]

Somewhat contradictory results were obtained by Moskowitz and Banerjee (1981) on $\approx 1\,\mu m$ size oxidized synthetic TM60 particles. Viscosity coefficients increased as z increased from 0 to 0.4 but declined with further oxidation. The difference in results between the two sets of experiments must be linked to the different grain sizes of the starting material. Coarser grains tend to crack during oxidation, and Moskowitz and Banerjee noted that their natural TM60 samples were moderately to highly cracked. M_{rs}/M_s remained constant at SD values for all of Özdemir and Banerjee's samples but decreased with oxidation for Moskowitz and Banerjee's samples, implying a decrease in grain size or possibly multiphase oxidation (phase-splitting to magnetite+ilmenite; §3.1).

10.2.4 Initial states and multidomain viscous magnetization

In multidomain magnetite, viscosity coefficients depend strongly on the sample's magnetic history. Tivey and Johnson (1981, 1984) found an order-of-magnitude decrease in the ability of very coarse MD grains to acquire VRM after their initial TRM was AF demagnetized. However, thermal demagnetization increased the acquisition coefficient. The effect was absent for SD magnetites. Halgedahl and Fuller's (1981) observations of Kittel-type lamellar domains in pyrrhotite after AF demagnetization but undulating, easily moved domain walls in the TRM or thermally demagnetized state are relevant here. Loosely-pinned wall segments presumably account for the high VRM capacity after thermal treatment. When these are eliminated by AF cleaning, the viscosity will drop. Lowrie and Kent (1978) documented a substantial decrease in VRM acquisition after the NRM's of submarine basalts containing coarse TM60 grains were AF demagnetized.

Lowrie and Kent's results are surprising because MD grains in any initial state, whether after TRM acquisition or thermal or AF cleaning, gradually lose their ability to acquire VRM when stored for increasing times in zero or weak fields (Plessard, 1971; Tivey and Johnson, 1981, 1984). It has been widely assumed that this 'wait-time' or Δt effect on VRM has the same origin as disaccommodation of short-term induced magnetization (decrease in susceptibility with wait time), namely diffusion of lattice defects to positions where they anchor walls. Moskowitz (1985) observed no correlation between measured disaccommodation and viscosity coefficients in natural and synthetic TM60 samples, and concluded that the case for diffusion after-effect is not compelling, particularly since SD grains also display a Δt effect (Tivey and Johnson, 1984).

Halgedahl (1993) discovered spectacular differences in the continuous thermal demagnetization behaviour of $1–2\,\mu m$ magnetite grains, depending on the initial state from which high-temperature VRM was acquired (Fig. 10.6a). VRM acquired from an AF demagnetized state demagnetized to the 5% level $< 50\,^{\circ}\mathrm{C}$ above the acquisition temperature of $225\,^{\circ}\mathrm{C}$. VRM acquired after ther-

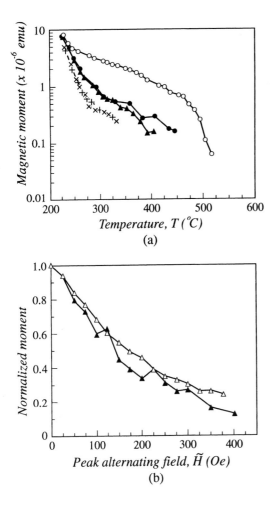

Figure 10.6
(a) Continuous thermal demagnetization curves of VRM's acquired by 1–2 μm glass-ceramic magnetites at 225 °C for 3 hr in $H_0 = 0.2$ mT starting from three different initial states, AF (crosses), TH (solid points) and TC (open circles) (see text for explanation of the states). VRM with an AF or TH initial state demagnetizes with ≤ 100 °C heating, but VRM with a TC initial state demagnetizes very slowly and has a tail of > 500 °C unblocking temperatures. (The effect does not appear with stepwise thermal demagnetization: see Fig. 10.11b) (b) Stepwise AF demagnetization at room temperature of VRM's similar to those of (a) with AF (open triangles) and TC (solid triangles) initial states. [After Halgedahl (1993) © American Geophysical Union, with the permission of the author and the publisher.]

mal demagnetization to T_C followed by zero-field heating to 225 °C (thermally heated or TH state) were slightly more resistant, requiring ≈ 100 °C heating for a similar VRM demagnetization level. But VRM acquired after zero-field heating to T_C and cooling to 225 °C (thermally cooled or TC state) demagnetized very slowly, with a demagnetization tail like that of MD pTRM (Fig. 9.7b) extending up to and above 500 °C. AF demagnetization at room temperature (Fig. 10.6b) revealed no obvious difference between the AF and TC states, however. Nor did viscous acquisition and decay curves at 225 °C after various zero-field wait times. And as we shall see later in this chapter, *stepwise* thermal demagnetization of high-temperature VRM, where measurements are made after cooling to T_0, reveal no great difference among AF, TH and TC initial states even in quite large (135 μm) magnetites (Fig. 10.11b).

10.3   Theory of single-domain VRM

10.3.1   Exact theory of viscous magnetization

The formal equation for acquisition of viscous magnetic moment m by N single-domain grains from a demagnetized initial state is

$$m(t) = \int\int NVM_{eq}(1 - e^{-t/\tau})f(V, H_{K0})dV \, dH_{K0}. \tag{10.4}$$

A similar equation describes viscous decay. If $f(V, H_{K0})$ is known (which is seldom the case), $m(t)$ can be found by numerical integration. For special forms of $f(V, H_{K0})$, analytic solutions are possible.

Walton (1980) assumed that the range of grain shapes, and thus H_{K0}, is much less than the range of V. Approximating M_{eq} (eqn. (8.2) or (8.11)) by $VM_s^2H_0/kT$, since H_0 is assumed to be small, (10.4) gives for the acquisition coefficient $S_a = \partial M/\partial\ln t = t(\partial M/\partial t)$

$$S_a = \int \frac{M_s^2 H_0}{kT}\frac{t}{\tau}(1 - e^{-t/\tau})NV^2f(V)dV. \tag{10.5}$$

The integrand is of the form $x(1 - e^{-x})$, with x itself exponential (eqns. (10.2), (10.3)). It peaks sharply at $x = 1$ or $t = \tau$, which is the blocking condition. All non-exponential factors can be assigned their blocking values and taken outside the integral. For example, $V \rightarrow V_B = 2kT \ln(2t/\tau_0)/\mu_0 M_s H_K$ (eqn. (8.26)). Then, assuming shape anisotropy, so that $H_K \propto M_s$

$$S_a \propto H_0\left(\frac{T}{M_s^2}\right)^2\left[\log\left(\frac{t}{\tau_0}\right)\right]^2 \tag{10.6}$$

if $f(V)$ is uniform over the range of ensembles affected in time t_a. If $f(V) \propto 1/V$, as for the larger grains in a log-normal distribution,

$$S_a \propto H_0\frac{T}{M_s^2}\log\left(\frac{t}{\tau_0}\right). \tag{10.7}$$

Either equation predicts a $\log t$ dependence for VRM intensity for times such that $\log t \ll -\log \tau_0 \approx 21$ (§8.5), i.e., for a few decades of t, as observed. Deviations from $\log t$ dependence will become obvious even at short times for very viscous samples and at longer times for all samples. Deviations will appear earlier if eqn. (10.6) applies rather than eqn. (10.7).

These equations also predict that VRM intensity is proportional to field strength H_0 in the weak-field range and that VRM acquisition (and decay) coefficients increase non-linearly as $T/M_s{}^2(T)$ to some power at high temperatures. Proportionality between VRM and field is well established experimentally. The inherent temperature dependence of viscosity coefficients tends to be obscured by the variability of the grain distributions, very different parts of which are sampled by the VRM 'window' at different temperatures. However, a good theoretical match to high-temperature data for SD magnetite (Fig. 10.7) was obtained by Walton (1983) using the known $f(V)$ of the sample (Dunlop, 1973a) and numerically integrating eqn. (10.4).

10.3.2 Blocking theory of VRM

Such success gives us some confidence in modeling viscous remagnetization beyond the range of laboratory experimentation. We would like to be able to predict how much VRM accumulates during the Brunhes epoch, for example, based on the measured magnetic properties of a particular rock. Clearly no laboratory zero-field storage test or VRM acquisition experiment can reproduce conditions in nature over a 1-Ma time scale. However, two factors work in our favour. First, as we mentioned earlier, eqns. (10.2) and (10.3) tell us that linear changes in variables like V, H_{K0}, T or H_0 are equivalent to *logarithmic* changes in τ and thus, under blocking conditions, in t. More than one-half the scale of $\log t$ from $\tau_0(\approx 10^{-9}$ s) to 1 Ma is accessible in the laboratory. Secondly, even the inaccessible part of the scale can be activated in mild heatings, well below the Curie point, because τ is such a strong function of T (cf. Fig. 8.10b).

Let us elaborate on the first point. Because we will be examining viscous changes over many decades of t or τ, the exact but rather cumbersome equations

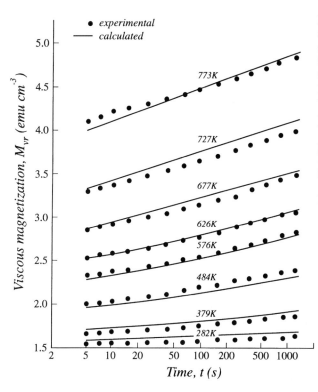

Figure 10.7 Viscous acquisition data for an SD magnetite (mean grain size 0.04 µm) at various temperatures (dots, after Dunlop, 1983b) compared to theoretical fits based on the theory of §10.3.1. [After Walton (1983), with the permission of the author. Reprinted with permission from *Nature*. Copyright © 1983 Macmillan Magazines Ltd.]

(10.4)–(10.7) are of no great benefit, and we shall return to Néel's blocking approximation, used with such success in Chapter 8. That is, we shall replace eqn. (10.1) by the step function

$$M = M_0, \qquad t < \tau$$
$$M = M_{eq}, \qquad t > \tau. \tag{10.8}$$

Clearly this is not as good an approximation as it was in modelling temperature changes in magnetization. Blocking (viscous acquisition) and unblocking (viscous decay) are fuzzy, drawn-out processes for small intervals of t, whereas they were quite sharp for small intervals of T (Fig. 8.10b). However blocking/ unblocking over equivalent intervals of T and $\log t$ are equally sharp, and this allows us to use Néel diagrams as an aid in visualizing changes over many decades in time.

An appropriate Néel diagram is given in Fig. 10.8, with blocking contours calculated from eqn. (10.2) by setting $\tau = t$. Equal increments of t would give rise to a set of contours identical to blocking temperature contours (Fig. 8.15). But whereas the spacing of contours increased with increasing T, it decreases with increasing t. In fact, equal areas swept out by the blocking curve, as sketched, are produced by equal increments of $\log t$. The intensity of VRM is therefore indeed proportional to $\log t$ if $f(V, H_{K0})$ is independent of V or H_{K0} or both over the small area swept out by the blocking curve in several decades of $\log t$.

Figure 10.8 demonstrates that about one-half the range of $\log t$ up to 1 Ma time scales is accessible in laboratory experiments. Even the inaccessible part can be activated by carrying out viscosity experiments at only mildly elevated temperatures. Comparing the contours in Fig. 10.8 with those in Fig. 8.15, we can estimate that for magnetite, heating to 200–250 °C should be sufficient to activate all τ up to 1 Ma. For minerals with lower Curie temperatures, like TM60 and pyrrhotite, correspondingly less heating is needed.

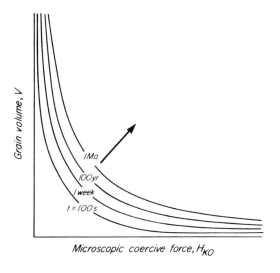

Figure 10.8 The effect of time t on single-domain blocking curves on the Néel diagram. VRM acquisition and viscous remagnetization can be modelled by sweeping a blocking curve across the grain distribution, as indicated by the arrow.

Figure 10.8 also makes it clear that single-domain VRM is carried by the same ensembles as, and is indistinguishable from, a partial TRM having the same upper blocking curve as the VRM and produced by the same field. The partial TRM in question occupies a small area on the Néel diagram, so that unless the grain distribution is unusually biased towards small V and H_{K0}, VRM over any reasonable time scale – up to 1 Ma in fact – will be a small fraction of the total TRM produced by a field of similar intensity. VRM is therefore normally a small fraction of NRM, if the NRM is a TRM.

10.3.3 Thermal demagnetization of VRM

VRM is also the *softest* fraction of the NRM (TRM). The upper contour bounding a weak-field VRM acquired in time t is approximately $VH_{K0} = C(T_0, t, 0)$, according to eqn. (8.33). The upper blocking-temperature contour of the equivalent partial TRM is $VH_{K0} = C(T, t_c, 0)$ where t_c is a characteristic time determined by the rate of heating or cooling. Therefore a room-temperature VRM is thermally demagnetized by heating to

$$\frac{T}{\beta^2(T)} = \frac{\ln(2t/\tau_0)}{\ln(2t_c/\tau_0)} T_0, \tag{10.9}$$

where $\beta(T) \equiv M_s(T)/M_{s0}$. If, for example, $t = 1$ Ma, $t_c = 100$ s and $\tau_0 = 10^{-9}$ s, $T \approx 2T_0 \approx 593$ K or $320\,°$C, neglecting $\beta(T)$. Using $\beta(T)$ for magnetite (Fig. 3.5a) reduces the required temperature to about $200\,°$C, as we estimated visually.

The analogy between T and $\log t$ is embodied in the relationship (Néel, 1955; Stacey, 1963)

$$S = \frac{\partial M}{\partial \ln t} = \frac{T_0}{\ln(2t_c/\tau_0)} \frac{\partial M}{\partial T} \approx 10 \frac{\partial M}{\partial T}, \tag{10.10}$$

which follows directly from (10.9) if T is close enough to T_0 that $\beta \approx 1$.

The fact that single-domain VRM is 'soft' as far as thermal demagnetization is concerned is of fundamental importance in paleomagnetism. It means that secondary NRM acquired viscously at or near T_0 can be erased by a very mild heating, leaving the primary NRM intact. In a Thellier paleointensity determination on SD material, VRM or viscous decay affect only the lowest-temperature steps (e.g., Fig. 8.14).

10.3.4 AF demagnetization of VRM

VRM affects only ensembles in the lowest part of the Néel diagram (Fig. 10.8). Therefore it is usually soft to AF as well as thermal cleaning. However, since the blocking curve for AF demagnetization (Fig. 8.16) does not exactly match the shape of the upper blocking contour bounding a VRM, there may be grains of small V and large H_{K0} that carry VRM because of their small size and consequent large $H_q(T_0, t)$, but are not easily demagnetized because of their large H_{K0}.

In more quantitative terms, the upper blocking contour bounding a VRM is specified by

$$VH_{K0} = \frac{2kT_0}{\mu_0 M_{s0}} \ln\left(\frac{2t_a}{\tau_0}\right),$$
(10.11)

according to eqn. (8.33). These are the last grains to be magnetized and they are also the most resistant to AF demagnetization (with a characteristic time $t_{af} \approx 1/f$). Their effective coercivity, from eqn. (8.24), is

$$H_c = H_{K0} - \left(\frac{2H_{K0}kT}{\mu_0 VM_{s0}}\right)^{1/2}\left[\ln\left(\frac{t_{af}}{\tau_0}\right)\right]^{1/2} = H_{K0}\left\{1 - \left[\frac{\ln(t_{af}/\tau_0)}{\ln(2t_a/\tau_0)}\right]^{1/2}\right\}$$
(10.12)

The resistance of a VRM to AF cleaning is proportional to H_{K0}, the factor inside brackets being independent of mineral properties.

Hard VRM of this origin is prominent in hematite-bearing red sediments and may occur also in some lunar rocks and soils. Although hard to AF demagnetize, VRM of this type is easily cleaned thermally, because the VRM and thermal demagnetization contours do match.

10.4   Theory of multidomain VRM

So far, VRM has been modelled solely in terms of SD grains so small as to be practically SP. It is a matter of common observation in paleomagnetic studies that coarse-grained rocks, which presumably contain much MD material, are more viscous than most fine-grained rocks. Multidomain VRM can be analysed using eqn. (10.3). Let us assume $\tau_0 = 10^{-10}$ s (Gaunt, 1977), which is similar to the value for domain rotation, $c = 1$, and $H_c = 5$ mT (50 Oe), a typical value for $\approx 1\,\mu$m magnetite grains. A Barkhausen jump cannot be less than about 10 Å or 1 nm, so that V_{Bark} for a 1 μm cube is at least 10^{-21} m^3. By substitution in eqn. (10.3), τ at room temperature turns out to be $\gg 1$ Ga. VRM is even more strongly prohibited in larger grains, because the increase in V_{Bark} more than offsets any decrease in H_c

On the face of it, MD grains should be entirely non-viscous. The flaw in the argument must be in the assumption that the wall is rigid and is activated as a unit, because no other parameter can be changed sufficiently to reduce τ to realistic values. Figure 5.18 shows that experimentally, domain walls in titanomagnetite will bend in spite of substantial magnetostatic opposition. There is every reason to expect that *segments* of a wall can undergo Barkhausen jumps, involving $V_{\text{Bark}} \ll 10^{-21} \text{m}^3$, past individual pinning centres (e.g., Gaunt and Mylvaganam, 1979).

It is possible to view VRM as resulting from the action of a 'viscosity field' (Néel, 1950, 1955) added to the applied field H_0, in much the same way we sketched earlier the blocking of single-domain TRM in the presence of a fluctuation field H_q (§8.7) and multidomain TRM in the presence of an analogous fluctuation field H_f (§9.4.2). In fact, the viscosity field is nothing more than $H_f(T_0, t)$. VRM is acquired if $H_0 > H_c - H_f(T_0, t)$. If the entire domain wall is activated, $H_f < 0.1$ mT (1 Oe), whereas H_f for small Barkhausen jumps of wall segments is substantial and allows VRM to be acquired. This point of view was developed in some detail by Néel (1950) and has been reviewed from the point of view of rock magnetism by Dunlop (1973c) and Bina and Prévot (1994).

Thermally activated transitions in domain structure have been proposed (Moon and Merrill, 1986a) as a possible cause of VRM's acquired over geologically long times, for example during ≈ 100 Ma superchrons of a single geomagnetic field polarity. Experimentally, superchron VRM's are more difficult to erase than simple Néel (1949) theory predicts, although it is now believed that much of the pervasive late Paleozoic remagnetization of Appalachian and Hercynian rocks on either side of the Atlantic, which coincided in time with the Permo-Carboniferous Kiaman reversed superchron, is of chemical origin, rather than thermoviscous. There is clear observational support for transdomain TRM, in which domains nucleate or denucleate during heating or cooling (§9.8). The corresponding process of transdomain VRM at constant temperature has yet to be observed. At ordinary temperatures, even the lowest energy barriers between LEM states with different numbers of domains are several thousand kT (Dunlop *et al.*, 1994), much in excess of the $(25–60)kT$ barriers that can be crossed with the aid of thermal energy on time scales varying from days to the age of the earth (cf. eqn. (8.4)).

10.5 Viscous noise problems

Viscous remagnetization or 'overprinting' of primary NRM through exposure to the present earth's field during the Brunhes epoch is a persistent contaminant of paleomagnetic information in rocks both young and old (Schmidt, 1993). Prévot (1981) estimates that an average of 25% of the NRM of mafic volcanic rocks is viscous noise. We noted in §10.3.4 that 'hard' VRM (Biquand and Prévot, 1971) is well known, particularly in hematite-bearing rocks. Any component of NRM directed near the present geomagnetic field direction at the sampling site is suspect, even if it is moderately resistant to AF demagnetization.

More insidious is VRM acquired in the laboratory during the course of measurements. It has been known for a long time that the viscosity coefficient S is dependent on initial state of a rock (§10.2.4, Fig. 10.6). Lowrie and Kent (1978) found very different rates of VRM acquisition before and after AF demagnetization of the NRM. It is also common experience that VRM acquisition is greatly enhanced when the NRM has been thermally demagnetized to low levels. Of course the very reduction of the signal (the primary NRM) tends to bring the viscous noise into prominence, but this is only part of the story. Plessard (1971) and Tivey and Johnson (1984) have observed that the viscosity coefficient changes markedly if a rock is allowed to relax for a time in zero field following thermal or AF demagnetization. The origin of this transitory effect seems to lie with domain walls that are loosely pinned (in shallow E_w minima) following demagnetization relaxing into deeper potential wells.

The cure for transitory VRM's is to make all measurements in a null-field environment. If specimens must be removed from null field between demagnetizer and magnetometer, they should be allowed time to relax before being measured. Superconducting quantum interference device (SQUID) magnetometers have sufficiently fast and continuous response that the relaxation can be monitored.

10.6   Thermoviscous remagnetization

10.6.1   Thermal demagnetization of thermoviscous overprints

The relationship between the time and temperature responses of magnetization is of considerable interest to paleomagnetists. If a remanence M_r has been produced by long exposure to a field or by very slow cooling in nature, what temperature in short-term laboratory thermal demagnetization will serve to destroy M_r? In §10.3.3, we dealt with viscous overprints acquired during the last 0.78 Ma at ordinary temperatures. In this section, we will turn our attention to thermoviscous overprints such as high-temperature VRM's and VpTRM's in ancient rocks. Thermoviscous magnetization is also important in the modern oceanic lithosphere (Arkani-Hamed, 1989; §14.2.9).

Néel SD theory provides a straightforward answer to the remagnetization question. The last grains to acquire remanence are also the last to be demagnetized. Assuming $H_0 \ll H_K$, eqn. (10.2) predicts that

$$\frac{T_L \ln(t_L/\tau_0)}{M_s(T_L)H_K(T_L)} = \frac{T_A \ln(t_A/\tau_0)}{M_s(T_A)H_K(T_A)} \qquad (10.13)$$

(Pullaiah et al., 1975), T_L, t_L and T_A, t_A being laboratory and ancient temperature and time, respectively. Dodson and McClelland Brown (1980) found an equivalent result for the demagnetization of a VpTRM acquired during slow cooling. Walton (1980), on the basis of the theory embodied in eqns. (10.5)–(10.7), has proposed the relation

$$\frac{T_{\mathrm{L}}[\ln(t_{\mathrm{L}}/\tau_0)]^2}{M_{\mathrm{s}}(T_{\mathrm{L}})H_{\mathrm{K}}(T_{\mathrm{L}})} = \frac{T_{\mathrm{A}}[\ln(t_{\mathrm{A}}/\tau_0)]^2}{M_{\mathrm{s}}(T_{\mathrm{A}})H_{\mathrm{K}}(T_{\mathrm{A}})} \qquad (10.14)$$

for equal intensities of VRM to be produced under conditions $(T_{\mathrm{L}}, t_{\mathrm{L}})$ and $(T_{\mathrm{A}}, t_{\mathrm{A}})$.

Enkin and Dunlop (1988) have argued that since VRM reactivates and replaces previous remanence rather than being superimposed on it, eqn. (10.13) is the correct answer to the paleomagnetists' question. Consider, for example, a VRM added at right angles to a prior remanence of equal intensity. In the Walton picture, the total remanence is a vector sum that exceeds the original remanence. If, however, the original remanence is gradually *replaced* by a new magnetization – not necessarily of the same magnitude – in a perpendicular direction, the final erasure temperature T_{L} of the original remanence is automatically detected by the vector analytical methods commonly used in paleomagnetism.

Williams and Walton (1988) and Worm and Jackson (1988) numerically calculated time–temperature relations based on Néel theory (eqn. (10.4)) for many different assumed grain-size distributions, $f(V)$. Both studies agree that the t–T contours are almost independent of the volume distribution chosen and that the curves agree in slope with eqn. (10.13), not eqn. (10.14). The theoretical contours in Fig. 10.9(a), for a log-normal distribution, agree very well with the Pullaiah *et al.* contours if samples are taken to be 'demagnetized' when only 5% of their original moments remain. For 1% and 0.1% demagnetization, higher-temperature, but parallel, contours are found.

Experimental data by Williams and Walton (1988) are in perfect accord with the Pullaiah *et al.* predictions (Fig. 10.9b). VRM's acquired by well-dispersed SD magnetite grains in a glass matrix in $t_{\mathrm{A}} = 10$ hr at $T_{\mathrm{A}} = 95\,^\circ\mathrm{C}$ and $233\,^\circ\mathrm{C}$ were reduced to $\approx 10\%$, 1% and 0.1% of their initial intensities in heatings for $t_{\mathrm{L}} = 10^3$ s at exactly the temperatures T_{L} predicted by eqn. (10.13).

Turning to field studies, eqn. (10.13) accurately predicts the temperature necessary to erase thermoviscous overprints in SD hematite (Kent and Miller, 1987) but it predicts cleaning temperatures that are too low in the case of magnetite. Figure 10.10 shows results of work by Kent (1985). Two thermoviscous overprints were studied. Overprint A is a room-temperature VRM acquired by glacial boulders whose orientations were randomized at the end of the last ice age. Thus $T_{\mathrm{A}} \approx 20\,^\circ\mathrm{C}$ and $t_{\mathrm{A}} \approx 10$ ka. The predicted erasure temperature T_{L} for this overprint in $t_{\mathrm{L}} = 30$ min heatings, according to eqn. (10.13) (assuming $H_{\mathrm{K}}(T) \propto M_{\mathrm{s}}(T)$, i.e., shape anisotropy) is $< 150\,^\circ\mathrm{C}$. The observed T_{L} required is $270\,^\circ\mathrm{C}$, and agrees more closely with Walton's relation (eqn. (10.14)), shown graphically by dashed contours, after Middleton and Schmidt (1982). Overprint B is an ancient secondary magnetization acquired on a time scale of $t_{\mathrm{A}} \approx 100$ Ma at $T_{\mathrm{A}} \approx 250\,^\circ\mathrm{C}$, as estimated from the alteration of conodont microfossils

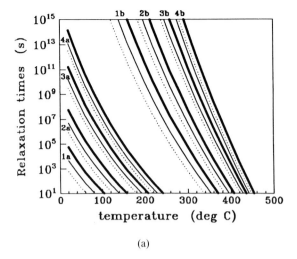

(a)

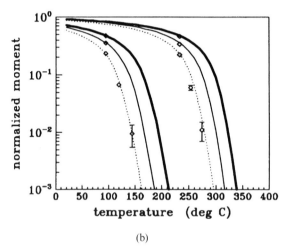

(b)

Figure 10.9
(a) Theoretical time–temperature contours for SD magnetites (V^{-1} size distribution assumed) that acquired VRM at 20 °C (left) or 300 °C (right) for times of 10 min, 10 days, 100 yr, or 10^5 yr. Times and temperatures for demagnetization to 1% (light curves) or 0.1% (heavy curves) of initial moment agree well with the Pullaiah *et al.* (1975) contours (eqn. (10.13); dotted curves). (b) Data on SD magnetites that acquired VRM during 10 hr at 95 °C (left) and 233 °C (right) and were demagnetized, first for times of $10–10^3$ s at a single higher temperature and then for a fixed time of 10^3 s at two further temperatures. The data agree perfectly with demagnetization curves predicted using Néel theory for heating times of 10, 10^2 and 10^3 s (heavy, light and dotted curves, respectively). [After Williams and Walton (1988) © American Geophysical Union, with the permission of the authors and the publisher.]

in the rocks. According to (10.13), $T_L < 400$ °C, but $T_L \approx 480$ °C is observed, again in reasonable accord with (10.14).

To sum up, exact numerical calculations of single-domain VRM result in time–temperature relations that are scarcely distinguishable from the simple Pullaiah *et al.* (1975) relations and are furthermore borne out by well-controlled laboratory experiments. The paleomagnetic 'experiments' over geological time scales are also relatively well-controlled in that (t_A, T_A) are known within fairly narrow limits, but in this case (Middleton and Schmidt, 1982; Kent, 1985; Jackson and Van der Voo, 1986) T_L disagrees significantly with eqn. (10.13). The

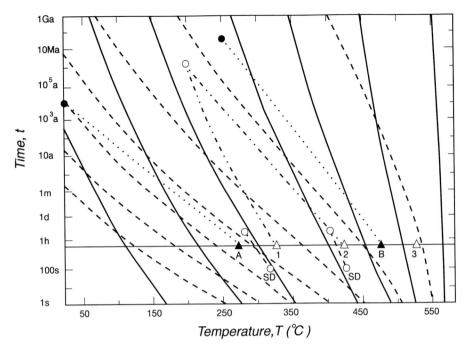

Figure 10.10 Single-domain blocking/unblocking time–temperature contours according to Pullaiah *et al.* (1975) (solid curves) and Walton (1980) (dashed curves). Each contour joins (t, T) values for which a particular SD ensemble (V, H_{K0}) is just blocked (remagnetized) or unblocked (thermally demagnetized). Data for SD magnetites (Dunlop and Özdemir, 1993) follow the Pullaiah *et al.* contours but the time range is relatively narrow. Among geological examples over greater time ranges, remagnetized limestones (solid circles and triangles; low-temperature and high-temperature viscous overprints: A and B, respectively; data from Kent, 1985) follow the Walton relation, but the remagnetized Milton monzonite (open circle and triangles; Middleton and Schmidt, 1982; Dunlop *et al.*, 1995a) agrees with the Pullaiah contours for SD samples (1), with the Walton contours for mixed SD and MD samples (2), and with neither for MD samples (3). [Contour plots after Pullaiah *et al.* (1975), reprinted with kind permission of the authors and Elsevier Science – NL, Sara Burgerhartstraat 25, 1055 KV Amsterdam, The Netherlands, and Middleton and Schmidt (1982) © American Geophysical Union, with the permission of the authors and the publisher.]

answer to this riddle lies in the very different thermal demagnetization responses of SD and MD grains.

Figure 10.11 contrasts experimental data on the stepwise thermal demagnetization of high-temperature VRM's acquired by SD and MD magnetites. The acquisition time t_A was 3.5 hr and the time t_L the sample was held at each demagnetization temperature was ≈ 100 s. VRM's in the SD magnetite sample demagnetized very sharply, mostly within $\pm 15\,^{\circ}$C of T_A (Fig. 10.11a). T_L for demagnetization to the 10% level was in excellent agreement with eqn. (10.13) for both runs (Fig. 10.10). Results on a 135 μm MD magnetite sample were com-

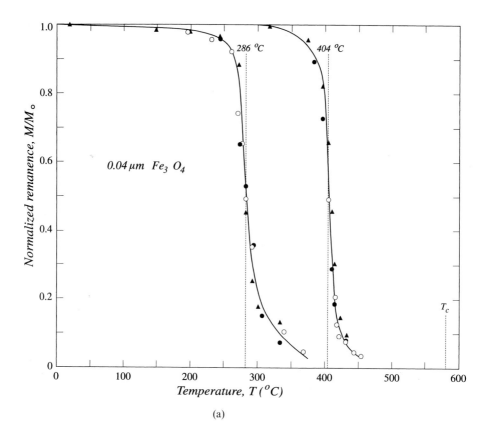

(a)

Figure 10.11 Stepwise thermal demagnetization of VRM's produced during 3.5 hr at the temperatures shown by $H_0 = 0.1$ mT (a) or 1 mT (b) in SD and large MD grains of magnetite, respectively. Most single-domain VRM demagnetizes within $\pm 15\,°C$ of the VRM acquisition temperature, whereas demagnetization of multidomain VRM extends over the entire range from T_0 to T_C. The high-temperature demagnetization tail agrees with pTRM theory (Chapter 9, eqn. (9.29)). There is no strong dependence on initial state in either case (the order AF, TH, TC of curves is actually opposite for stepwise demagnetization in (b) to the order TC, TH, AF in Fig. 10.6(a) for continuous demagnetization). [After Dunlop and Özdemir (1995) © American Geophysical Union, with the permission of the authors and the publisher.]

pletely different (Fig. 10.11b). Whether the initial state was AF, TH or TC, demagnetization continued over a broad temperature interval extending up to T_C. The thermal demagnetization tail was well described by the theory of demagnetization of MD partial TRM (§9.5, eqn. (9.29) with $n = 4$) and doubtless has the same cause: continuous re-equilibration of domain walls during heating due to self-demagnetization. The ultimate T_L in the MD case is close to T_C, much higher than predicted by either (10.13) or (10.14).

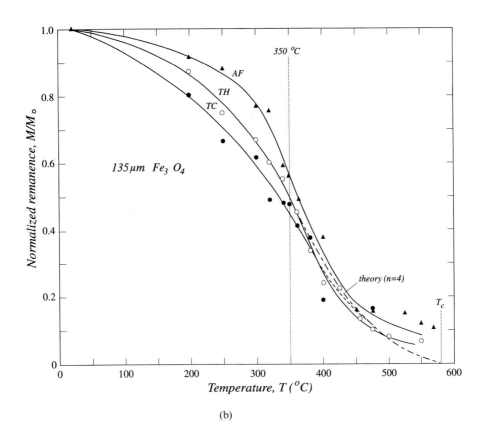

(b)

A restudy of the Milton monzonite, whose burial overprint first led Middleton and Schmidt (1982) to propose alternative t–T contours, has confirmed that the 'anomalously' high demagnetization temperatures originally reported are the result of mixed SD and MD carriers (Dunlop *et al.*, 1995a). In the restudy, samples from a single site contained almost exclusively SD magnetite; their T_L values agreed with the Pullaiah *et al.* contours (Fig. 10.10). Samples from two sites contained exclusively coarse MD magnetite; their T_L values were extremely high, much above the predictions of either set of contours. All other sites contained mixtures of SD and MD magnetite and approximately reproduced Middleton and Schmidt's T_L values.

We conclude that eqn. (10.13) gives the correct t–T relationship for SD grains. Thermoviscous overprints can be removed cleanly from SD samples at a prescribed maximum heating temperature T_L, leaving the surviving primary NRM untouched. However, in the case of MD grains, VRM's and VpTRM's, like multidomain pTRM's (§9.5), have a thermal demagnetization tail that continues to obscure the primary NRM virtually up to the Curie point. Thermal

demagnetization is not the optimal method of cleaning thermoviscous overprints
from MD material.

10.6.2   AF and low-temperature demagnetization of thermoviscous overprints

AF demagnetization efficiently removes low-temperature thermoviscous overprints in SD magnetite but becomes less and less effective as T_A increases (Fig.
10.12). VRM produced in 2.5 hr at 100 °C was completely erased by a peak AF
of 20 mT (200 Oe) but VRM's produced at 500 °C and above were almost as resistant to AF cleaning as the total TRM. They required AF's of 50 mT (500 Oe)
for removal; much of the primary TRM would be erased at the same time. This
is hardly astonishing, since higher-temperature VRM's magnetize areas of the
Néel diagram corresponding to higher-temperature pTRM's. The only way to
erase overprints in SD material without at the same time affecting the surviving
primary TRM is by thermal demagnetization.

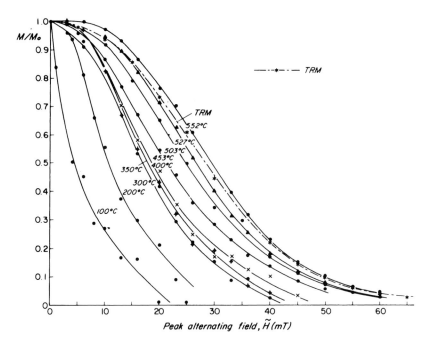

Figure 10.12   Room-temperature AF demagnetization curves of VRM's produced in 2.5-hr runs (50 μT field) at the temperatures shown for the same single-
domain magnetite sample used in Fig. 10.7. The AF demagnetization curve of total
TRM is similar to the curves for VRM's produced at 527 °C and 552 °C. [Re-
printed from Dunlop and Özdemir (1990), with kind permission of the authors and
Elsevier Science – NL, Sara Burgerhartstraat 25, 1055 KV Amsterdam, The
Netherlands.]

For MD material, thermal demagnetization of high-temperature VRM's and VpTRM's is clearly *not* the best strategy. Figure 10.13 shows that low-temperature demagnetization (LTD), which consists of zero-field cooling and reheating through the isotropic point of magnetite (§3.2.2) to unpin domain walls, or AF demagnetization to 10 mT (100 Oe) both make subsequent thermal demagnetization more effective. Either treatment has the effect of enhancing SD remanence. After LTD or partial AF demagnetization, the high-temperature VRM responded thermally in a more SD-like fashion. In particular, the high-temperature tail was reduced. In a paleomagnetic situation, this would allow more of the desired primary remanence to show through the blanketing overprint. Both LTD and partial AF demagnetization are quick procedures. They are recommended as standard pre-treatments prior to thermal demagnetization.

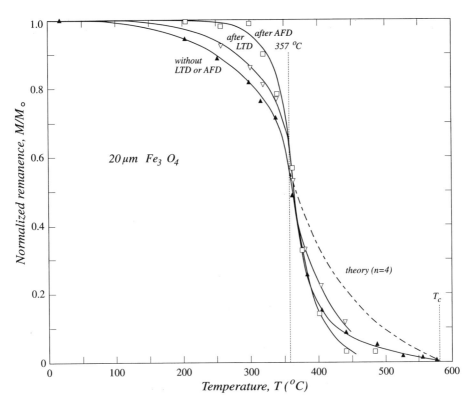

Figure 10.13   Stepwise thermal demagnetization of VRM's produced during 3.5 hr at 357 °C in a 20 μm magnetite sample, when the VRM's were pre-treated by low-temperature demagnetization (LTD) to 77 K or AF demagnetization (AFD) to 10 mT before thermal cleaning, and without pre-treatment. Both LTD and AFD render the thermal demagnetization more SD-like (cf. Fig. 10.11a) and reduce the high-temperature demagnetization tail. [After Dunlop and Özdemir (1995) © American Geophysical Union, with the permission of the authors and the publisher.]

10.7   Cooling rate dependence of TRM

We turn now to another practical question involving time–temperature relationships: how much is the intensity M_{tr} of TRM affected by the cooling rate? This question arises in determinations of the paleofield intensity recorded by primary TRM's. Dodson and McClelland Brown (1980) and Halgedahl *et al.* (1980) independently predicted that the intensity of TRM in SD grains should increase for longer times, possibly by as much as 40% between laboratory and geological settings. Fox and Aitken (1980) verified that there was about a 7% decrease in the intensity of TRM acquired by SD magnetite in a baked clay sample when the cooling time changed from 2.5 hr to 3 min, in excellent agreement with predictions (Fig. 10.14).

McClelland Brown (1984) theorized that for multidomain TRM due to pinned domain walls, slowly cooled TRM's should be *less* intense than rapidly cooled ones. This effect was confirmed in experiments on synthetic TM30 grains, but the magnitude of the intensity change greatly exceeded predictions. The titanomagnetites seemed to be two-phase mixtures, the lower T_C matrix apparently amplifying the TRM of higher-T_C inclusions.

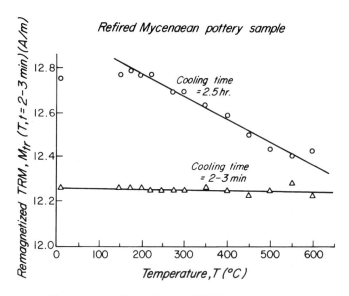

Figure 10.14   Dependence of TRM intensity on rate of cooling, illustrated for a pottery sample given TRM's by fast (2–3 min) and slow (2.5 hr) coolings, and partially remagnetized by fast coolings in the same field from a series of increasing temperatures, so as to replace part of each original TRM with a series of partial TRM's. [After Fox and Aitken (1980), with the permission of the authors. Reprinted with permission from *Nature*. Copyright © 1980 Macmillan Magazines Ltd.]

10.8   VRM as a dating method

Viscous remagnetization can be used as a means of determining the time at which the original magnetization of some geological material was demagnetized or randomized. Heller and Markert (1973) sampled dolerite blocks that had been reoriented in Roman times in building Hadrian's Wall in northern England. The VRM formed since that time was resolved from the surviving primary NRM by AF demagnetization, and the maximum coercivity H_c of the VRM was used to deduce $\ln(t/\tau_0)$, using an equation analogous to (10.12). Two of three blocks tested gave reasonable estimates of elapsed time: $t_a = 1.6$–1.8 ka.

The resolution of this method is determined by $\partial M/\partial t = (1/t)\,(\partial M/\partial \log t) = S_a/t_a$. There are two problems. First, S_a must be constant for time periods of geological length, contrary to usual laboratory observations (e.g., Fig. 10.3). The rock used must contain only a single magnetic phase, i.e., one mineral with constant domain structure, and the grain distribution $f(V, H_{K0})$ must be unusually uniform. The second problem is inherent in the method: the resolution is inversely proportional to the length of time to be measured. Thus, only geologically recent events are datable.

An interesting new approach is thermal demagnetization of viscous overprints. Borradaile (1996) has compared historical dates of limestone and chalk blocks from buildings ranging in age from Roman (AD300) to modern to ages predicted from $(T_{UB})_{max}$ of their VRM overprints, using both the Pullaiah *et al.* (1975) and Walton (1980)/Middleton and Schmidt (1982) contours (Fig. 10.10). Neither set of contours proved adequate. The Pullaiah *et al.* contours accounted for the low T_{UB} values of the younger VRM's, but older VRM's had higher T_{UB}'s than any SD theory would predict. MD magnetite grains may be responsible, although this remains speculative in the absence of hysteresis measurements. Borradaile established empirical $(T_{UB})_{max}$ vs. t curves suitable for dating these particular building stones, but the chalk curve was very different from that for the limestones.

10.9   Granulometry using magnetic viscosity

Viscous magnetization is a sensitive probe of narrow (V, H_{K0}) bands of the grain distribution. If H_{K0} has a narrow spread compared with V, inversion of eqn. (10.4) or its analog for viscous decay yields $f(\tau)$, i.e, $f(V)$. A cruder estimate, which is adequate in most applications, is to use a blocking approximation in eqn. (10.5) to obtain $f(V_B)$. Small V_B can be accessed by using the frequency dependence of susceptibility. Large V_B can be activated in viscous acquisition or decay experiments conducted above room temperature.

Chapter 11

Isothermal magnetization and demagnetization

11.1 Introduction

All magnetizations are produced by an applied magnetic field, but certain magnetization processes, for example isothermal remanent magnetization (IRM) and alternating-field (AF) demagnetization, are field-driven in a more restricted sense. They are produced isothermally, usually at or near T_0, over a time scale of at most a few minutes. They result, in other words, from the *sole* influence of an applied field. Even this definition is not entirely accurate: τ_0 is so much less than ordinary measurement times that thermal fluctuations play some role in practically all field-induced processes. For weak applied fields and small V, H_q or H_f are $> H_0$, and thermal excitation plays a major role (§8.7, 9.4.2).

Isothermal remanences do not carry useful paleomagnetic information. An IRM produced by the geomagnetic field is easily reset by later weak fields of similar magnitude: it lacks paleomagnetic stability. Only isothermal remanences due to strong fields, for instance saturation isothermal remanence (SIRM) and anhysteretic remanence (ARM) produced by the combination of a steady field and a strong but decaying AF, have the requisite stability, and these occur in nature only when outcrops have been struck by lightning and remagnetized. However, the stepwise acquisition or removal of magnetizations in the laboratory is used as a means of erasing NRM's of low stability (AF demagnetization) and of determining the composition and domain structure of mineral magnetic carriers (hysteresis, IRM acquisition, 'DC demagnetization', Preisach analysis).

In this chapter, we shall first review fundamental understanding of isothermal magnetization processes in SD and MD grains (§11.2–11.5). Then we shall deal with specialized methods and tests that use isothermal magnetic parameters,

288

such as the Lowrie–Fuller test (§11.6), the Wohlfarth relations and Henkel plots (§11.7), the Preisach diagram (§11.8), and hysteresis parameters (§11.9).

11.2 Single-domain coercivity spectrum and AF demagnetization

The effect of AF demagnetization on remanence carried by SD grains is most easily appreciated by using the Néel diagram (Fig. 11.1). As discussed in §8.11, the blocking contours for steady fields H_0 or alternating fields $\tilde{H}$ are obtained by setting $H_c = H_K - H_q$ (eqn. (8.24)) equal to H_0 or $\tilde{H}$. If thermal energy were of no significance in surmounting barriers, the magnetization would reverse at a field $\tilde{H}$ or $H_0 = H_K$ (dashed lines in Fig. 11.1a). This is approximately the case for large SD grains, but reversals in grains with small volumes V are in part thermally excited and have fluctuation fields H_q comparable to H_K. Small grains will therefore respond to steady or alternating fields H_0 or $\tilde{H}$ which are substantially less than their microcoercivities H_K. For this reason, the blocking contours are not vertical but hyperbolic.

AF demagnetization 'demagnetizes' ensembles of SD grains by leaving as many moments pointing in one direction along the easy axis (or axes) of anisotropy as in the opposite direction. The response of an SD grain depends on the angle between the AF and the easy axis along which its moment points (Fig. 8.7). Moments parallel to the AF, unless they reverse incoherently, have large

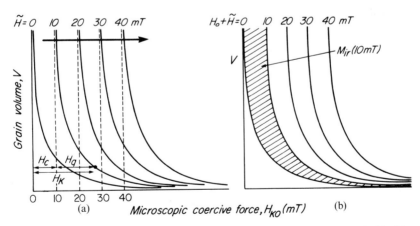

Figure 11.1 The effect of steady or alternating fields on single-domain blocking curves on the Néel diagram. (a) Modelling AF demagnetization. Note the strong effect of the thermal fluctuation field H_q on the effective microcoercivity of small grains. In the example (marked with a dot), $H_{K0} = 27\,\text{mT}$, but the grain reverses its magnetization in an AF of only 10 mT. (b) Modelling acquisition of IRM and total or partial ARM, or DC 'demagnetization' (remagnetization of remanence by stepwise application of back fields).

critical fields compared to moments at intermediate angles. Randomly oriented grains within a particular ensemble (V, H_{K0}) will reverse over a range of fields, the H_c values we calculate and show in Fig. 11.1 (for coherently reversing grains aligned with H_0) being upper limits. When samples are 'tumbled' in space at the same time the AF is slowly decaying, demagnetization is more homogeneous for individual ensembles, although the distribution $f(V, H_{K0})$ of ensembles should ensure overall random demagnetization even without tumbling. (The same is not true of MD grains. $180°$ domain walls perpendicular to the AF will not move (Fig. 8.7) and tumbling or multi-axis demagnetization is essential.)

Returning to Fig. 11.1, AF demagnetization to increasingly higher fields: 10 mT, 20 mT, etc., can be viewed as randomizing the moments of SD ensembles lying in the areas between the zero-field contour and the contours labelled 10 mT, 20 mT, etc. The **(micro)coercivity spectrum** or distribution of destructive fields of a particular remanence is determined by the distribution of magnetization in these different areas. Experimentally, the coercivity spectrum is determined by taking the point-by-point slope of the measured AF demagnetization curve. Conversely, the AF decay curve is the cumulative or integrated coercivity distribution. The median value of the spectrum, the field that reduces the remanence to one-half its initial value, is the **median destructive field** (MDF), $\tilde{H}_{1/2}$, of that remanence.

Because TRM, anhysteretic remanence (§11.4) and SIRM are carried by all grain ensembles, whatever their coercive forces, these remanences have approximately the same normalized coercivity spectrum and AF demagnetization curve. This intrinsic AF decay curve is labelled 'single-domain' in Fig. 11.2. It has an initial plateau, for magnetite extending to 10 mT (100 Oe) or so (cf. Figs. 8.17a, 10.12). Little or no demagnetization occurs in this range because magneto-crystalline anisotropy provides a minimum threshold field that must be overcome even in equidimensional grains with no shape anisotropy. For MD grains, the intrinsic spectrum and associated AF decay curve are dictated by barriers to wall motion. The characteristic exponential form of the curve (Fig. 11.2), which is much softer than the SD curve, will be discussed in §11.5 (see Fig. 11.13). Qualitiatively, the MD spectrum is softer than the SD spectrum because domain walls are less strongly pinned than most SD moments. In addition, H_d in MD grains assists in the demagnetizing process, as we observed in Chapter 9.

Remanences that affect only part of the Néel diagram, such as VRM's and partial TRM's, exhibit only part of the complete coercivity spectrum. Room-temperature VRM or low-temperature pTRM activate only the lowest part of the spectrum and are soft to AF demagnetization. High-temperature VRM's or pTRM's affect higher parts of the coercivity spectrum and may have larger average coercivities than total TRM (Figs. 8.17, 10.12).

The rationale behind AF demagnetization as a 'cleaning' procedure in paleo-magnetic studies is evident in Fig. 11.2. A rather low AF will serve to randomize

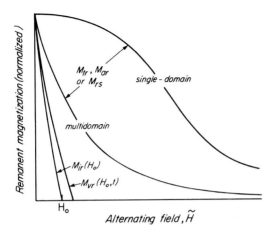

Figure 11.2 AF demagnetization curves for various types of remanence. TRM, ARM and SIRM, which affect the entire grain distribution, have similar demagnetization curves, which are much harder than those of weak-field IRM or VRM. Single-domain grains and large multidomain grains have sigmoid and quasi-exponential curves, respectively.

weak-field IRM or VRM acquired at ordinary temperatures (at least for magnetite), while scarcely affecting single-domain TRM. AF demagnetization to intermediate fields will begin to attack single-domain TRM, but much less than multidomain TRM, which is rapidly reduced to low levels. Highly anisotropic minerals like pyrrhotite, hematite and goethite require considerably higher cleaning fields than magnetite, but VRM and multidomain TRM will still be preferentially cleaned. In practice, commercial demagnetizers are often limited to AF's ≤ 100 mT (1 kOe). Complete AF cleaning of unwanted remanences in magnetically hard minerals is then not feasible, and thermal demagnetization must be used.

One way of combining the advantages of AF and thermal demagnetization is 'hot' AF demagnetization, illustrated in Fig. 11.3. This has never become a routine paleomagnetic method because the remanence must be remeasured at high temperature after each AF step, and no suitable equipment is available commercially. However, the technique does illustrate very graphically the contraction of the coercivity spectrum at high temperatures, due in part to $H_q(T)$ in the finer grains but mainly to $H_K(T)$, as shape and crystalline anisotropies both decrease with heating. Notice that the hardest fraction of SIRM (the 5% level in Fig. 11.3), which had coercivities > 50 mT at T_0, was cleaned by an AF of < 20 mT at 499 °C.

11.3   Acquisition and 'DC demagnetization' of IRM

Application and immediate removal of a steady field H_0 produces **isothermal remanent magnetization** (IRM). On the Néel diagram (Fig. 11.1b), IRM acquired in $H_0 = 10$ mT affects SD ensembles in the hatched area between the zero-field and 10 mT blocking contours. Since weak-field IRM affects only the

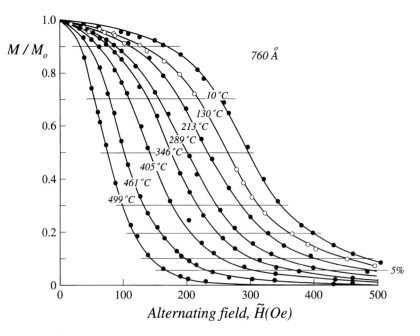

Figure 11.3 'Hot' AF demagnetization of SIRM carried by 0.076 μm SD magnetite grains. Both demagnetization and remanence measurements were carried
out at high temperature. The coercivity spectrum contracts dramatically as T rises.
Horizontal lines compare demagnetization fields necessary to reduce initial remanence M_0 to fixed levels (from 90% to 5%) at different temperatures. [After Dunlop
and Bina (1977), with the permission of the authors and the publisher, Blackwell
Science, Oxford, UK.]

lowest-H_c part of the coercivity spectrum, it is much softer and less intense than
TRM produced by the same field. Strong-field IRM approaches strong-field
TRM in intensity and hardness.

11.3.1   Lightning and drilling-induced IRM's

Lightning strikes cause IRM in nature. Within a few metres of a strike, the surface rocks experience a DC field of short duration and varying intensity that
remagnetizes all or part of the NRM (Rimbert, 1959: Graham, 1961). A lightning overprint is usually recognizable by its extreme intensity compared to the
NRM it replaces (Dunlop *et al.*, 1984). Directional stability during demagnetization is generally excellent, but lightning currents vary sufficiently on a centimetre
scale that directions of remanence for different specimens of the same paleomagnetic sample or core may diverge 20° or more. Lightly overprinted samples
some distance from a strike can be successfully AF cleaned, since weaker IRM's
affect only the low-coercivity grains.

   Drilling-induced remanent magnetization (DIRM) is an IRM produced by the
magnetic field H_0 of a magnetized steel core barrel during drilling. DIRM was

noticed during early Deep-Sea Drilling Project (DSDP) coring (Ade-Hall and Johnson, 1976) and is a persistent contaminant of samples from continental and oceanic boreholes. The intensity of DIRM generally increases by about a factor 10 from the centre of the drillcore to the core-barrel wall (Özdemir *et al.*, 1988; Audunsson and Levi, 1989; Pinto and McWilliams, 1990). DIRM tends to be a near-vertical remanence, probably because the horizontal component of the earth's field averages to zero while the rotating core barrel is acquiring its magnetization and also because the long axis of the cylindrical drillstring is the easy axis of shape anisotropy (§4.2.5, 4.3.3). Pinto and McWilliams found reasonable agreement between the theoretical profile of field H_0 across a 6-inch core due to a long, axially magnetized core barrel, the measured field profile, and the intensity of DIRM (Fig. 11.4).

DIRM carried by magnetite is largely demagnetized by AF cleaning to 20 mT or heating to $\approx 400\,°C$ (Özdemir *et al.*, 1988). The unblocking temperatures are higher than those expected for weak-field IRM or VRM produced at ordinary

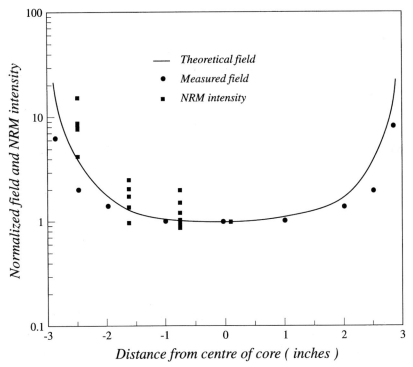

Figure 11.4 Comparison of theoretical and measured fields H_0 over a 6-inch diameter steel core barrel and the intensity of drilling-induced isothermal remanent magnetization (DIRM) measured in an edge-to-edge traverse across a deep continental drillcore. [After Pinto and McWilliams (1990), with the permission of the authors and the publisher, The Society of Exploration Geophysicists, Tulsa, OK.]

temperatures (§10.3, 10.6). The induced magnetization of the steel core barrel acts as an amplifier, producing a field substantially greater than the earth's field. Audunsson and Levi (1989) found that MD samples acquired DIRM more readily than SD samples. Fields of 5–15 mT produced IRM's with intensities and stabilities similar to those of the DIRM's.

Burmester (1977) reported that DIRM had a log t decay in the manner of VRM, but only a small part of the DIRM intensity could be accounted for by VRM in the ≈ 0.2 mT axial core-barrel field. However, the drillbit–rock interface could reach $\approx 200\,°C$, at which temperature submicron magnetites are about four times more viscous than at T_0 (Dunlop, 1983b; Fig. 10.3). The combination of field amplification by the core barrel and above-ambient temperatures seems to be responsible for producing IRM/VRM with the coercivities and unblocking temperatures characteristic of DIRM.

Drillcores are usually unoriented azimuthally. One method of orienting them is to use Brunhes-epoch VRM in the core as a north marker (Kodama, 1984). If DIRM is substantial, this method fails because the VRM is AF or thermally cleaned before the DIRM is completely removed. Jackson and Van der Voo (1985) found that a remanence analogous to DIRM is acquired even during *laboratory* drilling and sawing of paleomagnetic specimens. This DIRM was parallel to the ambient field H_0 during cutting and was preferentially carried by large MD grains. A fraction of this DIRM could not be removed by AF cleaning to 60 mT or thermal cleaning to 400 °C. High-coercivity or high-T_{UB} demagnetization 'tails' of this sort are characteristic of MD remanence (see §9.5, 11.5).

11.3.2   Laboratory acquisition and demagnetization of IRM

IRM acquired in stepwise fashion in the laboratory, after demagnetization of the NRM of a rock, is often used to measure the intrinsic coercivity spectrum (Fig. 11.5). Such an **IRM acquisition curve** successively magnetizes ensembles between the 0 and 10-mT, 10- and 20-mT, etc. blocking curves of Fig. 11.1(b). Since these contours are identical to those for AF demagnetization, the same coercivity spectrum should be found whether the IRM is built up in steps to the SIRM or the SIRM is demagnetized in steps to zero. In practice, this equivalence holds only for non-interacting SD grains (see §11.7). The IRM acquisition method of determining the coercivity spectrum has become popular in recent years because of the advent of automated hysteresis-loop plotters (alternating-gradient force magnetometer or AGFM; Flanders, 1988) that can also be programmed to measure IRM buildup. For highly anisotropic minerals like hematite and goethite, where AF demagnetization is inadequate to delineate the coercivity spectrum, IRM acquisition is the only viable approach (see Chapter 15). Lowrie (1990) has proposed a method in which the sample is twice rotated through 90° during

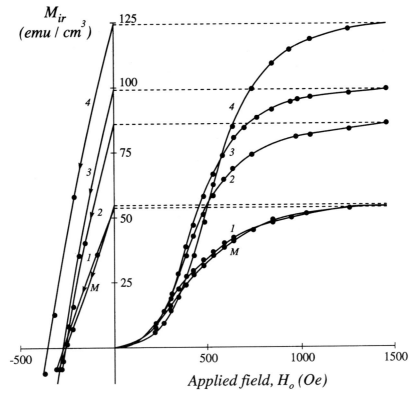

Figure 11.5   Partial remanent hysteresis curves for four SD or nearly SD magnetite samples (mean sizes: 1, 0.22 μm; 2, 0.10 μm; 3, 0.076 μm; 4, 0.037 μm). M
(Mapico magnetite) is similar in size and properties to 1. DC demagnetization
curves (left) cross the field axis at $-H_{cr}$, the coercivity of remanence. [Reprinted
from Dunlop (1986a), with kind permission of the author and Elsevier Science –
NL, Sara Burgerhartstraat 25, 1055 KV Amsterdam, The Netherlands.]

IRM acquisition, at field values marking the approximate limits of the coercivity
spectra of goethite, hematite, and softer minerals like magnetite, titanomagnetite
and maghemite. Thermal demagnetization of the three orthogonal components
of the composite SIRM then yields the relative quantities and unblocking temperature spectra of the phases.

Automated AGFM's also make easy the formerly laborious determination of
the **DC demagnetization curve** (Fig. 11.5). In reality, this is not a demagnetization curve but an IRM remagnetization curve. The sample, usually carrying a
saturation IRM, is exposed to reverse fields of increasing strengths. Its remanence decreases to zero at a field called the **coercivity of remanence** H_{cr} and then
grows to increasing negative values, culminating in $-M_{rs}$, the negative SIRM.
When a mixture of hard and soft phases is DC demagnetized, the ratio
$S = -M_{ir}(-300\,\text{mT})/M_{rs}$ is often used as a rough estimate of the relative quantities of the phases (Thompson and Oldfield, 1986), since 300 mT is sufficient to

reverse the remanence of even elongated SD magnetite or maghemite grains, but scarcely affects hematite or goethite remanence.

The DC demagnetization process consists of *reversing* the remanence of successive bands on the Néel diagram (Fig. 11.1b). Each incremental remanence is twice that in the IRM acquisition curve and is negative. Thus the DC demagnetization curve should be the mirror image of the IRM acquisition curve, scaled up by a factor 2. H_{cr} should be identical to H_{cr}', the coercivity of remanence acquisition or median IRM acquisition field. The DC demagnetization curve should also be identical to the AF demagnetization curve (in shape – no actual demagnetization or randomizing of moments takes place) but scaled up by a factor 2. Thus H_{cr} should also be identical to the MDF, $\tilde{H}_{1/2}$. These relationships prove to be true for non-interacting SD grains, but not otherwise (§11.7).

11.4   Anhysteretic remanence and other AF-related remanences

11.4.1   Anhysteretic remanent magnetization (ARM)

If the field applied during AF demagnetization is not perfectly symmetrical, the net effect is to superimpose a steady or directed field H_0 on a symmetric field $\tilde{H}$. Both fields decrease smoothly in intensity. As a result of this process (which is the writing process in audio or video magnetic recording), **anhysteretic remanent magnetization** (ARM) is acquired by the sample in the direction of H_0. Such an ARM may be an unwanted contaminant accompanying AF 'cleaning' if the demagnetizer waveform is distorted by higher harmonics. ARM is also acquired when the AF is perfectly symmetrical if the earth's magnetic field is not perfectly cancelled in the demagnetizing apparatus. ARM's of either type are serious in paleomagnetic work because they obscure the primary NRM rather than revealing it. ARM is magnetically hard: its coercive force is $\approx \tilde{H} + H_0$ (see Fig. 11.1b), not H_0.

The intensity of an ARM, M_{ar}, depends on the relative directions of H_0 and $\tilde{H}$. If H_0 is parallel to the axis of the AF, $\tilde{H}$, one polarity of $\tilde{H}$ is biased compared to the other. More than 50% of the moments block in the favoured direction, resulting in a net remanence parallel to H_0. If H_0 is perpendicular to $\tilde{H}$, there is no polarity bias, but the resultant field is no longer perfectly axial. Moments blocked in opposite orientations are not precisely antiparallel but have a net resultant, M_{ar}, in the direction of H_0. Experimentally, parallel ARM is about twice as intense as perpendicular ARM (Rimbert, 1959; Denham, 1979).

The '**anhysteretic susceptibility**', $\chi_a = dM_{ar}/dH_0$, is poorly predicted by SD theory: theoretically it should be infinite, i.e., even a vanishingly small biasing field should cause all moments to block parallel to H_0. Grain interaction fields in the SD case and internal demagnetizing fields in the MD case provide a distribution of biasing fields and make the anhysteretic susceptibility finite (Jaep, 1969; Shcherbakov and Shcherbakova, 1977; Veitch, 1984). In Fig. 11.6(a), the highest value of χ_a is that of plagioclase grains from the Lambertville diabase,

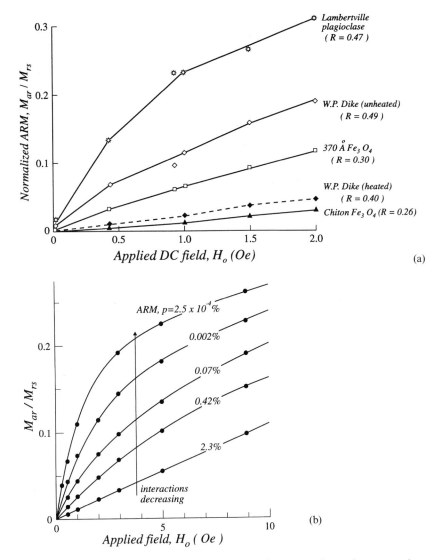

Figure 11.6   (a) Normalized ARM acquisition curves for various natural and synthetic samples. Grain interactions are smallest in the Lambertville plagioclase (well-separated needle-like inclusions of magnetite) and greatest in the chiton teeth (biogenic magnetite in chains of crystals). The 370-Å magnetite is sample 4 of Fig. 11.5. (b) ARM acquisition curves for samples containing dispersed magnetite grains in different concentrations. [Reprinted from Cisowski (1981) and Sugiura (1979), with kind permission of the authors and Elsevier Science – NL, Sara Burgerhartstraat 25, 1055 KV Amsterdam, The Netherlands.]

which contain well-separated SD magnetite crystals, and the lowest χ_a is for chiton teeth, containing chains of strongly interacting magnetite crystals. The effect of interactions on χ_a is shown directly in Fig. 11.6(b).

Experimentally, parallel ARM is less intense, usually by a factor 2 to 5, than TRM produced by the same steady field. However, in magnetite grains just around the critical size d_0, where SD, 2D and vortex states are all possible (§7.4, 7.5.2, 7.6), M_{tr}/M_{ar} rises to high values (Fig. 11.7). The cause is unknown but it is a useful 'fingerprint' of critical-SD-size magnetite grains.

ARM is not always an unwanted contaminant. Since ARM affects the entire grain distribution, it has often been argued that M_{ar} is analogous to M_{tr} produced by the same H_0, AF playing the role of a randomizing agent in the ARM case that is played by thermal excitations for TRM. The *coercivity spectra* of the two remanences are certainly similar (so is that of SIRM), and partial ARM's produced when H_0 is applied over only part of the range of $\tilde{H}$ are additive in the style of partial TRM's, but the argument is otherwise less than compelling. Be that as it may, ARM analog methods of estimating paleofield intensity use laboratory-induced ARM as a substitute for laboratory TRM (e.g., Banerjee and Mellema, 1974), thereby eliminating the need for heating and the risk of altering the rock chemically. This is a method of last resort (Bailey and Dunlop,

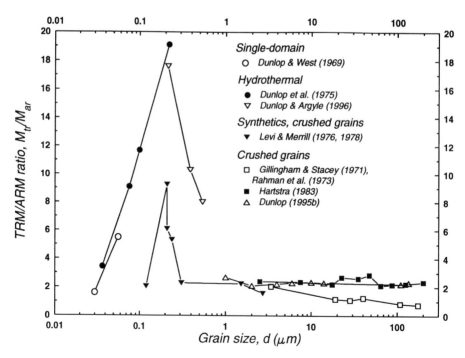

Figure 11.7   Observed peak of the TRM/ARM ratio around the critical SD threshold size d_0 in magnetite. [After Dunlop and Xu (1993) © American Geophysical Union, with the permission of the authors and the publisher.]

1977; Kono, 1978), when heating must be avoided at all costs: for example, with lunar rocks (Chapter 17), which equilibrated under very low oxygen fugacity conditions.

On safer ground is the use of ARM intensity as a normalizing factor in determining *relative* paleointensities in sediment cores (see Chapter 15). In this application, ARM intensity is used merely as a scaling factor to allow for differing magnetic mineral contents in different cores. However, the assumption that ARM intensity is simply proportional to the quantity of magnetite present in a core ignores the dependence of M_{ar} on grain size. In Chapter 12, we will see that M_{ar} is about an order of magnitude greater in 0.1μm magnetite grains than in 1μm magnetites. Other magnetic monitors must be used to ensure that grain size does not vary down-core, or to allow for size variations if they do occur.

Banerjee *et al.* (1981) and King *et al.* (1982) have suggested a comparison of anhysteretic susceptibility χ_a to DC initial susceptibility χ_0 as a suitable means of detecting grain-size variations (Fig. 11.8). The ratio χ_a/χ_0 can also be written as $Q_a = M_{ar}/\chi_0 H_0$, in which form it is immediately recognizable as the analog of the Koenigsberger ratio, $Q_t = M_{tr}/\chi_0 H_0$ for TRM (§9.3). Q_a is most sensitive as a grain-size indicator for magnetites $< 1\,\mu m$ in size, since this is the range in which χ_a is strongly size dependent (although to judge by Fig. 11.7, not as strongly size dependent as M_{tr} or Q_t over the same range). This is the size range

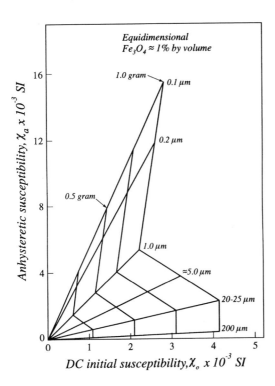

Figure 11.8 The field of variation of anhysteretic and DC susceptibilities, χ_a and χ_0, respectively, based on data for sized magnetites. The grain size and concentration of magnetite in natural samples can be determined by comparison with the diagram. [Reprinted from King *et al.* (1982), with kind permission of the authors and Elsevier Science – NL, Sara Burgerhartstraat 25, 1055 KV Amsterdam, The Netherlands.]

of greatest interest in deep-sea sediments, wind-transported loess, and many soils, silty sediments and muds.

Initial susceptibility χ_0 is probably the least size-dependent magnetic parameter over broad size ranges, being roughly similar for SD and MD magnetite alike (§5.7.1; Hunt et al., 1995a,, Fig. 1). It is this size independence that makes Q_a useful as a size indicator. Why not then simply use χ_0 as a normalizing factor to allow for differing magnetite contents in relative paleointensity determination? The objection usually raised is that χ_0 is determined by all magnetic material present, with the MD contribution if anything highlighted. Only the most stable part of NRM, resisting AF demagnetization in fields of ≈ 20 mT (200 Oe), is used for paleointensity determination. ARM, demagnetized to a similar level, is the best analog to this stable NRM. While this is true, it invalidates the use of χ_a/χ_0 as a grain-size proxy on the same sediments.

A final application of ARM is the use of partial ARM spectra to determine the coercivity spectrum (Jackson et al., 1988a). The reasoning is analogous to using partial TRM spectra as a measure of the blocking-temperature spectrum (Fig. 1.3, §8.8). The partial ARM spectrum agrees closely with the point-by-point slope of the AF demagnetization curve of total ARM and can be determined with less noise, since no numerical differentiation is required.

11.4.2   Gyromagnetic remanent magnetization (GRM)

Another laboratory remanence, acquired by fixed samples in alternating (or rotating) fields or by rotating samples in either steady or alternating fields, is called **gyromagnetic remanent magnetization** (GRM) (Stephenson, 1981). It originates in the gyroscopic response of SD moments to the torque of an applied field H_0. The moments precess about the direction of H_0 and only gradually relax into the direction of H_0. The net effect, especially pronounced for anisotropic rocks, is a spurious remanence M_{gr} in the plane perpendicular to H_0 or to the axis of the AF, $\tilde{H}$, *not* in the direction of H_0 or $\tilde{H}$. Because only SD moments have a gyroscopic response, domains and walls in MD grains should in principle be immune to GRM (Stephenson and Potter, 1995). It is a matter of common observation, however, that spurious remanences, probably GRM's, plague the highest steps when coarse-grained rocks are being AF demagnetized. It may be that the generally weak NRM of these rocks tends to highlight GRM moments carried by the fine grains.

Since the component of magnetization along the axis of AF demagnetization is unaffected by GRM, Dankers and Zijderveld (1981) proposed three separate measurements of remanence at each AF level, the sample being re-demagnetized in three orthogonal orientations with respect to the field axis. The cyclical permutation of the axis of demagnetization results in a cyclical displacement of the magnetization vector about the true NRM vector (Fig. 11.9), but the along-axis

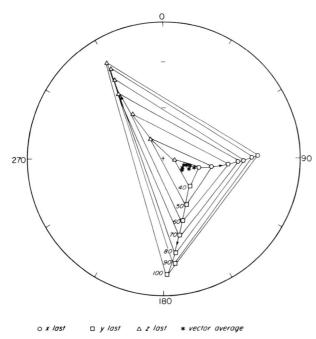

Figure 11.9 An example of gyromagnetic remanent magnetization (GRM) during static 3-axis AF demagnetization of a diabase sample. The equal-area stereographic projection illustrates the cyclic displacement of the measured NRM vector when the axis of demagnetization is cyclically permuted (x, y, z). Numbers are peak AF's in mT. The average vector at each level (stars) is stationary and estimates the true NRM direction, uncontaminated by GRM. [Unpublished data courtesy of L. D. Schutts.]

o *x last*   □ *y last*   △ *z last*   * *vector average*

components from the three steps give a good estimate of the uncontaminated NRM.

Another manifestation of GRM is a spurious remanence acquired parallel to the inner axis of rotation when a rock specimen is tumbled about two or three mutually orthogonal axes during AF demagnetization (Wilson and Lomax, 1972). This remanence is often called a **rotational remanent magnetization** (RRM), although it is not physically different from GRM (Stephenson, 1980). RRM reverses its direction when the sense of rotation is reversed or the frequency of rotation approaches that of the AF (Stephenson, 1976), and the RRM intensity has an unmistakable resonance peak (Fig. 11.10). The simplest cure for RRM is to select tumbling frequencies that are well off-resonance. A method of correcting for RRM is given by Hillhouse (1977).

11.5 AF demagnetization of SIRM of multidomain grains

The acquisition and demagnetization of remanence in MD grains involves two competing effects (§9.2–9.4): pinning of domain walls away from a demagnetized condition, leading to the idea of the microcoercivity distribution, $f(h_c)$, determined by $E_w(x)$; and the restoring force exerted by the internal demagnetizing field H_d. Except in 2D grains, self-demagnetization is not usually accomplished

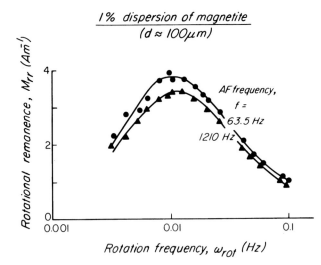

Figure 11.10 Intensity of rotational remanent magnetization (RRM) as a function of rotation frequency for synthetic coarse-grained magnetite. The resonance peak is here independent of the frequency f of the AF signal that produces the RRM. [After Edwards (1980), with the permission of the author and the publisher, Blackwell Science, Oxford, UK.]

by equal responses of all walls to H_d. It is often useful to imagine, as we did in §9.3, two populations of walls, hard and soft. The hard walls remain pinned in displaced positions, while the soft walls move freely in response to H_d so as to reduce the overall magnetization (magnetic screening).

11.5.1   A simple theory

A simple approach to AF demagnetization in the presence of screening is to write separate equations for the demagnetizing fields H_{df} and H_{dp} acting on the free and pinned walls, respectively (Xu and Dunlop, 1993; see also Shcherbakov and Markov, 1982):

$$H_{df} = -N_f M_f - N_m M_p, \qquad H_{dp} = -N_m M_f - N_p M_p. \qquad (11.1)$$

In (11.1), M_f, M_p are the magnetizations due to free and pinned walls, respectively, and N_f, N_p and N_m are interaction coefficients, analogous to demagnetizing factors, describing the self and mutual interactions between soft and hard walls. In the SIRM state, $H_0 = 0$ but the walls remain pinned at their maximum displacements by the microcoercivities h_c of their local energy wells. Making the simplifying assumption that all hard walls have the same microcoercivity $h_c = H_c$, while the soft walls have $h_c = 0$, we have by analogy with eqn. (9.14) for the descending branch of the hysteresis loop

$$N_f M_f + N_m M_p = 0, \qquad N_m M_f + N_p M_p = H_c. \qquad (11.2)$$

Solving and rearranging, we find

$$M_f = -\frac{N_m H_c}{N_f N_p - N_m^2}, \qquad M_p = \frac{N_f H_c}{N_f N_p - N_m^2}. \qquad (11.3)$$

The screening factor α (§9.3) is then

$$\alpha = \frac{M_p + M_f}{M_p} = 1 - \frac{N_m}{N_f}. \tag{11.4}$$

Keeping in mind that the intrinsic susceptibility is overwhelmingly due to the soft or free walls, it is easily shown that $\alpha \approx 1 - N_m \chi_i$. In §9.3, we found that the screening factor is $(1 + N\chi_i)^{-1}$, which is more readily evaluated than α and moreover gives us a way of estimating the mutual interaction coefficient N_m. Finally, combining (11.3) and (11.4), we have for the SIRM intensity

$$M_{rs} = \alpha M_p = \alpha \left(1 - \frac{N_m^2}{N_f N_p}\right)^{-1} \frac{H_c}{N_p}. \tag{11.5}$$

In AF demagnetization to a peak field $-\tilde{H}$, again by analogy with eqn. (9.14) but with $H_0 = -\tilde{H}$ rather than 0, we have that

$$N_f M_f + N_m M_p = -\tilde{H}, \qquad N_m M_f + N_p M_p = -\tilde{H} + H_c. \tag{11.6}$$

Solving for M_p, we obtain

$$M_p(\tilde{H}) = \left(1 - \frac{N_m^2}{N_f N_p}\right)^{-1} \frac{H_c - \alpha \tilde{H}}{N_p}. \tag{11.7}$$

At the negative peak field $-\tilde{H}$, the pinned walls are pushed closest to the demagnetized state. As the AF decays in amplitude, these walls merely oscillate within their local energy wells, and M_p scarcely changes. Thus after AF demagnetization, the SIRM is given by $\alpha M_p(\tilde{H})$, and by combining (11.5) and (11.7), we have for the normalized AF demagnetization curve of SIRM

$$\frac{M_{rs}(\tilde{H})}{M_{rs}(0)} = \frac{\alpha M_p(\tilde{H})}{\alpha M_p(0)} = 1 - \frac{\alpha}{H_c} \tilde{H}. \tag{11.8}$$

Equation (11.8) predicts that SIRM is only completely demagnetized when the AF peak field reaches a value H_c/α, which is $> H_c$ since $\alpha < 1$. Stacey and Banerjee (1974, p. 139) obtained an analogous result for ultimate demagnetizing field, namely $(1 + N\chi_i)H_c$. By setting $M_{rs}(\tilde{H})/M_{rs}(0) = 0.5$, we see that the median destructive field $\tilde{H}_{1/2}$ of SIRM is

$$\tilde{H}_{1/2} = H_c/2\alpha. \tag{11.9}$$

Thus the MDF is proportional to the average coercivity or bulk coercive force H_c, but is magnified by a factor $1/(2\alpha)$. In other words, a large apparent AF stability in MD grains can result either from large intrinsic microcoercivities h_c (firmly pinned walls) or from a large screening effect (i.e., a small value of α). In the latter case, the soft walls oscillate with large-amplitude displacements, producing a demagnetizing field H_{df} which to a great extent nullifies the effect of $\tilde{H}$ on the hard walls. It is the field $\tilde{H}$ that is in effect being screened.

11.5.2   Exponential AF demagnetization curves

The linear decrease of M_{rs} with increasing $\tilde{H}$ predicted by (11.8) is not in agreement with observed AF demagnetization curves for MD grains, which decrease
exponentially. Contrary to our assumptions in deriving (11.9), neither H_c nor α
remains constant during AF demagnetization. As $\tilde{H}$ increases in successive AF
steps, α will decrease (more effective screening) because more and more walls
are unpinned, while H_c will increase because the walls that remain pinned are
those with microcoercivities higher than the average. Both effects cause tailing
of the AF demagnetization curve to high coercivities.

To predict a realistic AF demagnetization curve, we must use numerical modelling with a full spectrum $f(h_c)$ of microcoercivities. In Figs. 9.1 and 9.2, we imagined a series of regular variations in $E_w(x)$ producing a set of energy barriers of
similar height and spacing. Such a picture is too simplistic for modelling real AF
demagnetization curves because identical barriers lead to a square hysteresis
loop with only a single microcoercivity, $h_c = H_c$ (§9.4), as in the analytic model
developed above. A more realistic microcoercivity distribution, one example of
many hundreds of numerical simulations, appears in Fig. 11.11. It represents the
variation of wall pinning force with position in a 10 μm magnetite grain resulting
from an array of 1000 randomly distributed screw dislocations (Xu and Dunlop,
1993). Both the heights and wavelengths of barriers vary in an irregular fashion.

In the absence of magnetic screening, the pinning force vs. position graph can
be transformed directly into a cumulative coercivity distribution or AF demagnetization curve. This is the case with 2D grains: each grain contains only a single

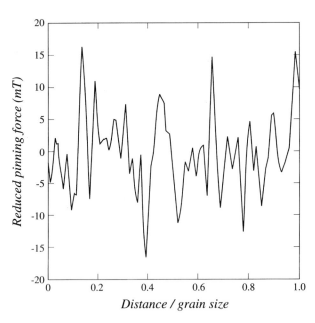

Figure 11.11   A
representative example of
the theoretical variation of
the pinning force, i.e.,
$h_c \propto |dE_w/dx|$ (eqn. (9.5)),
associated with an array of
1000 randomly distributed
dislocations in a 10 μm
magnetite cube. The
variation is much more
irregular than that given by
the idealized periodic
$E_w(x)$ functions of Figs.
9.1 and 9.2. [After Xu and
Dunlop (1993) ©
American Geophysical
Union, with the permission
of the authors and the
publisher.]

wall, and without additional walls, there can be no screening. As more and more walls are added, screening increases and the remanence becomes harder to demagnetize. Numerical modelling results, with screening included, are shown in Fig. 11.12 for an assumed dislocation density of 10^{13} m^{-2}.

The theoretical AF demagnetization curves of SIRM in Fig. 11.12 match quite well experimental curves of similar size magnetites if the grains contain on average between four and six domains in the 5–10 μm size range. This is a reasonable number according to Figs. 6.7(a), (b). All the curves have the exponential form characteristic of MD grains. As predicted, average coercivities increase as the number of domains increases and screening becomes more effective. At first sight, this result is counterintuitive. One must remember, however, that if *unnormalized* curves were plotted, the demagnetization tails would be similar in all cases, but the initial remanences would be weaker (because more effectively screened) for the larger numbers of domains.

Xu and Dunlop (1993) predicted that the screening effect can also result in *higher* MDF's when the dislocation density *decreases*. Weaker wall pinning results in lower microcoercivities, but microcoercivity becomes less important than screening when the level of internal stress drops. That is, α rather than H_c dominates in eqns. (11.8) and (11.9). The predicted effect is observed experimentally. In low-stress hydrothermal magnetites, Heider *et al.* (1992) measured AF

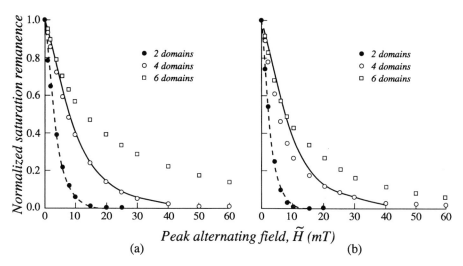

Figure 11.12 Model AF demagnetization curves (points) of SIRM for different numbers of domains in (a) 5 μm and (b) 10 μm magnetite cubes compared with measured curves (solid lines) from Bailey and Dunlop (1983). A moderately high dislocation density of 10^{13} m^{-2} was used in the modelling. The dashed curve is determined directly from the microcoercivity distribution and is appropriate for 2D grains. 4D and 6D grains have increasingly efficient screening by soft walls; the coercivities increase accordingly. [After Xu and Dunlop (1993) © American Geophysical Union, with the permission of the authors and the publisher.]

demagnetization curves that are almost size-independent and, for sizes $\geq 20\,\mu m$, have higher MDF's than corresponding AF curves for high-stress crushed magnetite grains.

Figure 11.13 illustrates the range of experimental shapes of AF demagnetization curves (in this case of TRM) over a very broad range of sizes, stretching from the SD range in magnetite to large MD sizes. The MD endmembers are distinctly exponential in form, while the SD endmembers are strongly sigmoid (cf. Fig. 11.2).

11.6   The Lowrie–Fuller test

Lowrie and Fuller (1971) proposed a test based on AF demagnetization which could be applied to NRM (of TRM origin) directly and would permit a rapid

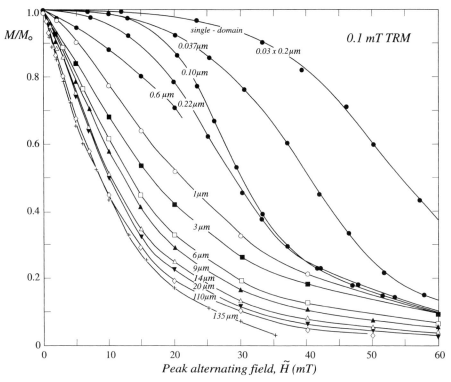

Figure 11.13 Comparison of AF demagnetization curves of a 0.1 mT TRM in magnetites with widely varying grain sizes. The shape of the demagnetization curve changes from sigmoid (S-shaped, with an initial plateau of no demagnetization and an inflection point at intermediate fields) to exponential as grain size increases from SD to large MD. [After Argyle *et al.* (1994) © American Geophysical Union, with the permission of the authors and the publisher.]

selection of igneous rocks with desirable SD carriers of NRM. The test is appealing because of its simplicity. It is based on the experimental observation that normalized AF demagnetization curves of weak-field TRM (i.e., NRM) and strong-field TRM (or SIRM, for experimental simplicity) have different relationships for SD and for large MD grains of magnetite. In large MD grains, SIRM requires larger destructive fields than weak-field TRM to reach the same normalized remanence level. In particular, the MDF of SIRM is larger than the MDF of weak-field TRM. In SD grains, the opposite is true. Laboratory investigations of the test often use weak-field ARM instead of weak-field TRM.

It was soon noticed that grains much larger than SD size d_0 gave an SD-type Lowrie–Fuller test result (Dunlop *et al.*, 1973a; Johnson *et al.*, 1975a; Bailey and Dunlop, 1983). This was not too disturbing because the changeover in the test result seemed to occur around 10–15 μm in magnetite, a size below which pseudo-single-domain (PSD) effects were believed to become important (see Chapter 12). However, a crisis arose when Heider *et al.* (1992) observed 'SD-type' behaviour for hydrothermal magnetites up to 100 μm in size (Fig. 11.14). Both the practical utility of the test and our understanding of why the observed responses occur came into question.

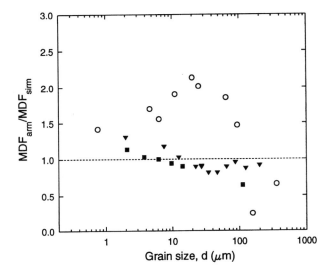

Figure 11.14   The result of the Lowrie–Fuller test expressed as the ratio between median destructive fields of weak-field ARM and SIRM, for low-stress hydrothermal magnetites (circles: data from Heider *et al.*, 1992) and crushed natural magnetites (triangles: Hartstra, 1982a; squares: Bailey and Dunlop, 1983). Points above and below the dotted line represent SD-type and MD-type test results, respectively. Hydrothermal magnetites have SD-type tests even in grains ≈ 100 μm in size. [After Xu and Dunlop (1995b) © American Geophysical Union, with the permission of the authors and the publisher.]

Based on the most recent theoretical work, the Heider *et al.* (1992) result is interpreted to be due to the increased importance of screening by soft walls in low-stress hydrothermal magnetites (cf. §11.5.2). The theory (Xu and Dunlop, 1995b) leads to the conclusion that the interplay between microcoercivity and screening can lead to opposite results of the Lowrie–Fuller test in high-stress and low-stress magnetites having the same size and number of domains. Thus the test result is not a simple indicator of grain size or domain state, as originally hoped.

In simple terms, the microcoercivity effect on AF demagnetization can be understood in the following terms (Lowrie and Fuller, 1971). A strong field H_0 pushes walls farther from a demagnetized state, and leaves them pinned by higher E_w barriers, than a weak H_0. If there were no self-demagnetization to contend with, this would certainly make strong-field remanence harder to AF clean than weak-field remanence. Bailey and Dunlop (1983), on the other hand, assumed that self-demagnetization is the more important effect. In a 2D grain, there is a very direct connection between the displacement of a wall and the resulting internal demagnetizing field: $H_d = -NM$. The larger the wall displacement, the larger M and H_d. In fact, as H_0 increases, pushing the wall farther from its demagnetized position, more and more of the remanence will be spontaneously demagnetized by H_d when $H_0 \rightarrow 0$. An exponential coercivity spectrum has the property that its form, after renormalization, is unaffected by truncation at the lower end: it gives a null Lowrie–Fuller result. A sub-exponential spectrum, which Bailey and Dunlop assumed to be typical of large MD grains, gives the MD-type result, while a superexponential spectrum (which is certainly characteristic of PSD and SD grains) gives the SD-type result.

The most recent theory (Xu and Dunlop, 1995b) embodies both the coercivity and self-demagnetization aspects. It yields much the same result Bailey and Dunlop argued for, except that exponential and Gaussian microcoercivity distributions, which are typical of large MD grains, yield an MD-type, not a null, test result (Fig. 11.15a). The predicted curves agree well with measurements for a sample containing crushed 110 μm magnetite grains (Fig. 11.15b).

In natural settings, internal stress levels as low as those in Heider *et al.*'s magnetites are not likely. The Lowrie–Fuller test may therefore still be useful as a grain-size discriminator for magnetite carrying NRM of TRM origin in igneous rocks. It should be kept in mind that the actual response is to the change in shape of the coercivity spectrum from exponential to more sigmoid shapes, not to domain state or grain size per se. In fact, the same information can be derived from the shapes of the AF curves directly (Fig. 11.13).

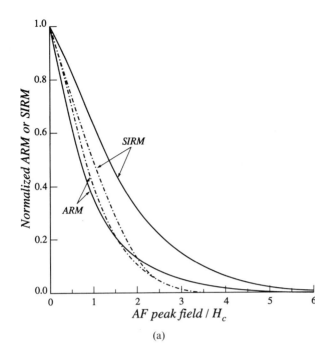

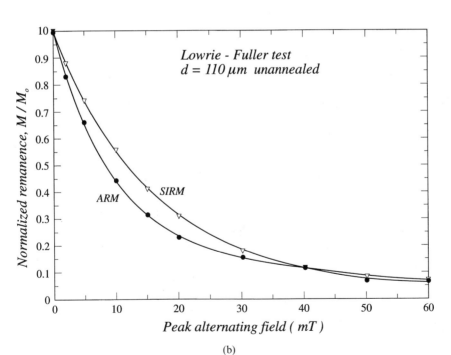

Figure 11.15 (a) Theoretical Lowrie–Fuller tests (comparison of AF demagnetization curves of weak-field ARM and SIRM) for MD grains with Gaussian (dot-dash) and exponential (solid curves) microcoercivity distributions. The test result is MD-type in both cases. (b) Experimental Lowrie–Fuller test for 110 μm magnetite grains, showing good agreement with the predictions of (a). [After Xu and Dunlop (1995b) and Argyle *et al.* (1994), both © American Geophysical Union, with the permission of the authors and the publisher.]

(a)

(b)

11.7   Grain interactions and the Wohlfarth relations

The interaction field H_{int} due to near neighbours of a single-domain grain plays in some respects an analogous role to the demagnetizing field H_d in multidomain grains (cf. §8.6). H_{int} increases in strength as the remanence level of the sample increases and is responsible, in the SD case, for the truncation effect described in the previous section. Isolated SD grains with no mutual interactions would yield a null Lowrie–Fuller result, i.e., weak-field and strong-field remanences would demagnetize in exactly the same way. The fact that these remanences are always observed to demagnetize differently is indirect evidence that SD grain interactions are not negligible.

Wohlfarth (1958) proposed formal relations among the IRM acquisition curve, $M_{ir}(H_0)$, the 'DC demagnetization' curve, $M_r(-H_0)$ (see Fig. 11.5), and the AF demagnetization curve, $M_r(\tilde{H})$. On the asumption that the three curves give identical measures of the coercivity spectrum for non-interacting SD grains,

$$M_r(-H_0) = M_{rs} - 2M_{ir}(H_0) \quad \text{and} \quad M_r(\tilde{H}) = M_{rs} - M_{ir}(H_0).$$

$$(11.10)$$

Deviation from these relations indicates particle interaction. When normalized to SIRM, M_{rs}, and related to $M_r(\tilde{H})$, which is most frequently determined in paleomagnetism, (11.10) becomes (Dunlop, 1986a)

$$\frac{M_r(\tilde{H})}{M_{rs}} = \frac{1}{2}\left[1 + \frac{M_r(-H_0)}{M_{rs}}\right] = 1 - \frac{M_{ir}(H_0)}{M_{rs}}.$$

$$(11.11)$$

The meaning of (11.10) and (11.11) is readily appreciated graphically (Fig. 11.16). The AF demagnetization curve has (approximately) the same shape as the DC demagnetization curve, reflected about the $H = 0$ axis and scaled down by a factor 2, and both are similar to the inverted IRM acquisition curve.

The fact that the rescaled and reflected or inverted curves in Fig. 11.16 are not identical indicates that the magnetite grains in this synthetic sample interact significantly, probably because they are clumped. If we compare median fields for the various curves, which are also median values of the corresponding coercivity spectra, dM_r/dH_0 or $dM_r/d\tilde{H}$, we find that $\tilde{H}_{1/2} < H_{cr} < H'_{cr}$. AF demagnetization always gives the lowest coercivities and IRM acquisition the largest coercivities, basically because the mean interaction field in a sample is a demagnetizing field which aids in AF demagnetization but opposes IRM acquisition. Dankers (1981) proposed the relation

$$H'_{cr} + \tilde{H}_{1/2} \approx 2H_{cr}, \quad \text{i.e., } H_{cr} = \tfrac{1}{2}(H'_{cr} + \tilde{H}_{1/2}).$$

$$(11.12)$$

Dankers' relation is obeyed quite well by the sample of Fig. 11.16. The AF demagnetization and inverted IRM acquisition curves flank the scaled-down DC demagnetization curve at about equal distances.

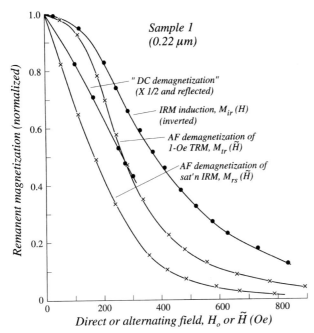

Sample 1
(0.22 μm)

"DC demagnetization"
(X 1/2 and reflected)

IRM induction, $M_{ir}(H)$
(inverted)

AF demagnetization of
1-Oe TRM, $M_{tr}(\tilde{H})$

AF demagnetization of
sat'n IRM, $M_{rs}(\tilde{H})$

Remanent magnetization (normalized)

Direct or alternating field, H_o or $\tilde{H}$ (Oe)

Figure 11.16
Comparison of DC 'demagnetization', IRM acquisition, and AF demagnetization curves, suitably reflected, rescaled or inverted for the purposes of testing the Wohlfarth relations (eqns. (11.10), (11.11)). The curves are all of approximately the same form, but the offsets between them are evidence of grain interactions. The sample contains magnetite grains somewhat above critical SD size. [Reprinted from Dunlop (1986a), with kind permission of the author and Elsevier Science – NL, Sara Burgerhartstraat 25, 1055 KV Amsterdam, The Netherlands.]

A popular way of comparing the AF and IRM acquisition coercivity spectra is by plotting $M_r(\tilde{H})/M_{rs}$ and $M_{ir}(H_0)/M_{rs}$ curves on the same graph (without inversion of the IRM curve) and examining the value R of M_r/M_{rs} at which the two curves cross (Cisowski, 1981). In the examples shown in Fig. 11.17, $R = 0.5$ for the non-interacting magnetites in the Lambertville plagioclase (cf. Fig. 11.6a), showing that the AF and IRM acquisition curves are mirror images of each other. For the interacting chiton magnetites, R is well below 0.5. This method is not well suited to an actual point-to-point comparison of the two curves and their coercivity spectra, and so rather under-utilizes the data. However, the curves are observed to cross at a field equal to H_{cr}, in agreement with (11.12), which shows that the average of $\tilde{H}_{1/2}$ and H'_{cr} is H_{cr}.

The internal demagnetizing field H_d in MD grains will have the same effect on the Wohlfarth relations as the average H_{int} in SD grains, and will result in graphs very much like Figs. 11.16 and 11.17(b). In fact, Dankers' (1981) investigation, on which (11.12) is based, was of MD magnetites. Neither type of plot is a definitive test for SD grain interactions, unless the grain size is known independently.

Another method of comparing IRM acquisition and DC demagnetization, utilized in magnetic recording research, is to plot remanence 'lost', $M_r(-H_0)/M_{rs}$, versus remanence gained, $M_{ir}(H_0)/M_{rs}$, with H_0 as a parameter. This graph, the Henkel (1964) plot, is the analog of the Arai plot in Thellier paleointensity determination (Figs. 8.14, 9.8, 9.9), with remanence gained/lost

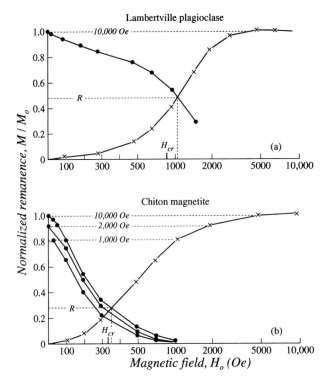

Figure 11.17
Comparison of IRM
acquisition and AF
demagnetization of SIRM
in (a) weakly interacting
magnetites (crossover or R
value 0.5) and (b) strongly
interacting magnetites
($R < 0.5$). In both cases,
the curves cross at a field
equal to H_{cr}, as predicted
by eqn. (11.12). The
samples are the same as in
Fig. 11.6(a). [Reprinted
from Cisowski (1981), with
kind permission of the
author and Elsevier
Science – NL, Sara
Burgerhartstraat 25, 1055
KV Amsterdam, The
Netherlands.]

at temperature T replaced by remanence acquired/demagnetized at field H_0.
Like the Thellier graph, a Henkel plot will sag below the ideal SD line (in this
case of slope -2) in the case of MD grains (cf. Fig. 9.8), and also in the SD case
if the mean grain interaction field is negative (Proksch and Moskowitz, 1994).

11.8   The Preisach diagram

A method of storing information about IRM acquisition, DC demagnetization,
ARM acquisition and SD grain interactions is the Preisach (1935) diagram.
Néel (1954) proposed the following interpretation. Suppose a local interaction
field H_i acts on an SD grain, due to nearby grains and independent of changes
in the overall magnetization of the sample. (This constant local field, which fluc-
tuates spatially – Néel's fluctuating field – but not temporally as M changes,
represents the fine structure of the interaction field H_{int} in an MD grain, whose
mean is the demagnetizing field H_d.) The effect of the local field H_i, if it acts parallel to H_0, is to offset the grain's hysteresis loop, producing unequal coercivities
$a = H_c - H_i$ and $b = -H_c - H_i$ (Fig. 11.18a). The Preisach diagram is a plot of

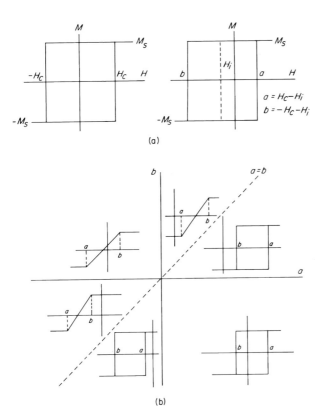

Figure 11.18 (a) Elementary hysteresis loops of an SD grain with and without a (constant) local interaction field H_i acting parallel to H_0 and the easy axis (cf. Fig. 8.6). H_i shifts the loop, making the critical fields a and b unequal. (b) Elementary reversible and irreversible magnetization cycles with (a, b) in different sectors of the Preisach diagram. [Reprinted from Dunlop et al. (1990b), with kind permission of the authors and Elsevier Science – NL, Sara Burgerhartstraat 25, 1055 KV Amsterdam, The Netherlands.]

b vs. a, with an associated grain distribution function $F(a, b)$. Only grains with (a, b) in the fourth quadrant carry a stable remanence (Fig. 11.18b).

$F(a, b)$ is the product of two independent distributions, $g(H_c)$ measured along the diagonal $a = -b$ and $f(H_i)$ measured perpendicular to $a = -b$ (Fig. 11.19a). Thus if we can determine $F(a, b)$, we can determine both the coercivity spectrum in the absence of interactions (by profiling along the diagonal) and the distribution of grain interactions H_i (by profiling perpendicular to the diagonal). AF demagnetization leaves the diagram polarized about the diagonal (Fig. 11.19b). IRM acquisition in field H_0 reverses the remanence in a triangular region above the diagonal (Fig. 11.19b), whereas DC demagnetization (not shown) reactivates a region bounded by a horizontal line, and changes part of the previously acquired IRM. By successive application of positive and negative fields, increasing in steps to eventual saturation, $F(a, b)$ can be worked out in squares on a grid (Hejda and Zelinka, 1990; Hejda et al., 1990, 1994). The automated alternating-gradient force magnetometer (AGFM) is well suited to such measurements (e.g., Proksch and Moskowitz, 1994).

$F(a, b)$ is widely used in magnetic technology as a kernel which efficiently stores information about the irreversible (and also reversible, if other quadrants are worked out) DC response of a sample. But, more excitingly, it can be used

(a)   Preisach distribution function F(a,b)

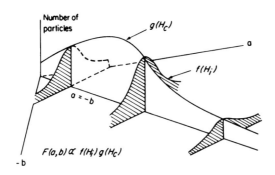

(b)   Magnetization processes on the Preisach diagram

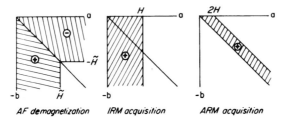

AF demagnetization   IRM acquisition   ARM acquisition

Figure 11.19   (a) The Preisach function $F(a,b)$ as a product of two independent distribution functions, $g(H_c)$ parallel to the diagonal $a = -b$ and $f(H_i)$ perpendicular to the diagonal. (b) Areas of positive and negative magnetization following AF demagnetization to peak field $\tilde{H}$, IRM acquisition in field $+H$, and ARM acquisition in a bias field H. The magnetization in blank areas depends on the previous magnetic history of the sample. [Reprinted from Dunlop *et al.* (1990b), with kind permission of the authors and Elsevier Science – NL, Sara Burgerhartstraat 25, 1055 KV Amsterdam, The Netherlands.]

to predict and interpret the results of quite different types of magnetic measurements. For example, ARM, since it results from a small bias field H_0 superimposed during AF demagnetization, polarizes the Preisach diagram in a strip of width $\sqrt{2} \times H_0$ parallel to the diagonal (Fig. 11.19b). Thus ARM acquisition amounts to overcoming interaction fields with a maximum strength $\sqrt{2} \times H_0$, and one can estimate $f(H_i)$ from the measured $M_{ar}(H_0)$ curve. Conversely, one can use the experimental Preisach function $F(a,b)$ to predict ARM acquisition and interaction fields (Petrovsky *et al.*, 1993).

Predicted and measured curves, based on Preisach diagrams and ARM acquisition measured independently at a variety of temperatures (Dunlop *et al.*, 1990b), are compared in Fig. 11.20. The sample contains interacting SD grains of magnetite. The agreement between the Preisach prediction and measured ARM acquisition is excellent at all temperatures. Notice that the ARM increases more rapidly with H_0, i.e., χ_a is higher, at high temperatures. This is because dipole interaction fields decrease in proportion to $M_s(T)$ (since $\mu = VM_s$) at high temperature (compare Figs. 11.6b and 11.20).

In Fig. 11.20, the TRM acquisition curve, $M_{tr}(H_0)$, is very similar to $M_{ar}(H_0)$ measured at 460 °C. In fact, there is a close analogy between ARM and TRM acquisition by interacting SD grains. Both can be regarded as resulting from the switching of grain moments as H_0 overcomes the effect of oppositely directed

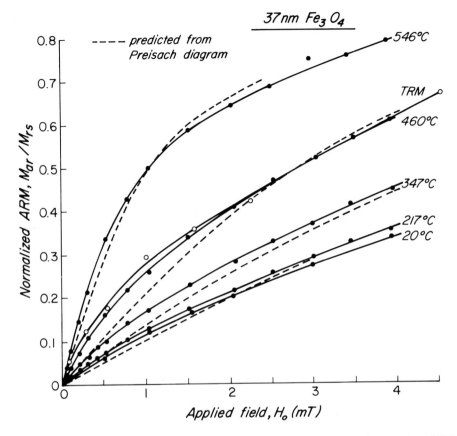

Figure 11.20 Experimental ARM acquisition curves measured at a series of high temperatures and the TRM acquisition curve, for an SD magnetite sample (sample 4 of Figs. 11.5 and 11.6a). Dashed curves are predictions based on the off-diagonal distribution $f(H_i)$ from experimental Preisach diagrams, determined by DC measurements (cf. Fig. 11.19). The TRM acquisition curve matches closely the acquisition of ARM at a temperature of 460 °C, suggesting a fundamental analogy between single-domain ARM and TRM. However, 460 °C is lower than typical TRM blocking temperatures (500–560 °C; Fig. 8.3a) for this sample. [Reprinted from Dunlop *et al.* (1990b), with kind permission of the authors and Elsevier Science – NL, Sara Burgerhartstraat 25, 1055 KV Amsterdam, The Netherlands.]

interaction fields, H_i (cf. §8.6). Theoretically, $M_{tr}(H_0)$ is the integrated interaction field spectrum at the blocking temperature T_B, just as the room-temperature ARM acquisition curve $M_{ar}(H_0)$ is $\int f(H_i)\, dH_i$ at ordinary temperatures (Dunlop and West, 1969; Jaep, 1971). Since average interaction strength increases with increasing concentration of magnetic grains, both ARM and TRM 'susceptibilities' have a concentration dependence (Sugiura, 1979; Shcherbakov and Shcherbakova, 1979).

11.9   Hysteresis and magnetic granulometry

11.9.1   Information from hysteresis loops

With the introduction of automated AGFM's and vibrating-sample magnet-ometers (VSM's), the measurement of hysteresis curves for even quite weakly magnetic rocks and sediments has become quick and reliable. Software packages automatically correct for the diamagnetic or paramagnetic signal of the rock matrix, which is isolated above ferromagnetic saturation. A hysteresis loop, if minor loops at intermediate fields are included, contains all the information about the coercivity spectrum that is usually obtained by AF demagnetization or IRM acquisition and in addition an equal wealth of information about reversible processes, e.g., the motion of unpinned domain walls. Most of this information at present goes unused in rock magnetic studies.

Some attempts have been made to characterize the entire hysteresis loop. Jackson *et al.* (1990) took an engineering approach, unfolding the loop into a pseudo-waveform and finding its frequency content by Fourier analysis. The Fourier coefficients are rather remote from the physical parameters controlling the magnetization process, but the loop can be fitted with arbitrary precision. Hejda (1985), on the other hand, used a smaller number of physically more meaningful parameters and obtained acceptable fits to symmetrical loops with one inflection point in each branch. The vast majority of measured loops falls in this category. With sufficiently precise fits, and by measuring minor loops as well as the major loop, the fourth quadrant of the Preisach diagram can be mapped out in detail (Zelinka *et al.*, 1987).

Usually only the parameters M_s, M_{rs}, H_c and H_{cr}, or ratios between them, are utilized from the hysteresis loop measurement. It is difficult to measure χ_0 in low fields because the AGFM and VSM both use electromagnets. Instead χ_0 is measured on a larger rock sample, e.g., a standard 2.5 cm cylindrical paleomagnetic sample, using a susceptibility bridge whose sensitivity is not cross-calibrated to that of the AGFM or VSM. Thus the ratio χ_0/M_s is usually not determined. This is unfortunate, because high values of this ratio are diagnostic of SP material (§5.7.2), which is otherwise hard to detect, since it has no remanence. The standard, rather cumbersome method of detecting SP grains is to compare hysteresis loops measured at different temperatures below T_0, looking for unblocking of remanence as evidenced by a drop in M_{rs}/M_s. More recently, with the introduction of automated instruments combining a superconducting quantum interference device (SQUID) detector and a superconducting solenoid, remanence unblocking can be tracked continuously by producing SIRM at very low temperature (5–10 K) and warming the sample to T_0. An example of such data, with a strong drop in remanence at low temperatures as grains become SP, appears in Fig. 3.6(b). Few rock magnetic laboratories are equipped with low-temperature SQUID magnetometers, and so room-temperature hysteresis parameters form the bulk of the published data.

11.9.2 M_{rs}/M_s and H_{cr}/H_c for ideal MD grains

Fortunately the remanence ratio M_{rs}/M_s and the coercivity ratio H_{cr}/H_c have diagnostic values for SD and for large MD grains. 'Large' or 'ideal' MD denotes a magnetization response to H_0 that is limited or offset by H_d to the greatest possible extent. A 2D or 3D grain cannot respond as effectively to self-demagnetization as an MD grain containing many walls, in which Barkhausen jumps of strongly pinned or hard walls (producing remanence) are largely screened or compensated for by negative displacements of loosely pinned or soft walls (with intrinsic susceptibility χ_i) in response to $H_d = -NM$. In ideal MD behaviour, the net magnetization is reduced to a fraction α or $(1 + N\chi_i)^{-1}$, which is the screening factor (see §9.3, 11.5.1).

There is no definite size above which grains are 'large' and behave in 'ideal MD' fashion. High internal stress reduces χ_i, because walls tend to be more strongly pinned. Ideal MD behaviour then occurs at larger grain sizes when more walls are present to respond to H_d. The nucleation of an entire domain and its bounding wall is a much larger event than small Barkhausen jumps of existing domain boundaries, and may be difficult to compensate by self-demagnetization. Failure of nucleation following saturation can leave a grain in a metastable few-domain state (e.g., SD or 2D), farther from ideal MD behaviour than it would be if it contained an equilibrium number (i.e., more) walls.

An ideal MD hysteresis loop was illustrated in Fig. 5.16. It has a slope $1/N$, which is the limit of $\chi_0 = \chi_i(1 + N\chi_i)^{-1}$ when χ_i is large (i.e., $\gg 1/N$). Since the loop is essentially linear over the small section between M_{rs} and $-H_c$, we have that $M_{rs} \approx H_c/N$ (eqns. (5.32), (9.26)). H_c is the average of the microcoercivity distribution $f(h_c)$ (see §9.2, 11.5). The value of H_c depends on the strength of wall pinning by dislocations and other lattice defects and thus in turn on the level of internal stress, but it is not likely to exceed $10\,\text{mT}$ ($8\,\text{kA/m}$ or $100\,\text{Oe}$) for MD magnetite. For equidimensional grains, $N_{MD} = \frac{1}{3}$ in SI or $4\pi/3$ in cgs emu (Fig. 5.4b), giving for magnetite ($M_s = 480\,\text{kA/m}$)

$$\frac{M_{rs}}{M_s} \approx \frac{H_c}{NM_s} \leq 0.05. \qquad (11.13)$$

To derive H_{cr}, we shall appeal to a comparison between the internal-field and external-field hysteresis loops, $M(H_i)$ and $M(H_0)$ as suggested by Néel. In §9.4.1, we saw that the intersection of $M(H_i)$ and a demagnetizing line of slope $-1/N$ yields the remanence M_r for $H_0 = 0$ (Fig. 9.3a). We can generate the entire externally observed loop $M(H_0)$ by considering intersections between $M(H_i)$ and a set of demagnetizing lines $M = (1/N)(H_0 - H_i)$ (eqn. (9.12)) for different values of H_0. The upright internal-field loops and the 'sheared' external-field loops were compared in Figs. 9.3(a) and (b). The inverse operation, in which $M(H_i)$ is obtained from the measured $M(H_0)$, is illustrated in Fig. 11.21. The hysteresis loop shown is actually intermediate between SD and true MD shapes and is typical of pseudo-single-domain (PSD) behaviour, to be discussed in Chapter 12. Most of the 0.22 μm magnetite grains are probably 2D, and there-fore not fully responsive to self-demagnetization, but we have followed Néel's procedure nonetheless.

The dashed internal-field loop (Fig. 11.21a) is more upright than the measured loop and has been obtained by 'unshearing' the measured loop. The values of H_c are the same in the sheared and unsheared loops, but M_{rs}/M_s is much larger when measured with respect to the net field $H_i = H_0 - H_d$ acting inside the grain.

A back field $-H_{cr}$ produces a magnetization $-M_D$. If instead of continuing to trace the major loop, $-H_0$ were reduced from $-H_{cr}$ to 0, zero remanence would result. Such a minor loop, with slope χ_i, from M_D to 0 is shown relative to the internal major loop in Fig. 11.21(b). In this plot, it is evident that the descending major loop is very steep between H_c and H_D (the *internal* field whose removal results in zero remanence – the 'true' coercivity of remanence). Approximating H_D by H_c, and noting that the line joining M_D and H_{cr} has slope $-1/N$ while the line joining M_D and the origin has slope χ_i, we have

$$M_D \approx \frac{H_{cr} - H_c}{N} \approx \chi_i H_c. \qquad (11.14)$$

Thus

$$H_{cr} \approx (1 + N\chi_i)H_c \quad \text{or} \quad H_{cr}/H_c \approx 1 + N\chi_i \approx \alpha^{-1}, \qquad (11.15)$$

where α is the screening factor (§9.3, 11.5.1).

Equations (11.14) and (11.15) have a number of important implications. First, (11.14) predicts that χ_i and H_c are closely related, which is hardly surprising since both are determined by the slopes dE_w/dx within potential wells (χ_i near the bottoms of wells and H_c at the point of escape or steepest slope). (A similar inverse relationship between χ_i and H_c holds for SD grains, e.g., Fig. 3.19.) Equa-tion (11.14) provides a formula for estimating χ_i, which is otherwise inaccessible to direct measurements.

Equation (11.15) demonstrates that H_{cr} is related to H_c by the inverse of the screening factor α. $H_c(\approx H_D)$ is the truest measure of average microcoercivity in a sample because it is measured when $M = 0$ and H_d, on average, is also zero. To measure H_{cr}, the walls must be driven away from the demagnetized state to magnetization M_D (Fig. 11.21) in the face of considerable opposition from H_d. Thus even though the walls spring back to a state of zero *remanent* magnetization when $H_0 \to 0$, H_{cr} is an inflated measure of average microcoercivity. Just how inflated can be judged by substituting a typical value $\chi_i = 10$ (≈ 0.8 in cgs), giv-ing $H_{cr} \approx 4H_c$ or $H_{cr}/H_c \approx 4$. This is often taken as the lower limit of H_{cr}/H_c for true MD grains.

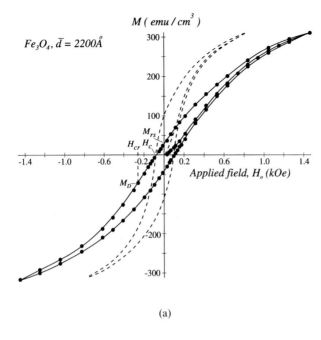

$Fe_3O_4, \bar{d} = 2200\text{Å}$

(a)

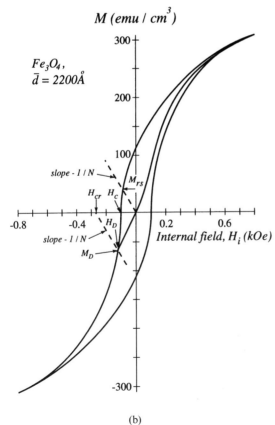

$Fe_3O_4,$
$\bar{d} = 2200\text{Å}$

(b)

Figure 11.21 (a) Measured hysteresis loop for a dispersion of small MD magnetite grains (probably containing 2 or 3 domains; see Fig. 11.26) and the corresponding internal-field hysteresis loop (dashed), simulated by 'unshearing' the measured loop, following Néel's procedure (see text). The internal-field loop is more upright than the external-field loop and has a higher M_{rs}/M_s. (b) The internal-field loop of (a) redrawn, with demagnetization lines of slope $-1/N$ (dashed), part of a minor loop (from M_D to the origin), and various parameters needed for the calculation of H_{cr}/H_c. [After Dunlop (1984) © American Geophysical Union, with the permission of the author and the publisher.]

H_{cr}/H_c can become very large as internal stress is reduced and walls become so loosely pinned that χ_i increases to values $\gg 10$. For example, the mm-size natural single crystal of magnetite illustrated in Fig. 11.22 has ideal MD hysteresis: a ramp-shaped $M(H_0)$ curve with slope $1/N$ and $M_{rs}/M_s = 0.003$. Its average microcoercivity is $H_c = 0.52\,$mT (5.2 Oe), but $H_{cr} = 12\,$mT (120 Oe) and $H_{cr}/H_c = 22.6$. Thus reduction of stress and wall pinning, resulting in increased magnetic screening, actually leads to an *increase* in remanence coercivity H_{cr}, and for similar reasons (as discussed in §11.5), in AF coercivities. This is the explanation of the quasi-exponential AF demagnetization tails measured for MD grains, and of the increase in AF coercivities with decreasing internal stress or increasing numbers of domains (Figs. 11.12, 11.14).

11.9.3 M_{rs}/M_s and H_{cr}/H_c for SD grains

An idealized SD hysteresis curve appeared in Fig. 5.19 and a theoretical curve for randomly oriented, uniaxial SD grains in Fig. 8.6. Following saturation by H_0, the magnetization M_s of each SD grain rotates back to the nearest easy direction as $H_0 \to 0$. The calculation of M_{rs}/M_s is thus purely geometrical: it is the average of $\cos\phi$ over a half-sphere (for uniaxial anisotropy) or over smaller-angle cones for $\langle 111 \rangle$, $\langle 100 \rangle$ or hexagonal easy axes. The results are (Dunlop, 1971; Joffe and Heuberger, 1974)

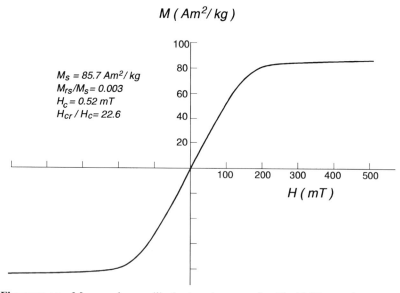

Figure 11.22 Measured ramp-like hysteresis curve of an ideal MD sample, a mm-size natural single crystal of magnetite. [After Özdemir *et al.* (1995) © American Geophysical Union, with the permission of the authors and the publisher.]

$$M_{rs}/M_s = \begin{array}{ll} 0.500 & \text{uniaxial} \\ 0.866 & \text{cubic, } \langle 111 \rangle \text{ easy axes (magnetite)} \\ 0.832 & \text{cubic, } \langle 100 \rangle \text{ easy axes (TM60)} \\ 0.500\text{--}0.637 & \text{uniaxial, easy plane (hematite)} \\ 0.750\text{--}0.955 & \text{triaxial, easy plane (hematite)} \end{array} \qquad (11.16)$$

For hematite, the lower M_{rs}/M_s values are appropriate if true saturation is achieved (M_s pulled out of the easy basal plane into parallelism with H_0) and the higher values are to be used if M_s remains pinned in the easy plane at the highest H_0 used. The distinction is not academic because saturation fields for hematite approach 3 T (30 kOe).

The estimation of H_{cr}/H_c is more problematic. For coherent rotation in randomly oriented grains of a single microscopic coercive force H_K, H_{cr}/H_K is 0.524 for uniaxial anisotropy (Wohlfarth, 1958), 0.204 and 0.333 for cubic anisotropy with $\langle 111 \rangle$ and $\langle 100 \rangle$ easy axes, respectively (Joffe and Heuberger, 1974), and 0.642 for uniaxial anisotropy within an easy plane (Dunlop, 1971). H_c has been calculated by averaging hysteresis loops at different angles (Stoner and Wohlfarth, 1948; Fig. 8.6), giving $H_{cr}/H_c = 1.04$–1.09 (Wohlfarth, 1958; Joffe and Heuberger, 1974). However, there are complications. A distribution of microcoercivities H_K can produce ratios as high as 2 (Gaunt, 1960). On the other hand, incoherent reversals reduce critical fields for grains at small angles to H_0 (Fig. 8.7), thereby reducing H_{cr}/H_c. However, H_{cr} can never be less than H_c, and so

$$1 \le H_{cr}/H_c \le 2 \qquad \text{uniaxial SD} \qquad (11.17)$$

must be true. A value $H_{cr}/H_c = 1.5$ has sometimes been used as an SD limit (Day et al., 1977) but this figure is purely arbitrary.

11.9.4   Correlation plot of M_{rs}/M_s versus H_{cr}/H_c

H_{cr}/H_c has less discriminating power than M_{rs}/M_s. Nevertheless it is common practice to use both parameters simultaneously in a plot of M_{rs}/M_s versus H_{cr}/H_c (Day et al., 1977). When data for both hydrothermally grown and crushed magnetites are plotted in this way (Fig. 11.23), they follow quite a narrow inverse relationship. Different levels of internal stress move points along the correlation line, but not off it. There are few data in the SD region ($M_{rs}/M_s \ge 0.5$, $H_{cr}/H_c \approx 1$) but rather more in the ideal MD region ($M_{rs}/M_s \le 0.05$, $H_{cr}/H_c \ge 4$). Most points fall between SD and ideal MD, in what is generally referred to as the PSD region (see Chapter 12).

The correlation line of Fig. 11.23 is not universally obeyed. Dunlop (1981) found that oceanic gabbros and peridotites of different provenances each followed a different inverse relationship between M_{rs}/M_s and H_{cr}/H_c. Figure 11.24 demonstrates that different limestones obey different correlation 'laws'

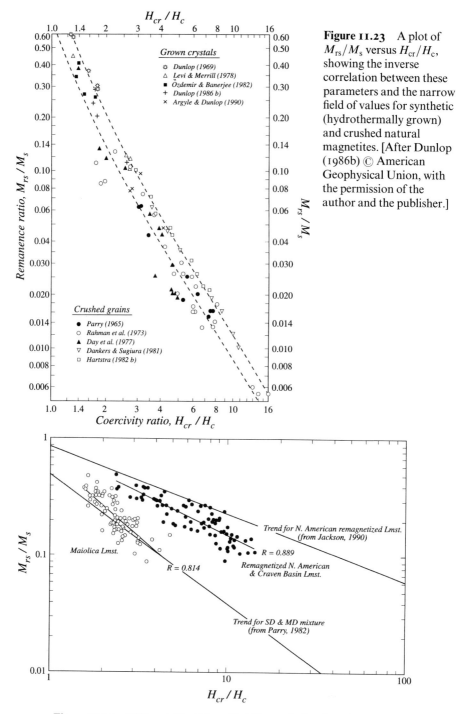

Figure 11.23 A plot of M_{rs}/M_s versus H_{cr}/H_c, showing the inverse correlation between these parameters and the narrow field of values for synthetic (hydrothermally grown) and crushed natural magnetites. [After Dunlop (1986b) © American Geophysical Union, with the permission of the author and the publisher.]

Figure 11.24 Different $M_{rs}/M_s - H_{cr}/H_c$ correlation lines for the unremagnetized Maiolica limestone and for remagnetized North American and European limestones. The SD & MD mixture line corresponds to the average trend in Fig. 11.23. [After Channell and McCabe (1994) © American Geophysical Union, with the permission of the authors and the publisher.]

(Channell and McCabe, 1994). The Maiolica limestone, which escaped late Paleozoic remagnetization (§1.2.5), follows the trend of Fig. 11.23, but remagnetized limestones from both sides of the Atlantic follow an unusual trend, with relatively high (SD to PSD) M_{rs}/M_s values but PSD to MD H_{cr}/H_c values of 3–10.

The precise carriers of NRM in the remagnetized limestones have proven difficult to identify but they are probably secondary magnetites finer in grain size than the primary NRM carriers in unremagnetized carbonate rocks. Jackson (1990) argued that unusually high H_{cr}/H_c values could result from a mixture of SP and stable SD grains. The high induced magnetization of SP grains in even a small negative field could cancel M_{rs} of the stably magnetized SD or larger grains, leading to an anomalously low value of H_c. H_{cr} would be unaffected because it depends only on remanence (in SD grains at least). Thus high values of H_{cr}/H_c are a hallmark of SP as well as MD grains.

Of course, M_{rs}/M_s should also be reduced by an admixture of SP material. The intercept of the remagnetization trend in Fig. 11.24 is >0.87, implying that the grains are nearly equidimensional and governed by cubic crystalline anisotropy (eqn. (11.16), line 2). If so, the measured M_{rs}/M_s values have been reduced substantially by SP admixture. Gee and Kent (1995) find similar evidence in their M_{rs}/M_s vs. H_{cr}/H_c plots for mid-ocean ridge basalts of dominant cubic anisotropy in nearly equidimensional SD grains.

11.9.5 Hysteresis of mixtures

We have just seen that rocks do not always, or even typically, contain grains in just one domain state. Nor do they necessarily contain only a single magnetic mineral. One striking manifestation of magnetic mixtures is a constricted, or 'wasp-waisted', hysteresis loop (Roberts et al., 1995; Tauxe et al., 1996; Fig. 11.25). The example shown was measured for a submarine basaltic glass but constricted loops are typical of remagnetized carbonate rocks as well. A negative induced magnetization of a sufficient quantity of SP magnetite will cancel the remanence of stable SD magnetite in very small reversed fields, as explained above, producing a constriction in the loop. At higher fields, the SD magnetization is less affected because the SP magnetization saturates rapidly (eqn. (5.44), Fig. 5.20) and the loop opens. Rather large quantities of SP material just below the SP–SD threshold size d_s are required.

A mixture of two minerals, one magnetically soft, like magnetite, and the other hard, like hematite, can produce constricted hysteresis if the two phases contribute comparable amounts of magnetization. Natural examples are not too common, because hematite must be about 100 times as abundant in a rock as magnetite to rival the latter's contribution to magnetization, especially in low fields.

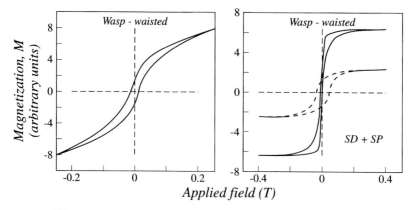

Figure 11.25   Constricted or wasp-waisted hysteresis loops, observed for submarine basaltic glasses, and modelled as a superposition of SD and SP magnetization curves (cf. Figs. 5.19, 5.20). [After Tauxe *et al.* (1996) © American Geophysical Union, with the permission of the authors and the publisher.]

Less extreme mixtures of mineral phases or domain states lead to less striking hysteresis loops, which nevertheless contain much useful information about the magnetic properties and relative quantities of the different phases. Soils and sediments are often magnetically heterogeneous. They contain detrital, diagenetic, authigenic and biogenic iron oxides and sulphides, such as magnetite, maghemite, hematite, goethite, pyrrhotite and greigite. Quite sophisticated methods have been developed to analyse bulk hysteresis data and extract the separate contributions of two or more mineral/grain-size fractions (e.g., Thompson and Peters, 1995). Similar methods have been developed to analyse IRM acquisition curves (e.g., Robertson and France, 1994). The great problem is to establish type curves for the various minerals in different domain states or combinations of domain states. There is such a variety of possible type curves that the 'deconvolution' problem seems insoluble without other independent information about the phases present.

11.9.6   Hysteresis as a function of temperature

SQUID magnetometers and AGFM's fitted with liquid-He dewars and VSM's equipped with either a dewar or a furnace are now commerically available for the measurement of hysteresis at both low and high temperatures. Hysteresis parameters at temperatures other than T_0 have not been widely used in rock magnetism and paleomagnetism, but they offer much information not accessible in other ways.

For example, one can detect changes in domain state at high or low temperature from measurements of M_{rs}/M_s, H_{cr}/H_c or H_c/M_{rs}. In magnetites and titanomagnetites of SD or nearly SD size, M_{rs}/M_s and H_{cr}/H_c are practically temperature invariant, except in the blocking range just below T_C, where M_{rs}/M_s falls and H_{cr}/H_c rises steeply as SP grains depress both M_{rs} and H_c (Özdemir and O'Reilly, 1981, 1982a; Dunlop, 1987). Another way of detecting changes in domain state at high temperatures is via the demagnetizing factor N, which depends on the number of domains (Fig. 5.4b). A reliable method of determining N has been much debated (Smith and Merrill, 1982; Dunlop, 1984; Hodych, 1986; Dunlop et al., 1987). The best method is that of Hodych (1986). Combining the MD relations $\chi_0(T) = \chi_i(T)/[1 + N\chi_i(T)]$ (eqn. (9.4)) and $\chi_i(T)H_c(T) \approx k'\beta(T)$ (from eqn. (11.14)) so as to eliminate χ_i, which is not directly measurable, we obtain

$$\chi_0^{-1}(T) = N + \frac{1}{k'}\frac{H_c(T)}{\beta(T)}, \tag{11.18}$$

where $\beta(T) \equiv M_s(T)/M_{s0}$. Thus a plot of $\chi_0^{-1}(T)$ versus $H_c(T)/\beta(T)$ for a single sample should be linear, with an intercept equal to N. Furthermore, SD grains should follow a similar linear relation but with zero intercept and a slope about $\frac{1}{3}$ as large (Dunlop et al., 1987, 1990c), providing a definitive test for MD versus SD structure.

The main problem in using (11.18) is that only rarely are $\chi_0(T)$ and $H_c(T)$ data available for the same sample (Hodych, 1986; Dunlop et al., 1987). A cruder method is to plot $H_c(T)/M_{rs}(T)$ versus temperature, since according to eqns. (5.32) or (9.26)

$$N(T) \approx H_c(T)/M_{rs}(T). \tag{11.19}$$

Figure 11.26 is an example of such a plot. Because these 0.39 μm and 0.54 μm magnetites are small enough that H_c is significantly reduced by thermal fluctuations at high temperatures, N determined from $H_K(T)/M_{rs}(T)$ is more reliable in the blocking range (H_K was calculated by thermal fluctuation analysis, as described in §8.7). Both samples seem to contain a mixture of 2D and 3D grains at all temperatures from $T_V \approx -150\,°C$ (the Verwey transition: §3.2) to around $500\,°C$, above which the average number of domains per grain may increase slightly. Gross changes in domain structure of submicron magnetites during heating or cooling are not suggested by these results. A similar analysis could be carried out for pyrrhotite using the $H_c(T)$ data of Menyeh and O'Reilly (1995).

Another interesting application is the use of $H_c(T)$ data to infer the mechanism of domain-wall pinning in magnetite (Fig. 11.27). Hodych (1982, 1990) showed that below T_0, $H_c(T) \propto \lambda(T)/M_s(T)$ for MD magnetite, where λ is magnetostriction constant (Fig. 3.8). The same dependence is found at high temperatures also (Dunlop and Bina, 1977; Heider et al., 1987). The explanation would seem to be pinning of domain walls by the stress fields of dislocations (Stacey and Wise, 1967), but a careful analysis shows that the pinning depends on a close match between stress wavelength and domain-wall width δ_w (Xu and Merrill, 1992). Since walls are expected to expand in width as T increases, the

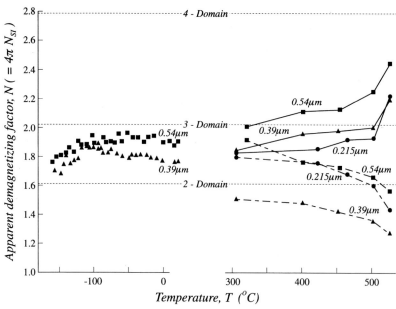

Figure 11.26 Apparent demagnetizing factor N, given by H_c/M_{rs} at lower temperatures or H_K/M_{rs} (solid curves) at high temperatures, for three samples containing < 1 μm magnetites. The high-temperature H_c/M_{rs} data (dashed lines) are not reliable measures of N because H_c at these temperatures is lowered by thermal fluctuations (§8.7). All samples seems to contain a mixture of 2D and 3D grains, with a possible increase in the number of domains above 500 °C. [After Argyle and Dunlop (1990) © American Geophysical Union, with the permission of the authors and the publisher.]

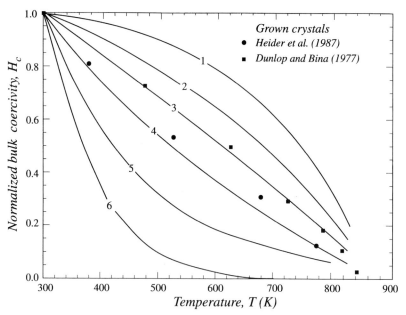

Figure 11.27 Comparison of experimental and theoretical temperature dependences of bulk coercive force H_c. Data for 12 μm and 1–5 μm magnetite crystals are from Heider *et al.* (1987) and Dunlop and Bina (1977), respectively. Theoretical curves: (1)–(4), positive and negative dislocation dipoles with various spacings; (5), (6), planar defects with exchange or anisotropy pinning, in a model 10 μm grain. [After Moskowitz (1993b) © American Geophysical Union, with the permission of the author and the publisher.]

$\lambda(T)/M_\mathrm{s}(T)$ dependence is not so obvious after all. Moskowitz (1993b), following Xu and Merrill (1990b), modelled wall pinning by many types of linear and planar crystal defects, including stacking faults, planar defects with exchange and/or anisotropy pinning, single dislocations, and pairs of dislocations (dislocation dipoles) with various spacings. Only an assemblage of dislocation dipoles with variable spacings can explain the $H_\mathrm{c}(T)$ data (Fig. 11.27).

Chapter 12

Pseudo-single-domain remanence

12.1 Introduction

Most magnetite grains in rocks are much larger than the critical SD size d_0 of $\approx 0.1\,\mu$m. Yet these rocks possess a TRM that is both harder and more intense than MD theory predicts. In magnetite there is no abrupt change from SD to MD TRM intensity at any grain size (Fig. 8.4). Instead TRM intensity decreases continuously above d_0, reaching MD levels around 10–20 μm. In high-titanium titanomagnetites, the corresponding range is ≈ 0.5–$35\,\mu$m (Day, 1977, Fig. 9). This **pseudo-single-domain** (PSD) (Stacey, 1963) size range incorporates most of the magnetite or titanomagnetite carrying stable TRM in igneous rocks. Therefore it is important that we understand the mechanism of PSD remanence.

The size dependence of TRM is not well documented except in titanomagnetites. However, strong-field remanence parameters like M_{rs} and H_c vary gradually over broad size ranges in a great many minerals, rather than changing sharply around d_0 (Fig. 12.1). In §11.9.4, we saw that most measured values of M_{rs}/M_s and H_{cr}/H_c are intermediate between SD and ideal MD values. Pseudo-single-domain behaviour seems to be an intrinsic feature of small MD grains rather than a special property of certain minerals.

The mechanism of PSD behaviour is still far from certain (for reviews of experimental data and theories, see Day, 1977; Dunlop, 1977, 1981, 1986b, 1990; Halgedahl and Fuller, 1983; Fuller, 1984; Halgedahl, 1987). Verhoogen (1959) proposed regions of deflected spins surrounding dislocations, while Stacey (1963) preferred 'Barkhausen discreteness' of domain-wall positions. In these models, the PSD moments are not independent of MD processes. For example, the moments can only reverse when domain walls are displaced. More recent

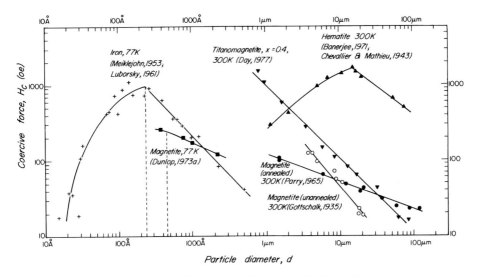

Figure 12.1 Grain-size dependence of bulk coercive force H_c for various minerals. (For a more complete data set for magnetite, see Fig. 12.4.) H_c peaks at the critical SD size d_0. The decrease in H_c below d_0 is due to thermal fluctuations (eqn. (8.28)). The decrease in H_c above d_0 is a PSD effect, with a similar size dependence in different minerals. [Reprinted from Dunlop (1981), with kind permission of the author and Elsevier Science – NL, Sara Burgerhartstraat 25, 1055 KV Amsterdam, The Netherlands.]

models emphasize truly SD moments, which can reverse independently of surrounding domains, e.g., the moments of domain walls (Dunlop, 1973b, 1977) and metastable SD grains in which walls have failed to renucleate following saturation (Halgedahl and Fuller, 1980, 1983). These models, however, are viable only over the lower part of the observed PSD range.

The mechanism is uncertain but the existence of some SD-like component in the magnetization of small MD grains, over and above their ideal MD remanence due to displaced walls, is widely accepted. In §12.2–12.5, we shall review the experimental evidence, while §12.6–12.9 will deal with theoretical models of PSD moments and diagnostic tests for their existence.

12.2 Lines of evidence for pseudo-single-domain behaviour

The earliest, and still the most important, evidence of a PSD effect is the high intensity of weak-field TRM in magnetite grains > 0.1–$0.2\,\mu m$ in size (Figs. 8.4, 12.2a). There is no sudden drop in TRM intensity marking an SD threshold but a continuous decrease over many decades of particle diameter. Above $\approx 5\,\mu m$, the TRM data diverge. Many data on crushed and sieved magnetites (Rahman

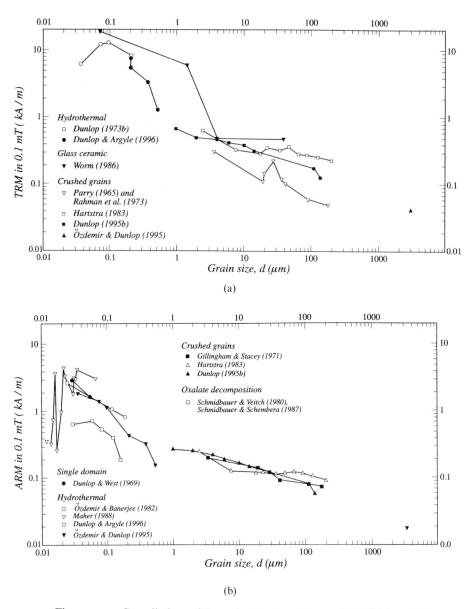

Figure 12.2 Compilations of the grain-size dependences of (a) TRM and (b) ARM intensities in magnetite. The size dependence below 1 μm is stronger than the size dependence above 1 μm, particularly for TRM. [After Dunlop and Xu (1993) © American Geophysical Union, with the permission of the authors and the publisher.]

et al., 1973; Dunlop and Xu, 1993; Dunlop 1995b) continue to decrease with increasing grain size and connect smoothly with the M_{tr} values for mm-size crystals (Levi, 1974; Özdemir and Dunlop, 1995). The data of Parry (1965) and Hartstra (1983), on the other hand, are almost size independent above 5 μm. They support Stacey's original idea of a PSD threshold above which SD-like moments are unimportant. One of the most pressing problems in rock magnetism is to measure the intrinsic size dependence of TRM intensity in the 5–300 μm range using strain-free magnetites grown by modern techniques.

The variation of TRM intensity M_{tr} with applied field H_0, for most rocks and titanomagnetites, is intermediate between theoretically predicted SD and MD dependences (Fig. 9.5). In §12.9.2, we shall show that superimposed SD and MD dependences account for at least some of the data. The unblocking of TRM in the PSD range has mixed MD and SD-like aspects. We saw in Fig. 9.7(b) that more than half of an intermediate-temperature partial TRM in 3 μm magnetite grains thermally demagnetized in SD fashion, i.e., $T_{UB} \approx T_B$, while the remainder had an MD demagnetization tail extending to T_C.

The coercivity spectrum of weak-field TRM also evolves from SD-like to MD-like over a broad size range (Fig. 11.13). High AF coercivities were considered by Verhoogen (1959) and Larson *et al.* (1969) to be important evidence for SD-like moments. But screening by soft walls in large MD grains can produce AF demagnetization tails, as well as high values of H_{cr} (see §11.5, 11.9.2). A more reliable indicator of SD-like behaviour is the shape of the AF demagnetization curve, which changes gradually from sigmoid, with an initial plateau, to quasi-exponential over the PSD range (Fig. 11.13).

Many other magnetic properties have grain-size dependences broadly similar to that of TRM. Examples are weak-field ARM, SIRM, and bulk coercive force H_c. These data will be reviewed in §12.3. As with TRM, the reality of a PSD threshold in the > 1 μm range remains uncertain (Heider *et al.* 1987; Worm and Markert 1987). Internal stress has a strong effect on the value of M_{rs} or H_c for grains of a particular size, although the overall size dependence of either parameter is about the same for low-stress or high-stress magnetites.

The Lowrie–Fuller test, which compares AF demagnetization of weak-field and strong-field remanences, changes from an SD-type result to an MD-type result in moderately stressed magnetite grains larger than 10–20 μm (Fig. 11.14). This changeover was long believed to mark a PSD threshold until Heider *et al.* (1992) showed that the changeover in low-stress magnetites occurs around 100 μm.

Low-temperature demagnetization (LTD), in which a sample is cycled in zero-field through the isotropic ($K_1 = 0$) temperature, $T_I = 120$–135 K, of magnetite and back to room temperature, serves to unpin domain walls in MD grains and destroy their remanence (§3.2.2; Merrill, 1970; Hodych, 1991). Single-domain remanence, if controlled by shape anisotropy, should survive as low-temperature (LT) memory. LTD experiments by Dunlop and Argyle (1991) showed unmistakably that separate SD-like and truly MD sources of remanence coexist in ≈ 0.5 μm magnetites. The LT memory was about 50% of the initial remanence for weak-field TRM and about 25% for SIRM. The shape of the AF decay curve after LTD was distinctively SD-like, whereas it was uninflected and MD-like before LTD. LTD data will be reviewed in §12.4.

Magnetites and titanomagnetites larger than any of the proposed PSD threshold sizes exhibit spontaneous increases in TRM and other remanence intensities in continuous thermal demagnetization to just below T_C (e.g., Rahman and Parry, 1975; Bol'shakov *et al.*, 1978; Hartstra, 1982c; Sugiura, 1988). In contrast to other manifestations of PSD moments, the percentage change in magnetization is greatest in the largest grains. A fuller discussion of the thermal demagnetization data will come in §12.5.

12.3   Grain-size dependence of magnetic properties

12.3.1   TRM and ARM

Weak-field (50–100 µT or 0.5–1 Oe) TRM and ARM have broadly similar grain-size dependences in magnetite (Fig. 12.2a,b). The TRM and ARM data of Hartstra (1983) are almost independent of grain size above $\approx 10\,\mu$m. Other data sets disagree, with TRM and ARM decreasing in similar fashion over the entire 1–200 µm range. The reality of a threshold size for PSD moments is not clear from these data.

Around 1 µm, there is a marked change in size dependences which may indicate a change in micromagnetic structure. Between 0.1 µm and 1 µm, $M_{tr} \propto d^{-1.4}$, while $M_{ar} \propto d^{-0.75}$ approximately, where d is grain diameter. The strong size dependence of TRM compared to ARM is responsible for the peak in M_{tr}/M_{ar} around 0.1 µm (Fig. 11.7). Below $\approx 0.05\,\mu$m or 50 nm, increasing admixtures of smaller SP grains cause a decrease in intensity, particularly in the ARM data.

Initial susceptibility χ_0 is one of the least size dependent of magnetic parameters for magnetite (Dunlop, 1986b, Fig. 3; Hunt *et al.*, 1995a, Fig. 1). Therefore the Koenigsberger ratio $Q_t = M_{tr}/\chi_0 H_0$, with $H_0 \approx 0.5$ Oe or 40 A/m, and the related ratio $Q_a = M_{ar}/\chi_0 H_0$ (§11.4.1) have grain-size dependences similar to those of M_{tr} or M_{ar}. The Q ratios normalize M_{tr} or M_{ar} for the quantity of magnetite in a rock or sediment core, which is generally undetermined.

12.3.2   Hysteresis parameters

Like ARM and TRM, SIRM is size dependent (Fig. 12.3). In this case, the remanent intensity M_{rs} (here normalized to M_s) varies more slowly with grain size, as $d^{-0.5}$ to $d^{-0.65}$, but continuously over a very broad size range, from 0.04 µm to at least 350 µm in synthetic hydrothermal magnetites and from 0.5 µm to 3 mm in natural magnetites (single crystals and sieve fractions of crushed grains prepared from them). This continuous size dependence is very different from the

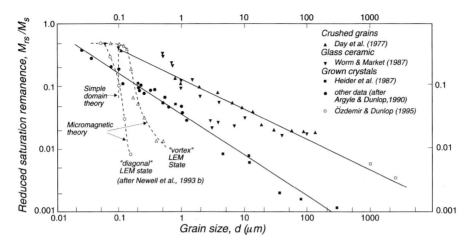

Figure 12.3 Grain-size dependences of the saturation remanence ratio, M_{rs}/M_s, in natural (mainly crushed) magnetites compared with data for synthetic glass-ceramic and hydrothermally grown magnetites. The continuous decreases observed are at odds with the abrupt decrease above d_0 predicted by micromagnetic and domain theories. [After Dunlop (1995a) © American Geophysical Union, with the permission of the author and the publisher.]

threshold behaviour predicted by simple domain theory or two-dimensional micromagnetic theory (Fig. 12.3). There is no indication of any change in size dependence around 1 µm. In this respect, SIRM behaves rather differently from TRM.

Quite apart from its implications for the origin of remanence, the regular and continuous variation of M_{rs}/M_s with grain size provides us with a reliable method of granulometry in magnetite. Unfortunately, some knowledge of the internal stress state in a sample is necessary before one can decide which line to use, the low-stress line defined by data for hydrothermally grown crystals (dislocation density $\rho \approx 10^{10}$ m^{-2}), the high-stress line defined by data for glass-ceramic and crushed magnetites ($\rho \approx 10^{13}$ m^{-2}), or some intermediate line.

The bulk coercive force H_c varies as $d^{-0.4}$ to $d^{-0.6}$ in magnetite above critical SD size (Fig. 12.4). This variation is similar to that of M_{rs}, not surprisingly in view of the MD relation $M_{rs} \approx H_c/N$ (eqns. (5.32), (9.26)). A size variation of H_c as $\approx d^{-0.5}$ can be accounted for if walls are pinned by ordered arrays of defects, such as dislocations (Stacey and Wise, 1967). The real puzzle is what causes coercive forces > 100 Oe or 10 mT, values much higher than expected for wall-pinning in magnetite, in grains up to several µm in size. As with the M_{rs}/M_s data, there are separate $H_c(d)$ trends for hydrothermal, glass-ceramic, and crushed natural grains.

Coercive force and SIRM data have been reported for crushed synthetic titanomagnetites with $x = 0.2$, 0.4 and 0.6 by Day *et al.* (1977) and for TM60 only by O'Donovan *et al.* (1986). The grain-size dependences were rather different in

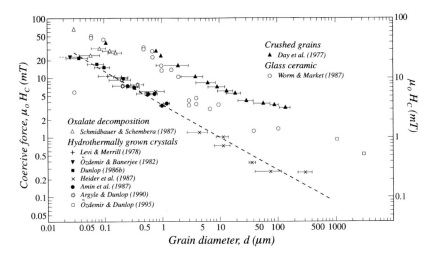

Figure 12.4 Compilation of bulk coercive force data for hydrothermally grown and glass-ceramic magnetites and one representative set of data for crushed magnetites. (For more complete crushed magnetite data, see Dunlop, 1986b or Hunt *et al.*, 1995). As with M_{rs}/M_s data, there is no sudden decrease in H_c above d_0 but a gradual decrease over many decades of grain size d. [After Argyle and Dunlop (1990) © American Geophysical Union, with the permission of the authors and the publisher.]

the two studies. Day *et al.*'s data over the 1–30 μm range, in which H_c is expected to be controlled by domain nucleation, follow a d^{-1} dependence approximately (see Fig. 12.1). Over the same range, O'Donovan *et al.*'s H_c data vary more as $d^{-0.5}$, and their M_{rs}/M_s data are less size dependent still. In both studies, grains around 1 μm in size have SD levels of M_{rs}/M_s (≈ 0.5) and coercive force (≈ 1000 Oe or 100 mT, presumably magnetoelastically controlled). (A single-domain threshold size $d_0 \approx 0.5$ μm is expected from the domain observations of Fig. 5.10.) Only when $d > 100$ μm do M_{rs}/M_s and H_c reach low values typical of ideal MD behaviour (§11.9.2).

Crushed synthetic pyrrhotites about 1.5 μm in size have SD values of M_{rs}/M_s and H_c similar to those of 1 μm titanomagnetites (Menyeh and O'Reilly, 1991). Above the SD threshold and up to 30 μm, H_c decreases by only a factor of 3 and M_{rs}/M_s remains essentially at SD levels, dropping from 0.48 to 0.40. Judging by these results, either d_0 is much larger than the 1.5 μm indicated by domain observations (Figs. 5.10, 6.2) or else PSD moments completely overshadow normal MD remanence in this size range.

Halgedahl and Fuller (1983) observed domain nucleation in a large number of polished natural pyrrhotite grains in the size range 5–40 μm. They found that $H_n = 120d^{-0.5}$, where H_n is in mT and d is in μm (Fig. 12.5). The size variations of H_c reported by Menyeh and O'Reilly (1991), and by Clark (1984) for crushed natural pyrrhotites in the 7–80 μm size range ($H_c = 430\ d^{-0.8}$), both give fairly

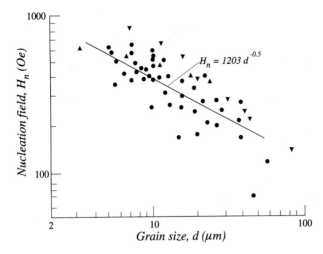

Figure 12.5 Grain-size dependence of the observed field H_n for primary nucleation of domain structure in pyrrhotite grains (dots) compared with measured bulk coercive force (H_c) data (inverted triangles, Clark, 1984; upright triangles, Menyeh and O'Reilly, 1991). Coercive force in pyrrhotite seems to be controlled by nucleation up to $\approx 50\,\mu m$. At larger sizes, H_n falls well below H_c. [After Halgedahl and Fuller (1983) © American Geophysical Union, with the permission of the authors and the publisher.]

similar numerical values to Halgedahl and Fuller's H_n data (Fig. 12.5). Coercivity seems to be nucleation controlled in pyrrhotite, at least up to $\approx 50\,\mu m$, where nucleation fields plummet. It is not clear why Menyeh and O'Reilly's H_c data are so much less size dependent, and Clark's H_c data so much more size dependent, than Halgedahl and Fuller's H_n data over similar size ranges.

Because of the effect of magnetic screening (§11.5, 11.9.2), H_{cr} and other measures of remanent coercivity do not give a true picture of the intrinsic size dependence of microcoercivity in MD grains. For this reason, the size variation of H_{cr} is much weaker than that of H_c (Dunlop, 1986b, Fig. 6; Hunt et al., 1995a, Fig. 5). The size dependence of H_{cr}/H_c is mainly determined by the variation of H_c. Since M_{rs}/M_s and H_c have similar size dependences, at least for titanomagnetites, one expects an inverse correlation between M_{rs}/M_s and H_{cr}/H_c. Such an inverse relation is the fundamental basis for correlation plots like Figs. 11.23 and 11.24.

12.4   Low-temperature demagnetization and memory

12.4.1   Low-temperature transition in magnetite

Low-temperature demagnetization (LTD) is the process of cooling a sample through the isotropic temperature, $T_I = 120\text{–}135$ K, of magnetite (§3.2, Fig.

3.7a), usually to liquid nitrogen temperature (77 K), and heating back to room temperature, all in zero field. Domain-wall width δ_w and energy per unit area γ_w depend directly on anisotropy K (eqns. (5.10), (5.11)). When $K_1 \to 0$ at T_I, the walls become very broad. Walls are most effectively pinned when their widths approximately match the wavelength of the stress fields created by lattice defects (Xu and Merrill, 1989, 1992). Thus walls tend to become unpinned and lose their remanence at T_I. Large, irregular Barkhausen jumps in magnetization during a cooling–heating cycle across T_I are illustrated in Fig. 12.6.

The amount of remanence demagnetized in crossing the transition is different for different grain sizes. The data shown in Fig. 12.7 are for SIRM produced at 10 K and taken in a zero-field heating–cooling cycle through T_I. About 75% of the initial remanence is lost by 135 µm unannealed magnetites but only 50% by unannealed 1 µm grains. These figures rise to 80% and 60%, respectively, when the same grains are annealed to reduce internal stress and weaken domain-wall pinning.

More germane to PSD behaviour is the remanence that survives LTD, called the **low-temperature (LT) memory**. LT memory is due in part to SD grains con-

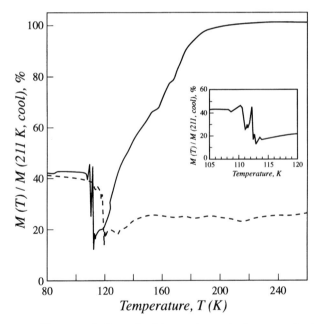

Figure 12.6 Large Barkhausen jumps in the magnetization of a natural octahedral crystal of magnetite during cooling (solid curve) and heating (dashed curve) through the isotropic temperature T_I, measured with a SQUID magnetometer. The inset is a magnified view of the jumps measured around T_I during cooling. [Reprinted from Halgedahl and Jarrard (1995), with kind permission of the authors and Elsevier Science – NL, Sara Burgerhartstraat 25, 1055 KV Amsterdam, The Netherlands.]

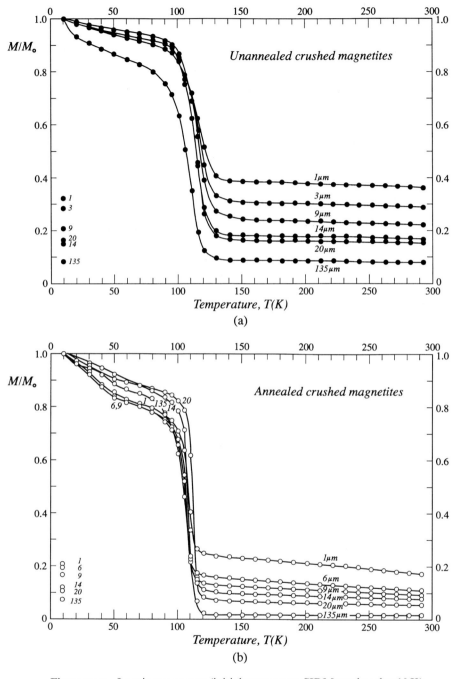

Figure 12.7 Loss in remanence (initial remanence: SIRM produced at 10 K) during zero-field heating across the Verwey transition T_V and isotropic temperature T_I (solid curves) and memory after zero-field recooling to 10 K (single points at left) for (a) unannealed and (b) annealed crushed natural magnetites. The loss in remanence is dependent on both grain size and internal stress. [After Dunlop and Xu (1993) © American Geophysical Union, with the permission of the authors and the publisher.]

trolled by shape anisotropy, or analogous SD moments in larger grains, and in part to walls that remain pinned because of unusually large local stresses. In Fig. 12.7(a), the LT memory at 293 K varies between 8% of the initial SIRM at 10 K for 135 μm grains and 37% of initial SIRM for the 1 μm grains. A second passage through T_I changes the memory rather little, as the final points at 10 K attest.

When the magnetites are annealed, the LT memory at T_0 decreases to about half its previous value (Fig. 12.7b), showing that roughly 50% of the previous memory was remanence pinned by stress. In addition, a certain amount of remanence that was lost in heating through T_I is now recovered in the second passage through T_I. The mechanism of this recovery is not clear (see Heider *et al.*, 1992) but a similar recovery often occurs in conventional LTD, i.e., in cooling–heating cycles through T_I (Ozima *et al.*, 1964; Kobayashi and Fuller, 1968; Halgedahl and Jarrard, 1995).

The memory ratio, R_I or R_T, is the fraction of initial SIRM or TRM that survives a complete cooling–heating cycle. Although many factors affect memory, for instance grain elongation and non-stoichiometry or surface oxidation in fine grains (Fig. 3.6), there is a clear grain-size variation in Fig. 12.8. R_I and R_T follow parallel trends, the TRM memory always being noticeably larger than SIRM memory. Among low-stress hydrothermal magnetites, R_I and especially R_T are very size dependent below 1 μm, less size dependent between 1 and 10 μm, and lack any clear size dependence at larger sizes. These breaks in the size variations are reminiscent of the breaks at similar sizes in TRM and ARM data (Figs. 8.4, 12.2).

Memory ratios for crushed magnetites are higher than ratios for low-stress magnetites of the same size. Similarly ratios for unannealed magnetites are higher than those for annealed magnetites of similar size. Stress as well as grain size plays a key role in LT memory.

12.4.2   AF demagnetization of low-temperature memory

LTD is an effective means of removing a large part of the remanence due to domain-wall pinning,thereby highlighting PSD remanence. Just how SD-like the residual remanence really is can be judged by stepwise AF demagnetization of the LT memory (Fig. 12.9; Dunlop *et al.*, 1995b). In annealed 1 μm grains, the softest part of TRM, ARM or SIRM has been erased. This fraction, with AF coercivities < 30 mT, presumably resided in pinned walls. The memory after LTD not only has higher average coercivities but also has the sigmoid shape characteristic of SD remanence (Fig. 11.13). Thus the memory possesses one of the hallmarks of PSD remanence.

More surprising is the fact that 135 μm magnetites, which by any criterion are much larger than PSD size, behave in exactly the same way as smaller grains

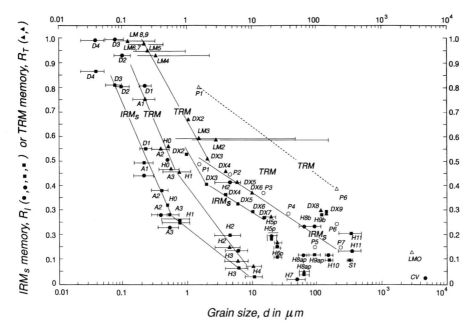

Figure 12.8 Low-temperature memory ratios of 0.1-mT TRM and SIRM in hydrothermal or annealed magnetites (squares) and unannealed, crushed or irregular magnetites (circles) as a function of grain size d. Data after: H, D, S, CV, Heider *et al.* (1992); A, Dunlop and Argyle (1991); LM, Levi and Merrill (1976, 1978); P, Parry (1979, 1980); DX, Dunlop and Xu (1993). The memory is dependent on type of remanence and internal stress as well as grain size. [After Heider *et al.* (1992) © American Geophysical Union, with the permission of the authors and the publisher.]

(Fig. 12.9b). The TRM and SIRM memories are smaller than in 1 μm grains, but the range of coercivities and the shapes of their AF demagnetization curves are distinctively SD-like. A 135 μm grain has a volume 10^{10} times that of an SD grain of critical size (≈ 0.1 μm) and should contain at least 10 domain walls (Fig. 6.7a,b). It is almost inconceivable that grains of this size could fail to nucleate any domain walls in both the TRM and SIRM states.

Yet the nature of the SD-like moments that comprise LT memory seems to be similar in grains of widely differing sizes. Certainly AF demagnetization curves of SIRM memory are very similar in shape for 1–135 μm crushed annealed magnetites and for 0.8–356 μm hydrothermal magnetites (Fig. 12.10a,b). What is more, the envelope of all these curves falls in the range of AF demagnetization curves measured for 0.04 μm and 0.1 μm SD magnetites (Fig. 12.10b). As far as AF demagnetization is concerned, SIRM memory is truly SD-like. The corresponding data for weak-field TRM are somewhat more grain-size dependent (they are also noisier, because the TRM's are much weaker than the SIRM's), but still less size dependent than almost any other magnetic property of magnetite.

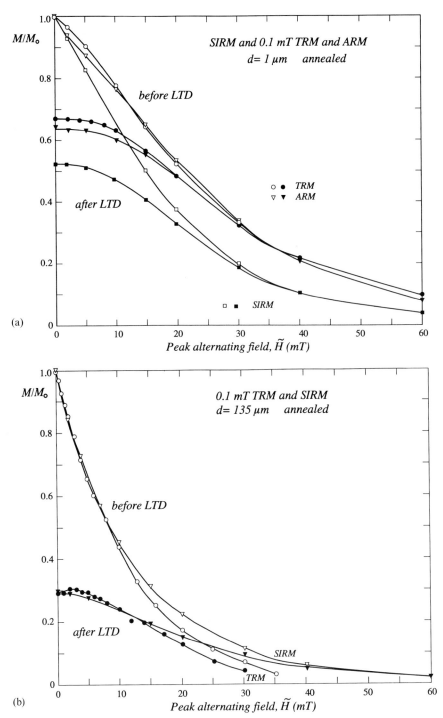

Figure 12.9 AF demagnetization curves of SIRM and of weak-field TRM and ARM, before and after low-temperature demagnetization (LTD) for (a) 1μm and (b) 135 μm annealed crushed magnetite grains. In both grain sizes, the LT memory of all three remanences has an SD-like AF curve with an initial plateau of little or no demagnetization. [After Argyle *et al.* (1994) © American Geophysical Union, with the permission of the authors and the publisher.]

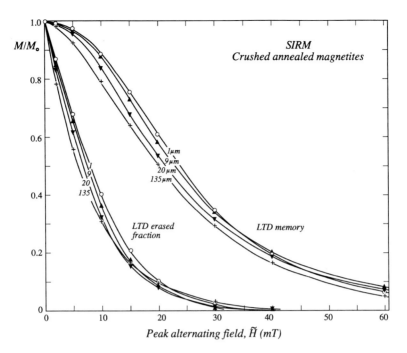

(a)

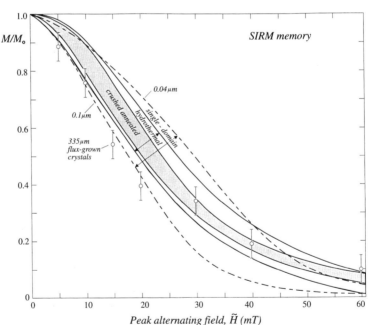

(b)

Figure 12.10 (a) AF demagnetization curves for the LT memory and LTD erased fractions of SIRM in annealed crushed magnetites of various sizes. Both sets of curves are almost independent of grain size. (b) Envelopes of AF demagnetization curves of LT memory for crushed and hydrothermally grown magnetites of widely varying sizes. Both envelopes agree well with measured AF curves of SIRM (without LTD) for 0.04 μm and 0.1 μm single-domain magnetites. [After Argyle *et al.* (1994) © American Geophysical Union, with the permission of the authors and the publisher.]

The fraction of SIRM that is erased by LTD also has an essentially grain-size independent AF demagnetization curve (Fig. 12.10a). Its shape is quasi-exponential, resembling the curves predicted for wall-pinning in 2D to 4D grains (Fig. 11.12). The overall AF demagnetization curve for various grain sizes is thus a mixture, in varying proportions, of distinct MD and SD curves which are themselves size independent. The grain-size dependence of AF demagnetization characteristics (e.g., Fig. 11.13) is not fundamental but a consequence of a higher proportion of SD-like moments in finer grains.

This conclusion has wide implications. TRM and other magnetic properties are not smoothly transitional between SD-like and MD-like when examined at the particle (or perhaps sub-particle) level. Instead the bulk properties are weighted averages of distinct SD and MD contributions, even in narrowly sized samples. As the grain size decreases, the SD-like moments become more prominent.

12.5 Thermal demagnetization of TRM and partial TRM

In §9.5 and Fig. 9.7(a),(b), we found that both total and partial TRM's of MD grains of magnetite must be heated almost to the Curie point before they are completely demagnetized. Just below T_C, a spontaneous increase in TRM intensity is sometimes observed (Fig. 12.11), presumably indicating a small, reversely magnetized fraction of TRM. Although the peaks appear large in Fig. 12.11, the high-temperature reversed moments are a small fraction of the total TRM. They are most easily seen in large (100–150 μm) grains because the background of multidomain TRM due to pinned walls is smaller, but the absolute intensities of the remanence peaks are similar in large and small (5–10 μm) grains alike.

If the peaks are the manifestation of SD inclusions oppositely magnetized to the MD regions in which they are embedded, then the total volume occupied by such PSD moments must be almost independent of overall grain size. This does not of itself imply that individual moments have similar volumes in large and small grains, but the AF demagnetization of LT memory (Fig. 12.10a,b) seems to require this. However, there are two points of apparent disagreement. First, the remanence peaks form a larger fraction of the total remanence as grain size increases, whereas the opposite is true of LT memory (Fig. 12.8). Second, the remanence peaks are reversely magnetized relative to the bulk of the remanence (otherwise they would not be detectable), while LT memory is in the same direction as the total remanence.

The issue is somewhat illuminated by stepwise thermal demagnetization of total and partial TRM's and their memories after LTD (Fig. 12.12). The magnetite sample has an average grain size of 135 μm and a similar size range

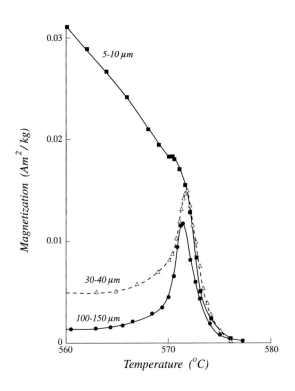

Figure 12.11 Continuous thermal demagnetization of weak-field (86 μT) TRM for three size fractions of crushed natural magnetite. Just below T_C, there is a spontaneous increase in TRM. This peak is more clearly seen in the larger grain sizes. [After Sugiura (1988), with the permission of the author and the publisher, Blackwell Science, Oxford, UK.]

(100–150 μm) to the sample with the most prominent remanence peak in Fig. 12.11. A number of salient points emerge.

1. The absolute intensity of the LT memory is almost the same for pTRM(T_C, 565 °C), pTRM(T_C, 400 °C) and TRM(T_C, T_0). Thus LT memory has mainly blocking temperatures $T_B > 565$ °C.

2. In all three cases, the main drop in the intensity of LT memory occurs above 550 °C. The LT memory has high unblocking temperatures $T_{UB} > 550$ °C, as well as high T_B's.

3. To a good approximation, $T_{UB} = T_B$ for the LT memory. This behaviour is SD-like.

4. In all three cases, there was a small increase in remanence intensity, amounting to ≈ 10% of the memory, during heating to 300 °C. The blocking temperature of this reversed component must be > 565 °C, since it is present in pTRM(T_C, 565 °C), but its unblocking temperatures are much lower. There was no sign of any reverse component unblocking around 570 °C, like that seen in the continuous demagnetization of total TRM in Fig. 12.11.

These results indicate that LT memory, as well as having SD-like AF demagnetization, also has SD-like thermal demagnetization: $T_{UB} \approx T_B$. Only about

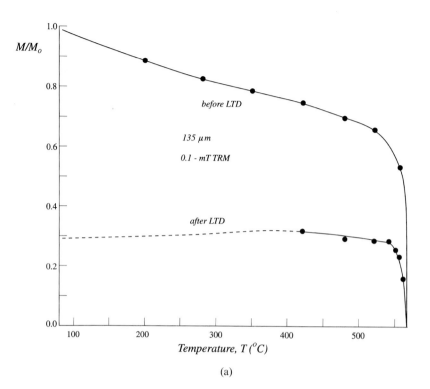

(a)

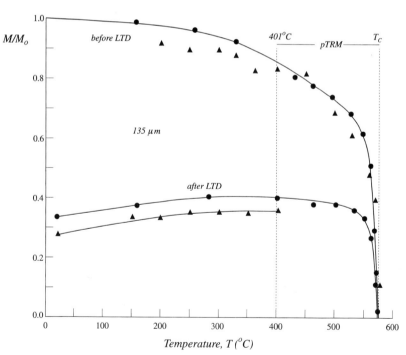

(b)

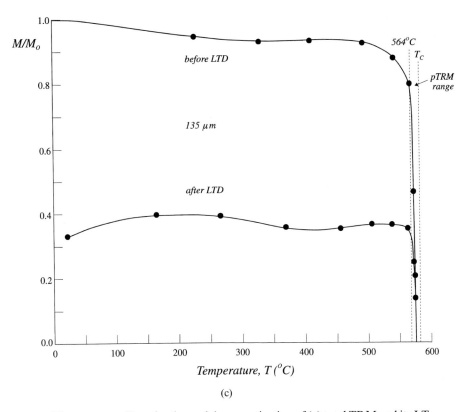

Figure 12.12 Stepwise thermal demagnetization of (a) total TRM and its LT memory, (b) pTRM(T_C, 401 °C) and its LT memory, and (c) pTRM(T_C, 564 °C) and its LT memory, all acquired in $H_0 = 0.1$ mT by 135 µm magnetite grains. In every case, the memory increases slightly with zero-field heating to 250–350 °C, but demagnetizes sharply above ≈ 550 °C. [After Dunlop and Özdemir (1995) © American Geophysical Union, with the permission of the authors and the publisher.]

10% of the memory is reversed to the total remanence, and this fraction has low rather than high unblocking temperatures.

The reversed fraction seems to be quite variable. In samples similar to the one illustrated in Fig. 12.12 but of larger or smaller grain size, there was *no* detectable reversed fraction. On the other hand, McClelland and Shcherbakov (1995) documented reversed fractions of pTRM(T_C, 565 °C), total TRM, and TRM memory in various fractions of crushed magnetites with grain sizes ≥ 20 µm. The reversed fraction was revealed by both AF and thermal demagnetization, and in some cases was so large as to cause the TRM memory to be self-reversed (i.e., opposite in direction to H_0 producing M_{tr}). A large self-reversed memory could account for the recovery of remanence described in §12.4.1 if the reversed fraction is preferentially erased in the second passage through T_I. Although interesting, self-reversed fractions of TRM or pTRM probably have little to do with the main issue of PSD remanence. Their T_{UB}'s are mainly low, even when their T_B is high, and their contribution to the total remanence is extremely variable.

12.6   Pseudo-single-domain models

12.6.1   Moments related to dislocations

Models of 'embedded' regions of unusually high magnetic moment and coercivity within otherwise soft MD grains have been popular over the years. Verhoogen (1959) proposed that subdomains of spins deflected from an easy axis of magnetization by the stress fields of dislocations could act as SD moments (Fig. 12.13). Favourably oriented dislocations produce subdomains whose moments are parallel to the dislocation line. The stress-controlled region is separated from surrounding body domain by a cylindrical domain wall.

There are three main objections to Verhoogen's model. First, Shive (1969) has shown that individual dislocations pin such small regions that the subdomain moments make a negligible contribution to total remanence. Second, such moments, like remanence of pinned walls, would be substantially screened by induced magnetization of the surrounding domains. The component of subdomain remanence parallel to the domain magnetizations creates a self-demagnetizing field that displaces loosely pinned walls (§9.3, 11.5, 11.9.2). The component of subdomain remanence perpendicular to the domains causes

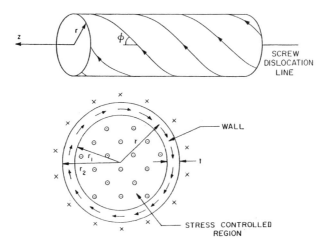

Figure 12.13   Verhoogen's (1959) model of a region of hard, SD-like magnetization pinned by the stress field surrounding a screw dislocation. [After Shive (1969), with the permission of the author and the publisher, Journal of Geomagnetism and Geoelectricity, Tokyo, Japan.]

rotation of the domain magnetizations, but this source of screening is less effective than parallel screening due to soft walls. The parallel screening factor, $(1 + N\chi_i)^{-1}$, is about 0.2 for magnetite and about 0.4 for TM60. Detailed analyses of screening (Stephenson, 1975; Moon and Merrill, 1986b; Xu and Merrill, 1990c) lead to similar values.

The third and most serious objection is that each dislocation moment is exchange coupled to its host domain. Its moment can reverse only when it is traversed by a domain wall, in low fields anyway. Such moments possess only MD-type reversal modes and cannot be the source of the SD-like AF demagnetization curves observed in Figs. 12.9 and 12.10.

Kobayashi and Fuller (1968) were guided in their theorizing by the observed important effect of internal stress on remanence loss and memory during LTD (Figs. 12.7, 12.8). The effect of a dislocation or a dislocation pileup on magnetization depends on the relative orientation of the magnetoelastic and magnetocrystalline easy axes. If the magnetoelastic easy axis is parallel to a $\langle 111 \rangle$ crystalline easy axis (an 'in-phase' stress centre), the anisotropy energies add and wall energy increases (eqn. (5.11)). Walls will be repelled from such regions, which can thus only change their magnetization by SD-type rotation in high fields. Because the moments of such regions are parallel to the host domains, they are screened quite effectively. Nevertheless, a fraction of the remanence will show through the opposing magnetization of the soft matrix.

If, on the other hand, the magnetoelastic easy axis is parallel to a $\langle 100 \rangle$ hard direction ('out-of-phase' stress centres), the magnetization will be deflected away from the host domain direction. Verhoogen's dislocation moments are of this type. Screening is somewhat less effective, although only for very large stresses would a subdomain moment be deflected perpendicular to the main domains. Out-of-phase stress centres attract and pin walls. They are therefore likely to reverse in MD fashion.

In crossing the isotropic point, K_1 changes sign and $\langle 100 \rangle$ become the magnetocrystalline easy axes. Then in-phase stress centres change to out-of-phase stress centres, and vice versa. Unpinning of walls is accounted for but the reason for memory is not so obvious. Kobayashi and Fuller suggest that the polarization of individual walls (Figs. 5.9b, 5.22) may be the key to renucleating a memory of the high-temperature state during heating back through T_1.

The Kobayashi and Fuller model improves on the Verhoogen model in a number of ways: by recognizing the existence of different types of dislocation moments, by showing that some have SD-like reversal modes while others attract walls, and by explaining remanence unpinning during LTD as a result of the two types of stress centres interchanging their roles. The *survival* of remanence in an LTD cycle is not obviously explained. Nor does their theory account for the part of memory that is independent of stress – the memory in annealed and in hydrothermally grown magnetites (Figs. 12.7b, 12.9, 12.10).

12.6.2   Surface and volume moments

The grain surface/volume ratio goes as d^{-1} and so, approximately, does the intensity of TRM of magnetite in the 0.1–1 μm range. PSD moments more or less uniformly distributed over grain surfaces would offer a natural explanation of the size dependence of TRM in the lower PSD range, and possibly of other magnetic properties as well (Banerjee, 1977).

Reverse spike domains, parts of closure domains, and near-surface domain-wall edges may be pinned by surface defects in a manner analogous to dislocation pinning of interior sub-domains. Because exchange and crystalline coupling are weakened by the lack of neighbouring atoms where the lattice terminates, near-surface spins are more easily deflected by a stress field than interior spins. The pinning can be quite strong. Primary nucleation of domains occurs preferentially at surface defects, but the reverse domain nucleus often remains pinned at its surface site before spreading into the grain interior at higher fields (secondary nucleation). Evidence for secondary nucleation includes asymmetric hysteresis and nucleation fields that depend on the orientation and strength of the 'saturating' field. Secondary nucleation following saturation was implied by the asymmetric succession of states illustrated in Fig. 6.1 for pyrrhotite. Surface domain trapping in magnetite is shown in Fig. 12.14.

The magnetization of a reverse nucleus like the one in Fig. 12.14 is small compared to the overall grain magnetization. In fact, the *failure* of nucleation,

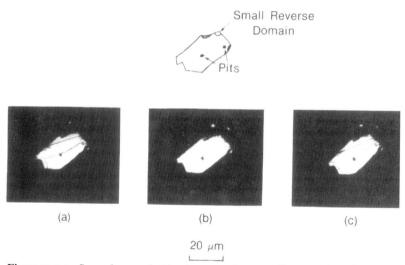

Figure 12.14   Secondary nucleation in magnetite: a small reverse domain pinned at its surface nucleating site in a 20 μm natural magnetite grain in the SIRM state (b). Two body domains form in the same grain after AF demagnetization to 100 mT (a) and after weak-field cycling about the SIRM state (c). [After Halgedahl and Fuller (1983) © American Geophysical Union, with the permission of the authors and the publisher.]

which leaves the grain in an almost SD state of anomalously high magnetization (see §12.8), is more significant than the moment of the trapped nucleus itself. However, in the case of closure domains, the pinned domains are part of an equilibrium domain structure, not a metastable phenomenon. Their moments, although small, persist even with field or temperature cycling that generally destroys metastable LEM states. Small closure domains abound in surface regions (Figs. 1.2, 5.8b, 5.11, 6.5), especially if altered, e.g., oxidized (Fig. 6.3). The closure domains have little freedom of adjustment and the orientation of their moments is largely unaffected by the mutual magnetostatic interaction that is important in the TRM and ARM of interacting SD grains (§8.6, 11.8). Furthermore, since their magnetization is oriented more or less perpendicular to M_s, in the interior body domains, they are poorly shielded.

However, it is difficult to imagine how surface moments of this sort could reverse independently of MD processes in the interior domains. Closure domains reverse only when an interior 180° wall passes their location, and domain-wall edges, if firmly pinned, do not change at all in small displacements of the main body of the wall (see Fig. 5.18 for an illustration of edge pinning and bending of walls). They do not possess SD-like reversal modes.

A different type of surface moment arises when low-temperature oxidation ('maghemitization') causes shrinkage and cracking of the surface of a magnetite or titanomagnetite grain (Fig. 3.10). Many of the individual crack-bound crystallites of maghemite or titanomaghemite are so small as to be SD or even SP at room temperature T_0. Evidence of this fact is the marked decrease during heating of SIRM acquired below T_0 (e.g., Fig. 3.6b, 10–50 K data) due to unblocking (SD→SP). These SD moments will interact magnetostatically, and they probably remain exchange coupled to the unoxidized grain below (see §13.3, Fig. 14.6), which will also have a magnetostatic shielding effect. Nevertheless surface crystallites remain a potent source of hard SD-like moments because surface oxidation is practically universal in natural titanomagnetites.

Interior regions, exchange decoupled from the rest of the grain by submicroscopic grain boundaries, were proposed by Ozima and Ozima (1965). The nature of such boundaries is difficult to imagine in single-phase material, since cracking due to oxidation occurs principally at the grain surface (Fig. 3.10). To explain high-temperature remanence peaks (Fig. 12.11), Sugiura (1988) proposed that magnetite can differentiate into two phases with slightly different Curie points. McClelland and Shcherbakov (1995) also favour this view, but there is no firm evidence for two chemically distinct phases in magnetite. All measured thermomagnetic curves show a single Curie temperature. Shcherbakov et al. (1993) proposed a different model in which inhomogeneous internal stress would effectively subdivide the grain into a large number of regions, each with a different easy axis of magnetization. This idea is akin to Kobayashi and Fuller's (1968) model, except that the stress-affected regions fill the entire grain volume and somehow act independently of one another. The existence and independence of such regions in magnetite is highly conjectural, although stress-dominated patterns in TM60 have led Appel and Soffel (1984) to a somewhat similar model for titanomagnetites.

On safer ground is partial exsolution of titanomagnetites into chemically distinct phases, invoked by Rahman and Parry (1975) and Hartstra (1982c) to explain remanence peaks. Oxyexsolution of titanomagnetites (§3.1, 14.4) produces a sandwich- or trellis-like pattern of magnetite regions separated by ilmenite lamellae in {111} planes (Figs. 3.2, 14.14). Oriented arrays of equant or elongated magnetite subgrains of possibly SD size can be produced in this way (Strangway *et al.* 1968) and could be the source of high AF stability in oxidized basalts (Larson *et al.*, 1969). Although individual SD-size regions could in principle reverse their moments in response to a field, the magnetostatic interaction in an oriented array will not allow this to occur independently of the other SD moments. A cascading reversal of many moments in unison or in rapid succession would likely occur.

12.6.3   Moments due to residual wall displacements

Stacey (1962, 1963) introduced the idea of Barkhausen discreteness (Fig. 12.15). In small grains at zero field, minimum total energy may not correspond to zero magnetization because variations in the wall energy have a wavelength that is

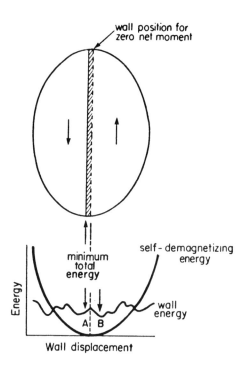

Figure 12.15 Stacey's (1963) model of spontaneous PSD moments arising from residual displacements ('Barkhausen discreteness') of domain walls in a minimum-total-energy or LEM state. A and B are two such LEM states. Note that they do not correspond to minima in wall energy E_w. The Barkhausen moment reverses if the wall is displaced from A to B. [After Dunlop (1977), with the permission of the author and the publisher, Journal of Geomagnetism and Geoelectricity, Tokyo, Japan.]

an appreciable fraction of the particle dimensions. The residual wall displacement is spontaneous and permanent because it corresponds to minimum total energy (a LEM state). The effect is not confined to 2D grains. In larger grains with more domains, the average residual Barkhausen moment of any wall increases with wall area as d^2 while the saturation moment of the grain increases as d^3. The intrinsic size dependence of individual moments is therefore as d^{-1}. However, larger grains contain more walls and therefore more moments.

On the other hand, these moments interact in such a way as to minimize their net magnetization. This screening or self-demagnetization is particularly effective because the Barkhausen discreteness moments are parallel to one another and to the domain magnetizations. Even in 2D grains, the residual Barkhausen moment is in effect self-shielded. The domain wall will not equilibrate at a position of minimum E_w. Rather, the internal demagnetizing field will drive the wall up the central barrier in the wall energy almost to the zero magnetization state (Fig. 12.15).

A related type of residual moment is due to domain imbalance in grains with odd numbers of domains (Craik and McIntyre, 1969; Dunlop, 1983a). In a 2D, 4D, etc. grain, apart from fluctuations in E_w, minimum energy is achieved when the positive and negative domains have equal sizes, yielding zero net moment. In a 3D, 5D, etc. grain, the two end domains are somewhat larger than half the size of the interior domains and they impart a net domain-imbalance moment to the grain. This effect can be seen in Fig. 5.4(a), where minimum energy for a 3D grain corresponds to a spontaneous remanence M_r about 12% of M_s. Even larger wall displacements and net moments are observed in practice, for example in Fig. 6.8(c). PSD moments of this type are intrinsic to MD materials and are no harder than ordinary remanences due to field-induced wall displacements. However, domain-imbalance moments are permanent. They cannot be removed, only reversed, by AF or thermal demagnetization, and they are not shielded by displacements of soft walls because the entire population of walls has an equilibrium net displacement. The domain-imbalance magnetization in a defect-free crystal decreases as the number of domains increases, but remains substantial. In 5D grains, for example, it is about 4% of M_s.

One problem with both Barkhausen discreteness and domain imbalance mechanisms is that the moments involve large regions, whose volumes increase as d^2. When one models TRM or ARM acquisition, either remanence is predicted to reach saturation in fields of a few Oe, contrary to observation (Fig. 9.5), except in grains approaching SD size (e.g., Stacey, 1963, Fig. 10). Another problem is that both types of moments reverse only by domain-wall displacement. To reverse the Barkhausen moment in Fig. 12.15, the wall need merely move from A to B over a local energy barrier. To reverse the domain-imbalance moment, all the domains must reverse. In neither case is the magnetization process decoupled from that of the main domains.

12.7 Domain-wall moments

The types of PSD moments considered in §12.6 suffer from various drawbacks. Most cannot be decoupled from their surroundings and can reverse only if traversed by a domain wall. They should have exponential AF demagnetization curves with mainly low coercivities. The source of their coercivity is not SD-type rotation and exactly how they could produce inflected SD-like AF demagnetization curves with initial plateaus (Figs. 12.9, 12.10) is enigmatic. Other major objections to most models are magnetostatic interaction among neighbouring regions in the same grain, coupling of the magnetization response of such regions to that of the main domains, and screening of the hard moments by the soft matrix. The models are also to a large extent untestable. In this section and the one that follows, we discuss two models of PSD moments with truly SD properties which are amenable to experimental testing.

A domain wall has a net moment perpendicular to the domain magnetization M_s, with a magnitude for a 180° wall of $\pm(2/\pi)M_s$ times the wall volume $V_w = \delta_w A$ (Fig. 12.16; see also §5.8, Fig. 5.22). For 71°, 90° or 109° walls, the moment is $\pm(2/\pi)M_{s\perp}V_w$, according to Fig. 5.9b. Domain-wall moments have a choice of two orientations or polarities, depending on the sense of spin rotation (called polarization or chirality) across the wall. Wall moments have truly SD behaviour, in that they are perpendicular to, and can reverse independently of,

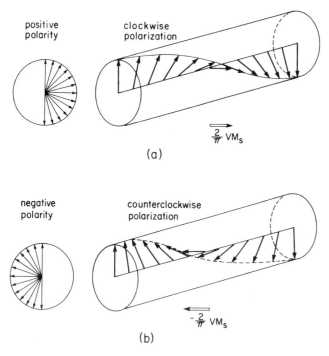

positive
polarity

clockwise
polarization

$\frac{2}{\pi}VM_s$

(a)

negative
polarity

counterclockwise
polarization

$-\frac{2}{\pi}VM_s$

(b)

Figure 12.16 Opposite polarizations or chiralities of spin rotation in a 180° domain wall, giving rise to opposite polarities of the wall moment m. Wall moments are independent of displacements of the wall. Conversely, m can reverse without affecting the domain magnetizations. Thus wall moments (or the equivalent vortex-line moments of single-vortex micromagnetic structures) have truly SD properties; they are 'psarks'. [After Dunlop (1977), with the permission of the author and the publisher, Journal of Geomagnetism and Geoelectricity, Tokyo, Japan.]

the domain magnetizations (Dunlop, 1977, 1981). When a wall is unpinned and moves, the wall moment is unchanged and is independent of any magnetization the wall may acquire (including Barkhausen or imbalance moments) because the wall and domain magnetizations are orthogonal. Similarly when the wall moment reverses (Fig. 5.22), the domains and domain-wall displacements are unaffected. Wall moments are only affected by surrounding domains through the torque they exert on the domain magnetizations, but as we noted earlier, perpendicular screening of this sort is relatively ineffective.

In grains containing more than one wall, demagnetizing energy is reduced if neighbouring walls have oppositely directed moments (Fig. 6.14). Grains with an even number n of domains will always have one uncompensated domain-wall moment, but the fraction it contributes to the total remanence decreases rapidly as n increases. In addition, wall width relative to particle diameter d decreases as d increases, for both equilibrium and non-equilibrium domain structures (e.g., Fig. 7.2), causing a further decrease in fractional remanence. Finally, self-demagnetization of wall moments by skirt formation (Figs. 5.6b, 7.2) also increases with increasing d. 2D grains just above equilibrium SD size have the most significant wall moments.

The mechanism by which a wall reverses its polarization, and thus its moment, can be deduced from the observations of DeBlois and Graham (1958) on iron (see also Hartmann and Mende, 1985; Proksch et al., 1994b). 180° Bloch walls are subdivided into a periodic structure of clockwise and counterclockwise wall segments, separated by Bloch lines with moments perpendicular to both wall and domain moments (Figs. 5.22, 6.17). The wall can change its magnetization gradually by Bloch-line displacement (an MD-like mode analogous to wall displacement) or abruptly at a critical field by reversal of unfavourably oriented segments. If the wall contains only one segment, i.e., has uniform polarization or chirality throughout its length (a 'single-domain wall'), rotation is the only possible response to a field. The rotation of spins may occur coherently or incoherently, e.g., by 'wall curling', resembling propagation of a vortex or a Bloch line along the wall (cf. Figs. 7.12, 7.14). Critical fields are comparable to those for reversal of an SD grain. Thus SD-like AF demagnetization curves (Figs. 12.9, 12.10) can be accounted for.

Pokhil and Moskowitz (1995), using the magnetic force microscope or MFM (§6.9), have observed subdivided walls in 10–20 μm magnetite grains. Williams et al. (1992b), also using an MFM, observed a single change of chirality in a 20 μm length of domain wall in a 5 mm polished magnetite crystal (Fig. 6.17). Judging from these observations, the upper size limit for a 'single-domain wall' seems to be a few μm in magnetite. Lorentz microscopy supports this figure (Fig. 6.14).

A theoretical calculation was given by Shtrikman and Treves (1960), who likened wall domains and Bloch lines to domains and walls in a thin film. Bloch lines occur because the alternation of wall-domain polarities reduces the magnetostatic energy of the wall, just as the alternating polarities of ordinary domains reduce the magnetostatic energy of the entire grain. However, Bloch lines themselves have energy, composed of exchange, crystalline anisotropy and demagnetizing energies. From a calculation like that of §5.3 for equilibrium wall width, Shtrikman and Treves found that the Bloch line has an equilibrium width

$$x = \left[\left(\frac{\mu_0}{4\pi} \right)^{-1} \frac{\pi^2 A^{3/2}}{8\sqrt{2}M_s^2 K^{1/2}} \right]^{1/3}. \tag{12.1}$$

Minimizing the sum of the Bloch line and wall energies (a calculation analogous to that of §5.4.1 for equilibrium domain width), they predicted the equilibrium wall-domain length l to be

$$l = \left(\frac{\pi}{2} \right)^{1/3} \left[\left(\frac{\mu_0}{4\pi} \right)^{-1} \frac{K}{M_s^2} \right]^{2/3} d, \tag{12.2}$$

assuming the walls extend across the full width d of the particle. In eqns. (12.1) and (12.2), the factor $\mu_0/4\pi$ is omitted in cgs.

Substituting for magnetite $A = 1.33 \times 10^{-11}$ J/m, $|K_1| = 1.35 \times 10^4$ J/m^3, and $M_s = 480$ kA/m, we obtain $x = 0.025\,\mu$m or 250 Å and $l = 0.82d$. The first figure suggests that Bloch lines will have difficulty in forming in grains just above critical SD size because the Bloch line and the wall have comparable dimensions. The second result implies that Bloch line formation will be inhibited in magnetite grains of *any* size, since the favoured number of wall domains is only 1–2. In the case of iron, where magnetostatic energy is more important because of the larger value of M_s, the predicted number of domains per wall is 3–4 (Shtrikman and Treves, 1960). In either case, the predictions are in reasonable accord with observations up to the 'PSD threshold' of 10–20 μm in magnetite. We can expect single-domain walls, or 'psarks' (Dunlop, 1977), to contribute significantly to remanence in magnetite grains up to perhaps 1 μm in size but not above.

12.8   Metastable SD grains

12.8.1   The observational evidence

A potent source of truly SD remanence is metastable SD structure in grains larger than equilibrium SD size. Metastable 2D, 3D,... states are of course also possible, but metastable SD grains have by far the largest moments. Theoretically, perfect crystals of magnetite will spontaneously nucleate a domain wall if they are larger than $\approx 0.3\,\mu m$ (Fig. 7.3). A vortex structure will form at even smaller sizes according to three-dimensional micromagnetic calculations (Enkin and Williams, 1994; Fabian *et al.*, 1996). Real magnetite grains with surface defects should nucleate walls more readily still (§6.3).

Experimentally, however, $>1\,\mu m$ grains of magnetite, pyrrhotite and TM60 in zero or small fields frequently lack visible walls (see Chapter 6). The frequency with which metastable SD states are observed depends on how the remanent state was achieved and decreases overall with increasing grain size. A metastable SD condition is most commonly observed following saturation, either in large fields at ordinary temperatures (SIRM) or in small fields near T_C (TRM). Clearly the controlling factor is nucleation or rather the failure of nucleation. The example shown in Fig. 12.14 suggests that secondary rather than primary nucleation is important in magnetite. A reverse domain has nucleated but remains pinned at its nucleation site.

Grains in metastable SD states can have giant moments, orders of magnitude greater than the moments of small grains in equilibrium SD states. However, the metastable state is typically rather unstable and a full domain structure tends to develop during cycling in small fields (Fig. 12.14). These grains therefore do not necessarily possess high AF stabilities. They lose virtually all their remanence quite abruptly when the nucleation field H_n is exceeded and domain structure develops. An initial SD-like plateau, as observed in the AF demagnetization of LT memory (Figs. 12.9, 12.10), can only be explained if most H_n values exceed 5 mT (50 Oe) and at least some exceed 20 mT (200 Oe). The plateau cannot be a result of SD-style rotation because metastable SD structures transform rather than rotating in reverse fields. The reverse metastable SD state is only generated following reverse saturation.

Some have argued that some metastable SD states may be an observational artefact: walls or other non-uniform micromagnetic structures may be present but not clearly imaged by Bitter patterns (Ye and Merrill, 1991; Williams *et al.*, 1992a; see also §7.7). Others have argued that nucleation failure is the norm, and that grains of all sizes typically contain fewer than the equilibrium number of domains (Fig. 12.17). However, the one-dimensional modeling used to generate the theoretical curve in Fig 12.17 ignores closure domains, which greatly reduce the demagnetizing energy in magnetite and result in fewer, wider body domains (§5.4.2; Fig. 7.7). The theoretical curve should be moved to the right by almost an order of magnitude in grain size. It was long believed (e.g., Worm *et al.*, 1991) that closure domains do not form in magnetite. This has been shown decisively to be untrue (Özdemir and Dunlop, 1993a; Özdemir *et al.*, 1995). When crystals are polished so that {110} is the viewing plane, surface closure domains are invariably observed (Figs. 1.2, 5.8b, 5.11, 6.5). $180°$, $109°$ and $71°$ walls are all sharply resolved, and the numbers and widths of domains agree quite well with theoretical predictions (Figs. 5.12, 6.7a,b). Magnetic colloid

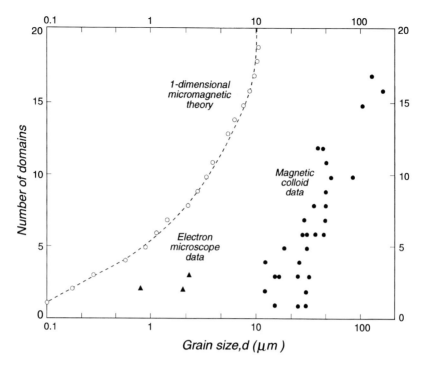

Figure 12.17   Discrepancy between the number of domains in a magnetite grain of given size predicted by 1-dimensional micromagnetic theory and domain observations (mainly by the Bitter method on unoriented surfaces). The disagreement largely disappears when 2-dimensional or 3-dimensional theories, which permit closure domains, are compared with Bitter patterns on {110} oriented surfaces, which display plentiful closure domains. [Reprinted from Worm *et al.* (1991), with kind permission of the authors and Elsevier Science – NL, Sara Burgerhartstraat 25, 1055 KV Amsterdam, The Netherlands.]

observations on unoriented surfaces (Fig. 12.17) disagree with the newer data and seem to be untrustworthy. It is interesting that extrapolation of the data in Fig. 6.7(a),(b) implies $d_0 = 0.06$–$0.2 \, \mu$m, which is consistent with observed and predicted values (Table 5.1).

We can conclude that nucleation failure is not the norm, but a relatively rare occurence. At the same time, there is little doubt that most observations of metastable SD states are reliable, since walls are clearly resolved in the same grains after field cycling. A third consideration is that the same grain, when repeatedly given a TRM under identical conditions, sometimes nucleates a domain structure and sometimes does not (Halgedahl, 1991; Fig. 5.14). Nucleation is a statistical event, even in individual grains. Furthermore, metastable SD states and states with pronounced domain imbalance seem to be divorced from the main distribution of LEM states (Fig. 5.14). Some sporadic inhibition factor preventing nucleation seems to be implied.

12.8.2   The Halgedahl and Fuller theory

Taking the observations as a guide, Halgedahl and Fuller (1980, 1983) developed a theory of PSD magnetization based on the idea that nucleation sites are randomly distributed and failure of a grain to nucleate at least one wall is a rare event. The probability that a grain whose equilibrium number of walls is λ actually contains only $w(<\lambda)$ walls was assumed to follow the Poisson distribution

$$P(w) = \lambda^w e^{-\lambda}/w! \tag{12.3}$$

Figure 12.18(a) shows that (12.3) is a good predictor of the observed number of domains in 73 pyrrhotite grains 6–11 μm in size if the equilibrium state is taken to be 2D over this size range.

Metstable SD grains ($w = 0$) account for most of the remanence. Their probability is $P(0) = e^{-\lambda}$. According to eqn. (5.17), the equilibrium number n of domains $(= \lambda + 1)$ in a grain of size $d(= D$ as used in Chapter 5) is

$$n = \left(\frac{\mu_0 N M_s^2 d}{2\gamma_w}\right)^{1/2} = \left(\frac{d}{d_0}\right)^{1/2} \tag{12.4}$$

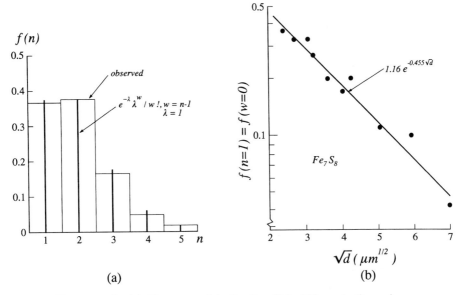

(a)   (b)

Figure 12.18   (a) Histogram of the fraction $f(n)$ of 73 pyrrhotite grains between 6 and 11 μm in diameter which contain n domains in their SIRM state. The equilibrium number is $n_{eq} = 2$ domains for this size range. The observed $f(n)$ is explained almost perfectly by a Poisson distribution (vertical bars) with $\lambda = n_{eq} - 1 = 1$. (b) Observed fraction of metastable SD states ($w = 0$ or $n = 1$) as a function of grain size in pyrrhotite grains carrying SIRM. [After Halgedahl and Fuller (1983) © American Geophysical Union, with the permission of the authors and the publisher.]

noting that $n = 1$ at the critical SD size d_0. Thus the probability of a metastable SD state $(w = 0)$ in a grain of size d, whose equilibrium number of walls is $\lambda = n - 1$, is

$$P(0) = e^{-(n-1)} = \exp[-(d/d_0)^{1/2} + 1] = 2.72 \exp[-(d/d_0)^{1/2}]. \qquad (12.5)$$

Experimentally, when 297 pyrrhotite grains in the 5–50 μm size range were classified into groups of similar d, the fraction $f(n = 1)$ of metastable SD grains in each group was well described by $f(n = 1) = 1.16 \exp[-(d/4.8)^{1/2}]$, with d in μm (Halgedahl and Fuller, 1983; Fig. 12.18b). The functional dependence on d is as expected from (12.5), but the observed numbers of SD grains at any size are about half those predicted, and the implied critical SD size of 4.8 μm is about three times that observed for pyrrhotite (Fig. 5.10, Table 5.1).

The reduced remanence following saturation due to metastable SD particles should be $(M_{rs}/M_s)_{SD}$, which is 0.5 for uniaxial anisotropy (magnetite with shape anisotropy, pyrrhotite) or 0.832 (TM60, magnetocrystalline controlled), times $P(0)$:

$$
\begin{aligned}
M_{rs}/M_s &= 0.500 \exp[-(d/d_0)^{1/2} + 1] \\
&= 1.36 \exp[-d/d_0)^{1/2}] \qquad \text{(uniaxial)} \\
&= 0.832 \exp[-(d/d_0)^{1/2} + 1] \\
&= 2.26 \exp[-(d/d_0)^{1/2}]. \qquad \text{(cubic)}
\end{aligned}
\qquad (12.6)
$$

The theory does not predict the size dependence of the nucleation field H_n, but experimentally, for a large number of pyrrhotite grains in which primary nucleation was observed, H_n in mT was given by

$$H_n = 120d^{-1/2}, \qquad (12.7)$$

as illustrated in Fig. 12.5.

12.9   Testing the models

12.9.1   Testing AF demagnetization behaviour

Models invoking magnetically hard regions embedded in softer MD surroundings have been criticized on many grounds but the only quantitative test is by Metcalf and Fuller (1987a). Using a SQUID magnetometer, they were able to measure IRM acquisition and backfield and AF demagnetization of SIRM for individual large ($\approx$ 50–100 μm) grains of natural TM60 and synthetic TM70. Grains with mainly straight parallel walls and grains of comparable size with mainly complex, stress-controlled patterns (Fig. 6.10) had very similar IRM growth and decay curves. In all cases, the AF decay curve was of characteristic MD exponential form (Fig. 12.19). There was no indication of any SD-like inflected demagnetization curve.

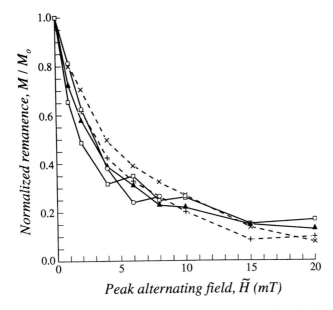

Figure 12.19   AF demagnetization curves of SIRM of individual ≈ 100 μm TM60 or TM70 grains with simple lamellar (solid curves) or complex stress-controlled (dashed curves) surface Bitter patterns, measured with a SQUID magnetometer. In both cases, the AF curves are of exponential MD shape (cf. Fig. 11.13), implying that the interior domain structure is simple even in grains with complex surface patterns. Such surface structures are not likely sources of PSD behaviour. [After Metcalf and Fuller (1987a) © American Geophysical Union, with the permission of the authors and the publisher.]

Metcalf and Fuller remarked that stress-controlled regions appeared to be a surface phenomenon. They were unresponsive to fields, but planar walls could sometimes be seen moving beneath them. Metcalf and Fuller concluded that displacements of simple planar walls are the source of most remanence in large titanomagnetite grains. The chaotic regions of intricate walls proposed by Appel and Soffel (1984) as sources of PSD remanence are indeed hard, but they seem to carry little coherent remanence, perhaps because directions of internal stress and magnetization vary rapidly over short distances. This source at least of surface moments cannot explain initial plateaus in AF demagnetization curves of LT memory.

12.9.2   Tests of TRM acquisition

Stacey (1963) proposed that the TRM of PSD moments with possible orientations at angles ϕ or $\pi - \phi$ to a field $\boldsymbol{H}_0$ is given by

$$M_{\text{tr}} = \frac{m}{V}\cos\phi\,\tanh\left(\frac{\mu_0 m(T_{\text{B}})H_0\cos\phi}{kT_{\text{B}}}\right), \tag{12.8}$$

m and *V* being the PSD moment and volume. This function, labelled PSD in Fig. 12.20(a), saturates rapidly with increasing H_0. Equation (12.8) is a strictly SD formulation, written down directly from Néel SD theory (eqn. (8.21)). It allows for no intermediate states with moments between $\pm m$, such as the 'collapsed' state of a metstable SD particle on those occasions when it does nucleate domain structure, or for alternative states with a quasi-continuum of *m* due to pinning of displaced walls (Barkhausen and domain imbalance moments). Nor are sub-domain moments that cannot be decoupled from the main domains permitted. The PSD moment must reverse freely above T_B between its two equilibrium states. Otherwise a potential energy of coupling to the host domains would enter the Boltzmann factors which are the source of the tanh function in eqn. (12.8). Only wall moments satisfy all these requirements and can be considered canonical PSD moments, or 'psarks'.

Added to the TRM of PSD origin is the TRM due to pinned walls, labelled MD in Fig. 12.20(a). It is approximately linear in H_0 up to 1 mT (10 Oe) and saturates only when $H_0 > 10$ mT (Fig. 9.5). The sum of these contributions, with two adjustable parameters (the PSD/MD ratio and the value of *m*), gives a close fit to measured TRM acquisition curves for submicroscopic magnetites (Fig. 12.20b). It is in this size range just above d_0 where SD, 2D and vortex structures are the likely LEM states, that 'psarks' should be most important. We noted in Chapter 7 that the single-vortex state resembles a 2D structure with two closure domains; the vortex-line moment is analogous to the wall moment and, like the wall moment, has two possible polarities.

For the two intermediate-size magnetites, *m* from the curve fits of Fig. 12.20(b) corresponds to the saturation moment of the entire grain. That is, some grains are SD, while others are 2D: the PSD and MD contributions come from different grains. However, in the largest magnetites (average size 0.22 μm), *m* corresponds to the volume of a 0.1 μm grain, much smaller than any of the grains in the sample. In this case, each individual grain makes both PSD and MD contributions, associated with the intrinsic moment and the displacement, respectively, of the domain wall. Independent estimates of the PSD volume from measured TRM unblocking temperatures (Fig. 8.3, eqn. (8.19)) and thermal fluctuation analysis (TFA) of $H_c(T)$ data (Fig. 8.13b, eqn. (8.29)) lead to the same conclusions (Dunlop, 1977, Fig. 9). Furthermore, TFA of $H_c(T)$ data for larger magnetites (up to 0.54 μm) gave PSD volumes that are almost independent of grain size (Argyle and Dunlop, 1990) and considerably less than the grain volume. A grain-size independent PSD volume is incompatible with metastable SD behaviour.

The TRM of metastable SD grains cannot be considered a partition between $\pm m (= \pm V M_s)$ states in the manner of eqn. (12.8) because the grain cannot switch directly between the + and − states. The dependence of TRM on H_0 has a different cause in this case. Metcalf and Fuller (1988) observed that the proportion of

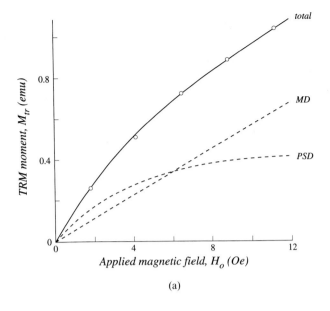

(a)

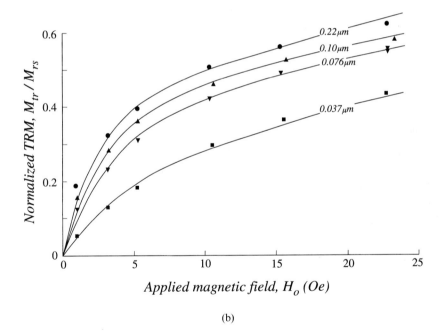

(b)

Figure 12.20 (a) Superposition of a rapidly saturating TRM acquisition curve for PSD moments and a slowly saturating MD curve for pinned walls to give observed total TRM curves. (b) TRM acquisition curves synthesized as in (a) compared with TRM data for four submicron magnetites. [Reprinted from Dunlop *et al.* (1974), with kind permission of the authors and Elsevier Science – NL, Sara Burgerhart-straat 25, 1055 KV Amsterdam, The Netherlands.]

metastable SD grains (i.e., $P(0)$) in 2–50 μm TM60 grains increased with an increase in the field H_0 used to impart TRM. On the simple assumption that the moments of the metastable SD grains overshadow all other sources of TRM, they synthesized a reasonable $M_{tr}(H_0)$ function by plotting $P(0)$ observed for the state TRM(H_0) normalized to $P(0)$ for the SIRM state (Fig. 9.11).

12.9.3   Testing predicted grain size dependences

Domain-wall moments and metastable SD grains are compared as probable sources of SIRM in <0.5 μm magnetites in Fig. 12.21. The log(M_{rs}/M_s) vs. $\sqrt{d}$ plot is convenient because eqn. (12.6) is then linear. The moments of 2D grains with wall overshoot or skirting (Fig. 7.2) decrease too rapidly with increasing d to explain the data for any but the finest grains. Predicted metastable SD moments, using $d_0 = 0.07$ μm, are higher than observed moments, but do give about the observed size dependence. The best fit results from an equipartition of

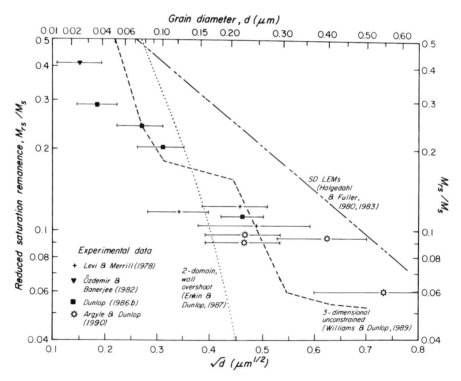

Figure 12.21   Experimental M_{rs}/M_s data for submicron magnetites compared with theoretical grain-size dependences. Metastable SD grains have the correct size variation but predict too high values. An ad-hoc combination of 3-dimensional micromagnetic states explains the data but is physically unsatisfying. [After Argyle and Dunlop (1990) © American Geophysical Union, with the permission of the authors and the publisher.]

various LEM states predicted by Williams and Dunlop's (1989) three-dimensional micromagnetic modelling, but since the states have quite different energies, it is not obvious that they should occur with equal probabilities.

A better fit to M_{rs}/M_s data in the 0.1–0.7 µm size range for magnetite results from three-dimensional micromagnetic simulations of hysteresis (Williams and Dunlop, 1995; §7.6). Although not all field orientations were tested, the extremes of H_0 parallel and perpendicular to the easy axis probably bracket the hysteresis behaviour. The same simulations produce a very encouraging fit to $H_c(d)$ data, although the predicted H_c values tend to be on the low side (Fig. 12.22). The success of this type of modelling, in which the grain selects its own preferred LEM state during the hysteresis cycle (in general it is a vortex state: Fig. 7.9), suggests that both metastable SD structures and domain-wall moments (indeed, domain walls as recognizable entities) may be something of an abstraction in the size range from d_0 to 0.7 µm in magnetite.

M_{rs}/M_s data for synthetic TM60, 40, 20 and 0 (i.e., magnetite) grains in the 1–15 µm size range (Day et al., 1977) have, as predicted by eqn. (12.6), a $\sqrt{d}$ dependence on grain size (Fig. 12.23a). The coefficients for TM60, TM40 and TM20 (1.12, 0.823 and 0.707) are lower than the 1.36–2.26 predicted. The probability of metastable SD states is thus somewhat lower than expected, as was

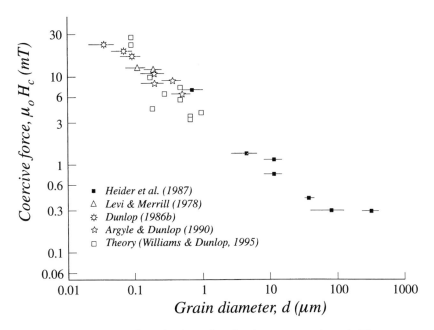

Figure 12.22 Coercive force data for the same samples as in Fig. 12.21 explained by theoretical H_c values from 3-dimensional micromagnetic simulations of hysteresis. [After Williams and Dunlop (1995) © American Geophysical Union, with the permission of the authors and the publisher.]

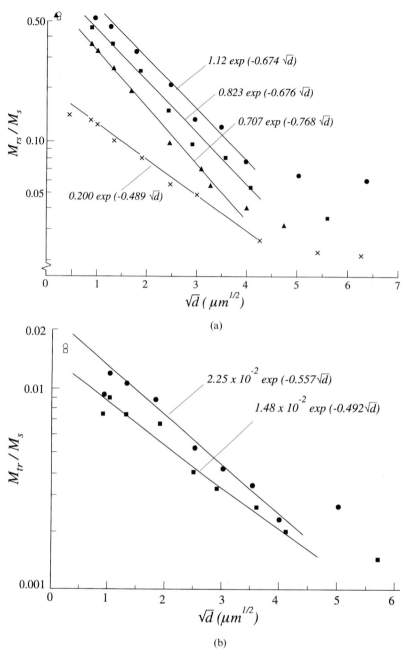

Figure 12.23 (a) Measured grain size dependences of M_{rs}/M_s for synthetic crushed titanomagnetites $Fe_{3-x}Ti_xO_4$ with $x = 0$ (crosses), 0.2 (triangles), 0.4 (squares) and 0.6 (circles) fitted in the manner of metastable SD theory (eqn. (12.6), solid lines). Data after Day (1973), Day *et al.* (1977), Özdemir (1979) and Özdemir and O'Reilly (1981). (b) Similar semilogarithmic plot, with line fits, of TRM data for crushed synthetic TM40 (squares) and TM60 (dots). Data after Day (1977), Özdemir (1979) and Özdemir and O'Reilly (1982c). [After Halgedahl and Fuller (1983) © American Geophysical Union, with the permission of the authors and the publisher.]

the case also for direct measurements of the metastable SD fraction in pyrrhotite grains of similar size (§12.8.2). The implied d_0 values are 1.7–2.2 μm, 3–4 times higher than experimental values (Table 5.1). This discrepancy is similar to that in the pyrrhotite data. The size dependence of M_{tr} data for TM60 and TM40 in the same size range gives fits of similar quality (Fig. 12.23b), with even higher implied d_0 values (3.2–4.1 μm).

Above ≈ 15 μm, the data can no longer be fitted by a $\sqrt{d}$ function. Fuller (1984) invokes a separate size-dependent MD contribution to remanence in these larger grains to produce a match to M_{rs} and M_{tr} data. O'Donovan *et al.* (1986) take a similar phenomenological approach. Although the theoretical fits are improved by adding more parameters, one loses the great virtue of the original formulation – its simplicity.

The M_{rs}/M_s data for magnetite follow a $\sqrt{d}$ relation (Fig. 12.23a) but the metastable SD probability is unreasonably low (the coefficient is only 0.2) and the implied value of d_0, 4.2 μm, is much too high. Nevertheless metastable SD states are observed for magnetite in this size range (Fig. 12.14; Boyd *et al.*, 1984). An even more problematic value of d_0, 450 μm, results from a fit of M_{rs}/M_s data for 1.5–30 μm pyrrhotites by Menyeh and O'Reilly (1991, Fig. 8).

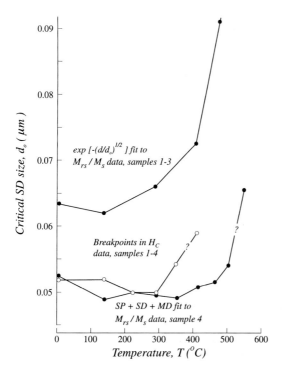

Figure 12.24 Estimates of the critical SD size d_0 in magnetite at high temperatures, based on experimental measurements of M_{rs}/M_s in submicron magnetites, and with less certainty, on H_c data. Samples 1–4 are the 0.22 μm, 0.10 μm, 0.076 μm and 0.037 μm samples, respectively, of Fig. 12.20(b). An increase in d_0 at high temperature was predicted theoretically in Fig. 5.15(b). [Reprinted from Dunlop (1987), with kind permission of the author and Elsevier Science – NL, Sara Burgerhartstraat 25, 1055 KV Amsterdam, The Netherlands.]

12.9.4   Testing the temperature dependence of d_0

A final test involves M_{rs}/M_s data measured at a variety of temperatures. The samples are the four magnetites of Fig. 12.20(b). The 0.037 μm sample contains mainly SD grains and its size distribution is known from transmission electron microscopy. One can crudely divide the size spectrum into an SD fraction, with $M_{rs}/M_s = 0.5$, and an MD fraction with negligible remanence, in order to estimate the critical SD size d_0 separating the two fractions. Using this method, d_0 is estimated to be about 0.05 μm from T_0 up to 350 °C and then to increase at higher temperatures (Fig. 12.24). A strong increase in d_0 at high temperatures was predicted in §5.5.1 and Fig. 5.15(b).

M_{rs}/M_s data for the large magnetites were fitted in the manner of Fig. 12.23(a). The implied d_0 values have a similar, somewhat more extreme temperature dependence (Fig. 12.24), all the values being somewhat higher than those determined experimentally for the 0.037 μm grains. A test of this type is important because it shows that the metastable SD model behaves in a reasonable manner (insofar as size dependence of hysteresis parameters is concerned) at high temperatures, where TRM is produced, as well as at room temperature.

Chapter 13

Crystallization remanent magnetization

13.1 Introduction

Crystallization remanent magnetization (CRM) results from the formation of a new magnetic mineral in the presence of a magnetic field, either by nucleation and growth to a stable blocking volume V_B (single-phase or growth CRM) or through alteration of an existing magnetic phase (two-phase or parent–daughter CRM). The commonly used term chemical remanent magnetization is not always strictly accurate, e.g., in the $\gamma Fe_2O_3 \rightarrow \alpha Fe_2O_3$ (spinel $\rightarrow$ rhombohedral) transformation, where no chemical change occurs, only a restacking of the lattice.

CRM is usually thought of as being blocked when grains grow from superparamagnetic (SP) to thermally stable SD size at $V = V_B$. This simple picture breaks down in the case of two-phase CRM because the growing daughter phase is influenced not only by an external field H_0 but also by its magnetic parent phase, to which it may be magnetostatically or exchange coupled with varying degrees of efficiency. Exchange coupling can occasionally result in self-reversed CRM (§13.4.6, 14.5.3).

CRM is usually regarded as a contaminant by paleomagnetists because the time of secondary mineral formation is difficult to date. Unfortunately CRM is not always easy to recognize because its unblocking temperatures and coercivities overlap those of primary TRM and DRM.

Many processes generate growth CRM in low-temperature sedimentary environments: precipitation of hematite cement from iron-rich solutions in red beds (Larson *et al.*, 1982); microbially mediated production of authigenic magnetite in the iron-reducing zone of recent marine sediments (Karlin *et al.*, 1987);

367

biogenic magnetite production in calcareous sediments that eventually form limestones (Chang *et al.*, 1987); and inorganic authigenesis of magnetite in soils (Maher, 1986; Maher and Taylor, 1988). These processes will be discussed in Chapter 15.

In igneous rocks (Chapter 14), subsolidus recrystallization of titanomagnetites (Tucker and O'Reilly, 1980b) or precipitation of magnetite in olivine and other iron-rich silicates (Brewster and O'Reilly, 1988) may continue below the Curie temperature during initial cooling (deuteric alteration). Secondary magnetite may form at lower temperatures through hydrothermal alteration of the titanomagnetites and silicates in continental and oceanic lavas, dikes and plutons (Hall, 1985; Hagstrum and Johnson, 1986; Smith and Banerjee, 1986; Hall and Fisher, 1987). Subsolidus reactions often go to completion during laboratory heatings of rock samples (e.g., Kono, 1987; Özdemir, 1987; Holm and Verosub, 1988), generating unwanted CRM's.

Finally, during deep burial alteration and metamorphism, CRM processes abound (Chapters 15 and 16). Widespread remagnetization of Appalachian rocks in eastern North America (§1.2.5) has been traced to secondary magnetite, some as spherules replacing pyrite, in carbonate rocks (McCabe *et al.*, 1983; Jackson *et al.*, 1988b; Suk *et al.*, 1990, 1992; Sun and Jackson, 1994) and hematite formed during deep burial diagenesis of red beds (Hodych *et al.*, 1985). CRM formed in natural environments has been reviewed by Merrill (1975), Henshaw and Merrill (1980) and Levi (1989).

In all these natural examples, direct evidence of the nature of the CRM is lacking. We shall focus attention in the present chapter on a relatively few laboratory studies under controlled time–temperature conditions in which CRM response to field direction and strength, and to prior remanence in the parent phase, was carefully monitored.

13.2   Single-phase or grain-growth CRM

13.2.1   Superparamagnetism

We have seen in Chapters 8, 9 and 11 that small changes in temperature or applied field can change the relaxation time τ for domain rotation or wall displacement from very small to very large values. A remanent magnetization (TRM, ARM or IRM) is blocked as a consequence. A small increase in grain size has the same effect on τ (Fig. 8.10a) and in the presence of a field H_0, results in **grain-growth CRM**.

A crystal so small that $\tau \ll t$ at room temperature is superparamagnetic (SP) or thermally unblocked. It has zero remanent magnetization (Fig. 5.20) and, since $H_q > H_K$ in eqn. (8.24), zero coercive force. The critical blocking volume

V_B at which such a crystal, growing in a weak field, passes from an SP to a thermally stable SD condition follows from eqn. (8.18):

$$V_B = \frac{2kT \ln(2t/\tau_0)}{\mu_0 M_s H_K},$$ (13.1)

giving for the room-temperature SP threshold size

$$d_s = V_B^{1/3} \approx \left(\frac{25kT}{K}\right)^{1/3}$$ (13.2)

if $t = 60$ s, since $H_K = 2K/\mu_0 M_s$ for uniaxial anisotropy. This is the justification for eqn. (5.29). Values of d_s for common minerals are given in Table 5.1.

Substituting (13.1) in eqn. (8.24), the coercive force measured in hysteresis or the destructive field in AF demagnetization of a single-domain ensemble will be (Kneller and Luborsky, 1963):

$$H_c(V, H_K) = H_K[1 - (V_B/V)^{1/2}].$$ (13.3)

Equation (13.3) describes a slow increase from zero coercive force when $V \leq V_B$ to $H_c = H_K$ at large V. The SP–SD transition size d_s can be determined from the grain-size dependence of H_c in the single-domain range using (13.3), a useful method when data for very small grains are lacking (Fig. 8.13a). Figure 13.1 illustrates the step-like increase in M_{rs} of iron–cobalt particles at d_s compared to the gradual increase in H_c throughout the SD range, which follows eqn. (13.3) quite closely.

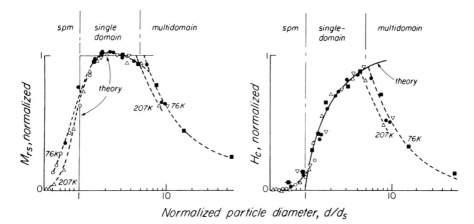

Figure 13.1 Saturation remanence M_{rs} and bulk coercive force H_c at 76 K and 207 K as a function of grain size d (normalized to the critical SP size, d_s) for electrodeposited spherical grains of iron–cobalt. Superparamagnetic, single-domain, and multidomain size ranges are well expressed, especially in the M_{rs} data. The sharpness of the subdivision is limited by size dispersion in the experimental samples. [After Kneller and Luborsky (1963), with the permission of the publisher, The American Institute of Physics, Woodbury, NY.]

The null remanent magnetization of an SP grain does not imply that exchange coupling of spins has broken down. SP grains retain a spontaneous magnetization M_s, as evidenced by their very large induced magnetization (Fig. 5.20). The most clearcut indication of SP grains in a rock is, in fact, an anomalously high susceptibility.

What does in effect vanish is anisotropy. Thermal fluctuations overcome the energy barriers to domain rotation (Figs. 8.5, 8.9) or wall displacement (Fig. 9.1). The smaller the number of coupled spins, the more likely it is that they can be excited coherently by a thermal impulse of energy $\approx kT$. The same situation prevails above the blocking temperature of a ferromagnetic grain of any size: exchange coupling remains essentially unaffected until the Curie temperature. The blocking temperature is nothing more than the temperature at which a grain becomes SP.

13.2.2   CRM acquisition by grain growth

Figure 13.2 traces the development of CRM in a ferromagnetic crystal growing in an applied field. Immediately after nucleation, the crystal is so small as to be SP. During its growth, it reaches the critical volume V_B at which $\tau \approx t$. Growing above V_B, τ becomes $\gg t$. There is an abrupt onset of CRM (Fig. 13.2, stage II)

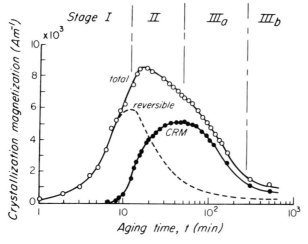

Figure 13.2   Induced and remanent crystallization magnetizations and their difference (the reversible magnetization) for spherical cobalt grains growing in a nonmagnetic matrix in an applied field $H_0 = 5\,mT$. The reversible magnetization peaks at the end of stage I when most of the grains are SP. The CRM intensity peaks at the end of stage II when most grains are SD, and decreases with growth into the MD range (stage III). Notice the similarity of the CRM growth curve in this figure and the size dependence of SIRM in Fig. 13.1. [After Kobayashi (1962a), with the permission of the author and the publisher, The Physical Society of Japan, Tokyo, Japan.]

and a gradual increase in H_c (Fig. 13.1). Later in the grain's growth (Fig. 13.2, stage III), M_{cr} and H_c fall off to PSD and then MD levels.

Figure 13.3 models on the Néel diagram the acquisition of single-domain CRM by crystal growth. CRM is frozen in when the representative points of growing grains cross the fixed blocking curve $VH_{K0} = C(T_0, t, H_0)$. When grain growth is complete, the representative points of grains carrying CRM cover a large part of the Néel diagram. Like TRM and SIRM, grain-growth CRM has a broad spectrum of unblocking temperatures and coercive forces.

CRM intensity was first considered by Kobayashi (1962b), who proposed, by analogy with eqn. (8.21),

$$M_{cr} = M_{rs} \tanh \left(\frac{\mu_0 V_B M_s H_0}{kT} \right). \tag{13.4}$$

Here T and M_s refer to the temperature at which CRM is acquired and V_B is given by eqn. (13.1). For the weak fields of interest in paleomagnetism, tanh can be replaced by its argument and eqn. (8.19) for T_B simplifies. Substituting (8.19) and (13.1) in (8.21) and (13.4) and rearranging, one obtains

$$\frac{M_{cr}}{M_{tr}} = \frac{H_K(T_B)}{H_K(T)} = \frac{K_B}{K} \left(\frac{M_s}{M_{sB}} \right), \tag{13.5}$$

as found by Stacey and Banerjee (1974, eqn. (9.17)).

Equation (13.5) is more of an idealization than (13.4). Whereas all grains growing at a constant T acquire CRM at the same volume V_B, it is unlikely that they will all eventually attain an identical terminal volume V and have a single T_B. Equation (13.5) does not explicitly depend on V, but it does so implicitly through T_B. It is clear from (13.5) that theoretically $M_{cr} < M_{tr}$, in agreement with the results of most experiments (see next section). A precise value for the

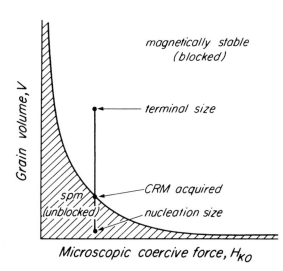

Figure 13.3 Modelling the acquisition of grain-growth CRM on the Néel diagram. The contour is the room-temperature blocking curve $VH_{K0} = C(T_0, 60\,\text{s}, H_0 \approx 0)$. CRM is acquired when a grain growing from its nucleation size to its terminal size crosses the blocking curve.

CRM/TRM ratio is not easy to predict, since M_s, K and H_K change rapidly over the typical range of T_B's just below T_C.

Grain-growth CRM, as sketched here, is the simplest possible CRM process. If a ferromagnetic phase crystallizes at the expense of an existing magnetized phase, the situation may be more complex, particularly if the old and new phases are exchange coupled across crystallization boundaries. The exchange forces outweigh any externally applied field and the CRM direction may be controlled by a pre-existing remanent magnetization. These complexities will be considered in §13.3 and 13.4.

13.2.3   Examples of grain-growth or single-phase CRM

Pure growth CRM uninfluenced by a magnetic parent is comparatively rare. Two classic examples are CRM resulting from high-temperature precipitation and growth of cobalt in a Co–Cu alloy (Kobayashi, 1961; Fig. 13.2) and CRM of magnetite crystallized from solution at high pressure and temperature (Pucher, 1969). Both reactions are fairly distant analogs of processes occurring in nature, but they serve to illustrate the physics involved.

Kobayashi (1961) measured the magnetizations of Co crystals growing at 750 °C after times of 20 s to several hr. He showed clearly the existence of three grain-size ranges in the growing crystals: SP with high induced magnetization but zero remanence, SD with a comparatively sharp onset of strong CRM as particles grew through d_s, and MD in which both reversible and irreversible magnetizations decreased rapidly (Fig. 13.2, stages I, II, III). The onset of remanence occurred at exactly the same stage of growth whether weak or strong fields H_0 were applied. CRM blocking is basically volume, rather than field, controlled.

Partial TRM spectra measured by cooling from 750 °C after various stages of growth, confirmed that blocking temperatures increase steadily as growth proceeds. The changeover from mainly low to mainly high T_B's coincided with the onset of CRM blocking. The intensity of CRM increased more or less linearly with H_0 up to about 4 mT.

Pucher (1969) crystallized SD-size magnetites in sealed gold capsules in 1–4 hr runs at $T = 300$ °C and $P = 200$ MPa, using HCl as a mineralizer. A similar technique is used to grow large magnetite crystals by hydrothermal recrystallization (Heider and Bryndzia, 1987). After demagnetization of each CRM, Pucher produced TRM's by cooling from above T_C. With one exception, CRM's and TRM's produced in the same magnetites by the same field H_0 had very similar intensities and AF and thermal responses. One would anticipate similar stabilities, since TRM and CRM after terminal grain growth are carried by exactly the same distribution of magnetite grains. The similarity in intensities is unexpected, since the ultimate TRM blocking temperature T_B when grains have grown well past V_B is considerably greater than T at which CRM was frozen in. Equation (13.5) then predicts $M_{cr} < M_{tr}$.

Hoye and Evans (1975) produced magnetite by oxidizing synthetic olivines at 500 °C for times up to several days. The onset of CRM and its eventual decrease in intensity when magnetite crystals grew beyond SD size were similar to Kobayashi's observations, although the kinetics were slower. In contrast to Pucher's findings, at any stage of grain growth, CRM had an intensity about one-half that of TRM but was significantly harder to AF demagnetize than TRM.

A third method of growing magnetite is by the decomposition of green rust (a ferrous ferric hydroxide) under reducing conditions. The process can be carried out at ordinary temperatures and bears some relationship to magnetite formation in soils. Using this procedure, Pick and Tauxe (1991) measured a magnetite growth CRM whose direction was parallel to H_0 with an intensity proportional to H_0 up to about 1 mT. The approach to saturation at higher fields was much slower than predicted by (13.4), however.

Stokking and Tauxe (1987) grew SD (0.1–1 μm) hematite (αFe_2O_3) and goethite ($\alpha FeOOH$) crystals for 6 hr at 95 °C from a ferric nitrate solution on to porous alumina wafers, which were oriented in an applied field. The experiments were very similar to those of Hedley (1968), who precipitated goethite, hematite and lepidocrocite ($\gamma FeOOH$) over a period of almost one year from ferrous bicarbonate solution at ≈ 50 °C on to oriented glass discs. The two sets of experiments lead to similar conclusions. M_{cr} was in all cases parallel to H_0 acting during crystallization. Stokking and Tauxe (1990) reported that M_{cr} was proportional to H_0 for fields ranging from 10 μT to 0.8 mT and approached saturation intensity by 1.5 mT. An interesting observation is that the samples have an overall magnetic anisotropy of $\approx 20\%$. Thus H_0 can influence the orientation of growing crystals as well as the orientations of their domains.

13.3 CRM during low-temperature oxidation

Even in Stokking and Tauxe's and Hedley's experiments, it is not certain that pure single-phase CRM has been isolated, since hematite and goethite are often finely intergrown and goethite can carry an appreciable remanence below 100 °C (Dekkers, 1989a; Özdemir and Dunlop, 1996a). In the two-phase transformations to be described in this section, there is no doubt that the magnetic parent phase influences the CRM of the crystallizing daughter phase, and a simple blocking theory of CRM no longer holds. Because of the paleomagnetic implications, the direction of CRM is of more concern than CRM intensity. In modern experiments, the parent remanence and the applied field are non-collinear, usually perpendicular, so that the influence of each on CRM can be measured separately.

13.3.1   Low-temperature oxidation by oxygen addition

The low-temperature oxidation reaction $Fe_3O_4 \rightarrow \gamma Fe_2O_3$ and the corresponding oxidation of titanomagnetite to titanomaghemite are ubiquitous in surface-weathered continental rocks and altered seafloor rocks (§14.2), respectively. An inverse spinel lattice is preserved during oxidation, except at very high z values (Fig. 14.6). The lattice parameter changes by $< 1\%$, e.g., from 8.396 Å for Fe_3O_4 to 8.337 Å for γFe_2O_3 (§3.3). Thus one anticipates that exchange coupling will be unbroken across the magnetite–maghemite phase boundary. For SD-size crystals, oxidation should be no different in its effect on initial remanence than adding further layers of magnetite of TM60 to the original crystal. Even after a high degree of oxidation, the CRM should be indistinguishable from the initial remanence, apart from a decrease in intensity due to the change in M_s.

This predicted preservation or inheritance of remanence direction is found when SD magnetite (Johnson and Merrill, 1974; Heider and Dunlop, 1987; Özdemir and Dunlop, 1989) or titanomagnetite (Özdemir and Dunlop, 1985; Brown and O'Reilly, 1988; Nishitani and Kono, 1989) oxidizes in air to a cation-deficient spinel. An exception , in which M_{cr} was parallel to H_0, was reported by Nguyen and Pechersky (1987a,b), but here the reaction took place in solution and the grains were not immobilized. Oxidation of MD-size magnetite or titanomagnetite results in a field-controlled CRM (Johnson and Merrill, 1972, 1973). This result is logical since only the domains immediately adjacent to the phase boundary are directly exchange coupled. Other more distant domains can, and obviously do, adjust their sizes and orientations in response to the applied field.

Figure 13.4 gives some results of experiments by Özdemir and Dunlop (1985). Since TM60 has a Curie point of 150 °C, while oxidation proceeds at a reasonable rate only around 200 °C, they used Al-doped TM40 with $T_C = 270$ °C. The unoxidized parent material (grain size < 0.05 μm) was given an initial TRM M_{tr} in 50 or 100 μT while in sealed evacuated capsules. The capsules were then opened to permit oxidation under varying time and temperature conditions in a field H_0 perpendicular to M_{tr}. For $z < 0.65$, the oxidation product was a cation-deficient spinel and M_{cr} remained parallel to M_{tr} even when AF demagnetized to 100 mT. When $z > 0.7$, increasing amounts of rhombohedral hematite were detected by X-ray analysis among the oxidation products and M_{cr} consisted of two superimposed vectors. The softer of these (i.e., the one first AF demagnetized) was deflected 30–40° towards H_0, while the harder vector remained close to the direction of M_{tr}. In very similar experiments on ≈ 0.1 μm TM60, 90% of the CRM remained in the initial TRM direction even when z was as high as 0.8 (Brown and O'Reilly, 1988). Only when a change of lattice structure effectively breaks exchange coupling between parent and daughter phases can the external field begin to play a role in CRM acquisition.

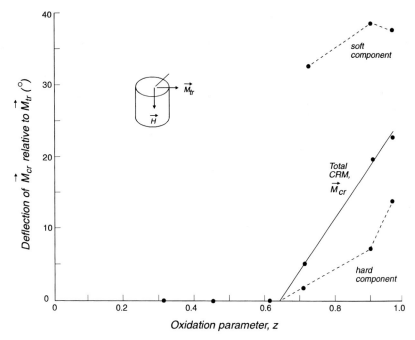

Figure 13.4   The direction of the CRM vector M_{cr} at different stages of in-air oxidation of SD titanomagnetite relative to the direction of the original TRM M_{tr}. For $z < 0.65$, the oxidized phase is exchange coupled to the parent mineral and M_{cr} inherits the direction of M_{tr}. For $z \geq 0.65$, there are multiphase oxidation products, exchange coupling is broken, and M_{cr} is increasingly deflected towards the field H_0, which was applied perpendicular to M_{tr}. [After Özdemir and Dunlop (1985) © American Geophysical Union, with the permission of the authors and the publisher.]

Özdemir and Dunlop (1985) concluded that when SD titanomagnetite undergoes single-phase oxidation, its original TRM is inherited by the resulting titanomaghemite. The external field acting during oxidation has no influence on the CRM direction unless $z \geq 0.7$. This was the first experimental demonstration that the original TRM direction could survive low-temperature oxidation of TM60 in submarine basalts, explaining the preservation of patterns of magnetic stripes over the oceans.

13.3.2   Low-temperature oxidation by iron removal

Kelso *et al.* (1991) used a similar configuration of M_{tr} and H_0 but their samples consisted of larger PSD-size TM45 and TM60 grains, immobilized in a silicate glass matrix through which warm (90 °C) acidic water was percolated to simulate oxidation under oceanic hydrothermal conditions. In this process, iron diffusing out of the crystal during maghemitization is leached by the solution rather than nucleating new spinel unit cells at the crystal surface (O'Reilly, 1983, 1984, p. 15; Worm and Banerjee, 1984; Brown and O'Reilly, 1987).

In contrast to the results of Fig. 13.4, a CRM component appeared in the direction of H_0 from the earliest stages of oxidation, gradually shifting the total remanence towards the field direction as oxidation proceeded . It is not clear whether the partial field control of CRM is attributable to the larger grain size and more complicated domain structure of the titanomagnetites or to the oxidation mechanism (iron loss rather than oxygen addition).

13.3.3   Low-temperature oxidation on the seafloor and magnetic stripes

It is crucial to resolve this question. Oxidation of titanomagnetites is the norm in seafloor older than 1 Ma, but its precise effect on the initial TRM is still unclear, despite more than two decades of study (e.g., Marshall and Cox, 1972; Beske-Diehl and Soroka, 1984; Smith, 1987). If CRM 'remembers' its parent TRM direction, magnetic stripe anomalies (§1.2.1, 14.1), which are the main record of seafloor spreading and lithospheric plate motions, will be unaffected except in intensity. If CRM tracks the periodically reversing geomagnetic field at the time oxidation occurs, the stripe pattern will be shifted in proportion to the time lag and may be blurred beyond recognition if oxidation is inhomogeneous in space and time.

Raymond and LaBrecque (1987) have estimated that as much as 80% of seafloor magnetism is 'CRM' (by which they mean remanence that does not remember the primary field direction) and only 20% is 'TRM' (i.e., CRM that has inherited a memory of the primary field). This is about the ratio required to explain the decrease in magnetic anomaly amplitudes between spreading ridges where fresh, TRM-bearing lavas are erupted and old, cold seafloor that is fully oxidized. However, their calculations are suspect because they fly in the face of evidence (like that of Fig. 13.4) that CRM due to maghemitization of fine-grained titanomagnetite preserves the original TRM direction. This evidence includes observations of NRM directions in variably oxidized pillow lavas from the seafloor (Soroka and Beske-Diehl, 1984) as well as laboratory experiments.

13.4   CRM with a change of lattice

13.4.1   Change of crystal structure and phase coupling

Many naturally occurring oxidation and alteration processes transform the crystal structure of the magnetic phase. Important examples in which the accompanying CRM has been studied are the dehydration of various isomers of FeOOH to either αFe_2O_3 or γFe_2O_3 (Hedley, 1968; Özdemir and Dunlop, 1993b), oxidation of Fe_3O_4 to αFe_2O_3 (Heider and Dunlop, 1987; Gie and Biquand, 1988;

Walderhaug, 1992), reduction of αFe_2O_3 to Fe_3O_4 (Haigh, 1958; Kobayashi, 1959; Gie and Biquand, 1988), inversion of γFe_2O_3 to αFe_2O_3 (Porath, 1968a; Özdemir and Dunlop, 1988; Walderhaug, 1992), the corresponding inversion of titanomaghemite to a multiphase spinel–rhombohedral intergrowth (Marshall and Cox, 1971, 1972; Bailey and Hale, 1981; Nguyen and Pechersky, 1987a,b), high-temperature oxidation of TM60, which also involves phase-splitting (Creer et al., 1970; Petherbridge, 1977; Nguyen and Pechersky, 1987a), and partial oxidation of pyrrhotite to magnetite (Bina and Daly, 1994).

In many cases, the nature of the crystallization process, whether homogeneous, inhomogeneous with many nuclei, or surface-dominated (as in the case of maghemitization), is not well documented except by inference from textures following alteration. A knowledge of the configuration of the phase boundary or boundaries during alteration is vital to understanding phase-coupling, but this knowledge is largely lacking.

Two-phase (or multiphase in the case of intergrown daughter minerals) CRM is often referred to as phase-coupled CRM, implying that coupling of the phases during CRM production is a demonstrated fact. Phases in intimate contact must at least interact magnetostatically, but this does not guarantee a coupling that is effective in preserving memory of the parent remanence. The evidence for effective phase coupling when the crystal lattice transforms is contradictory, ranging from assertions that M_{cr} is always parallel to H_0 (Nguyen and Pechersky, 1987a) to observations of partially deflected CRM's (Bailey and Hale, 1981; Heider and Dunlop, 1987; Walderhaug et al., 1991; Walderhaug, 1992) to demonstrations of self-reversal of CRM (Hedley, 1968; McClelland and Goss, 1993) or of subsequently acquired TRM (Creer et al., 1970; Bina and Daly, 1994).

13.4.2   Oxidation of magnetite to hematite

CRM's with stable directions intermediate between the direction of the applied field H_0 and the (perpendicular) direction of the parent remanence, which was an ARM, M_{ar}, were found by Heider and Dunlop (1987) following oxidation of $\approx 0.2\,\mu m$ (small PSD) cubic and octahedral magnetites in 2.5-hr runs at 500 °C. The strengths of both H_0 and M_{ar} were varied systematically. Some typical results are shown in Fig. 13.5. For initial ARM intensities ranging from 0.2% of M_{rs} (KA1) to 5% of M_{rs} (KA5), M_{cr} vectors were progressively deflected away from H_0 (0.1 mT in the runs illustrated) towards M_{ar}.

Upon AF demagnetization, the CRM's remained relatively stable in direction (Fig. 13.6). The CRM's in some cases (e.g., KA3, KA5) consisted of two vectors of slightly different intermediate orientations, but neither vector was close to either M_{ar} or H_0. The CRM vectors also remained in stable intermediate directions when thermally demagnetized to above 600 °C to erase any residual magnetite remanence.

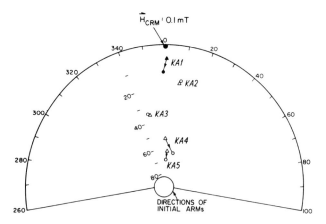

Figure 13.5 A stereographic projection of vectors M_{cr} of CRM acquired by $\approx 0.2\ \mu m$ magnetite after oxidizing for 2.5 hr at 500 °C under the simultaneous influence of $H_0 = 0.1$ mT (applied horizontally) and vertical initial ARM's M_{ar} ranging in strength from 0.2% (KA1) to 5% (KA5) of SIRM. Triangles and circles are the same CRM before and after thermal cleaning to temperatures above the Curie temperature $T_C = 580$ °C of magnetite. In each case, M_{cr} is a stable univectorial or nearly univectorial remanence (see Fig. 13.6) in a direction intermediate between H_0 and M_{ar}. [Reprinted from Heider and Dunlop (1987), with kind permission of the authors and Elsevier Science – NL, Sara Burgerhartstraat 25, 1055 KV Amsterdam, The Netherlands.]

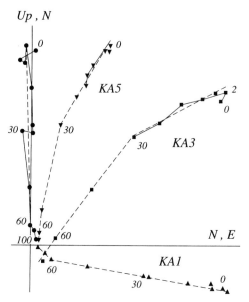

Figure 13.6 Orthogonal vector projections of AF demagnetization data for three of the CRM's of Fig. 13.5. The horizontal-plane (N vs E) data are similar for all three CRM's (circles). The vertical-plane (Up vs N) data (triangles and squares) are increasingly displaced away from H_0 (horizontal) toward M_{ar} (vertical) as the strength of M_{ar} increases (from KA1 to KA5). [Reprinted from Heider and Dunlop (1987), with kind permission of the authors and Elsevier Science – NL, Sara Burgerhartstraat 25, 1055 KV Amsterdam, The Netherlands.]

The deflected directions must have resulted from magnetostatic rather than exchange coupling; otherwise the coupling could not have been overcome by the rather small fields used. Heider and Dunlop postulated that the average internal field generated by M_{ar} is proportional to, and in the same direction as, M_{ar}. Then the resultant field acting during CRM production is

$$H_{res} = H_0 + \beta M_{ar} = H_0 \hat{i} + \beta M_{ar} \hat{k}, \tag{13.6}$$

for the orientations of H_0 and M_{ar} in Fig. 13.5. M_{cr} is produced in the direction of H_{res}. Thus its inclination I should be given by

$$\tan I = \beta M_{ar}/H_0. \tag{13.7}$$

The experimental data follow this relation quite well, except for one sample with a very high value of M_{ar}/H_0 (Fig. 13.7). The effective interaction field βM_{ar} deduced from the slope of the graph is somewhat greater than H_d for uniformly dispersed grains, implying some clustering of grains.

Walderhaug et al. (1991) and Walderhaug (1992) also obtained stable intermediate directions of CRM when magnetite regions of SD to PSD size in natural magnetite–ilmenite oxyexsolution intergrowths from a basalt were oxidized to hematite at 525 °C. There was some systematic deflection of M_{cr} away from the initial NRM direction towards H_0 as the field strength was increased from 0 to 60 µT. However, in similar experiments on coarser-grained magnetite from a dolerite dike, M_{cr} was parallel to H_0 for field strengths ranging from 12.5 to 75 µT and was uninfluenced by the pre-existing NRM. The dolerite magnetite

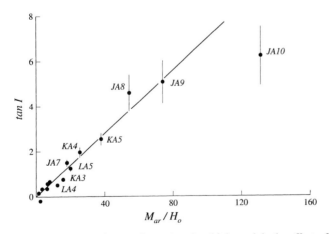

Figure 13.7 A test of eqn. (13.7), which models the effect of parent remanence M_{ar} on CRM in terms of an internal field whose strength is proportional to M_{ar}. The linearity of the graph of $\tan I$, where the inclination I is the angle between M_{cr} and the horizontal field H_0, versus M_{ar}/H_0 verifies the theory and gives a value for the internal field coefficient β. [Reprinted from Heider and Dunlop (1987), with kind permission of the authors and Elsevier Science – NL, Sara Burgerhartstraat 25, 1055 KV Amsterdam, The Netherlands.]

was profoundly maghemitized, exhibiting shrinkage cracks and granulation (§3.4, 3.6, 14.2.4), so that the CRM process is probably more analogous to the inversion of maghemite (§13.4.4) than to the direct oxidation of magnetite to hematite.

13.4.3   Reduction of hematite to magnetite

The earliest experiments in which the influences of both a parent remanence and of the applied field were monitored were those of Haigh (1958). He reduced hematite powders of unspecified grain size to magnetite in 45-min heatings at $\approx 300\,°C$ in a weak field (50 μT). H_0 was applied either parallel or antiparallel to the initial hematite remanence. CRM was of comparable intensity to the original remanence, which considering the great contrast in M_s values between the two minerals, implies a rather inefficient CRM process. M_{cr} was always parallel to H_0, leading Haigh to dismiss phase coupling. However, CRM acquired in the antiparallel configuration was only half as strong as that in the parallel config-uration, suggesting some control by the parent remanence.

Kobayashi (1959) also studied the reduction of hematite (grain size $< 1\,\mu m$) around $300\,°C$. Although he was not able to measure CRM while it was being acquired, as Haigh had done, he noted that CRM was relatively weak, about 10% of the TRM subsequently acquired by the daughter magnetite in the same field. CRM and TRM had similar resistances to AF demagnetization. Haigh had shown that their thermal unblocking is also similar. Gie and Biquand (1988) in recent detailed experiments found comparable results, except that CRM of the newly formed magnetite was about 25 times weaker than TRM.

13.4.4   Inversion of maghemite to hematite

Not since the early days of Haigh (1958) and Kobayashi (1961) had there been serious attempts to follow the development of CRM during a phase transforma-tion until the work of Özdemir and Dunlop (1988). Intrigued by the results of Porath (1968a), who reported anomalously high CRM intensities in maghe-mite–hematite intergrowths when the hematite content was about 90–95%, they investigated intermediate stages of the $\gamma Fe_2O_3 \rightarrow \alpha Fe_2O_3$ inversion in runs at temperatures from $300\,°C$ to $600\,°C$. Each run was 3 hr long in a field H_0 of 50 μT. Conversion to hematite ranged from $\approx 4\%$ at $300\,°C$ to ≈ 70–75% by 550–560 °C. In a first series of runs, 'pure' CRM's were generated in AF demag-netized samples by heating in H_0 with no initial remanence. In the second series of runs, pure ARM's M_{ar} were heated at the same temperatures in zero field, thereby progressively thermally demagnetizing them.

In the final runs, M_{ar} and H_0 acted perpendicular to each other, generating remanences in intermediate directions. However, since these remanences were produced at intermediate stages of the transformation, with both parent and daughter phase still present, the total remanence M_T is the vector sum of M_{cr} and surviving M_{ar}. The vector diagrams determined during AF demagnetization are continuously curving, showing that M_{cr} is not a stable, intermediate-direction CRM (Fig. 13.8a). In higher temperature runs, where M_{ar} is weaker, M_T is deflected closer to H_0.

In all but the highest temperature run, it was possible to synthesize the AF behaviour of M_T in the third set of runs almost perfectly by adding, at right angles, the AF demagnetization trajectories of pure CRM's and pure ARM's as determined in the first and second set of runs (Fig. 13.8b). In other words, the CRM, M_{cr}, produced in the presence of both H_0 and M_{ar} is identical (in its AF demagnetization characteristics) to pure CRM produced in the presence of H_0 alone. There is no phase coupling between initial ARM in the maghemite phase and CRM in the hematite phase.

Stable intermediate-direction CRM's did not occur in this $\gamma Fe_2O_3 \rightarrow \alpha Fe_2O_3$ transformation. Rather M_{cr} formed parallel to H_0, adding vectorially to any surviving initial remanence M_{ar}. Thus maghemite and hematite are not exchange or magnetostatically coupled during inversion. This conclusion is diametrically opposed to Porath's (1968a) conclusion that exchange coupling of maghemite and hematite accounted for the remanence peak in his 540 °C runs. Özdemir and Dunlop pointed out that since the ferromagnetic moment in hematite is perpendicular to the magnetic sublattices, exchange coupling should not in any case lead to a CRM parallel to the initial remanence. They interpreted the remanence peak to be a thermal activation effect, dictated by temperature, not phase-coupling (see §13.5).

13.4.5   Oxyexsolution of titanomagnetite

CRM's resulting from high-temperature oxyexsolution of titanomagnetite and from low-temperature oxidation plus inversion of titanomaghemite were first studied in detail by Creer et al. (1970) and by Marshall and Cox (1971), respectively. For compositions around TM60, the final daughter products of the two subsolidus reactions are a Ti-poor spinel (near-magnetite) with lamellae of paramagnetic ilmenite exsolved in {111} planes of its host (§3.1, 14.4).

During transformation, the parent spinel evolves towards magnetite and its Curie point rises (cf. Fig. 3.11), as Ti is partitioned to the rhombohedral phase. Although the overall structure of the grain will be disrupted by lamella development, exchange coupling should remain unbroken as far as individual daughter spinel regions are concerned. Indeed, as first reported by Wilson and Smith (1968), the original remanence direction in basalts and basaltic dikes is commonly preserved during heating well above T_C of the starting titanomagnetites. On the other hand, magnetostatic interaction favours alternating magnetic pola-

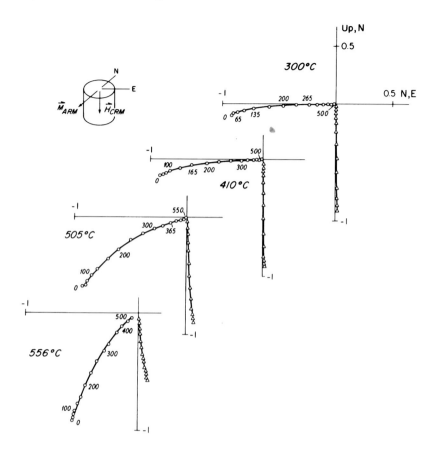

Figure 13.8 (a)

rities of the finely interleaved spinel regions (Stacey and Banerjee, 1974, pp. 63–65), greatly reducing the potential CRM intensity.

Experimentally, phase coupling exists but is rather weak. After 1–40 μm natural TM46 grains were oxidized at 400 °C for long times in a 45-μT field and cooled in zero field to T_0, the remanence during subsequent thermal demagnetization showed a reversible hump centred on the Curie point of the mother phase (Creer *et al.*, 1970). CRM of the higher-T_C daughter phase was negatively coupled to, and partially screened by, spontaneous TRM of the residual mother phase (Petherbridge, 1977). The equivalent interaction field was only 15 μT and must be of magnetostatic origin.

Bailey and Hale (1981) inverted natural seafloor titanomagnetites in 24-hr heatings at 515 °C. NRM's, M_{nr}, were at various orientations relative to the applied field H_0, whose strength was varied systematically from 10 to 50 μT. A complete set of control experiments was also performed on demagnetized sister samples. In demagnetized samples, M_{cr} was always parallel to H_0. In samples

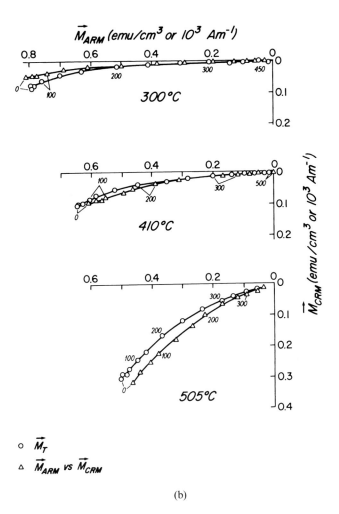

$\overline{M}_{ARM}$ (emu/cm³ or 10³ Am⁻¹)

300°C

410°C

505°C

$\overline{M}_{CRM}$ (emu/cm³ or 10³ Am⁻¹)

○ $\overrightarrow{M}_T$

△ $\overrightarrow{M}_{ARM}$ vs $\overrightarrow{M}_{CRM}$

(b)

Figure 13.8 (a) Orthogonal vector projections of experimental AF demagnetization trajectories for CRM's produced at intermediate stages during the inversion of maghemite to hematite at temperatures from 300 °C to 556 °C. The projections in a vertical plane containing the parent remanence M_{ar}, and the field H_0 applied during CRM production (circles) are continuously curving trajectories, rather than univectorial (straight) trajectories. Therefore the measured remanence M_T is not a single vector of intermediate direction, but a sum of M_{ar} and M_{cr} vectors with different resistances to AF demagnetization. (b) A demonstration that the measured data M_T (circles) can be synthesized by adding orthogonally the AF trajectories of pure CRM's and pure ARM's measured in separate experiments (triangles). Thus M_{cr} is produced parallel to H_0 and not in an intermediate direction. [After Özdemir and Dunlop (1988) © American Geophysical Union, with the permission of the authors and the publisher.]

with an initial remanence, M_{cr} was in a direction intermediate between M_{nr} and H_0 if H_0 was 20 μT or less. Furthermore M_{cr} remained in a stable intermediate direction when AF demagnetized to 100 mT. It was *not* a vector resultant of initial remanence that somehow survived inversion and a CRM along the applied field.

In Marshall and Cox's (1971) original experiments, in which H_0 was applied perpendicular to M_{nr}, intermediate direction CRM's may have been present but unrecognized as such. In their analysis of results, the component of magnetization along the direction of M_{nr} was assumed to be surviving original TRM and the component along H_0 was assumed to be the newly created CRM. Until Bailey and Hale's work, it was not realized that preconceived ideas had been built into the interpretation of the data.

13.4.6   Lepidocrocite dehydration and maghemite inversion

Phase-coupling of some type also occurs when γFe_2O_3 is produced by dehydrating lepidocrocite ($\gamma FeOOH$) and this maghemite then inverts to αFe_2O_3. Özdemir and Dunlop (1993b) studied CRM produced by a 50 μT field during these successive phase transformations in runs at temperatures from 200 °C to >600 °C. The thermomagnetic curve exhibits a sharp increase in M_s around 250 °C during heating, as the orthorhombic lattice of paramagnetic lepidocrocite ($T_N = 70$ K) dehydrates and converts to the spinel lattice of ferrimagnetic maghemite (Fig. 13.9a). Beginning around 450 °C, the maghemite inverts to hematite and M_s drops. The cooling curve shows that inversion is complete after heating to 600 °C.

The intensity of CRM shows an interesting correlation with the thermomagnetic behaviour. CRM first peaked at 250 °C, during the $\gamma FeOOH \rightarrow \gamma Fe_2O_3$ transformation, then decreased, and finally reached a second peak around 400 °C during the earliest stages of the $\gamma Fe_2O_3 \rightarrow \alpha Fe_2O_3$ inversion (Fig. 13.9b). The interpretation was that antiphase domains with their M_s vectors coupled antiparallel formed in the maghemite in the 275, 300 and 350 °C runs; growth of hematite on the antiphase boundaries, starting around 400 °C, broke the negative exchange coupling, allowing maghemite antiphase domains to align individually with H_0.

In the range ≈ 500–560 °C, the CRM is weak but has the interesting property of being either self-reversed to H_0 (Hedley, 1968; McClelland and Goss, 1993) or approximately perpendicular to H_0, with an internal self-reversal (Özdemir and Dunlop, 1993b; Fig. 13.10). During the inversion of maghemite, hexagonal platelets of αFe_2O_3 grow in {111} planes of γFe_2O_3 octahedra (Fig. 3.3). The M_s vector of hematite is confined to the basal plane of the platelets, perpendicular to one set of $\langle 111 \rangle$ easy axes containing the M_s vector of maghemite. Of course, there are eight different {111} faces and a corresponding number of $\langle 111 \rangle$ directions, so that many different geometrical relationships between the

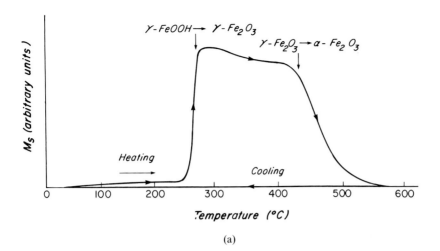

(a)

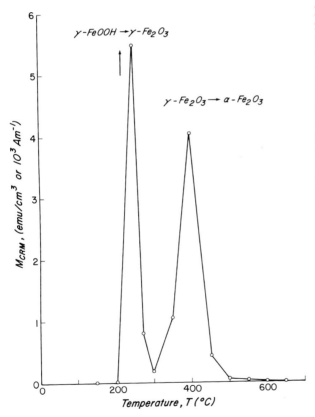

(b)

Figure 13.9 (a) Thermo-magnetic curve showing the transformations $\gamma FeOOH \rightarrow \gamma Fe_2O_3 \rightarrow \alpha Fe_2O_3$. Only γFe_2O_3 is ferrimagnetic and has a large M_s. (b) Intensities of CRM produced by $H_0 = 50\ \mu T$ in 2.5-hr runs at a series of 13 temperatures from 200 °C to > 600 °C. The peaks in CRM intensity correlate with the phase transformations in (a). For an interpretation, see text. [After Özdemir and Dunlop (1993b) © American Geophysical Union, with the permission of the authors and the publisher.]

maghemite and hematite M_s vectors are possible. Why one should be favoured over another is unclear, but there is certainly the potential for different relative orientations of CRM and initial remanence in experiments under slightly different conditions or with different starting samples of lepidocrocite.

Because of the variety of directions of phase coupling possible in this system, including internal self-reversals of CRM at a large angle to H_0 (Fig. 13.10), one must be cautious about inferring self-reversals of hematite CRM (M_{cr} anti-parallel to H_0) in nature (e.g., Wilson, 1961; McClelland, 1987). The NRM components may be reversed to each other without either one necessarily recording the paleomagnetic field direction.

13.4.7   Partial oxidation of pyrrhotite to magnetite

Self-reversal of a different sort can occur when natural pyrrhotite is partially oxidized to magnetite at 500 °C (Bina and Daly, 1994). The CRM of the magnetite

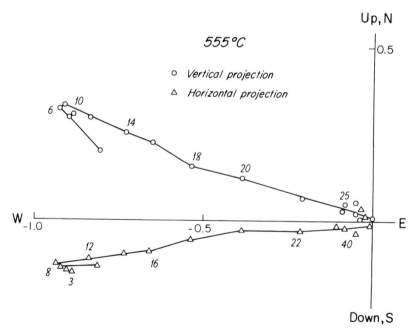

Figure 13.10 Orthogonal vector projections of AF demagnetization data for the CRM produced in the $\gamma Fe_2O_3 \rightarrow \alpha Fe_2O_3$ transformation at 555 °C, for the same material as in Fig. 13.9. The CRM M_{cr} is approximately horizontal, i.e., perpendicular to the vertical field H_0 which produced M_{cr}. The low-coercivity fraction of the CRM, up to 8 mT, is self-reversed to the high-temperature fraction, which forms the bulk of the CRM. This is an *internal* self-reversal: neither CRM vector is representative of the *external* field direction. [After Özdemir and Dunlop (1993b) © American Geophysical Union, with the permission of the authors and the publisher.]

is normal, i.e., parallel to H_0, but upon cooling in zero field, the residual pyrrhotite acquires a self-reversed TRM below its Curie temperature of 320 °C. This spontaneous TRM is sufficiently strong to reverse the total remanence. Although the microstructure of the partially oxidized grains is not known, Bina and Daly postulate that each grain has a magnetite rim and a largely unaltered pyrrhotite core and that negative magnetostatic coupling between the two is responsible for self-reversal. An exchange-coupling mechanism is unlikely because synthetic pyrrhotites treated in the same way do not self-reverse.

13.5   Chemicoviscous remanent magnetization (CVRM)

The theory of growth CRM (§13.2) makes it clear that the blocking of CRM due to growth through a critical blocking volume V_B is no different in principle from thermoviscous processes like the freezing in of TRM at a blocking temperature T_B or the acquisition of VRM in times $t \approx \tau$. In fact, if the terminal sizes V of growing particles are not much larger than V_B, or equivalently if their ultimate blocking temperatures T_B are not much larger than the temperature T at which growth occurs, the CRM will evolve viscously during and after grain growth.

It was recognized by early experimenters that VRM (called by them time-dependent IRM) would enhance the CRM due to volume-blocking. Haigh (1958) corrected for this viscous effect by allowing his samples to decay in zero field for times $\geq$ the CRM acquisition time before cooling to T_0. 30–50% of the remanence was lost as a result. Kobayashi (1959) corrected remanences produced during the reduction of hematite to magnetite at 340 °C by subsequently sealing the magnetites in capsules under vacuum, reheating them to 340 °C, and allowing them to acquire VRM for 20 hr in the same field used to produce CRM. This VRM, which amounted to about 20% of the original remanence, was subtracted to obtain the corrected CRM. Later workers have been less concerned with the effect, although some (e.g., Pucher, 1969) have noted the necessity of short runs to minimize viscous effects. Creer et al. (1970) found that certain degrees of high-temperature oxidation produced daughter-phase regions which were potently viscous, although no attempt was made to separate the effects of CRM, spontaneous (i.e., zero-external-field) TRM, and VRM. Beske-Diehl (1990) observed a VRM acquired by fine-grained titanomaghemites in oxidized seafloor basalts which was difficult to remove by AF demagnetization.

Özdemir and Dunlop (1985, 1988) pointed out that in some cases, what has conventionally been thought of as CRM may be largely, rather than just partly, VRM. In experiments that generated rhombohedral hematite from a spinel parent (§13.4.4), the magnetic moment of hematite was inadequate to explain the CRM intensity. Much of the remanence had to be a property of spinel regions which were shrinking, not growing, and VRM was the obvious mechanism.

Özdemir and Dunlop (1989) carried out parallel experiments on CRM production in the oxidation of SD-size magnetite to maghemite and VRM production in the same magnetite when sealed in evacuated capsules. Heatings were in a 50 μT field to 13 different temperatures. CRM and VRM rose to nearly identical peaks just below the respective magnetite and maghemite Curie points, and CRM fell sharply above 600 °C, even though much of the maghemite was still inverting to hematite (Fig. 13.11). Remanence peaks of this type are thus due to thermal activation of a single mineral species and not to phase coupling of two minerals as proposed by Porath (1968a).

In low-temperature runs, VRM comprised about one-half of CRM and was only slightly softer than CRM. At high temperatures, CRM and VRM intensities were virtually identical, as were their AF demagnetization characteristics (Fig. 13.12). The CRM was therefore called chemicoviscous remanent magnetization (CVRM). High-temperature VRM and CVRM were both slightly harder than total TRM. Viscous and chemicoviscous overprints in rocks of paleomagnetic interest, if produced at high temperatures, cannot necessarily be separated from primary TRM by AF cleaning. This is a matter of serious practical concern.

Similar experiments by Gapeev *et al.* (1991) used slightly larger magnetites (0.1–0.2 μm) and compared the AF demagnetization and time decay of CRM (or CVRM), acquired during oxidation of the parent magnetite in a field H_0, with AF demagnetization and time decay of pure VRM acquired by the maghemite daughter phase in the same H_0. Times and temperatures, as well as fields,

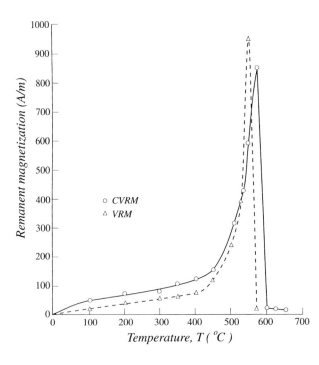

Figure 13.11   A comparison of the intensities of CVRM of oxidized or partially oxidized SD-size magnetite with intensities of VRM of unoxidized magnetite after 2.5-hr heatings in $H_0 = 50\,\mu$T at the same temperatures. [Reprinted with permission from Ö. Özdemir and D. J. Dunlop, *Science*, **243**, 1043–1047 (1989). Copyright © 1989 American Association for the Advancement of Science.]

were identical in the CRM and VRM experiments. As Fig. 13.13 shows, CRM was 2–3 times larger than pure VRM in the daughter phase, but both CRM and VRM decayed in zero field in proportion to log t over 4 decades of time, and at

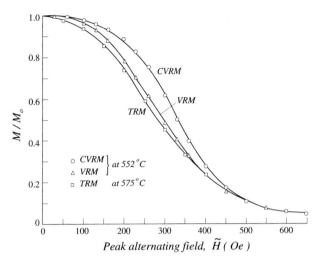

Figure 13.12 Comparison of AF demagnetization curves of VRM in unoxidized magnetite and CVRM in oxidized magnetite, both produced in 2.5-hr heatings at 552 °C and of total TRM in the unoxidized magnetite. A field of 50 μT was applied in all cases. High-temperature VRM and CVRM are just as resistant to AF demagnetization as total TRM. [Reprinted with permission from Ö. Özdemir and D. J. Dunlop, *Science*, **243**, 1043–1047 (1989). Copyright © 1989 American Association for the Advancement of Science.]

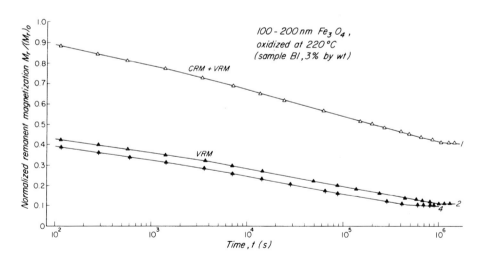

Figure 13.13   Zero-field decay of CVRM produced by oxidizing 0.1–0.2 μm magnetite at 220 °C in $H_0 = 100$ μT for 30 hr, compared with the decay of VRM produced under the same conditions of field, time and temperature in the oxidized daughter product (maghemite). Both remanences decay in proportion to log t over 4 decades of time, and at almost the same rate. CVRM seems to consist of CRM (time-independent) + VRM (time-dependent) fractions, each forming about one-half of the total remanence. [After Gapeev *et al.* (1991), with the permission of the authors and the publisher, Blackwell Science, Oxford, UK.]

almost the same rates. About half of what is nominally CRM in these samples seems to be indistinguishable from VRM.

The AF demagnetization curves of CRM and pure VRM, either before or after zero-field viscous decay, were very similar to each other (Fig. 13.14). They were also very hard and SD-like, comparable to the demagnetization curves of total TRM's in SD or nearly SD magnetites (cf. Fig. 11.13). Considering the low temperature of the CRM and VRM runs (220 °C), the high stability is astonishing.

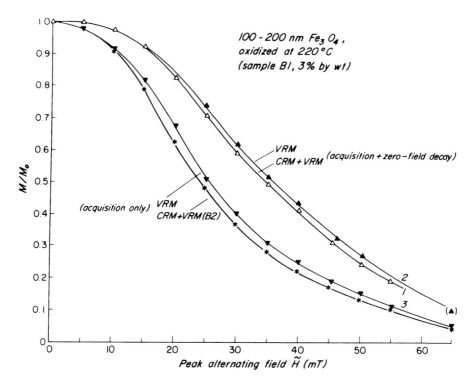

Figure 13.14 Comparison of AF demagnetization curves of CVRM and VRM, both immediately after acquisition (30-hr runs) and after a further 11.5 days of zero-field decay. In either case, CVRM and VRM have very similar AF demagnetization behaviour, which is remarkably SD-like. [After Gapeev *et al.* (1991), with the permission of the authors and the publisher, Blackwell Science, Oxford, UK.]

Chapter 14

Magnetism of igneous rocks and baked materials

In spite of the generality of the physical principles outlined in previous chapters, the natural remanent magnetization (NRM) of rocks is enormously variable in its intensity, its time stability and its resistance to thermal and AF demagnetization. This chapter and the two that follow will study magnetism in the context of the formation and subsequent alteration of igneous, sedimentary, and metamorphic rocks. In each case, the environment determines which magnetic minerals form, their composition and grain size, and microstructure such as exsolution intergrowths and internal stress.

14.1 The oceanic lithosphere and linear magnetic anomalies

Our treatment of igneous rocks begins with the oceanic lithosphere, the largest continuous igneous body available for paleomagnetic study. Linear anomalies in the geomagnetic field over the oceans ('magnetic stripes') are replicas of the regular pattern of NRM in the seafloor and provide tangible evidence of seafloor spreading (§1.2.1). In turn, it was the obvious fidelity of the seafloor paleomagnetic record that, in the minds of most earth scientists, proved the validity of the paleomagnetic method as applied to much older continental, and even extraterrestrial, rocks.

14.1.1 The Vine and Matthews' model

Figure 1.4 summarizes the Vine and Matthews' (1963) model. A similar interpretation of magnetic stripes was put forward about the same time by L.J. Morley

391

(see Cox, 1973, p. 224). At a mid-ocean ridge, basaltic magma rising from shallow depths in the upper mantle is extruded as pillow lavas (seismic layer 2A, 0.5–1 km thick), intruded at shallow depths as sheeted dikes and sills (layer 2B, 1–2 km) or cools at depth as massive gabbroic plutons (layer 3, 2–5 km). This crustal structure, illustrated in Fig. 14.1, was first inferred from ophiolite suites, slices of ancient seafloor obducted on to the continents, and only later confirmed by direct sampling in the Deep Sea Drilling Project (DSDP) and the Ocean Drilling Program (ODP). As the rocks cool, TRM is acquired parallel to the geomagnetic field (Johnson and Tivey, 1995). As we shall see later, the TRM is largely replaced within at most a million years of formation by CRM of reduced intensity, but the overall picture is unchanged.

The newly created seafloor, carrying its record of the geomagnetic field, spreads away from the ridge. Each reversal of the geomagnetic field gives rise to a reversal of the NRM acquired by strips of seafloor paralleling the ridge and symmetrical about it. For a uniform spreading rate, the strips and their magnetic images, the linear magnetic-field anomalies ('magnetic stripes') at the ocean surface, have widths proportional to the durations of geomagnetic polarity epochs. Profiled in the direction of seafloor spreading, the anomalies give a precise spatial replica of the geomagnetic polarity time scale.

Magnetic stripes endure until the seafloor is subducted at a trench. The oldest seafloor in the present oceans, and the oldest magnetic stripes, date back to ≈175 Ma. The record of the geomagnetic field during the earlier 90% of earth history is to be found only in the paleomagnetism of continental rocks. The sole

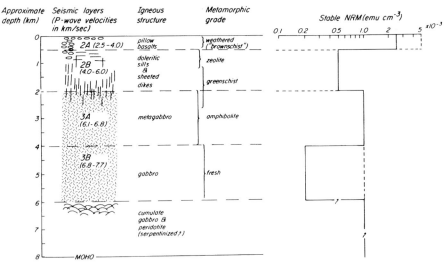

Figure 14.1   The petrologic, seismic and magnetic structure of typical oceanic crust. The solid and dashed lines indicate alternative possible NRM variations.

exception occurs when young oceanic crust is obducted on to the continental crust at a convergent plate margin, forming ophiolite complexes. Only young, relatively buoyant oceanic crust can be obducted, and so typically ophiolites are derived from marginal basins rather than from deep oceanic crust.

14.1.2 The nature of the magnetic anomaly source

Seafloor spreading is so far-reaching in its implications that a careful test of Vine and Matthews' model is in order. The principal unknown is the nature of the sub-oceanic magnetic source. (Sub-continental magnetic structure is just as uncertain: see Shive and Fountain, 1988.) How deep in the crust does it extend? What rock types constitute the source and what are their typical NRM intensities? Why is the crust beneath the ridge axis more strongly magnetized than the oceanic crust elsewhere? Are NRM inclinations consistent with the paleolatitude predicted by seafloor spreading?

Agreement between measured and predicted NRM inclinations is not particularly impressive (Fig. 14.2). The sampling is, of course, hardly representative of the ocean crust. DSDP and ODP cores for the most part have sampled only the top few metres of the basalt layer at a few widely separated locations. The magnetic stripes give a more representative picture of the NRM of the ocean crust at all depths. Unfortunately, inverting the anomalies to determine the magnetic moment of the source yields the product of NRM intensity and thickness of the magnetized layer, not either parameter separately (Harrison, 1976).

One constraint on the thickness of the source layer is the depth of the Curie point isotherm (Fig. 14.3). The geothermal gradient beneath mid-ocean rifts may be as high as $100\,°C/km$ and the primary titanomagnetites in submarine

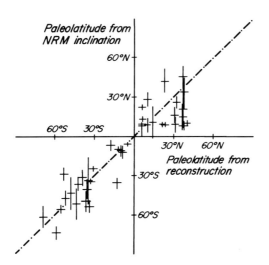

Figure 14.2 Paleolatitudes determined from the inclination of NRM at 40 DSDP sites compared with the paleolatitudes expected from plate reconstructions. [After Lowrie (1977), with permission of the author and the publisher, The Geological Society, London, UK.]

Paleolatitude from NRM inclination

Paleolatitude from reconstruction

basalts have Curie points of 150–200 °C. The Curie point isotherm for titano-magnetite is therefore only a few km deep. However, other magnetic minerals with higher Curie points occur at greater depths (§14.3).

Early dredged samples of layer 2A basalts (Irving *et al.*, 1970; Talwani *et al.*, 1971) from the North Atlantic were so strongly magnetized (10–100 A/m or 10^{-2}–10^{-1} emu/cm^3) that a thin veneer, only 200–500 m thick, seemed sufficient to produce the observed anomalies. However, the first deep (>0.5 km) drilling of layer 2A during DSDP Leg 37 revealed weak (10^{-1}–1 A/m or 10^{-4}–10^{-3} emu/cm^3) NRM's with highly non-uniform directions over the entire depth of the hole (Hall, 1976). A deeper source seemed inescapable. Subsequently, DSDP Hole 504B, which has been re-entered and drilled to greater depths in a number of ODP legs, has encountered higher NRM intensities in layer 2A and the upper part of layer 2B (Smith and Banerjee, 1985, 1986). Useful discussions of early DSDP and manned submersible results are to be found in Johnson and Atwater (1977) and Prévot *et al.* (1979).

The depth extent of the source of magnetic stripes is still uncertain. The sheeted dikes of layer 2B are rather weakly magnetized but mafic and ultramafic rocks of layer 3 and the uppermost mantle often carry strong and stable magneti-zations residing in magnetite (§14.3), sufficient to contribute significantly to mag-netic stripes. As the newly-formed lithosphere spreads away from the ridge, deeper rocks cool more slowly than near-surface rocks. Thus cooling fronts and magnetic polarity boundaries in the anomaly source are not vertical, as Vine and Matthews imagined them, but curve away from the ridge at depth (Fig. 14.3). Arkani-Hamed (1988) has shown that deeper sources with curved bound-aries are indeed required to explain the skewness of the magnetic anomalies.

14.2   Submarine basalts

Layer 2A basalts drilled or dredged from the seafloor are tholeiites of rather uni-form composition. They are very rich in iron and titanium and may contain as much as 5% titanomagnetite, although 1–2% is more typical. Their magnetic properties are dictated by the composition of the magma, particularly its Fe/Ti ratio and high sulphur content, the rapid chilling and consequent fine grain size of the extruded lava pillows, and the pervasive presence of circulating seawater at depth in the hot and highly fissured axial rift zone.

14.2.1   Primary homogeneous titanomagnetite

The most abundant primary magnetic mineral in submarine basalts is titano-magnetite, $Fe_{3-x}Ti_xO_4$, with $x \approx 0.6$ (TM60), often containing Al and Mg as impurities. TM60 is a metastable mineral, preserved by quenching from high

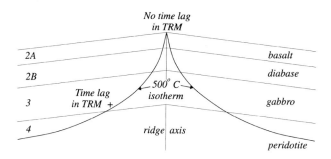

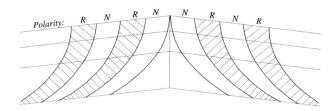

Figure 14.3 Sketch of vertical structure and the 500 °C isotherm in oceanic crust near a spreading centre. The normal (N) and reverse (R) bands of magnetized sea-floor, which are the sources of magnetic stripe anomalies, are sickle-shaped in cross-section because cooling fronts are curved in vertical cross-section.

temperatures where magnetite, Fe_3O_4, and ulvöspinel, Fe_2TiO_4, form a complete solid-solution series (Fig. 3.1). The grains may be skeletal or preserve other quench textures, but they are homogeneous. They do not unmix into two phases during the lifetime of an ocean basin nor during laboratory heatings in vacuum to temperatures well above the Curie point of 150–200 °C (Figs. 3.12a, 14.4a). For a detailed description of the formation, structure and properties of titanomagnetite and other iron–titanium oxides, the reader is referred to Haggerty (1976), Lindsley (1976, 1991) and O'Reilly (1984, Chapters 2 and 7).

Oxidation of titanomagnetite during initial cooling is prevented by rapid quenching and by the excess of sulphur over oxygen once the oxides and silicates have crystallized. Rapid cooling is also responsible for the fine grain size which makes oceanic basalts excellent recorders of the paleomagnetic field. In the sulphur-rich environment, iron sulphides are also common. Antiferromagnetic pyrite is abundant, but ferrimagnetic pyrrhotite is rare. Titanomagnetite therefore dominates the NRM.

14.2.2   Structure and cation distribution of titanomagnetite

Titanomagnetite is cubic with inverse spinel structure (Fig. 3.4). Cations on octahedral or B sites are surrounded by six O^{2-} ions at the corners of an octahedron. Cations on tetrahedral or A sites have four nearest neighbour O^{2-} ions in

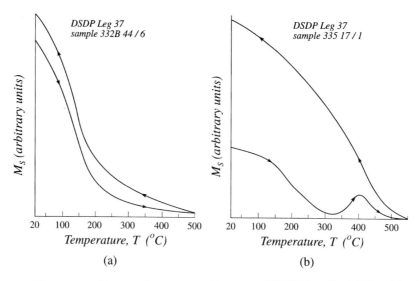

Figure 14.4 Vacuum thermomagnetic curves of DSDP basalts. (a) Unoxidized basalts have reversible $M_s(T)$ curves due to titanomagnetite. (b) Oxidized basalts have irreversible curves due to the inversion of titanomaghemite. [After Dunlop and Hale (1977), with the permission of the authors and the publisher, The National Research Council of Canada, Ottawa, Canada.]

tetrahedral coordination. Negative exchange coupling of the magnetic moments of A and B site cations, via intermediate O^{2-} ions, gives rise to A and B magnetic sublattices. Sixteen B sites and eight A sites comprise one unit cell.

Distribution of the cations between the sublattices determines the net ferrimagnetic moment. In magnetite, the net moment is that of the Fe^{2+} ion $(4\mu_B)$ because the Fe^{3+} ions are equally distributed between the sublattices. The cation distribution for intermediate titanomagnetites $(Fe^{2+}_{1+x}Fe^{3+}_{2-2x}Ti_xO_4)$ is less certain. The variation of magnetic moment with composition for different models and the experimental observations were compared in Fig. 3.11. There is a steady decrease in moment from $4\mu_B$ for Fe_3O_4 to 0 for Fe_2TiO_4. TM60 has about $\frac{1}{3}$ the spontaneous moment of magnetite. Its value of M_{s0} is ≈ 125 kA/m.

14.2.3   Identification and magnetic properties of titanomagnetite

Composition is usually determined by measuring Curie temperature or unit cell edge, which vary in a regular way with x (Fig. 3.11), or by direct measurement of the Fe/Ti ratio by electron microprobe. A Curie point of $\approx 150\,°C$ is characteristic of pure $Fe_{2.4}Ti_{0.6}O_4$. Al and Mg impurities lower T_C, while oxidation (or maghemitization; §14.2.4) raises T_C.

Synthetic SD titanomagnetites with large x may have coercive forces of 100 mT (1 kOe) or more due to crystalline or magnetoelastic anisotropies (Fig.

3.16), which increase sharply as x increases. Such high coercivities are not observed in submarine basalts, however. One reason is that, except in recently erupted pillows, TM60 grains small enough to be SD ($<0.6\,\mu m$; Table 5.1) tend to be oxidized (§14.2.4). Coarse MD grains resist oxidation but typically have coercivities $<20\,mT$ (200 Oe) (Figs. 3.16, 14.5).

14.2.4 Low-temperature oxidation of titanomagnetite

The thermomagnetic signature of an oxidized titanomagnetite with $x \approx 0.6$ is a Curie point of 200–400 °C during in-vacuum heating and irreversible increases in M_s and T_C upon cooling (Figs. 3.12b, 14.4b). The (pre-heating) mineral is *titanomaghemite*, a titanium-bearing analog of maghemite, γFe_2O_3, and the low-temperature oxidation of titanomagnetite is often called maghemitization. Titanomaghemite is metastable at ordinary temperatures but inverts to other minerals when it is heated in vacuum or in air. Titanomaghemite with $x \approx 0.6$ inverts to a multiphase mixture including ilmenite ($FeTiO_3$) and magnetite. It is the creation of strongly magnetic magnetite that causes the 'hump' in the heating curves in Figs. 3.12(b) and 14.4(b) and the irreversible increase in M_s upon cooling.

The oxidation of magnetite to maghemite has been thoroughly studied. The process is limited to temperatures below 250 °C and is accelerated by fine grain size (Nishitani and Kono, 1982). The oxidation mechanism is diffusion of Fe^{2+}

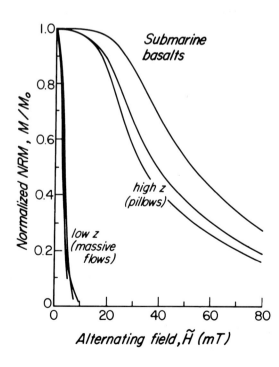

Figure 14.5 Typical AF decay curves of NRM in DSDP basalts, showing that oxidized pillows have the high coercivities expected of SD titanomaghemite, whereas unoxidized massive flows contain mainly low-coercivity coarse MD titanomagnetites. [After Dunlop and Hale (1976) © American Geophysical Union, with the permission of the authors and the publisher.]

ions to the surface of the crystal, leaving the inverse spinel structure unchanged but $\frac{1}{9}$ of the octahedral sites vacant. For this reason, titanomaghemites are called cation-deficient spinels. The remaining Fe^{2+} ions are converted to Fe^{3+} by electron hopping. Maghemite thus has the formula $Fe_{8/3}\square_{1/3}O_4$, where $\square$ indicates a lattice vacancy. The net moment per formula weight Fe_2O_3 is $\approx 2.4 \, \mu_B$ (Table 3.2) and $M_{s0} \approx 380 \, kA/m$, compared to 480 kA/m for magnetite.

Because it requires diffusion of cations, maghemitization is above all a surface phenomenon (Fig. 3.10). The interiors of large grains may remain unoxidized indefinitely. The fine grains in quenched pillows, in contact with seawater and warmed through burial beneath later eruptions, oxidize rapidly, certainly within 0.5 Ma (Irving, 1970; Ozima, 1971; Johnson and Atwater, 1977; Johnson and Pariso, 1993) and perhaps within 20 ka (Gee and Kent, 1994). Because of the range of grain sizes in any real sample, hysteresis properties like M_s and M_{rs} are often controlled by the less oxidized larger grains with low Curie points, while the NRM is carried by the more oxidized fine grains with Curie points as high as 550 °C (Kent and Gee, 1994).

Because the lattice parameters of magnetite and maghemite are slightly different (Fig. 3.9, Table 3.1), maghemitization creates strain. Accumulated lattice strain is eventually accommodated by microfracturing (Johnson and Hall, 1978; Petersen and Vali, 1987; Fig. 3.10), thereby reducing the effective grain size and promoting further oxidation.

14.2.5   Properties of titanomaghemite

On the ternary diagram of Fig. 3.1, the degree of maghemitization is measured by the oxidation parameter z. Unoxidized titanomagnetites on the magnetite–ulvöspinel join have $z = 0$. Fully oxidized titanomaghemites with $z = 1$ lie on the γFe_2O_3–Fe_2TiO_5 join. Curie temperature T_C (where this lies below the inversion temperature) increases with oxidation (Figs 3.1b, 3.17), whereas spontaneous magnetization decreases with increasing z (Fig. 3.18). M_{s0} is in the range 75–105 kA/m for $x \approx 0.6$ and $0.3 < z < 0.7$.

Natural titanomaghemites from oceanic pillows have SD-like AF demagnetization curves (cf. Fig. 11.13), with initial plateaus and relatively high coercivities (Fig. 14.5). Unoxidized coarse grains from massive flows have much lower coercivities. Their curves are quasi-exponential in shape, as expected for MD grains.

14.2.6   Inversion of titanomaghemite

Maghemite and Ti-poor titanomaghemites invert to weakly magnetic hematite, αFe_2O_3, usually above 250 °C (Table 3.3). Titanomaghemites with $x \approx 0.6$ invert to a multiphase intergrowth of magnetite, ilmenite and other minerals (Readman and O'Reilly, 1972; Özdemir, 1987), at temperatures as low as

200 °C if the heating is prolonged. The endproduct may be up to ten times as magnetic as the original titanomaghemite (Fig. 14.4b).

Over geological times, a similar inversion may take place at the temperatures of 100–200 °C prevailing in layer 2B and the upper part of layer 3 (Banerjee, 1980). Magnetite is the principal magnetic mineral in the sheeted dikes of DSDP Hole 504B (Smith and Banerjee, 1985, 1986). It originates both as a product of oxyexsolution or inversion and as a result of secondary recrystallization by convecting fluids (Shau *et al.*, 1993).

14.2.7 CRM of titanomaghemite

Oxidation of titanomagnetite and inversion of titanomaghemite do not merely change the magnetic mineral that carries NRM. They cause the primary TRM to evolve into one or more generations of CRM. The crucial question is whether the CRM is randomly oriented, acquired parallel to the ambient geomagnetic field during oxidation or inversion (which may have reversed), or is phase-coupled parallel to the primary TRM. The first two possibilities would invalidate the Vine and Matthews' model. Magnetic stripes would become weak and blurred or would lag changes in the field by a variable amount.

It is fortunate that CRM due to maghemitization of SD-size titanomagnetites appears to parallel the original TRM (Johnson and Merrill, 1974; Özdemir and Dunlop, 1985; see §13.3.1). Figure 14.6 is a schematic model of the CRM process occurring during the oxidation of magnetite to maghemite. First an Fe^{2+} ion is freed from a near-surface B site and diffuses to the surface. Exchange coupling is broken between the vacant site and its O^{2-} neighbours, but all other Fe^{2+} and Fe^{3+} ions remain exchange coupled. Two B-site Fe^{2+} ions are converted to Fe^{3+} by electron hopping. Their moments increase from 4 to 5 μ_B but exchange coupling is unbroken. The surface layer of Fe_3O_4 has now been converted to

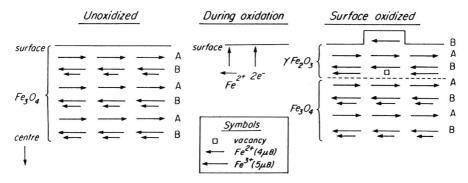

Figure 14.6 A schematic representation of phase-coupled CRM accompanying the oxidation of magnetite to maghemite. A and B lattice sites are depicted as alternate layers (for the true structure, see Fig. 3.4).

γFe_2O_3, whose net moment remains parallel to the B-site ions. Across the γFe_2O_3–Fe_3O_4 interface, lattice mismatch is accommodated and exchange coupling is little, if at all, affected. Finally, the surface Fe^{2+} ion, now converted to Fe^{3+}, is exchange coupled antiparallel to a surface A-site Fe^{3+} ion. It can form the nucleus for a new surface layer of γFe_2O_3, whose eventual moment will again parallel the B-site ions.

This picture of the development of phase-coupled CRM during the $Fe_3O_4 \rightarrow \gamma Fe_2O_3$ transformation is very different from the grain growth CRM model of §13.2.2, in which a nucleating phase grows through its critical blocking volume V_B and passes from SP to SD. During maghemitization, because of the nearly compatible lattices, old and new phases are exchange coupled. The 'exchange field' outweighs the geomagnetic field as the daughter phase grows through its critical SP size, d_s.

The model will fail if the grain, or crack-bounded regions at its surface (Fig. 3.10), are so near SP size that one of the coupled phases becomes thermally unpinned at some stage. The model will also fail for a large MD grain, whose NRM is due to relatively small displacements of walls, some of them loosely pinned. Internal stress may be relieved, or at any rate pinning centres relocated, as a result of cation diffusion. Momentarily unpinned wall segments can respond to an applied field even though the bulk of each domain is firmly exchange coupled across the phase boundary. The net result is a CRM in the direction of the geomagnetic field (Johnson and Merrill, 1973; Kelso et al., 1991).

14.2.8   Intensity and stability of CRM due to maghemitization

The titanomagnetite→titanomaghemite transformation in submarine basalts in the main affects SD and small PSD grains whose CRM, according to the model of Fig. 14.6, should pseudomorph the parent TRM, except for a decrease in intensity due to the change in M_s. In reality, because some domain walls unpin or lattice mismatch leaves an occasional cation without a neighbour, CRM is much less than 100% efficient. Decreases in anomaly amplitude and NRM intensity away from spreading ridges probably result from maghemitization (Figs. 14.7, 14.8) (Irving, 1970; Johnson and Atwater, 1977). On a smaller scale, Marshall and Cox (1972), Ryall and Ade-Hall (1975) and Beske-Diehl and Soroka (1984) have traced a similar decrease in intensity (but an increase in coercivity) of NRM from the fresher core to the altered rim of large pillows.

The increase in coercivity of $x = 0.6$ titanomagnetites with maghemitization observed in individual pillows seems to be true generally of submarine basalts (Fig. 14.5). The magnetocrystalline anisotropy of TM60 is due to Fe^{2+} single-ion anisotropy (§2.8.2) and should decrease with the decrease in Fe^{2+} ions with oxidation. An *increase* in H_c measured at T_0 is actually observed in moderately oxidized ($z = 0.1$–0.4) SD-size TM60 (Özdemir and O'Reilly, 1982a), but this

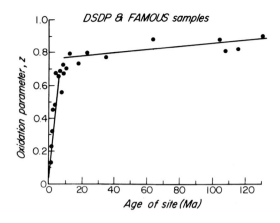

Figure 14.7 Mean oxidation parameter, z, at various DSDP and manned-submersible sites as a function of the age of the site. Maghemitization of primary titanomagnetites occurs rapidly, within at most 1 Ma after extrusion. [After Bleil and Petersen (1983), with the permission of the authors. Reprinted with permission from *Nature*. Copyright © 1983 Macmillan Magazines Ltd.]

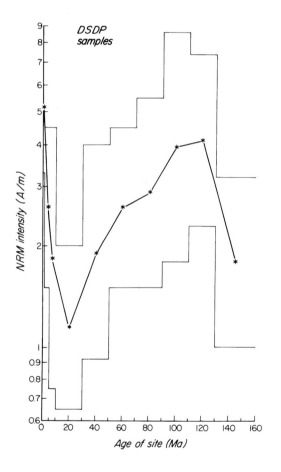

Figure 14.8 Average NRM intensity (stars and solid line) with upper and lower bounds at various DSDP and manned-submersible sites as a function of the age of the site. The decrease in NRM intensity in the first 5–10 Ma is a result of both the reduction in M_s during oxidation and the replacement of TRM by CRM. Subsequent oxidation is minor; the recovery of NRM intensity with increasing age must have an independent cause. [After Bleil and Petersen (1983), with the permission of the authors. Reprinted with permission from *Nature*. Copyright © 1983 Macmillan Magazines Ltd.]

does not indicate that K_1 has increased. The Curie temperature rises rapidly with oxidation (Fig. 3.17), with the result that T_0 is no longer close to T_C, and the fluctuation field H_q at T_0 drops sharply. H_K also decreases, as confirmed by measurements at 77 K (Özdemir and Banerjee, 1981), where H_q is negligible. At room temperature, however, the percentage decrease in H_q with oxidation outweighs the decrease in H_K, producing an *increase* in H_c (cf. eqn. (8.24)).

In nature, the coercivity of titanomaghemites is higher than that of unoxidized titanomagnetites even for high z values. This has a simple explanation. It is fine SD or nearly SD grains with high coercivities that tend to oxidize, and fracturing makes them finer and harder still. The coarser MD titanomagnetite grains which resist oxidation are magnetically soft (Fig. 3.16). The contrast in coercivity between oxidized fine-grained pillows and coarse-grained massive flows is striking (Fig. 14.5).

14.2.9   Viscous remagnetization of submarine basalts

Partial viscous remagnetization by the present earth's field of basalts originally magnetized in an earlier, reversed field will also decrease NRM intensity away from a ridge. Zones of normal magnetization are unaffected. Figure 10.3 illustrates the strong viscous magnetization changes that characterize some basalts containing largely unoxidized titanomagnetite grains (Prévot and Bina, 1993; Bina and Prévot, 1994). The changes will be larger at depth where the temperature is higher (Fig. 10.3). Unoxidized basalts sampled at the ridge axis are relatively non-viscous (Plessard and Prévot, 1977), as are some moderately maghemitized basalts. SD-size titanomaghemites become much more viscous at high values of z, however (Fig. 10.5).

Whatever the mechanism, the viscous magnetization of some submarine rocks is exceptionally large, even over periods of a few hours. Occasionally (e.g. Lowrie and Kent, 1978), the projected VRM that would be acquired during a geomagnetic polarity epoch would completely replace primary TRM or any subsequent CRM. However, Hall *et al.* (1995), in a systematic study of viscous magnetization in basalts, metabasalts, dikes and mafic and ultramafic intrusives from the Troodos (Cyprus) ophiolite (representing a complete cross-section of oceanic crust), estimate that long-term VRM forms at most one-third of the NRM and typically much less than this.

14.2.10   Summary

Submarine basalts are not the simple TRM carriers envisaged in Vine and Matthews' original model. The primary titanomagnetites carry a soft TRM that is sometimes readily remagnetized viscously in later reversed fields. Moreover all but the coarser grains have been oxidized to titanomaghemite and the TRM has been replaced by relatively hard CRM of reduced intensity.

Because the NRM is no longer a TRM, most submarine basalts are unsuitable for paleointensity determination (see, however, Grommé *et al.*, 1979). However, for moderate oxidation of SD-size grains, the CRM *direction* is the same as that of the TRM (Özdemir and Dunlop, 1985; Fig. 13.4), so that magnetic stripe anomalies should be unchanged, except in amplitude.

Deeper in the crust, titanomaghemite may invert to magnetite and other oxides. Oxyexsolution of titanomagnetite and serpentinization of olivine (§14.3) are other possible sources of magnetite (Wooldridge et al., 1990; Shau et al., 1993). In each case, the resulting CRM is a potent source of magnetic anomalies.

14.3 Oceanic intrusive and plutonic rocks

We have limited knowledge of the magnetic properties of rocks from the deeper oceanic crust – sheeted diabase dikes (seismic layer 2B), gabbro (layer 3), and peridotite (layer 4). The sampling bias is enormous. Measurements are largely confined to tectonically emplaced ophiolite suites (e.g., Banerjee, 1980; Swift and Johnson, 1984; Hall et al., 1991; Horen and Dubuisson, 1996), altered dredgehaul or manned submersible samples from mid-ocean rift valleys and fracture zones (Wooldridge et al., 1990), and limited sections of DSDP and ODP drill holes in which the layer 2B, 3 and 4 rocks have been brought near the surface by block faulting. We have very few samples of fresh, young, tectonically unaffected oceanic intrusives (Smith and Banerjee, 1985; Pariso and Johnson, 1989, 1991, 1993a,b; Kikawa and Ozawa, 1992).

Figure 14.9 shows that in the Macquarie Island ophiolite, the sheeted dikes have NRM intensities at least an order of magnitude below those of overlying pillow lavas. The same is true of the Troodos ophiolite, but fresh sheeted dikes from DSDP Hole 504B have NRM's almost as strong as those of the overlying basalts (Smith and Banerjee, 1985). The sheeted-dike NRM is not necessarily soft to AF cleaning, nor is it significantly more viscous than the NRM of the volcanics (Hall et al., 1995). A reversible in-vacuum thermomagnetic (M_s–T) curve with a single 580 °C magnetite Curie point characterizes all the Macquarie and Hole 504B dikes. In Hole 504B, both primary magnetite preserving deuteric oxyexsolution textures and secondary magnetite resulting from hydrothermal recrystallization are observed (Shau et al., 1993).

The Macquarie dikes are metamorphosed to at least greenschist facies (see §16.2.2). Some degree of metamorphism is inevitable at depth in the oceanic crust because of the high geothermal gradient beneath mid-ocean rifts (Cann, 1979; Wooldridge et al., 1990). The freshness of the 504B dikes suggests, however, that the grade of metamorphism in ophiolites is not typical of oceanic crust in situ. For example, although gabbros from ODP Hole 735B are metamorphosed to amphibolite facies in the upper part of the section, gabbros in the lower section are fresh and unaltered (Kikawa and Ozawa, 1992; Fig. 14.1).

Oceanic gabbros and peridotites in ophiolite suites or from dredgehauls and deep drill holes have reversible thermomagnetic curves, with Curie points of

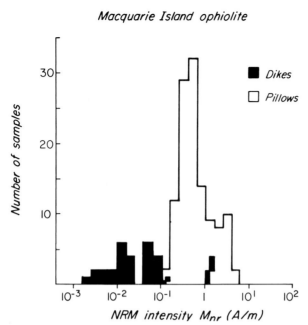

Figure 14.9 NRM intensities of pillow basalts and sheeted dikes in the Macquarie Island ophiolite complex. [After Levi and Banerjee (1977), with the permission of the authors and the publisher, Journal of Geomagnetism and Geoelectricity, Tokyo, Japan.]

520–580 °C (Kent *et al.*, 1978; Fig. 14.10). The intensity of NRM, however, varies enormously depending on whether or not the rocks are serpentinized (Bina and Henry, 1990). Serpentinization of olivine occurs in the presence of seawater at low temperatures and shallow depths very soon after initial cooling of the rock. Abundant magnetite is a by-product of serpentinization: serpentinized gabbros and peridotites possess NRM intensities approaching those of fresh layer-2A basalts (Fig. 14.11). The magnetite inclusions may remain of SD or pseudo-SD size (Horen and Dubuisson, 1996), in which case the NRM they carry is hard and stable, or the inclusions may grow to MD size, imparting soft magnetic properties (Fig. 14.12).

The extent to which serpentinization takes place in deep oceanic crust is unknown. If layer-3 or layer-4 rocks, in situ, are serpentinized to a significant extent and are coherently magnetized (magnetite inclusions of SD or pseudo-SD size), the deep crust may be an important source of magnetic stripe anomalies (Harrison, 1976; Banerjee, 1984).

14.4 Subaerial basalts and andesites

Oceanic magnetic anomalies are an integrated signal, observed several km from the source. Geomagnetic field reversals are detected, but the details of the NRM of submarine rocks and the information they record about short-period variations of the field are lost. Subaerial lavas, on the other hand, can be sampled

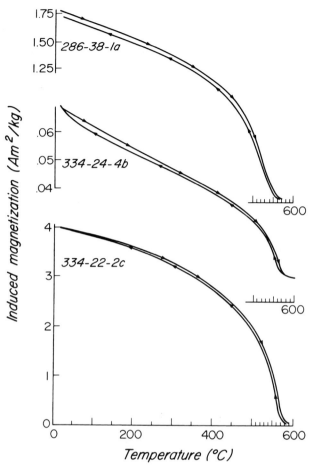

Figure 14.10 Typical vacuum thermomagnetic curves of DSDP gabbro, serpentinized gabbro, and serpentinized peridotite. [After Dunlop and Prévot (1982), with the permission of the authors and the publisher, Blackwell Science, Oxford, UK.]

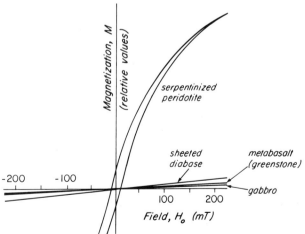

Figure 14.11 Hysteresis of serpentinized and unserpentinized units of the Ordovician Betts Cove ophiolite, Newfoundland. [Unpublished data of D. J. Dunlop; samples courtesy of K. V. Rao and E. R. Deutsch.]

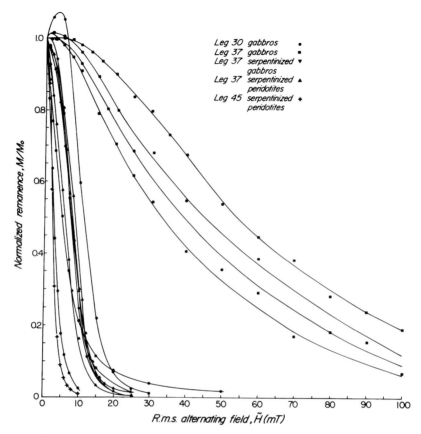

Figure 14.12 Typical AF demagnetization curves of NRM in DSDP intrusive and plutonic rocks. [After Dunlop and Prévot (1982), with the permission of the authors and the publisher, Blackwell Science, Oxford, UK.]

flow by flow. Since each flow cools in a few years at most, the field is fossilized at, geologically speaking, an instant in time. By studying a sequence of flows, a detailed picture of the geomagnetic field can sometimes be obtained (e.g., Prévot *et al.*, 1985), limited mainly by the uncertainty in ages of the flows and the variable frequency of eruption.

The conditions under which subaerial lavas form and subsequently alter differ in several significant ways from the submarine environment:

1. Instead of quenched pillows, massive flows with slowly cooling interiors are common.

2. Oxygen fugacity is frequently high, leading to deuteric oxidation of the lava at high temperature during initial cooling.

3. There is less external water; consequently maghemitization of those flows that escaped deuteric oxidation is slow.

Magnetic properties, as a result, are distinctively different from those of submarine basalts. We shall consider first the magnetic properties of subaerial basalts and andesites. Basalts are the typical products of volcanic activity at rifted continental margins, of rift valleys and grabens in continental interiors, and of oceanic islands like Iceland and Hawaii. Andesites and associated volcanoclastic rocks are typical of volcanoes at convergent plate boundaries above subducting oceanic crust. The magnetization of dacites and other felsic volcanics will be described in §14.5.

14.4.1   High-temperature oxidation of titanomagnetite

Fresh subaerial basalts, heated in vacuum, may exhibit a low Curie point (100–200 °C) due to homogeneous titanomagnetite with $x \approx 0.6$–0.7, a high Curie point (500–580 °C) due to oxyexsolved low-Ti titanomagnetite near magnetite in composition, or, rarely, both Curie points (Fig. 14.13). The thermomagnetic curves are practically reversible. **High-temperature oxidation** occurring deuterically as the lava cools from about 900 to 500 °C (Grommé *et al.*, 1969; Tucker and O'Reilly, 1980b) is responsible for the appearance of the high-Curie-point phase. It is this process, especially when carried to completion with the production of fine-grained hematite, that causes subaerial basalts and andesites to be such good recorders of the field.

Titanomagnetite heated in air above ≈ 300 °C does not oxidize to titanomaghemite, but instead forms an intergrowth of cubic magnetite with rhombohedral ilmenite exsolved in {111} planes of its host (Fig. 14.14). The reaction products are similar to the inversion products of titanomaghemite of the same x

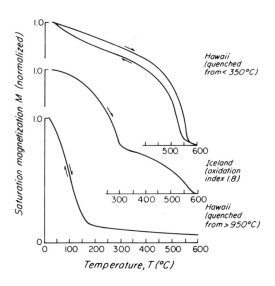

Figure 14.13 Characteristic vacuum thermomagnetic curves of subaerial basalts with high, moderate but inhomogeneous, and low oxidation states. [After Grommé *et al.* (1969) © American Geophysical Union and Lawley and Ade-Hall (1971), reprinted with kind permission of the authors and the publishers, The American Geophysical Union and Elsevier Science – NL, Sara Burgerhartstraat 25, 1055 KV Amsterdam, The Netherlands.]

value (Özdemir, 1987). As oxidation proceeds, the bulk composition moves along the oxidation lines in Fig. 3.1(a).

With continued oxidation, the magnetite-ilmenite intergrowth evolves to mixtures of hematite (αFe_2O_3), pseudobrookite (Fe_2TiO_5) and rutile (TiO_2) (Buddington and Lindsley, 1964), and certain silicates, notably olivine, generate magnetite and hematite (Haggerty and Baker, 1967). Highly oxidized basalts have thermomagnetic curves with both magnetite and hematite (675 °C) Curie points.

The oxide mineral assemblage varies from point to point within a single lava because of differences in oxygen fugacity, temperature and cooling rate. The *interiors* of flows, as well as their margins, may become oxidized by their contained fluids. In fact, Watkins and Haggerty (1967) found that the highest oxidation states are achieved only in very thick flows. Here the cooling rate is sufficiently slow that the various subsolidus reactions go to completion. On the other hand, among samples taken from cooling Hawaiian lava lakes (Grommé *et al.*, 1969), only those quenched from very high temperatures escaped oxidation entirely and had single low Curie points (Fig. 14.13). Intermediate deuteric oxidation states are the norm.

Wilson and Watkins (1967) and Watkins and Haggerty (1967) proposed the convenient classification of oxidation states reproduced in Table 14.1. Colour photomicrographs of grains in these classes are given by Wilson and Haggerty (1966). The **oxidation index** of a sample is the weighted average of the observed oxidation states of several hundred opaque mineral grains.

Not considered in the Wilson–Watkins–Haggerty classification but of importance paleomagnetically are iron oxides produced by deuteric oxidation of sili-

Table 14.1   Oxidation classes of deuterically oxidized titanomagnetites

Class	Characteristics
I	Homogeneous (single-phase) titanomagnetites
II	Titanomagnetites contain a few 'exsolved' ilmenite lamellae in {111} planes
III	Abundant ilmenite lamellae: equilibrium two-phase intergrowths
IV	Ilmenite lamellae oxidized to rutile+hematite
V	Residual titanomagnetite, as well as ilmenite, oxidized to rutile+hematite
VI	Total oxidation, to pseudobrookite as well as hematite and/or rutile

cate minerals such as olivine (Hoye and Evans, 1975) and micas (Borradaile, 1994a). Because of the fine particle size of the oxides produced, the magnetization may be very hard and stable. Also of significance in porphyritic lavas of basic and intermediate composition are fine exsolution blades or needles of magnetite (and less frequently hematite) in porphyroblasts of calcic plagioclase and pyroxene. Oxide inclusions in silicates are of paramount importance in plutonic rocks and will be discussed in detail in §14.6.

14.4.2 Thermochemical remanent magnetization (TCRM)

The NRM of subaerial basalts has long been taken to be TRM par excellence. But is this really true? Carmichael and Nicholls (1967) pointed out that, given the oxygen fugacities characteristic of basaltic lavas, many oxidized mineral assemblages could form at equilibrium temperatures as low as 600 °C. Precipitation of magnetite in olivine could occur at still lower temperatures. Only the presence of pseudobrookite, which cannot form below 585 °C, guarantees that any magnetite present carries a true TRM.

In practice, equilibrium conditions are seldom achieved. The subsolidus oxidation reactions are slow and lag behind falling temperatures in the lava. In cooling lava lakes, oxyexsolution of titanomagnetite to magnetite and ilmenite was observed to continue 100–400 °C below equilibrium temperatures for the coexisting compositions observed (Grommé et al., 1969). In one case, magnetite was still forming below 500 °C.

Under these circumstances of simultaneous slow cooling and volume growth of magnetite (or another magnetic phase) below its Curie temperature, the NRM produced is termed a **thermochemical remanent magnetization** (TCRM). Upon reheating during thermal demagnetization, unblocking temperatures of TCRM will be higher than the temperatures at which magnetization was blocked during initial cooling because the magnetic phases continued to grow after the TCRM was frozen in. In rapidly chilled lavas, on the other hand, reactions tend to be quenched above the Curie points of magnetite or hematite and true TRM is acquired.

For successful paleointensity determination, magnetic properties, including blocking temperatures, must not change irreversibly during laboratory heating. An equilibrium oxidized mineral assemblage carrying TCRM fulfills this requirement very well. The laboratory blocking and unblocking temperatures will be identical, although both will be somewhat higher than the blocking temperatures in initial cooling. Since the intensity of TRM or TCRM is not acutely sensitive to small variations in blocking temperature (eqns. (8.2), (8.21)), equilibrium TCRM may serve as a reasonable analog of TRM in paleointensity determination (Kellogg et al., 1970). The TRM of a quenched, non-equilibrium assemblage may actually be less suitable, because the initially incomplete oxidation reactions may be driven toward completion by laboratory heatings (Kono, 1985), even in the relatively low oxygen fugacity of a vacuum furnace.

Since TCRM is acquired during initial cooling, there is no doubt about directional fidelity. Even when TCRM is produced long after initial cooling, for example in the course of laboratory thermal demagnetization of unoxidized titanomagnetite, the newly formed magnetite seems to 'remember' the TRM direction (Wilson and Smith, 1968; see §13.4.5). However, the same cannot be said for the inversion of titanomaghemite (Bailey and Hale, 1981; Özdemir and Dunlop, 1985; Fig. 13.4, $z > 0.65$), although the reaction products are similar.

14.4.3   Effective grain size of oxidized titanomagnetite grains

We saw earlier that fracturing reduces the effective grain size and hence increases the coercivity of titanomaghemite grains. The exsolution texture of deuterically oxidized titanomagnetites has a similar effect (Larson *et al.*, 1969; Evans and Wayman, 1974; Manson and O'Reilly, 1976; O'Donovan and O'Reilly, 1977). During oxidation, the growth of ilmenite lamellae on two or more sets of spinel {111} planes subdivides the crystal into sheet-like or rod-like subgrains of magnetite (Strangway *et al.*, 1968). The elongated subgrains, if small enough, will be interacting SD particles with high shape anisotropy and hence high coercivity. Electron micrographs of magnetite–ilmenite intergrowths of this sort appear in Figs. 3.2 and 14.14.

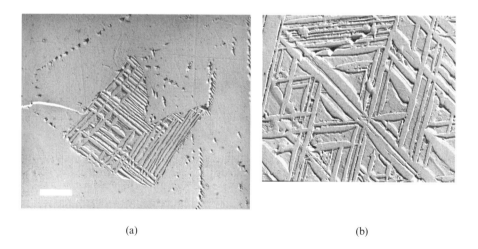

(a)   (b)

Figure 14.14   Electron micrographs of etched polished surfaces, showing magnetite–ilmenite intergrowths produced by high-temperature (600 °C) oxidation of homogeneous titanomagnetite. Lamellae of ilmenite form on {111} planes of magnetite. The magnification is ×4200 (scale bar: 2 μm) in (a) and ×7000 in (b). [After Davis and Evans (1976) © American Geophysical Union, with the permission of the authors and the publisher.]

Stacey and Banerjee (1974, pp. 63–65) have questioned the SD-like behaviour of such strongly interacting arrays. Davis and Evans (1976) confirmed experimentally that the magnetostatic interactions are strong: measured hysteresis loops of oxide grains like those in Fig. 14.14 were 'sheared' in the same way the hysteresis curve of an MD ferromagnet is sheared by the internal demagnetizing field (Fig. 11.21a). However, after 'unshearing' the loops, Davis and Evans found $M_{rs}/M_s = 0.5$, which is an SD value (eqns. (5.41), (11.16)). The coercivities were also SD-like, predominantly 10–60 mT (100–600 Oe). Thus oxyexsolved grains of TM60 have PSD behaviour, including the high stability necessary to preserve a primary NRM and with it a faithful record of the paleomagnetic field.

Figure 14.15 illustrates how AF coercivity increases with high-temperature oxidation in the laboratory or in nature. Index I basalts (similar to Davis and Evans' laboratory starting material) contain MD grains of unoxidized

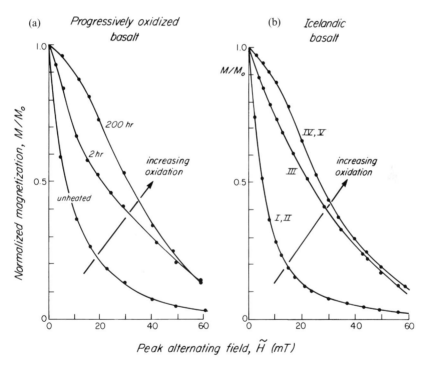

Figure 14.15 Increasing AF coercivity of remanence (here weak-field ARM) with increasing degree of high-temperature oxidation of titanomagnetite. (a) Basalt containing initially homogeneous titanomagnetite after oxidation at 600 °C for times up to 200 hr. (b) Icelandic basalts with oxidation indices from I to V. [After Davis and Evans (1976) © American Geophysical Union and Dunlop (1983c), reprinted with kind permission of the authors and the publishers, The American Geophysical Union and Elsevier Science – NL, Sara Burgerhartstraat 25, 1055 KV Amsterdam, The Netherlands.]

titanomagnetite and are soft to AF cleaning. Index III basalts in which lamella growth is complete (Davis and Evans' final state) are relatively hard. Further oxidation (indices IV through VI) is accompanied by further hardening, probably due to hematite (Beske-Diehl, 1988; Li and Beske-Diehl, 1991).

In field sampling, unaltered chilled margins with their fine grain size are at first sight appealing candidates for high-stability NRM. However, they commonly contain unoxidized titanomagnetites with low unblocking temperatures and low coercivities. It is the slowly cooled centres of lava flows or sills, containing exsolved titanomagnetites, and fully oxidized flow tops, containing hematite, which have the smallest effective grain sizes and the highest NRM unblocking temperatures and coercivities.

14.5   Subaerial felsic volcanics and pyroclastics

14.5.1   Titanomagnetite in felsic volcanic rocks

The magnetic mineralogy and properties of felsic volcanic and pyroclastic rocks have not been studied in as great detail as those of basaltic lavas. Like basalts and andesites, felsic lavas contain primary titanomagnetites. The initial composition of the titanomagnetite before deuteric oxidation tends to be less titanium-rich than in basalts: $0.3 < x < 0.6$. Consequently the unoxidized primary titanomagnetites tend to have Curie temperatures of 200–$400\,^\circ$C, higher than in basic volcanics. However, iron–titanium oxides are less abundant, generally constituting less than 1% of the rock. Pale-coloured dacites and rhyolites often have low NRM intensities (10^{-3}–10^{-2} A/m) and relatively low coercivities, making them unpromising candidates for paleomagnetic work. However, felsic volcanics that contain deuterically oxidized titanomagnetites may be excellent field recorders. These rocks are often visibly reddened because of the presence of hematite. Felsic lavas also commonly contain phenocrysts of pyroxene, micas and feldspar which can be partially or entirely altered under hydrothermal conditions following eruption. Magnetite and hematite are common secondary alteration products, imparting high unblocking temperatures and coercivities.

14.5.2   Titanohematite

Although felsic volcanics are usually dominated, magnetically speaking, by titanomagnetites and their oxidation products, they differ from rocks of more basic composition in having ferrimagnetic titanohematite (or hemoilmenite), $Fe_{2-y}Ti_yO_3$ with $0.5 \le y \le 0.7$ (Fig. 3.23), as a common primary mineral. Ferrimagnetic titanohematites have the remarkable property of **self-reversal**. When exchange coupled to an intergrown titanohematite with $y < 0.5$, they can acquire a TRM that is reversed in direction to the field in which they cooled. Because of its great physical interest, we will describe the self-reversing property of titanohematites and the phenomenon of self-reversal generally in some detail, although documented self-reversals of TRM are relatively rare.

Titanohematite, $Fe^{3+}_{2-2y}Fe^{2+}_yTi^{4+}_yO_3$, is a rhombohedral mineral with corundum structure (Fig. 3.22). Cations are arranged very nearly in basal planes perpendicular to the ternary or c-axis, with O^{2-} ions in intervening layers. Neighbouring Fe^{2+} or Fe^{3+} ions in adjacent cation planes are negatively exchange coupled, but exchange coupling within each plane may be positive or negative.

For $0 \le y \le 0.5$, the cation distribution is 'disordered': Ti^{4+} and Fe^{2+} ions are equally distributed among all c-planes. Titanohematite in this compositional range is antiferromagnetic with a superimposed weak 'parasitic' ferromagnetism. In the range $0.5 \le y \le 1$, the structure, in principle, is 'ordered' and ferrimagnetic: Ti^{4+} is confined to alternate cation planes and Fe^{2+} to the intervening planes. This ordered structure is approximately achieved around $y = 0.7$, giving a spontaneous moment close to the theoretical $2.8\,\mu_B$ per formula weight (Fig. 3.23). However, the Curie temperature is barely above room temperature T_0 for $y = 0.7$ and rises only to $\approx 200\,°C$ when $y = 0.5$ (Fig. 3.23). For this latter composition, $M_{s0} \approx 100\,kA/m$.

As with titanomagnetite, homogeneous titanohematite of intermediate composition is not an equilibrium phase at ordinary temperatures. If cooling is slow enough, the homogeneous mineral exsolves towards an equilibrium intergrowth of Ti-rich and Ti-poor titanohematites. This process is analogous to the sluggish titanomagnetite→magnetite+ulvöspinel exsolution, rather than the oxidation reaction titanomagnetite→magnetite+ilmenite. When the bulk composition of titanohematite is near $y = 0.5$, exchange coupling between the exsolved ferrimagnetic ($y > 0.5$) and antiferromagnetic ($y < 0.5$) phases can give rise to self-reversal of TRM.

14.5.3   Self-reversal mechanisms

There are at least four distinct self-reversal mechanisms. The first two apply to homogeneous grains, while the other two require two-phase intergrowths.

In **N-type ferrimagnetism** (Néel, 1948), the moments M_A, M_B of the magnetic sublattices have sufficiently different temperature dependences that the resultant spontaneous magnetization, $M_s = M_A - M_B$, changes sign during heating (Fig. 2.7). Only two natural examples are known (Schult, 1968, 1976; Kropáček, 1968).

Self-reversal by **ionic re-ordering** occurs during extreme low-temperature oxidation of high-Ti (large x) titanomagnetites (O'Reilly and Banerjee, 1966; Readman and O'Reilly, 1972). As explained in §14.2.7 and Fig. 14.6, the maghemitization of SD magnetite should preserve the original TRM direction. The same is not true of single-phase titanomaghemites with very large x and z

values, however. Furthermore Ozima and Sakamoto (1971) have produced single and double self-reversals in *multiphase* oxidation products of low-temperature oxidation. Such phenomena are really reversed CRM's rather than self-reversed TRM's. They cannot be reproduced by giving the rock a TRM in the laboratory. Reversals of this sort may possibly occur in highly maghemitized submarine basalts, although this remains to be demonstrated.

Magnetostatic interaction (Néel, 1951) is the only mechanism for self-reversal in physically decoupled magnetic phases, e.g., magnetite regions in oxyexsolution intergrowths. Interaction fields in such intergrowths are strong enough to outweigh the earth's field (Davis and Evans, 1976) and certainly influence TRM but positive and negative coupling are equally probable. There are no undisputed natural examples of magnetostatic self-reversal. In the laboratory Tucker and O'Reilly (1980c) reported self-reversed weak-field TRM after high-temperature oxidation of synthetic single-crystal TM60 (Fig. 14.16a), and Bina and Daly (1994) discovered what appears to be spontaneous (i.e., zero-field) self-reversed TRM in pyrrhotites which had been partially oxidized to magnetite (§13.4.7).

Negative exchange coupling (Néel, 1951) does not necessarily lead to reversal. In Fig. 14.6, negative exchange coupling across a phase boundary reproduced,

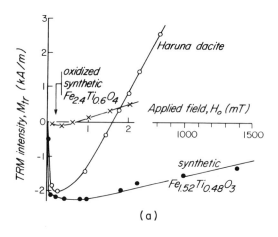

(a)

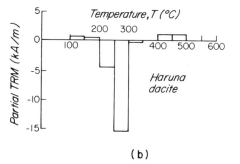

(b)

Figure 14.16   (a) Self-reversing TRM as a function of applied field for oxidized synthetic titanomagnetite, synthetic titanohematite, and the Haruna dacite. (b) Spectrum of weak-field partial TRM's for the Haruna dacite. [After Nagata (1961) Figs. 5-33, 5-37 and Tucker and O'Reilly (1980c), with the permission of the authors and the publishers, Maruzen Co. Ltd. and Journal of Geomagnetism and Geoelectricity, both Tokyo, Japan.]

in the maghemite, the magnetic sublattices of the original magnetite, albeit with a modified arrangement of cations. The net moments of the phases were always coupled parallel. A magnetite–maghemite intergrowth (if it were thermally stable) could not produce self-reversed TRM. However, in partly unmixed titanohematites, the situation is quite different.

14.5.4   Self-reversal of TRM in titanohematite

In titanohematite exsolution intergrowths, the cation distribution and net moment of the ferrimagnetic (ordered or partially ordered) phase are reasonably well understood, but the direction of the weak 'parasitic' moment of the nearly antiferromagnetic disordered phase is uncertain. Cation directions throughout one phase are certainly linked to cation directions throughout the other phase through negative exchange coupling of cation planes of either phase at their interface. Fields of 2T (20 kOe) cannot suppress self-reversal in some synthetic titanohematites (Fig. 14.16a). Only exchange forces could possibly resist so large a field. 'Exchange anisotropy' in the hysteresis of self-reversing samples (Haag *et al.*, 1990) is also evidence of this strong phase coupling.

In hematite, canting of cation moments (Fig. 2.8) to give a weak ferromagnetic moment perpendicular to the sublattices is relatively easy because the cation moments lie in the basal plane (Fig. 3.22a), whose anisotropy is much weaker than that along the c-axis. In two-phase titanohematite intergrowths, cation moments in the disordered phase should be exchange coupled parallel or antiparallel to cation moments in the ordered phase, which lie along the c-axis (Fig. 3.22b). The origin of the weak ferromagnetic moment of the Fe-rich phase is then difficult to explain (e.g., Hoffman, 1992).

Whatever the mechanism by which it arises, the experimental evidence implies that the antiferromagnetic phase does possess a ferromagnetic moment and favours the view that this moment is *antiparallel* to the much stronger moment of the ferrimagnetic phase. At normal temperatures, the weak ferromagnetic moment is completely masked by the ferrimagnetic moment, and the question of how, or even if, it exists is academic. But above the Curie temperature of the ordered phase, the disordered phase remains magnetic (Fig. 3.23). It is to this circumstance that self-reversed TRM owes its existence.

Let us take as a specific example the celebrated dacite tuff of Mt. Haruna, Japan (Nagata *et al.*, 1952; Uyeda, 1958). Other dacitic and andesitic pyroclastics have the same self-reversing property (Kennedy, 1981; Heller *et al.*, 1986; Lawson *et al.*, 1987; Ozima *et al.*, 1992). Figure 14.16(b) shows the partial TRM spectrum of the titanohematite separate from one of these dacites. Titanomagnetite was also present but was found to be irrelevant to the self-reversal phenomenon. The blocking temperatures have two peaks below inferred Curie points of

450–500 °C (antiferromagnetic $Fe_{1.8}Ti_{0.2}O_3$) and 250–300 °C (ferrimagnetic $Fe_{1.5}Ti_{0.5}O_3$).

When cooled from above 500 °C, the rock first acquires a normal TRM of low intensity carried by the disordered phase. Around 300 °C, the ordered phase becomes ferrimagnetic. At interfaces between the phases, the cation moments in the ordered phase are fixed, not in response to the applied field, but by exchange coupling with the adjacent cations in the disordered phase.

If the ordered regions are of MD size, only their boundary domains are pinned antiparallel to the high-temperature TRM. The interior domains respond to the applied field and the TRM is normal. But if the ordered and disordered regions are both of SD size, the strong moments of ordered domains are coupled antiparallel to, and greatly outweigh, the weak (but hard) moments of the disordered domains, resulting in self-reversed TRM.

Since exchange coupling provides the main aligning force, a small external field is sufficient to saturate self-reversed TRM (Fig. 14.16a). Quite high fields may be required to produce normal TRM, particularly in synthetic titanohematites, since the coupled antiferromagnetic phase must be switched below its blocking temperature. However, self-reversing rocks usually acquire a normal TRM in fields > 1–2 mT (Fig. 14.16a; Haag et al., 1988).

Because of the requirement that the coupled regions be of SD size, microintergrowths of ordered and disordered phases promote self-reversal. Haag et al. (1993) and Hoffmann and Fehr (1996) have observed regions a few μm wide at the boundary of self-reversing titanohematite grains which electron microprobe analysis showed to be Ti-poor, and therefore of higher T_C than the body of the grain. These observations confirm the existence of Ishikawa and Syono's (1963) 'X-phase' (the antiferromagnetic disordered phase).

It is interesting that *quenched* titanohematites with $y \approx 0.6$ have the strongest reversed TRM of all (Fig. 14.17) and also the strongest phase-coupling (Ishikawa and Syono, 1963). The suggestion is that exsolution into compositionally distinct phases is almost incidental to self-reversal. The quenched grains are only partially, and *inhomogeneously*, ordered. There need be only a slight compositional difference between ferrimagnetic regions of high order and antiferromagnetic regions of low order to produce the non-overlapping blocking temperature spectra required for self-reversal (Hoffman, 1975). What does seem to be important is large numbers of antiphase domains in the ordered phase (Lawson et al., 1981; Nord and Lawson, 1989, 1992). If the ordered phase were homogeneous, iron and titanium layers would alternate throughout each grain, as in Fig. 3.22b. In reality, in cooling from high temperatures, ordering begins in numerous separate regions. When these regions meet, they can only coalesce into a single phase if an iron layer in one region is adjacent to a titanium layer in another. Where this is not the case, the neighbouring regions are chemically antiphase. Hoffman (1992) proposes that the antiphase boundaries between these domains

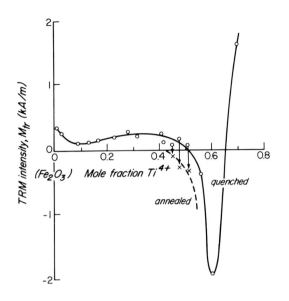

Figure 14.17 Self-reversed and normal TRM of quenched and annealed synthetic titanohematites as a function of composition. [After Nagata (1961) Fig. 5-36, with the permission of the publisher, Maruzen Co. Ltd., Tokyo, Japan.]

act as the disordered X-phase, whose weak moment serves to magnetically couple the ferrimagnetic antiphase domains.

In view of this discussion, it is possible that the $y \approx 0.2$ phase in the Haruna dacite is not essential for self-reversal. On the other hand, titanohematites with $y \approx 0.1-0.2$ in the Allard Lake anorthosite (Carmichael, 1961) and the Bucks diorite (Merrill and Grommé, 1969) themselves have self-reversed partial TRM's over some temperature intervals.

Annealing non-self-reversing titanohematites with $0.4 < y < 0.6$ may induce self-reversal (Fig. 14.17). Ordering, exsolution, and growth of grains or antiphase domains within grains to SD size are all promoted. Over-annealing or too fine a grain size suppresses self-reversal (Westcott-Lewis and Parry, 1971; Westcott-Lewis, 1971). The phases must be neither above nor below SD size.

Self-reversing titanohematites have a decided fascination. They are rare in nature, however, because natural titanohematites usually have $y > 0.7$ or <0.1, outside the self-reversing range. In addition, the contribution of titanohematites to NRM is usually secondary to that of accompanying titanomagnetites.

14.6.   Continental intrusive and plutonic rocks

14.6.1   Introduction

Hyperbyssal and plutonic rocks contain primary opaque minerals similar to those found in their extrusive equivalents, but cooling is slower and the grains tend to be larger. Magmatic fluids are retained within the cooling body and the magnetic oxides are often deuterically oxidized. However, the extreme oxidation, accompanied by wholesale production of hematite, that sometimes occurs in lavas is absent in intrusive rocks. Low-Ti titanomagnetite, approaching pure magnetite in composition, is the principal magnetic mineral, but pyrrhotite (§3.10) occurs frequently. Exsolved titanohematite, occasionally exhibiting partial self-reversal of TRM (e.g., Carmichael, 1961), is common in anorthosites and diorites.

The grain size of the opaque minerals varies from a few μm to a few tens of μm in thin dikes and sills to hundreds of μm in large plutons. It is surprising, therefore, that stable magnetizations should occur at all in intrusive rocks. It is true that there is usually a prominent soft and viscous MD fraction of NRM, but against this background of soft, unstable remanence, more stable SD or PSD components of NRM can often be isolated by AF or thermal cleaning. Fine-grained magnetite or hematite, a product of unmixing, oxidation or alteration (e.g., Hagstrum and Johnson, 1986), is one source of hard NRM. Another, very important, source of stable NRM is what are sometimes referred to as 'magnetic silicates'. These contain magnetite and, less frequently, hematite, either as crystallographically oriented, exsolved inclusions or as products of alteration filling cracks and voids in the crystal.

Magnetic silicates are not particularly conspicuous in extrusive rocks. In mafic and intermediate intrusive rocks, on the other hand, they appear to be ubiquitous. The inclusions are products of either high-temperature ($>800\,°C$) deuteric oxidation of calcic plagioclase (especially labradorite), pyroxene and hornblende or of secondary alteration of micas and olivine. Specific examples will be considered in §14.6.2 and 14.6.3.

As a result of these magnetic inclusions in silicates, many mafic and intermediate intrusive rocks possess unexpectedly intense and hard NRM's. The discrete opaque minerals in these rocks are for the most part paleomagnetically irrelevant, since they contribute only a soft viscous magnetization. Felsic intrusives like granites, which owe their NRM's entirely to large opaque grains (since the potassic or sodic feldspars they contain do not possess magnetic inclusions), are generally poor paleomagnetic recorders.

14.6.2   Continental mafic dikes and sills

Diabase dikes have been widely and successfully used in paleomagnetism, particularly in establishing apparent polar wander paths in Precambrian time (Buchan and Halls, 1990). Their grain size is comparable to that of slowly cooled lavas and although oxidation tends to be less extreme, subdivision of titanomagnetite grains by ilmenite lamellae (class III oxidation) is common. Hence, at least part of the NRM possesses SD-like stability.

Another source of stable remanence is cloudy feldspars (Zhang and Halls, 1995). The clouding is due to submicroscopic magnetite with SD/PSD properties exsolved during slow cooling at relatively deep crustal levels (see next section). Dikes that cooled more rapidly at shallow depths are more deuterically altered but have clear feldspars that lack magnetic inclusions.

Strangway (1961) pointed out that the internal field (and with it the NRM) within a strongly magnetized thin sheet, such as a dike or sill, will be deflected away from the external field H_0 into the plane of the sheet. The in-plane component of NRM produces no self-demagnetizing field H_d, but the transverse component, $(M_{nr})_\perp$, produces a field $H_d = -(M_{nr})_\perp$ (or $-4\pi(M_{nr})_\perp$ in cgs) that reduces the transverse component of H_0 within the sheet. (The demagnetizing factors 0 and 1 (4π in cgs) for a thin sheet were given in §4.2.5 and Table 4.1.) For an NRM intensity of 1 A/m (10^{-3} emu/cm^3), $(M_{nr})_\perp$ is only a few percent of $H_{0\perp}$, but the deflection is important if M_{nr} is ≥ 10 A/m (10^{-2} emu/cm^3) (Coe, 1979; Dunlop and Zinn, 1980).

Macroscopic shape anisotropy of this sort does not seem to be important in diabase dikes (Evans, 1968), which rarely have $M_{nr} > 1$ A/m, but it must be significant in many lavas flows. Anomalously shallow NRM inclinations in submarine basalts (e.g., Hall, 1976) may originate in this way. A related effect is deflection of the magnetic field *outside* a strongly magnetized lava flow, which will produce spurious directions of TRM in subsequently erupted lavas (Baag *et al.*, 1995).

14.6.3   Continental plutonic rocks

A wide variety of rocks, from gabbros and anorthosites through diorites and granodiorites to tonalites and syenites, come under this heading. These rocks, particularly the more felsic ones, have been avoided in paleomagnetic work because the large oxide grains they contain should have truly MD behaviour, implying an NRM that is weak, soft, and apt to be contaminated by viscous noise.

There are surprising features to the NRM of certain plutonic rocks, however. A case in point is the Michikamau anorthosite (Murthy *et al.*, 1971, 1981), a rock containing practically no discrete opaque minerals. Nevertheless, the NRM is as strong as that of most diabases and some volcanics. Moreover the NRM is exceptionally hard and SD-like. The source of the NRM is needles of magnetite, of SD or near-SD size, crystallographically oriented in pyroxene. Similar magnetite needles in plagioclase generate a very hard but rather weak NRM in unserpentinized oceanic gabbros (Davis, 1981; Fig. 14.12). On the other hand, SD-like behaviour of the Jimberlana norite (McClay, 1974) has been traced to fine magnetite–ilmenite intergrowths, magnetite in pyroxene or plagioclase playing only a minor role.

The Michikamau anorthosite is particularly striking because it entirely lacks the soft NRM typical of plutonic rocks. But there are many cases in which a stable and hard NRM due to magnetic silicates underlies MD-type TRM and VRM. For example, the Michael gabbro has an unusual Lowrie–Fuller test (§11.6), characteristic of bimodal mixtures of soft and hard phases (Fig. 14.18). The SIRM is due mainly to the coarse opaques and AF demagnetizes exponentially, in MD fashion (cf. Fig. 11.13), whereas the weak-field ARM is carried in large part by the SD-like inclusions and has a much harder AF decay curve.

The best-documented example of a bimodal plutonic rock is the Modipe gabbro. Evans and McElhinny (1969) showed that the most stable fraction of NRM, with AF coercivities as high as 150 mT (1.5 kOe), was associated with pyroxene. Subsequently, Evans and Wayman (1970) examined replicas of the etched polished surfaces of pyroxene grains under the electron microscope. They discovered that the size distribution of opaque inclusions that had been etched out of the pyroxene extended well into the $<1\,\mu m$ range, perhaps to truly SD sizes (Fig. 14.19; Evans, 1977). These microcrystals were sufficiently elongated to have high coercive forces due to shape anisotropy, and they responded to etching in the same manner as micron-size crystals that could be positively identified as magnetite.

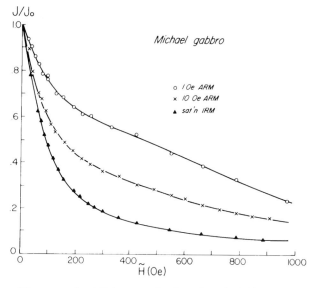

Figure 14.18 AF demagnetization of weak-field and strong-field remanences (Lowrie–Fuller test) for the Michael gabbro, which contains a bimodal size distribution of magnetite grains. Large MD magnetites carry most of the SIRM and demagnetize at low fields. Small SD or PSD magnetite inclusions in plagioclase and pyroxene carry much of the 1-Oe ARM and produce much harder AF behaviour. [After Dunlop (1983c), with kind permission of the author and Elsevier Science – NL, Sara Burgerhartstraat 25, 1055 KV Amsterdam, The Netherlands.]

Figure 14.20 illustrates the AF demagnetization of 'reconstituted NRM's' of several silicate mineral fractions in the Lambertville diabase (Hargraves and Young, 1969). IRM acquisition and AF demagnetization for the plagioclase fraction of this rock were shown in Fig. 11.17. The 'reconstituted NRM's' were

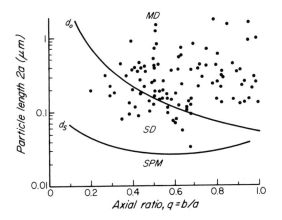

Figure 14.19 Measured sizes and shapes of 115 etched magnetite inclusions in pyroxene crystals from the Modipe gabbro. At least some of the inclusions are of SD size. [After Evans and Wayman (1970), with kind permission of the authors and Elsevier Science – NL, Sara Burgerhartstraat 25, 1055 KV Amsterdam, The Netherlands.]

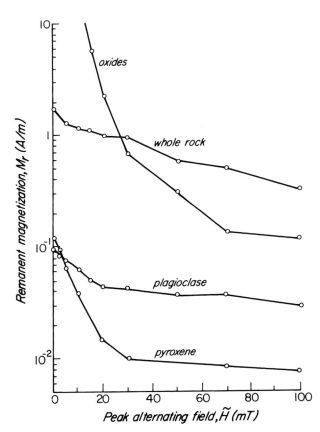

Figure 14.20 AF demagnetization of reconstituted NRM's of separated oxide, plagioclase and pyroxene fractions (laboratory depositional remanences, 3-mT field) and of the original whole-rock NRM for the Lambertville diabase. [After Hargraves and Young (1969), with permission of the authors. Reprinted by permission of American Journal of Science.]

laboratory depositional remanent magnetizations (see Chapter 15) produced by sedimenting crushed separated mineral fractions in a weak field. Plagioclase carries the strongest and hardest 'NRM', but all the silicates are appreciably magnetic. The weighted sum of the individual mineral curves is a reasonable replica of the whole-rock AF demagnetization curve.

Wu *et al.* (1974) measured the NRM's of individual *oriented* crystals cut out of the Tatoosh granodiorite. This **magnetic microanalysis** technique revealed, with a high degree of certainty, that stable NRM resides in plagioclase, while unstable MD remanence characterizes biotite and hornblende grains (Fig. 14.21). In the Stillwater and Laramie intrusions of Wyoming, Geissman *et al.* (1988) have characterized the silicate inclusions responsible for stable NRM using both electron microscopy and magnetic microanalysis.

Opaque inclusions have been identified or inferred in virtually all silicate minerals except quartz and potassic and sodic feldspars. They are particularly abundant in calcic plagioclase and pyroxene, where they are a product of phase disequilibrium during initial cooling or of deuteric oxidation. Their fine size, frequently elongated form, and crystallographic orientation in the host mineral make them ideal SD or PSD NRM carriers, although it is often uncertain whether the NRM is TRM or CRM.

14.6.4   Pyrrhotite

Pyrrhotite ($Fe_{1-x}S$, $0 < x < 0.125$) is a common magnetic mineral in syenites and gabbros (Fig. 14.22), where it occasionally carries a larger fraction of the NRM than magnetite. It occurs not uncommonly in other intrusive rocks, including ultramafics and submarine plutonic rocks, in lavas, and as an altera-

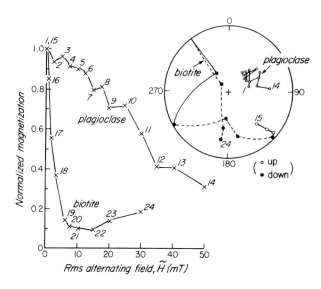

Figure 14.21 Microanalysis of NRM of the Tatoosh granodiorite: AF demagnetization of the NRM's of individual oriented mineral crystals cut from the rock. Plagioclase has a directionally stable primary NRM (see stereoplot). Biotite carries mainly viscous noise. [After Wu *et al.* (1974), with kind permission of the authors and Elsevier Science – NL, Sara Burgerhartstraat 25, 1055 KV Amsterdam, The Netherlands.]

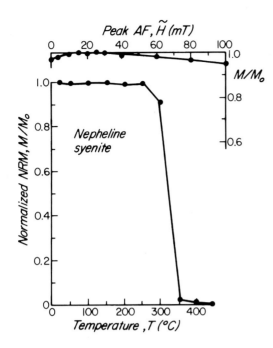

Figure 14.22 Thermal and AF demagnetization of NRM carried by fine-grained pyrrhotite in a nepheline syenite sample. [After Dunlop (1985), with the permission of the author and the publisher, The National Research Council of Canada, Ottawa, Canada.]

tion product in sediments. In these cases, its role in NRM is usually secondary to that of magnetite.

Pyrrhotite seems to be a mixture of phases like Fe_7S_8 and Fe_9S_{10}, some of which are ferrimagnetic, rather than a member of a solid-solution series with antiferromagnetic troilite (FeS) (Schwarz and Vaughan, 1972). The ferrimagnetic Curie point of 320 °C for Fe_7S_8 (Fig. 3.27) is, however, very close to the Néel point of troilite. With a spontaneous magnetization $M_{s0} \approx 80$ kA/m, pyrrhotite is a potent source of NRM. The ferrimagnetism arises, as in γFe_2O_3, from the preferential location of lattice vacancies on one of the magnetic sublattices (Fig. 3.26).

SD pyrrhotite is magnetically hard. The critical SD size d_0 is about 1.6 µm (Table 5.1). The pyrrhotite in rocks is therefore quite likely to be SD or SD-like. The relatively high magnetocrystalline anisotropy of pyrrhotite accounts for its high SD coercivities (Fig. 14.22). Pyrrhotite's only drawback is that it decomposes to magnetite around 500 °C in routine thermal cleaning runs. If the field in the furnace is not perfectly zeroed, the resulting laboratory TCRM can obscure the NRM of the primary magnetite over its main unblocking temperature range.

14.7 Bricks, pottery and other baked materials

The archeomagnetic record of the earth's field in the historical past is based on the study of man-made materials that acquire TRM in the same manner as igneous rocks: pottery, bricks, and the walls and floors of the kilns in which they were fired. The clays from which bricks and pottery are made contain detrital grains of magnetite, maghemite and hematite (Le Borgne, 1955), many of them $< 10\,\mu\text{m}$ in size. After firing at high temperatures, the grains are converted to either hematite or magnetite, depending on whether the supply of air to the kiln produces oxidizing or effectively reducing conditions. Country rocks, sediments or soils baked by igneous intrusions, which are of importance in the baked contact test (§16.1.1), share many of the attributes of artificially baked materials.

Bricks and pottery should be ideal materials for paleointensity determination because they are strongly magnetized and the minerals have been stabilized, physically and chemically, well above their Curie temperatures before initial cooling. The NRM is a pure TRM residing principally in SD or PSD grains, and no chemical changes should occur during laboratory heating under appropriate oxidizing or reducing conditions (see, however, Walton, 1984). This is not always so for Indian and aboriginal hearths or other baked earths, which may have been inadequately heated or have suffered later chemical or textural alteration during weathering (Barbetti *et al.*, 1977).

Other problems of which experimenters should be aware are fabric anisotropy in pottery (Rogers *et al.*, 1979) and 'magnetic refraction', the change in the direction, and also the intensity of TRM due to the shape anisotropy (§14.6.2) of kiln floors and walls (Aitken and Hawley, 1971; Dunlop and Zinn, 1980; Lanos, 1987). The basic theory, methodology and results of paleointensity studies using baked archeological materials have been reviewed by Creer *et al.* (1983).

Chapter 15

Magnetism of sediments and sedimentary rocks

15.1   Introduction

The erosion products of igneous, sedimentary or metamorphic rocks are sources of detrital particles that go to make new sedimentary rocks. The **detrital** and **post-depositional remanent magnetizations** (DRM and PDRM) acquired when sediments are deposited and consolidated are no more than a reconstitution of the NRM's of detrital magnetic grains from the source rock (or rocks). Depending on the size and remanence mechanisms of grains in the source rocks, DRM may inherit the temperature and time stability and AF hardness of TRM, CRM, TCRM or VRM of SD, PSD, or MD grains.

DRM and PDRM are inherently weak, often $< 10^{-3}$ A/m (10^{-6} emu/cm^3). There are two reasons for this low intensity. First, dense oxide grains are less readily transported than silicate grains of similar size. Secondly, DRM represents only a partial realignment of original NRM vectors. With the advent of cryogenic magnetometers (Collinson, 1983), measuring the weak magnetizations of sediments and sedimentary rocks is no longer a major problem.

Red sedimentary rocks – so-called red beds – possess, in addition to depositional remanence, a CRM (10^{-3}–10^{-1} A/m) carried principally by hematite pigment and cement. The CRM is useful paleomagnetically if the time at which the hematite formed is known.

The great attraction of sedimentary rocks is the comparative continuity of their magnetic record. Major igneous activity and metamorphism occur intermittently, sometimes at long intervals. Although a long (several Ma) sedimentary sequence is unlikely to be continuous and uniform over its entire length, it is usually easier to find a sedimentary sequence covering a given time span than

425

an igneous one. It is from sediments and sedimentary rocks that our detailed knowledge of fine-scale variation of the geomagnetic field, e.g., paleosecular variation or PSV (§1.2.3), has come.

Depending on sedimentation rate and post-depositional reworking, the time resolution possible with a given sequence may be quite fine. In oceanic sediments, standard 2.5 cm cores, representing about 1000 yr of sedimentation, have readily measurable NRM's. With lake sediments, 100 yr and sometimes even 10 yr resolution is possible. In undisturbed glacial varves, annual layers can be distinguished and sometimes measured individually.

In the last decade, measurements on sedimentary sequences have been used to infer extensive and detailed records of polarity epochs and events and PSV of the geomagnetic field. The reliability of such records turns on the fidelity of sediments as magnetic recorders, a subject that will occupy us throughout this chapter.

15.2   Detrital and post-depositional remanent magnetizations

15.2.1   Theory of detrital remanent magnetization (DRM)

A spherical or near-spherical grain of diameter d with remanence M_r, as it rotates in response to the earth's field H_0 during settling in calm water, is subject to balanced inertial, viscous and magnetic aligning torques:

$$I \frac{d^2\phi}{dt^2} + \pi d^3 \eta \frac{d\phi}{dt} + \frac{\pi}{6} \mu_0 d^3 M_r H_0 \sin\phi = 0. \qquad (15.1)$$

Here ϕ is the angle between M_r and H_0, I the moment of inertia of the grain, and η the viscosity of water at ordinary temperatures ($\approx 10^{-3}$ Pa s or 10^{-2} poise). In addition, when the grain reaches the bottom, we must consider the mechanical torque exerted on the grain by the surface on which it settles. This torque tends to misalign M_r with H_0.

Figure 15.1 compares typical values of the various torques. The calculations of inertial and viscous torques presuppose the time constants for alignment to be calculated below (eqn. (15.2)). The mechanical torque plotted is the maximum torque experienced by a spheroid of dimensional ratio 1.5:1 settling on a flat surface. Viscous and magnetic torques have identical d^3 dependences except in the PSD range where they go approximately as d^2, since M_r, if it is a TRM carried by magnetite, goes approximately as d^{-1} (Figs. 8.4, 12.2a) and $d\phi/dt \propto M_r$ (eqn. (15.2)). Similarly $d^2\phi/dt^2 \propto M_r^2$, so that inertial torques have an approximately d^3 dependence in the PSD range and a d^5 dependence elsewhere. Except in mm-size grains, inertial torques are negligible.

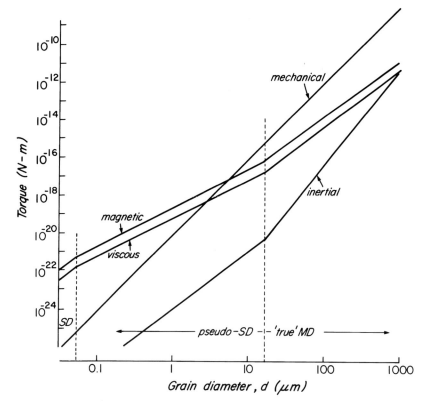

Figure 15.1 Torques exerted on a near-spherical magnetite grain settling in still water.

For grains larger than 1 mm, misaligning mechanical torques are overwhelmingly influential (Fig. 15.1) and magnetic alignment achieved by grains during settling is destroyed. The magnetic aligning torque is also weakened by the fact that 'magnetic' grains this large are composites, containing non-magnetic silicates as well as magnetic oxides and sulphides. The paleomagnetic conglomerate test is based on the expectation of randomized DRM in extremely coarse-grained sediments. Lack of any systematic direction of NRM in a conglomerate is a positive result of the test, indicating no post-depositional CRM in the conglomerate and a stable, non-viscous NRM in the source rock. Conglomerate pebbles are rarely transported any great distance, and the source rock and conglomerate may be found close together.

Sediments containing micron-size and submicron magnetic grains do acquire DRM. According to Fig. 15.1, we may drop the inertial term in eqn. (15.1). Then, replacing $\sin \phi$ by ϕ and solving, we have

$$\phi(t) = \phi_0 \exp\left(-\frac{\mu_0 M_r H_0}{6\eta} t\right) = \phi_0 e^{-t/\tau}, \qquad (15.2)$$

provided the initial value ϕ_0 is small. Even if ϕ_0 is not small, ϕ decreases monotonically and eqn. (15.2) describes the approach to perfect alignment.

Alignment occurs quite rapidly during settling. Assuming that M_r varies between 10^2 and 10^4 A/m (0.1–10 emu/cm^3) over the PSD range (0.1–15 µm in magnetite), $\tau = 6\eta/\mu_0 M_r H_0$ is 10^{-2}–1 s when $H_0 = 50$ µT (0.5 Oe). Magnetostatic coagulation of the particles (Fukuma, 1992) will tend to reduce M_r and should cause τ to increase. Yoshida and Katsura (1985) have verified that the magnetization relaxes exponentially with about the predicted time constants after suppressing the aligning field applied to dilute suspensions of clays and limey muds.

Stacey (1972) has shown that, according to Stokes' law, the distance a grain settles before achieving essentially perfect alignment with H_0 is

$$h = \frac{d^2(\rho - \rho_0)g}{3\mu_0 M_r H_0},$$

(15.3)

where $\rho - \rho_0$ is the density contrast between the grain and the water and g is gravitational acceleration. Even grains as large as 100 µm should exhibit saturation DRM if they settle for ≈ 1 s through several cm of still water in a weak field of the order of 0.1 mT (1 Oe). Times and distances are much less for smaller grains.

Shive (1985) has demonstrated that perfect alignment of the moments of synthetic magnetic particles in suspension is achieved in fields of 30–100 µT (0.3–1.0 Oe), as the theory above would lead one to expect. However, in many laboratory redeposition experiments using natural sediments, fields ≈ 1 mT (10 Oe) are necessary to saturate the DRM (Fig. 15.2) and DRM produced by a field of the order of the earth's field is relatively weak.

The intensity of DRM is grain-size as well as field dependent. Relatively strong (i.e., well-aligned) DRM is confined to a range of sizes between approximately 0.5 and 15 µm in magnetite and maghemite dispersions (Fig. 15.3; Shive, 1985). DRM intensity decreases steadily for sizes smaller or larger than the optimum range.

15.2.2 Inclination and other errors in DRM

We shall consider first > 1 µm size grains, whose remanence vectors achieve near-perfect alignment while settling but are misoriented by mechanical torques when the grains touch bottom. In fact, mechanical torques so far outweigh magnetic torques in > 10 µm magnetic grains (Fig. 15.1) that the initial degree of alignment is almost incidental. Two models of mechanical misalignment have been proposed. The first model (King, 1955) considers elongated particles which, mechanically, prefer to settle with the long axis horizontal, and magnetically, by virtue of shape anisotropy (§4.2.4), prefer to have their remanence M_r along the long axis. A model of this sort led to the mechanical torque curve in

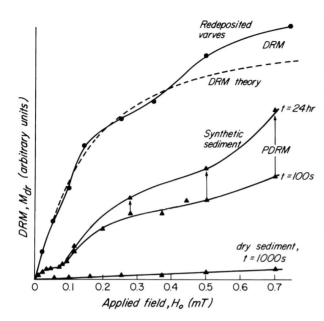

Figure 15.2 DRM and PDRM as a function of applied field for redeposited varved clays and a synthetic sediment. The theoretical field dependence of DRM follows eqn. (15.4). [After Johnson *et al.* (1948) © American Geophysical Union and Tucker (1980), with the permission of the publishers, The American Geophysical Union and Blackwell Science, Oxford, UK.]

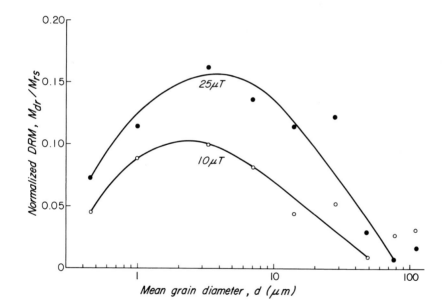

Figure 15.3 Particle size dependence of DRM intensity for artificially sedimented magnetites. [Data of C. Amerigian, cited by Dunlop (1981) Fig. 13 and reprinted with kind permission of Elsevier Science – NL, Sara Burgerhartstraat 25, 1055 KV Amsterdam, The Netherlands.]

Fig. 15.1. The effect is certainly real. Susceptibility in the bedding plane of a sediment does exceed susceptibility transverse to the plane (Fig. 15.4), but the anisotropy is normally only a few percent. The second model (Griffiths *et al.*, 1960), in which spherical grains roll when they settle on an irregular surface, reduces the overall DRM without introducing a systematic anisotropy of susceptibility.

Although grain rotations resulting from either cause are essentially random and have a random effect on the declination D of M_r vectors, they result in a systematic decrease in the inclination I. This **inclination error** as it is called can amount to 10–20°. It has been documented in artificial sediments deposited in the laboratory and in some fresh-water sediments (Fig. 15.5; see also Fig. 15.11).

If deposition occurs on a sloping plane, there may be systematic changes in D as well as I. Since spheres roll farther downhill than up, this **bedding error** should

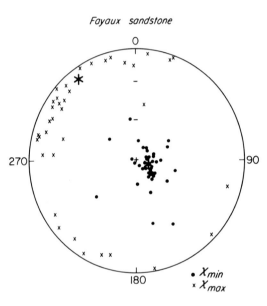

Figure 15.4 Anisotropy of susceptibility in 38 samples of the Fayaux sandstone. Directions of maximum susceptibility (mean indicated by the large star) tend to lie in the bedding plane, reflecting the tendency for elongated magnetite grains to settle with their long axes horizontal. [After Channell *et al.* (1979), with the permission of the authors and the publisher, Eclogae Geologicae Helvetiae, Zürich, Switzerland.]

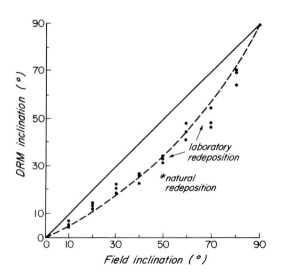

Figure 15.5 Inclinations of DRM in laboratory and natural riverbed deposition of detrital hematite. The DRM vector is shallower than the applied field. This inclination error is most pronounced in middle magnetic latitudes. [After Tauxe and Kent (1984), with the permission of the authors and the publisher, Blackwell Science, Oxford, UK.]

exceed the slope of the plane, as is observed in redeposition experiments (Griffiths *et al.*, 1960).

Deposition in flowing or turbulent water brings into play systematic viscous torques due to the velocity gradient in the boundary layer of laminar fluid flow at the water–sediment interface (Granar, 1958). Both D and I may be changed systematically by as much as several tens of degrees due to this **current rotation effect** (Rees, 1961). DRM intensity is also significantly weakened.

15.2.3 Effect of thermal fluctuations on DRM

Submicron grains are subject to all the mechanical effects discussed above, but the magnetic torque is sufficiently powerful by comparison (Fig. 15.1) that alignment of grain moments achieved during settling should be preserved. Non-saturation of DRM intensity in $< 1\,\mu m$ grains presumably reflects non-achievement of alignment during settling.

Very small grains suspended in water experience thermal fluctuations in their orientations. These fluctuations are not readily represented by an average torque that can be plotted in the manner of Fig. 15.1., but the average alignment of M_r vectors can easily be calculated. The ensemble of grain moments acts like a classical paramagnetic 'gas', obeying Boltzmann statistics, albeit the response time for rotations is 10^{-2}–10^{-1} s (eqn. (15.2)) rather than $\approx 10^{-9}$ s (e.g., eqns. (8.16)). The predicted intensity of DRM is (Collinson, 1965a; Stacey, 1972)

$$M_{dr} = (M_{dr})_{sat} L\left(\frac{\mu_0 V M_r H_0}{kT}\right) \leq (M_{dr})_{sat} \frac{\mu_0 V M_r}{3kT} H_0. \tag{15.4}$$

Here $L(\alpha)$ is the Langevin function $\coth(\alpha) - 1/\alpha$ and V is the volume of each moment. $(M_{dr})_{sat}$ is the saturation DRM observed when all M_r vectors are perfectly aligned. Implicit here is the assumption that the field H_0 is never large enough to change the magnitude M_r of NRM of any grain. DRM that is 'saturated' in the sense that all moments are aligned remains much weaker than the saturation remanence, M_{rs}.

King and Rees (1966) noted that the saturation predicted by eqn. (15.4) is very sensitive to changes in grain moment $V M_r$. Grains $> 1\,\mu m$ in size (large V) and most SD grains (large M_r) have such large moments that, in theory, saturation is achieved in the earth's field. Only PSD moments in submicron grains are sufficiently small to be significantly misaligned by thermal fluctuations during weak-field sedimentation. The decrease in DRM intensity below $0.5\,\mu m$ (Fig. 15.3) is presumably due to thermal fluctuations.

By integrating eqn. (15.4) over a uniform distribution of PSD moments, Stacey (1972) obtained a reasonable fit (Fig. 15.2) to Johnson *et al.*'s (1948) data for redeposited varves. Stacey's average PSD moment, corresponding to $M_r = 5$–$10\,kA/m$ or emu/cm^3 in 0.1–$0.2\,\mu m$ magnetite grains, is at first sight reasonable (cf. Fig. 8.4). However, M_r in Fig. 8.4 represents the vector sum of misaligned individual grain moments, which for weak fields is much less than their arithmetic sum. When M_r for *individual* PSD moments is used in eqn. (15.4), saturation of DRM is predicted to occur in much smaller fields than are observed experimentally. The thermal-fluctuation theory of DRM is thus inadequate in exactly the same way as the thermal-fluctuation theory of single-domain TRM (eqn. (8.21), §8.6).

15.2.4   Post-depositional remanent magnetization (PDRM)

Detrital remanence is not finally fixed in intensity and orientation until the sediment has been compacted by the overburden of later deposits, dewatered , and lithified. After deposition, wet unconsolidated sediments are often disturbed by mechanical slumping or through bioturbation by burrowing organisms. In either case, DRM should lose most of its directional coherence if it is not entirely destroyed. In practice, however, slumped beds frequently exhibit a uniform magnetization (Irving, 1957). They must have been remagnetized since deposition and slumping occurred.

   To account for this observation, Irving proposed that in a water-saturated sediment, magnetic particles would remain free to rotate in response to the earth's field until compaction and reduction in water content eventually restricted their movement (Fig. 15.6b; Blow and Hamilton, 1978). This mechan-

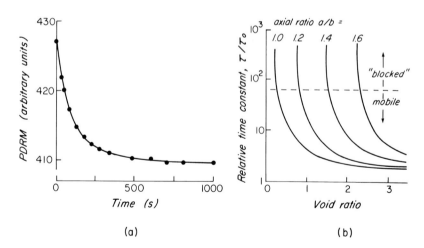

Figure 15.6   (a) Time-dependent realignment of PDRM following reorientation of the field applied to an artificial magnetite slurry. [After Tucker (1980), with the permission of Blackwell Science, Oxford, UK.] (b) Theoretical time constants for realignment of PDRM in sediments with various void ratios and axial ratios of the magnetic particles. The time constant τ_0 for a very dilute suspension is used as a normalizing factor. Compaction of the sediment below a critical void ratio blocks particle rotations. [Reprinted from Hamano (1980), with kind permission of the author and Elsevier Science – NL, Sara Burgerhartstraat 25, 1055 KV Amsterdam, The Netherlands.]

ism of **post-depositional remanent magnetization** (PDRM) was shown to be physically plausible by Irving and Major (1964). It is illustrated in Fig. 15.6(a). PDRM in a reworked deep-sea sediment is documented in Fig. 15.7(a).

PDRM reestablishes the remanence of a deformed or reworked sediment and reduces or eliminates inclination error in an undeformed, unreworked sediment. Indeed the best evidence for the widespread occurrence of PDRM is the absence of inclination error in many classes of sediments, especially deep-sea sediments (Fig. 15.7b). Using a large data set from 185 deep-sea sediment cores, Schneider and Kent (1990) have confirmed that average NRM inclinations do approximately match expected dipole field inclinations. However, there is significant inclination shallowing at equatorial and southern latitudes (see also Gordon, 1990; Tarduno, 1990).

Since PDRM may continue evolving until a sediment is thoroughly compacted and its grains cemented in place, the theory of DRM developed in previous sections may be irrelevant in many natural settings. PDRM is most efficient if the magnetic grains are significantly finer than the silicate grains and can rotate readily in pore spaces (Fig. 15.6b). The density differential during sedi-

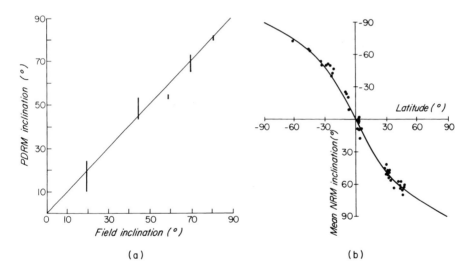

(a)   (b)

Figure 15.7 Evidence that deep-sea sediments record the field direction without major inclination error. (a) PDRM of a deep-sea sediment redeposited in the laboratory. Vertical bars indicate the spread of replicate measurements. PDRM inclination equals field inclination within experimental error. [After Kent (1973), with permission of the author. Reprinted with permission from *Nature*. Copyright © 1973 Macmillan Magazines Ltd.] (b) Average NRM inclinations for deep-sea sediment cores, each spanning the 0.78-Ma Brunhes epoch, compared to the dependence on latitude λ, $\tan I = 2 \tan \lambda$, expected for a geocentric axial dipole field (solid curve; this relation is easily derived from eqn. (2.2)). [Reprinted from Opdyke and Henry (1969), with kind permission of the authors and Elsevier Science – NL, Sara Burgerhartstraat 25, 1055 KV Amsterdam, The Netherlands.]

ment transport seems to ensure that this situation generally holds, at least in deep-sea sediments (e.g., Opdyke and Henry, 1969). The ultimate determining factor in the acquisition or non-acquisition of PDRM is water content. When water content drops below a critical value, remagnetization ceases (Verosub *et al.*, 1979). Slow deposition (e.g., in the deep oceans) and fine sediment particle size favour PDRM, because they promote high water content and delay compaction. A review of this and many other topics in sedimentary magnetism is given by Verosub (1977).

Compaction rotates grains into the bedding plane and is itself a source of inclination error (Anson and Kodama, 1987; Celaya and Clement, 1988; Arason and Levi, 1990a,b; Deamer and Kodama, 1990). Anisotropy of susceptibility (Fig. 15.4) and of remanence (§16.4.1), for example ARM, can be used to measure the resulting planar magnetic fabric. It may perhaps also be possible to correct for the inclination error by using anisotropy of ARM (Jackson *et al.*, 1991; Kodama and Sun, 1992; Hodych and Bijaksana, 1993; Collambat *et al.*, 1993).

PDRM to some extent restores the time resolution (e.g., in recording polarity reversals) that is lost when a sediment is reworked. A time lag remains, however, between the time of deposition and the time PDRM is acquired. Løvlie (1976) simulated a polarity reversal by reversing the applied field H_0 applied during laboratory redeposition of marine sediments. He found that PDRM was acquired at depths 2–7 cm below the sediment–water interface at the time of field reversal (Fig. 15.8). The grains carrying PDRM were locked in place while the sediment was only partially compacted. Extrapolating from Løvlie's laboratory results, for a deposition rate of ≈ 1 cm/ka, the time lag would be ≈ 2–7 ka. By way of comparison, a few tens of ka of delay in the fixing of PDRM in deep-sea sediments were inferred by deMenocal *et al.* (1990). A pos-

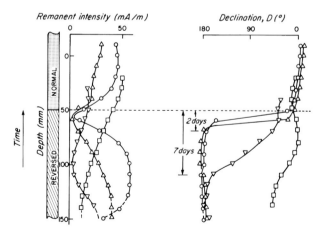

Figure 15.8 PDRM in sediments continuously deposited over a period of many days in a 42 µT laboratory field. Reversal of the field was recorded by sediments deposited 2–7 days earlier. Fixing of PDRM required 1–5 days, blurring the magnetic record of the sharp field reversal. [Reprinted from Løvlie (1976), with kind permission of the author and Elsevier Science – NL, Sara Burgerhartstraat 25, 1055 KV Amsterdam, The Netherlands.]

sible explanation of the discrepancy between laboratory and natural lock-in depths/times is Quidelleur *et al.*'s (1995) observation, in redeposition experiments similar to Løvlie's, that major changes in inclination (marking final compaction and associated inclination error) occurred 13 cm below the position of the field reversal.

In Løvlie's and Quidelleur *et al.*'s experiments, the laboratory field reversal was not recorded as a sharp event. PDRM changed gradually over a 1–5 cm interval. Reversal events of ≈ 1–5 ka duration in the original sediments would thus be obscured (Tauxe and Badgley, 1984) and the record of paleosecular variation (PSV) between reversals smoothed out (Lund and Keigwin, 1994). In addition, the coexistence of grains with opposite NRM polarities at the same depth could create spurious intermediate directions of total remanence, which might be wrongly interpreted as recording transitional geomagnetic field directions (Quidelleur and Valet, 1994; Quidelleur *et al.*, 1995; see also Langereis *et al.*, 1992).

Hyodo (1984) and Hyodo and Yaskawa (1986) have proposed two methods of overcoming the smoothing problem. The first and more general method is deconvolution of the magnetization-depth record using a moment fixing function $r(t)$ which describes the time (or in a continuously deposited sediment, the depth) dependence of PDRM and an assumed half fixing depth $z_{1/2}$, which will depend on such factors as grain size and shape, void ratio and rate of compaction. An exponential $r(t)$ is indicated by data like those of Fig. 15.6. A priori knowledge of $z_{1/2}$ is usually lacking, but Hyodo *et al.* (1993) found that assumed values of 10–50 cm in different parts of their marine and lacustrine cores gave an optimal match between deconvolved PSV records and a detailed master curve. The second method is AF demagnetization. It makes the assumption that the fraction of PDRM with the highest AF coercivities is acquired over a small depth interval and therefore provides a sharp record of field variations. A good match between the AF demagnetized magnetization-depth record of a shallow marine core and the deconvolved record of the same core seems to justify the method (Hyodo and Yaskawa, 1986).

15.2.5   Paleointensity determination using sediments

The earliest DRM experiments were aimed at determining paleofield intensity. The principle is simple and analogous to that of paleointensity determination using the TRM of igneous rocks (§1.2.4, 8.9). Assuming that the sediment is relatively unconsolidated and can be restored to liquid suspension, comparison of the intensity of DRM produced in known fields when the sediment is redeposited in the laboratory to the measured NRM intensity determines the paleofield in which the NRM was acquired.

Although the experiments of Tucker (1981) provide some guidelines, such paleointensity estimates must be regarded with caution. We do not understand which factors govern the intensity of weak-field DRM and PDRM in nature nearly well enough to have confidence that we can reproduce either process in the laboratory. Johnson *et al.* (1948) noted that natural DRM's in their varved clays approximately matched the laboratory DRM due to redeposition in the earth's field (Fig. 15.2), but only within a factor 2–4. As a means of *absolute* paleointensity determination, laboratory simulation of DRM gives order-of-magnitude estimates only.

Relative values of paleofield intensity can be determined with considerably more certainty (Levi and Banerjee, 1976), but only if one compares the NRM intensities of sediments of similar mineralogy, grain size and provenance (for a review, see Tauxe, 1993). DRM/PDRM presumably formed in the same manner in these related sediments, so that a single DRM or PDRM intensity versus applied field curve should apply to all, although this type curve is itself unknown. Figure 1.5 is an example of a high-quality relative paleointensity record over a 0.6 Ma time span, determined from deep-sea sediments cored at three different ODP sites. Field intensity variations at and between reversals are clearly resolved. Other high-quality records have been published by Tauxe and Wu (1990), Tric *et al.* (1992), Meynadier *et al.* (1992, 1994) and Valet and Meynadier (1993). Thibal *et al.* (1995) report a continuous record of field intensity between 4.7 and 2.7 Ma deduced entirely from downhole magnetic measurements.

Variations in magnetic mineral content cause a proportional change in NRM intensity. To correct for these variations, NRM intensity must be normalized to some independent parameter which is also proportional to the quantity of magnetic minerals, for example susceptibility χ_0, ARM or 'anhysteretic susceptibility' χ_a (§11.4.1), or SIRM measured for the same sample. This method has been used in Fig. 1.5. Improved long-core measurement techniques, such as continuous measurement of U-channel samples with cryogenic magnetometers (Weeks *et al.*, 1993), facilitate the process. Unfortunately, all these parameters (and the DRM intensity as well) are also sensitive to variations in grain size, so that grain-size differences down a sediment core will lead to spurious apparent variations in relative paleointensity.

Since the different parameters vary with grain size in different ways, it is possible to use two of the parameters jointly, e.g., χ_0 and χ_a (King *et al.*, 1982, 1983), to detect grain-size changes (see discussion in §11.4.1). However, *correcting* the relative paleointensity values to a common datum (i.e., a fixed grain size) is a much riskier business. The common assumption that ARM can be used as a renormalization factor that will allow for variations in grain size and domain structure, as well as variations in magnetic mineral concentration, is unjustifiable. There is little resemblance between the grain-size dependences of ARM and DRM intensities (Figs. 12.2 and 15.3).

15.3   Oceanic and continental sediments and sedimentary rocks

15.3.1   Deep-sea sediments

Layer 1 of the oceanic crust consists of up to 0.5 km of clastic and chemical sediments, the thickness increasing with age of the underlying igneous rocks, i.e., with distance from the spreading centre. Haggerty (1970), Opdyke (1972), Lowrie and Heller (1982), Freeman (1986) and Vali *et al.* (1989) report that the principal lithogenic magnetic minerals are titanomagnetites in a variety of oxidation states and detrital hematite. If the titanomagnetites were eroded from submarine basalts, they would be expected to have $x = 0.6$–0.7 (cf. §14.2.1). Instead, they are near magnetite in composition and, in the North Pacific at least, are thought to be wind-transported volcanic ash and continental erosion products (Johnson *et al.*, 1975b). The grains tend to be fine; submicron particles are abundant (Yamazaki and Katsura, 1990).

In addition to detrital magnetic minerals, authigenic and diagenetic iron oxides, hydroxides and sulphides commonly form in pelagic and hemipelagic environments (Verosub and Roberts, 1995). In slowly deposited oxic sediments with little organic input, Fe–Mn oxides and hydroxides form and may obscure any primary DRM or PDRM (Henshaw and Merrill, 1980). In more rapidly deposited sediments, for example continental slope or mid-ocean ridge deposits, fine-grained magnetite is formed just above the iron reduction zone (Karlin *et al.*, 1987; Karlin, 1990), perhaps by the metabolic activity of dissimilatory iron-reducing bacteria (e.g., GS15: Lovley *et al.*, 1987). This authigenic magnetite is responsible for the strong and stable NRM of suboxic sediments.

Deeper in the sediment column in the iron reduction zone, fine-grained magnetite and other oxides of detrital or biogenic origin dissolve and reform as iron sulphides, particularly pyrite (FeS_2) (Karlin and Levi, 1985; Leslie *et al.*, 1990; Bloemendal *et al.*, 1992). Magnetic monitoring of cycles of magnetite dissolution is of environmental importance because the fluctuating redox conditions that produce the cycles are climatically driven and correlate with glacial cycles (Tarduno, 1994, 1995). As a result of sulphide diagenesis, anoxic sediments are often magnetically barren. However, under sulphate-reducing conditions, e.g., in muds and some rapidly deposited marine sediments, pyrrhotite and greigite (Fe_3S_4) may be preserved (Tric *et al.*, 1991; Roberts and Turner, 1993; van Velzen *et al.*, 1993; Reynolds *et al.*, 1994). Magnetite can also be produced by sulphate-reducing bacteria (Sakaguchi *et al.*, 1993).

'Magnetofossils', the remains of chains of magnetosomes produced by magnetotactic bacteria and other organisms (§1.2.6, 3.13), are major, perhaps dominant, sources of detrital fine-grained magnetite and occasionally greigite in many deep-sea sediments (Kirschvink and Lowenstam, 1979; Petersen *et al.*,

1986; Stolz *et al.*, 1986, 1990). For reviews of magnetofossils and magnetite bio-mineralization, see Chang and Kirschvink (1989) and Moskowitz (1995). Perfect crystals of SD size with distinctive cubic, octahedral and bullet-like shapes, often in intact chains, are abundant in both siliceous and calcareous pelagic sediments (Vali *et al.*, 1987, 1989; Yamazaki *et al.*, 1991). They are absent from pelagic clays, which generally carry only weak and unstable NRM's. Moskowitz *et al.* (1993) and Oldfield (1994) propose rock magnetic criteria for assessing the relative roles of biogenic and lithogenic magnetic minerals in sediments.

Not all deep-sea cores preserve a faithful record of the geomagnetic field. Figure 15.9 compares the variations of inclination (after AF cleaning to 15 mT) with depth in two cores of similar lithology and length from the North Pacific. One core preserves a clean reversal record covering the Brunhes and part of the Matuyama polarity epochs (see Fig. 1.4). The other core has a chaotic record containing numerous spurious reversals. Directional stability of remanence during AF demagnetization is also poorer for the second core than for the first. CRM produced during in situ maghemitization of magnetite probably caused the confused magnetic record of the second core.

Maghemitization of titanomagnetites in deep-sea sediments is frequent (Vali *et al.*, 1989) and tends to render the NRM soft and viscous (Kent and Lowrie, 1974). By way of contrast, maghemitization of submarine basalts tends to generate a hard and directionally stable CRM (§14.2.8). The difference in behaviour must result from the very fine initial size of magnetite grains in the sediments (Yamazaki and Katsura, 1990). As a result of cracking during oxidation (Fig. 3.10), a large proportion of the maghemite grains behave superparamagnetically.

CRM acquired during diagenesis, bioturbation of the sediment by organisms, and viscous remagnetization (Yamazaki and Katsura, 1990) are the major in

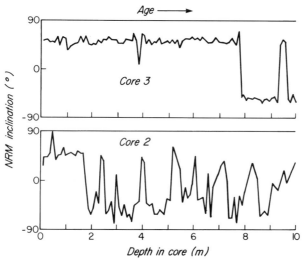

Figure 15.9 NRM inclination following 15-mT AF cleaning for two deep-sea sediment cores from the North Pacific Ocean. [After Johnson *et al.* (1975b), with the permission of the authors and the publisher, The Geological Society of America, Boulder, CO.]

situ sources of magnetic noise in deep-sea cores. Much of the secondary remanence can be AF cleaned, but sometimes thermal cleaning is necessary (Yamazaki, 1986). Post-coring errors arise from disturbance of the sediment by the
piston coring process and partial remagnetization by the present field during drying of the core. 'Drying remanence' (a problem also with lake sediment cores) is
in part a PDRM due to mechanical rotation of grains and in part a VRM. The
'drying remanence' is softer than the pre-drying NRM but nevertheless has
some coercivities in excess of the AF's of 10–15 mT (100–150 Oe) commonly
used to clean deep-sea sediments (Johnson *et al.*, 1975b).

15.3.2 Fresh-water and marginal-sea sediments

A variety of coarse-grained to fine-grained late Tertiary, Quaternary and Recent
sediments from continental and continental margin depositional environments
have been used in paleomagnetism. It is clear from studies of lake sediments
that lithofacies and environment of deposition correlate with the strength and
fidelity of the recorded DRM and with rock magnetic properties (e.g., see review
by Verosub and Roberts, 1995). Conditions during deposition may range from
oxidizing (shallow-water flood-plains), producing red beds (§15.5), to reducing
(deep still lakes with abundant organic matter), producing black shales.

Biogenic magnetite and greigite produced by magnetotactic bacteria are common in anaerobic shallow-water environments (Kirschvink, 1983; Petersen *et
al.*, 1989; Snowball and Thompson, 1990; Snowball, 1994). The magnetosomes
in these bacteria are a potent source of CRM or DRM because their sizes and
shapes fall in the SD range (Fig. 15.10). Vali *et al.* (1987) observed a variety of
sizes, shapes and numbers of magnetosomes in freshwater bacteria but comparative uniformity in marine bacteria. Fossil magnetosomes as old as Jurassic were
morphologically similar to those in living bacteria, apart from some surface cor-
rosion of the crystals.

The high rate of sedimentation in continental and continental margin environments, and consequent rapid dewatering compared to deep-ocean sediments,
appear in some cases to inhibit PDRM or else arrest it early in its development.
Verosub (1975) showed that certain deformed glacial varves had not been remagnetized since folding, which occurred soon (within three years) after deposition.
On the other hand, Irving (1957) and McElhinny and Opdyke (1973) demonstrated post-slumping remagnetization of sandstones, and Graham (1974), Barton and McElhinny (1979) and Barton *et al.* (1980) produced laboratory
PDRM in redeposited slurries of organic muds.

DRM with attendant inclination error appears to be the norm in glacial
deposits. An inclination error averaging 10–15° is seen in Swedish glacial
varves (Fig. 15.11a). The standard for comparison is geomagnetic inclination
from observatory records. Mean site inclination (average inclination for all

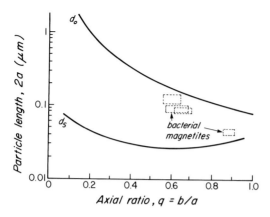

Figure 15.10   Sizes and shapes of biogenic magnetite crystals produced by magnetotactic bacteria. The bacterial magnetites fall in the stable single-domain field
between d_0 and d_s and are potentially important carriers of remanence in some
sediments. [After Butler and Banerjee (1975b) © American Geophysical Union
and Kirschvink (1983), with the permission of the authors and the publishers, The
American Geophysical Union, Washington, DC and Plenum Publishing, New
York, NY.]

samples from a single site) for lake sediments also tends to be ≈ 10° lower than
the expected dipole-field inclination, although in some cases non-dipole features that persisted longer than the averaging time at the site may be recorded.

Direct comparisons of the fidelity of volcanic rocks and contemporaneous
lake deposits in recording the paleofield (Biquand and Prévot, 1972; Reynolds,
1979) show that the primary NRM's of the sedimentary beds, isolated by AF
and thermal cleaning, are in some cases faithful field recorders but in other cases
have an inclination error of 10–20° (Fig. 15.11b). In many lake sediment studies,
it is standard practice to blanket clean cores at an AF level determined by the
response of a few pilot specimens. Under these conditions, it is doubtful if the primary NRM is isolated at all. Apart from other sources of magnetic noise, lake
sediments readily acquire VRM, and this VRM typically is erased only by thermal demagnetization to ≈200 °C or AF demagnetization to several tens of mT
(Biquand, 1971).

Paleomagnetic inclination records are compared in Fig. 15.12 for cores from
the Aegean and Black Seas. Dispersion in the data, a reflection of noise in the
magnetic recording process, is considerable, particularly near the bottom of
either core. In conventional paleomagnetic studies, averaging of data at the sample and site levels pinpoints mean directions and allows a statistical estimate of
significance to be made. In continuous sediment core studies, in the effort to
resolve the details of PSV, each measurement is treated as a significant datum.
The danger in this approach is that systematic errors in the magnetic recording
may be mistaken for real trends in the geomagnetic field.

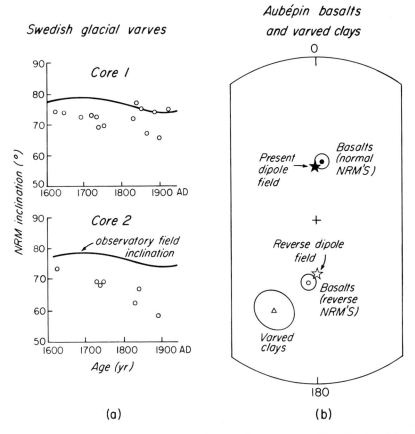

Figure 15.11 Inclination errors in some Quaternary and late Tertiary lake sediments. (a) NRM inclinations of Recent glacial varves from Sweden compared to geomagnetic observatory records. [After Granar (1958), with the permission of the author and the Royal Swedish Academy of Sciences.] (b) Average cleaned remanence directions with circles of 95% confidence for contemporaneous Pliocene volcanic and sedimentary rocks from France. The basalts accurately record the geomagnetic field, while the sediments have NRM vectors that are 10–20° shallow. [After Biquand and Prévot (1972), with the permission of the authors and the publisher, Editions BRGM, Orléans, France.]

It is clear that some smoothing of the data is needed, by the use of running means for example, to enhance the periodicities in the record, but whether all the periodicities are significant, i.e., record real geomagnetic variations, remains unclear. Lack of correlation between the inclination variations in geographically nearby cores (e.g., Fig. 15.12) is unfortunately typical of marginal-sea and lake-sediment studies. On the other hand, Verosub and Banerjee (1977) found that two geomagnetic excursions, at 10 and 15 ka, were common to many North American lake sediment records. Correlation of cyclic changes in declination in cores from three English lakes has been reported by Thompson (1975). However,

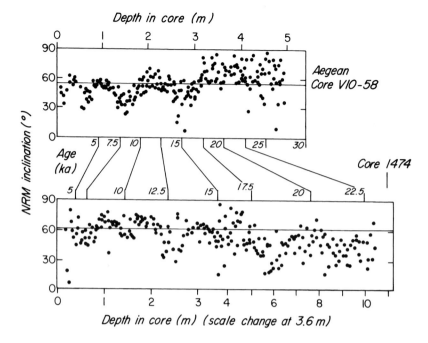

Figure 15.12   NRM inclinations after 10-mT AF cleaning for individual sub-samples of piston-cored sediments from the Aegean and Black Seas. Paleosecular variation about the dipole field inclination (solid lines) is evident in the upper half of each core but is obscured by between-sample scatter of NRM directions in the lower half of the cores. [Reprinted from Creer (1974) and Opdyke *et al.* (1972), with kind permission of the authors and Elsevier Science – NL, Sara Burgerhart-straat 25, 1055 KV Amsterdam, The Netherlands.]

the same recorded feature in different cores from a single lake can have ampli-tudes that vary from 30° to 90° (Fig. 15.13).

Such behaviour clearly stems from the shortcomings of sediments as magnetic recorders. Van Hoof and Langereis (1992) concluded that most of the 'fine struc-ture' recorded by Pliocene marls in southern Sicily, including a 60° excursion, was determined by the remanence acquisition process, which is complicated by authigenic formation of magnetic minerals at varying times in response to cyclic paleoredox conditions. Until we thoroughly understand these problems and how to deal with them, our knowledge of geomagnetic field behaviour based on sediment records will remain tentative and uncertain. Recent reviews of sediment magnetism and remanence acquisition processes are given by Lund and Karlin (1990) and King and Channell (1991).

15.3.3   Soils and loess

The main magnetic minerals in soils and loess are maghemite, goethite and hema-tite (Mullins, 1977; Maher, 1986; Schwertmann, 1988; Liu *et al.*, 1993) and occa-

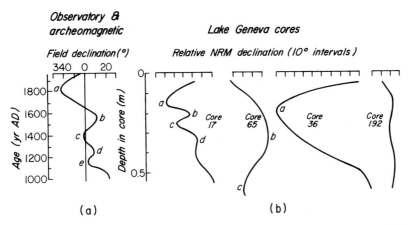

Figure 15.13 A comparison of paleosecular variation in geomagnetic declination according to (a) observatory records, and (b) NRM measurements on sediment cores from Lake Geneva. The fidelity of the magnetic recordings is quite variable. [Reprinted from Creer *et al.* (1975), with kind permission of the authors and Elsevier Science – NL, Sara Burgerhartstraat 25, 1055 KV Amsterdam, The Netherlands.]

sionally magnetite (Longworth *et al.*, 1979). They may be water- or wind-transported erosion products or they may form in situ biogenically (Fassbinder *et al.*, 1990; Evans and Heller, 1994) or by organic or inorganic pedogenesis (Schwertmann, 1971; Taylor *et al.*, 1987; Fine *et al.*, 1995). The oxidation and reduction of magnetic minerals in soils reflect climatic conditions and are therefore of great interest in global paleoclimate studies. In addition to changes in chemical composition, the evolution from rock through subsoil to soil involves changes in grain size of the iron oxides, which are reflected in magnetic properties (Özdemir and Banerjee, 1982).

The A (upper) horizon in soils has enhanced magnetic susceptibility. Le Borgne (1955, 1960b) proposed that the high susceptibility results from reduction of goethite or hematite to magnetite followed by oxidation to maghemite, and suggested two mechanisms: fires and organic fermentation. The burning of organic material produces both increased temperature and a reducing atmosphere (Longworth *et al.*, 1979). Fermentation, the decay of organic matter under anaerobic conditions achieved during wet periods, can also reduce hematite. Oxidation occurs during subsequent dry aerobic conditions. Cyles of oxidation and reduction may thus record variations in rainfall (Mullins, 1977). This is particularly evident in Mediterranean soils formed on a permeable limestone substratum (Tite and Linington, 1975).

Other suggested mechanisms for high susceptibility in surface soils are preferential accumulation of lithogenic minerals (Fine *et al.*, 1989) and in-situ production of magnetic oxides by soil-forming processes (Maher and Taylor, 1988). Fine pedogenic grains can be dissolved by citrate–bicarbonate–dithionite

(CBD) treatment (Mehra and Jackson, 1960), leaving the coarser lithogenic oxides intact (e.g., Hunt *et al.*, 1995b; Sun *et al.*, 1995).

Magnetic susceptibility variations in loess/paleosol sequences correlate well with oceanic oxygen isotope stages and are of considerable interest as indicators of past climate changes. A typical loess stratigraphy, e.g., in the loess plateau of China, consists of alternating loess and paleosol layers. The loess was deposited during cold and dry (glacial) periods and, being wind-blown, was initially relatively barren in magnetic constituents. Paleosols developed in warm and humid (interglacial) periods and are magnetically much richer. Pedogenesis is the main cause of the high susceptibility of the paleosols (Liu *et al.*, 1993; Evans and Heller, 1994) and the magnetic enhancement is a proxy for rainfall. Past rainfall rates can be estimated (Heller *et al.*, 1993), and the local climate signal can even be separated from the regional monsoon record (Banerjee *et al.*, 1993). For a recent review, see Heller and Evans (1995).

15.3.4   Sedimentary rocks

Secular variation, excursions and reversals of the geomagnetic field are recorded by continental and oceanic sediments for the last few million years. Fully lithified sedimentary sequences offer the same advantages of comparative continuity, resolution and datability (by fossil or radiometric methods) in establishing the polarity record of the earth's field throughout the Phanerozoic (e.g., Lowrie and Alvarez, 1977; Lowrie *et al.*, 1980). The inverse process, the use of distinctive magnetic events to date and correlate sedimentary horizons, is the science of magnetostratigraphy.

Sedimentary rocks are prey to the same ills as the sediments from which they originated: inclination and other errors in primary DRM; partial destruction of DRM by bioturbation and slumping; and remagnetization by CRM or VRM. Burial by later deposits and eventual uplift add further possibilities of overprinting by CRM and VRM. We shall elaborate on some of these problems in §15.5 and 16.3.

Thermal demagnetization is difficult for wet, unconsolidated lake and marine sediments, but it is the standard method of magnetically cleaning sedimentary rocks. The unblocking temperature spectrum so obtained, supplemented by Curie point data from thermomagnetic analysis, is the principal means of identifying the magnetic minerals that carry NRM. AF demagnetization is relatively ineffectual in cleaning sedimentary NRM because the frequently occurring minerals goethite (αFeOOH) and hematite (αFe$_2$O$_3$) have coercivities that in large part exceed 300 mT (3 Oe), the practical maximum field of AF demagnetizing equipment.

However, the coercivity spectrum can still be estimated, from the incremental IRM vs applied field curve (Fig. 15.14; cf. §11.3). Applied fields up to 2 T

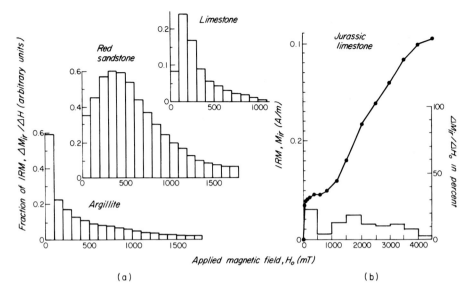

Figure 15.14 Coercivity spectra determined from the rate of acquisition of IRM M_{ir} with applied field H_0. (a) The dominant magnetic minerals in the limestone and red sandstone samples are detrital and pigmentary hematites. Magnetite with coercivities < 100 mT is important in the argillite. (b) IRM acquisition curve and corresponding differential spectrum for a Jurassic limestone containing magnetite, hematite, and goethite with roughly equal magnetic expressions. [After Dunlop (1972) and Heller (1978), with the permission of the authors and the publishers, Blackwell Science, Oxford, UK and Deutsche Geophysikalische Gesellschaft, Köln, Germany.]

(20 kOe) are routinely produced by means of electromagnets and fields of 5 or 6 T are obtainable with superconducting solenoids. In the examples of Fig. 15.14, coercivities < 100 mT (1 kOe) are mostly due to magnetite and maghemite, those between 100 and 300 mT are due to coarse detrital hematite, and those > 300 mT to authigenic hematite and goethite. There is, however, considerable overlap in the coercivity ranges of these phases. Rotational hysteresis (Brooks and O'Reilly, 1970) is an alternative method that distinguishes minerals by their differing anisotropy constants, but few laboratories are equipped for these measurements.

Identifications by either coercivity spectra or rotational hysteresis are normally checked by determining unblocking temperatures and Curie points of the various phases. Lowrie (1990) has devised a method in which three separate IRM's are produced along orthogonal axes, with coercivity limits corresponding to magnetite, hematite and goethite, respectively. Thermal demagnetization then yields the T_{UB} spectra of the three constituents separately. A practical difficulty is ensuring that the three IRM's are acquired at exactly 90° to one another, and that the measurement axes correspond exactly to the IRM axes. If this is not the case, the smaller signal of one constituent IRM may be contaminated or even masked by a component of the largest IRM.

Experimenters should be forewarned that irreversible chemical changes frequently occur during thermomagnetic analysis or thermal demagnetization (van Velzen and Zijderveld, 1992). The irreversible dehydration of goethite to hematite around 300 °C is distinctive (Hedley, 1968; Fig. 15.15). Clay minerals tend to partially break down between 600 and 700 °C, producing magnetite. In argillites, the new mineral may be produced in sufficient quantities to swamp the NRM (Dunlop, 1972). Reduction of hematite to magnetite can occur during vacuum heatings, although not usually at such low temperatures as in the example of Fig. 15.15.

15.3.5   Chemical remagnetization

A useful review of remagnetization of North American sedimentary rocks is given by Elmore and McCabe (1991). As well as chemical remagnetization, it

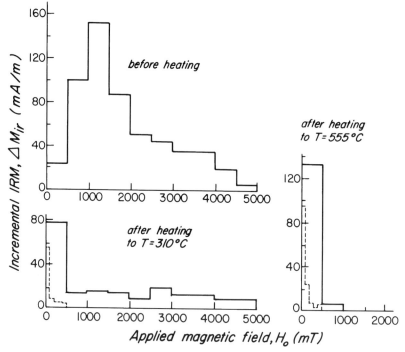

Figure 15.15   IRM coercivity spectra (dashed lines are fine-scale spectra for 100 mT intervals below 500 mT) for a limestone sample originally containing mainly goethite. Heating to 310 °C generated fine-grained hematite, by dehydration of the goethite, and some magnetite. Heating to 555 °C (in air) converted hematite to magnetite. [After Heller (1978), with the permission of the author and the publisher, Deutsche Geophysikalische Gesellschaft, Köln, Germany.]

covers thermoviscous overprinting (Chapters 10 and 16) and remagnetization related to stress and deformation (Chapter 16).

Remagnetization of carbonate rocks and red beds during the late Carboniferous (≈ 300 Ma ago) is widespread on both sides of the Atlantic, not only in the Appalachian and Hercynian orogenic belts but also inland in stable platform areas. Remagnetization may have been caused by basinal fluids driven onshore during the plate convergence (McCabe and Elmore, 1989). NRM overprints in undeformed, mildly heated platform carbonates were at first believed to be thermoviscous overprints (§10.6, 16.3) but were subsequently found to be low-temperature CRM's carried by authigenic or diagenetically altered magnetites (McCabe et al., 1989; Jackson, 1990). A direct connection between CRM acquisition and alteration by basinal fluids in and around mineralized veins has been established for the Viola limestone of Oklahoma by Elmore et al. (1993a).

Not all limestones are remagnetized. Magnetic properties are distinctively different for remagnetized and unremagnetized limestones (Fig. 11.24; Jackson et al., 1992, 1993a). The probable controlling factors are original magnetite grain size, governed by depositional environment (pelagic, shelf, reef, lagoonal: see Borradaile et al., 1993) and paleotemperature during uplift and diagenesis (Jackson et al., 1988b). Some of the authigenic magnetites observed in these rocks have exotic and beautiful forms, for example botryoidal or spheroidal polycrystalline aggregates 10–100 μm in size composed of 0.5–2 μm crystallites (McCabe et al., 1983). In some cases, the spherules formed by magnetite replacement of pyrite framboids (Suk et al., 1990). Some of the spherules have SD to PSD hysteresis properties but these are more similar to those of unremagnetized than remagnetized whole rocks (Suk and Halgedahl, 1996; cf. Fig. 11.24). Alternative candidates for the carriers of CRM are ultrafine nearly superparamagnetic magnetite grains (Jackson et al., 1993a; Channell and McCabe, 1994).

On a more local scale, an exciting possibility, with exploration potential, is concentration of authigenic magnetite (Elmore et al., 1987; Kilgore and Elmore, 1989; Elmore and Leach, 1990) or pyrrhotite (Reynolds et al., 1990) in clastic and carbonate rocks above or near hydrocarbon reservoirs. These rocks have been altered by migrating hydrocarbons and their associated fluids. The exact mechanism of magnetite or pyrrhotite authigenesis is often not clear, but the association of ferrimagnetic minerals (with accompanying aeromagnetic anomalies) and hydrocarbon diagenesis is well documented, and could be the basis of a relatively inexpensive exploration technique. For summary reviews, see Elmore and McCabe (1991) and Elmore et al. (1993b).

Ore genesis may also result in chemical remagnetization of adjacent host rocks. Prime examples are Mississippi Valley-type Pb–Zn deposits in largely undeformed platform carbonates (Pan and Symons, 1993; Symons et al., 1993). Magnetite inclusions in sphalerite and galena account for the remanence of the ores, while authigenic magnetic minerals, probably magnetite, in spheroidal or

other morphologies carry CRM in the altered host rocks. As in late Carboniferous (Alleghenian) remagnetization, the mechanism seems to be migration of
basinal brines from adjacent orogens, but the orogenies concerned range from
Acadian (Devonian, ≈ 370 Ma) to Laramide (late Cretaceous, ≈ 70 Ma). Paleomagnetic dating using CRM directions agrees with these inferred ages of ore
genesis.

15.4   Magnetic properties of hematite

Before discussing red sedimentary rocks, we shall review the properties of the
mineral hematite, to which these rocks owe their distinctive magnetic behaviour.
Antiferromagnetic titanohematite ($Fe_{2-y}Ti_yO_3$, $y \leq 0.5$) is a frequent constituent of rocks of all types, but only in oxidized sedimentary rocks does its weak
parasitic ferromagnetism dominate the NRM. The titanohematites in red beds
are close in composition to pure hematite (αFe_2O_3), whose distinctive magnetic
properties have been exhaustively studied (e.g., Flanders and Remeika, 1965;
Fuller, 1970; Dunlop, 1971; Jacobs et al., 1971; Bucur, 1978; Hejda et al., 1992).

15.4.1   Antiferromagnetism of hematite

The susceptibility and saturation magnetization of single crystals of hematite
(Fig. 15.16) indicate a Curie temperature of 675 °C (identical to the antiferromagnetic Néel point) and a pronounced discontinuity, the Morin transition, at
$T_{Morin} \approx -15$ °C (Liebermann and Banerjee, 1971). The same transition temperatures are observed in fine (0.2–0.7 μm) grains as well (Fig. 3.20; Nininger
and Schroeer, 1978), although grains smaller than ≈ 0.1 μm have no Morin transition (Bando et al., 1965). Small amounts of titanium likewise depress T_{Morin}
and eventually suppress the transition. Red beds usually exhibit a Morin transi-

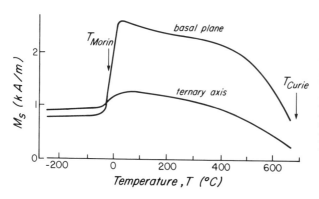

Figure 15.16 Saturation magnetization of a large single crystal of hematite measured parallel and per-pendicular to the ternary or c-axis. [After Néel and Pauthenet (1952), with the permission of the pub-lisher, l'Institut de France, Paris, France.]

tion just below room temperature, indicating that they contain nearly pure αFe_2O_3 in $> 0.1\,\mu m$ grains.

When $T < T_{\text{Morin}}$, cation moments parallel the crystallographic c-axis, as in ilmenite (Fig. 3.22b). Antiferromagnetic balance of the Fe^{3+} moments is, in principle, exact. Above the Morin transition, all spins lie in basal cation planes perpendicular to the c-axis (Fig. 3.22a). Compared to the strong c-axis coupling below T_{Morin}, anisotropy in the basal plane above T_{Morin} is weak, although still sufficient to impart coercivities $\approx 100\,mT$ (1 kOe) to SD grains (§15.3.4).

15.4.2   Parasitic ferromagnetism of hematite

Above T_{Morin}, even in the absence of an external field, there remains what amounts to a $\approx 1.5\,T$ (15 kOe) field, in the basal plane, perpendicular to the sublattice moments (§2.7). The moments are deflected a fraction of 1° from perfect antiparallelism, giving rise to weak ferromagnetism with $M_s \approx 2.5\,kA/m$ coupled perpendicular to the antiferromagnetism (Fig. 2.8). This phenomenon of **spin-canting** was first proposed and accounted for by Dzyaloshinsky (1958).

This **'parasitic' ferromagnetism** of hematite is 200 times weaker than the ferromagnetism of magnetite. If the remanences of SD hematite and MD magnetite are compared, the contrast is reduced to about a factor of 10. Hematite therefore does not usually dominate the uncleaned NRM. However, its NRM is eventually isolated by either AF or thermal cleaning, since hematite has higher coercivities than magnetite and most of its unblocking temperatures are above 580 °C, the Curie point of magnetite.

Hematite also assumes importance as an NRM carrier because its SD range is broad and encompasses many of the grain sizes typical of hematite in red beds. Figure 15.17 illustrates the transition from soft MD behaviour to SD hysteresis with $H_c > 0.1\,T$ (1000 Oe) that occurs between 240 and 15 µm. Although 'hard' ferromagnetism seems to disappear below about 1 µm in the samples illustrated, this is not typical. Submicron grains are usually very hard (e.g., Dunlop, 1971), and the superparamagnetic size is not reached until about 0.03 µm (Banerjee, 1971; Table 5.1).

In addition to the anisotropic (i.e., basal plane only) spin-canted ferromagnetism that vanishes below T_{Morin}, hematite possesses an underlying isotropic ferromagnetism that is unaffected by the Morin transition (Figs. 3.20, 3.21b, 15.16). The isotropic ferromagnetism is largest in large crystals (compare Figs. 3.20 and 15.16). It is for this reason that M increases steadily with grain size in Fig. 15.17. The isotropic component can be increased by neutron irradiation or reduced by annealing (Gallon, 1968). Since its origin seems to lie in chemical impurities (local magnetite-like regions) or in lattice defects, it is called the **defect moment**. Being stress and temperature sensitive, its remanence may be untrustworthy if the rock has been even mildly metamorphosed (see Chapter 16).

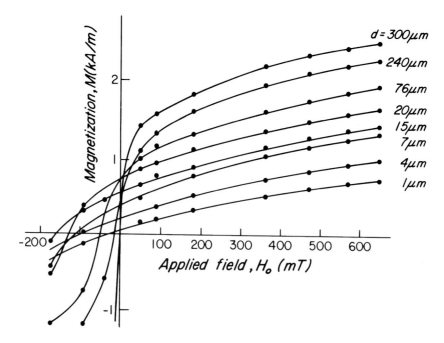

Figure 15.17   Magnetic hysteresis of synthetic hematites of various grain sizes. [After Chevallier and Mathieu (1943), with the permission of the publisher, Les Editions de Physique, Les Ulis, France.]

15.4.3   Low-temperature memory

The spin-canted moment itself behaves in an unexpected manner in cooling–heating cycles across the Morin transition. Below T_{Morin}, spin-canted remanence vanishes and only the isotropic defect remanence remains. But on reheating in zero field, a fraction or **memory** of the spin-canted remanence is recovered, its direction unchanged (Haigh, 1957; Fig. 3.21b). (The analogous phemonenon during low-temperature cycling of magnetite across the isotropic point or Verwey transition was described in §12.4.1.) It is not clear to what extent the defect moment is necessary to renucleate the spin-canted moment during reheating, but the memory of the spin-canted moment decreases when the defect moment is partially annealed out (Gallon, 1968). The two types of moment are therefore not entirely independent.

Cycling across the Morin transition is a chemically non-destructive alternative to thermomagnetic analysis for the detection of hematite (Fuller and Kobayashi, 1964). It is *not* recommended as a method of low-temperature cleaning because it enhances the defect moment at the expense of the probably more reliable intrinsic ferromagnetism. In any case, surface rocks at latitudes $> 40°$ or so, or at high elevation, have been repeatedly cycled through the transition in nature.

15.4.4   Coercivity of hematite

Coercivity due to shape anisotropy is proportional to M_s (eqns. (5.37), (8.8)) and is negligible in hematite. Coercive forces due to crystalline anisotropy and magnetostriction vary from ≈ 1 mT (10 Oe) in large MD grains to many hundreds of mT in SD hematite. The triaxial crystalline anisotropy in the basal plane is weak, certainly not more than $10\text{--}100\,\text{J/m}^3$ ($10^2\text{--}10^3\,\text{erg/cm}^3$), torque curves on large crystals indicating values more like $1\,\text{J/m}^3$ (Flanders and Schuele, 1964). The main source of the observed high coercivities in SD hematite must be magnetoelastic anisotropy arising from internal stress (Porath, 1968b). The magnetostriction constant of $\alpha\text{Fe}_2\text{O}_3$ is 8×10^{-6} (Table 3.1), so that an internal stress $\sigma = 100\,\text{MPa}$ ($10^9\,\text{dyne/cm}^2$ or 1 kb) produces a coercive force $\mu_0 H_c = 3\lambda\sigma/2M_s \approx 500\,\text{mT}$ (5 kOe).

Figure 15.18 illustrates how sensitive the properties of SD (in this case $0.5\text{--}2\,\mu\text{m}$) hematite are to annealing. There are two effects of annealing. First, the defect moment is annealed out and so M_s decreases. Second, internal stress is reduced and M_{rs}/M_s increases from near 0.5 (reflecting the predominance of *uniaxial* magnetoelastic anisotropy due to internal stress) to >0.7 (residual *triaxial* anisotropy, perhaps due to crystal twinning (Porath and Raleigh, 1967)). The residual spin-canted moment, pinned by a triaxial anisotropy, is extremely hard in these SD grains. It would be isolated by AF demagnetization, as well as by annealing. In large MD crystals, on the other hand, where the spin-

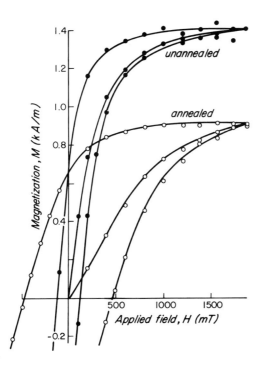

Figure 15.18   Effect of annealing on the hysteresis of $0.5\text{--}2\,\mu\text{m}$ synthetic hematite grains. [After Dunlop (1971), with the permission of the author and the publisher, Gauthier-Villars, Montrouge, France.]

canted moment has coercivities < 10 mT (100 Oe), it is the defect moment that is revealed by AF demagnetization (Smith and Fuller, 1967).

15.5   Red sedimentary rocks

In the early 1950's, when instruments capable of measuring the weak magnetizations of sedimentary rocks had just been developed, the first paleomagnetic studies of these rocks focused on fossil-bearing strata that were well-dated geologically. Their NRM's, however, usually proved to be incoherent or else directed close to the present field.

A systematic study of different sedimentary rock types showed coherent directions of NRM only in fine-grained red sedimentary rocks. This discovery led to a concentration of effort on red beds. A disadvantage of red beds is that they contain few or no fossils, being predominantly continental deposits laid down in high-energy depositional environments. They can be dated accurately only if interbedded with fossiliferous sediments.

Red beds owe their distinctive colour to fine-grained, chemically produced hematite pigment. Larger black crystalline hematite grains of detrital origin are found in both red and non-red sediments. Such detrital hematite grains, often referred to as specularite (Irving, 1957; Van Houten, 1968), are either derived from erosion of earlier red beds or volcanic terrains or form by oxidation (martization) of magnetite during erosion and transport (Steiner, 1983). Red beds typically have NRMs of 10^{-3}–10^{-1} A/m (10^{-6}–10^{-4} emu/cm^3), stronger than the typical DRM or PDRM of non-red sediments. As well as being relatively intense, the NRM of many red beds is extremely hard and stable.

15.5.1   CRM of hematite pigment

Hematite pigment in red beds may form by the oxidation of magnetite, the inversion of maghemite, or the dehydration of goethite (Fig. 15.15):

$$2Fe_3O_4 + O_2 \rightarrow 3\alpha Fe_2O_3$$
$$\gamma Fe_2O_3 \rightarrow \alpha Fe_2O_3$$
$$2\alpha FeOOH \rightarrow \alpha Fe_2O_3 + H_2O$$

The source of the goethite in the third reaction is the breakdown of hydrous clay minerals and ferromagnesian silicates (Van Houten, 1968; Özdemir and Deutsch, 1984). All the reactions are slow at room temperature but accelerate with mild heating, e.g., during burial in a sedimentary pile.

The lattices of parent and daughter phases are incompatible in all cases. We would therefore expect a growth CRM controlled by the ambient field to accompany each reaction. This seems to be so in the second and third reactions (Hedley,

1968; Özdemir and Dunlop, 1988), but is not always the case in the first reaction. When synthetic magnetites were oxidized to hematite, the CRM often appeared in a spurious direction intermediate between the applied field H_0 (= 50–200 µT) and a pre-existing ARM, M_{ar} (Heider and Dunlop, 1987; see §13.4.2 and Fig. 13.5). The CRM could be made parallel to either H_0 or M_{ar} by increasing one of these factors at the expense of the other.

Chevallier (1951) was the first to note that dehydration of goethite generally produces a hematite that lacks hysteresis. The dehydrated grains are polycrystalline and the crystallite size is apparently <0.03 µm. This phenomenon is probably the cause of superparamagnetic (Creer, 1961) or incoherently magnetized (Collinson, 1969) hematite in red beds, even in grains as large as 0.25 µm (Strangway et al., 1967).

Since all traces of goethite disappear before hematite crystallites grow to stable SD size, all memory of the NRM of goethite (which is even weaker than that of hematite (Banerjee, 1970; Hedley, 1971)) is lost. With continued growth of the crystallites, part at least of the pigment eventually acquires a true growth CRM (Hodych et al., 1985). The burning questions in red bed paleomagnetism are 'How much of the NRM is CRM carried by pigmentary hematite?' and 'How long after the rock formed did the pigment develop?' (see Purucker et al., 1980; Larson et al., 1982; Channell et al., 1982).

15.5.2   Grain growth CRM

Non-exchange-coupled or growth CRM was modelled in §13.2.2. In the case of hematite, it is reasonable to assume that the daughter phase does not grow beyond SD size. Then, by analogy with Néel's (1949) expression for single-domain TRM, the CRM intensity should be (Kobayashi, 1962b)

$$M_{cr} = M_{rs} \tanh\left(\frac{\mu_0 V_B M_{s0} H_0}{kT_0}\right). \tag{15.5}$$

Here V_B is the critical blocking (or superparamagnetic) volume at which CRM is frozen in. Growth is assumed to take place at room temperature T_0. If H_0 is small enough that the hyperbolic tangent can be replaced by its argument, the ratio of CRM/TRM intensities (cf. eqn. (8.21) for TRM) is

$$\frac{M_{cr}}{M_{tr}} = \frac{V_B M_{s0} T_B}{V M_{sB} T_0} = \frac{H_{KB}}{H_{K0}}, \tag{15.6}$$

substituting from eqn. (8.19) for T_B and from eqn. (13.1) for V_B.

Equation (15.6) should not be taken too literally, since eqn. (8.21) did not predict TRM intensity particularly well. It does seem clear that growth CRM in a given applied field will be of comparable intensity to the TRM the grains would acquire if cooled in the same field from high temperature after having grown to their terminal sizes. Coercivities will be identical: growth CRM, like TRM, tends to be hard. In fact, where SD hematite is the carrier, it is ultra-hard.

15.5.3   Hard VRM in red beds

VRM is normally a soft remanence, removed by AF cleaning in at most a few tens of mT. Avchyan and Faustov (1966) and Biquand and Prévot (1970) independently discovered that in some red beds, hematite's high coercivities may be imparted to VRM. Hardness of the VRM increases with longer exposure to the field (Fig. 15.19) and with decreasing grain size (Dunlop and Stirling, 1977). In fact, hard VRM is carried by grains that because of their small volumes, V, are affected by thermal fluctuations and acquire VRM but which resist AF demagnetization because their intrinsic microcoercivities, H_K, are high (Figs. 10.8, 11.1).

In more quantitative terms, when exposed to a weak field for a long time t, the last grains to acquire VRM are those whose product VH_{K0} just satisfies the blocking condition (8.33)

$$VH_{K0} = \frac{2kT_0}{\mu_0 M_{s0}} \ln\left(\frac{2t}{\tau_0}\right).   \tag{15.7}$$

These are also the grains whose VRM is most resistant to AF demagnetization. Their effective AF coercivity, from eqn. (8.24), is

$$H_c = H_{K0}\left\{1 - \left(\frac{2kT_0}{\mu_0 M_{s0} VH_{K0}}\right)^{1/2}\left[\ln\left(\frac{t_{af}}{\tau_0}\right)\right]^{1/2}\right\}$$

$$= H_{K0}\left\{1 - \left[\frac{\ln(t_{af}/\tau_0)}{\ln(2t/\tau_0)}\right]^{1/2}\right\}   \tag{15.8}$$

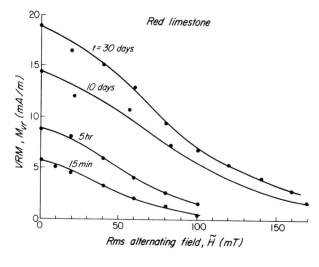

Figure 15.19   AF demagnetization of 'hard' VRM acquired by a hematite-bearing limestone after various exposures to a 500 µT field. [After Biquand and Prévot (1971), with the permission of the authors and the publisher, Deutsche Geophysikalische Gesellschaft, Köln, Germany.]

substituting from eqn. (15.7). Here $t_{af} = 1/f$ is a short time characteristic of AF demagnetization at a frequency f.

Equation (15.8) shows that H_{K0} is the sole material property that determines the hardness of a VRM. The AF coercivities of VRM's acquired under identical conditions of time and field by SD hematite and magnetite will be in the ratio of their respective microscopic coercivities: $H_{K0} \approx 500$ mT (5 kOe) for SD hematite compared to ≈ 50 mT (500 Oe) for SD magnetite, or about 10:1.

The increased hardness for longer exposure times evident in Fig. 15.19 follows directly from eqn. (15.8). The grain-size dependence is less obvious, but follows from eqn. (15.7). If H_{K0} is large enough to render VRM hard, V must be very small to satisfy eqn. (15.7). Figure 15.20 shows that for a nominal microscopic coercivity H_{K0} of 1 T (10 kOe), only 0.06–0.075 μm grains can acquire VRM over the entire time span 1 hr to 1 Ma. The AF coercivities of these VRM's range from $H_c = 100$–350 mT (1.0–3.5 kOe).

The important point to remember is that the hardness of VRM is proportional to the intrinsic coercivity of the mineral concerned. Any NRM component directed near the present earth's field, no matter how hard, should be suspected of being a VRM if hematite is an important constituent of the rock.

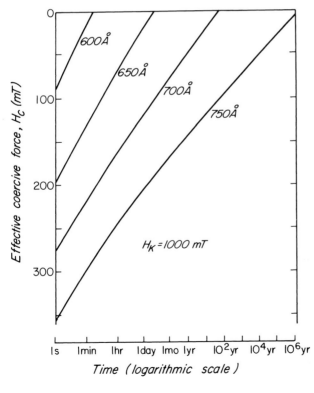

Figure 15.20 Theoretical time dependence of AF coercivity for SD hematite of various grain sizes. If $H_{K0} = 1000$ mT, grains with short-term coercivities as high as 300 mT have $H_c \leq 0$ on a 1-Ma time scale and can acquire 'hard' VRM. [After Dunlop and Stirling (1977) © American Geophysical Union, with the permission of the authors and the publisher.]

15.5.4   The best method for cleaning red beds

Since AF demagnetization is standard practice in paleomagnetism, it is worth considering what this method can and cannot achieve with hematite-bearing rocks. If the equipment is available, AF cleaning to $\approx 300\,mT$ (3 kOe) will demagnetize all NRM carried by magnetite and maghemite, much of the NRM of detrital hematite, and practically all VRM, hard VRM included, carried by fine pigment grains (cf. Fig. 15.20). It has the disadvantage that it enhances CRM at the expense of the primary, usually more interesting, DRM or PDRM.

Thermal demagnetization is the usual method of cleaning sedimentary rocks (Irving *et al.*, 1961; Chamalaun and Creer, 1964; Irving and Opdyke, 1965). It efficiently removes hard VRM (Fig. 15.21). It cannot resolve NRM's of specularite and pigment on the basis of their Curie temperatures, which are identical, but pigment, being finer-grained, has the lower unblocking temperatures (eqn. (8.19)). However, since the M_s–T curve of hematite is very 'blocky' (Fig. 3.20), most of the decrease in M_s occurring in a 50 °C range below T_C, there is a strong grouping of unblocking temperatures in this range for grains of all sizes (Fig. 15.22b). Very small temperature steps (5–10 °C) may be required to resolve hematites of detrital and chemical origins.

Some red beds alter when heated. Goethite and maghemite that have not changed to hematite in nature do so in laboratory heating around 300 °C, acquiring a spurious CRM if any stray field is present. The primary hematite is, of course, *chemically* stable during in-air heating, but may alter *physically*, e.g., by further growth of SP pigment crystallites to SD size. More serious than either of these changes is the spontaneous generation of magnetite described in §15.3.4.

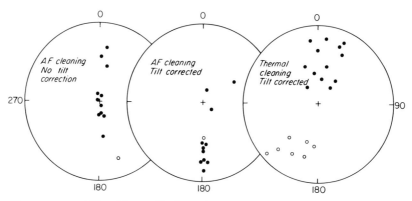

Figure 15.21   AF and thermally cleaned remanence directions in folded red marls from southern France. Tilt-corrected results show that thermal cleaning has successfully isolated a bipolar pre-folding remanence. AF cleaning fails to remove a secondary VRM, which before tilt correction is steep and positive like the present-day field in the area. [Reprinted from Ouliac (1976), with kind permission of Elsevier Science – NL, Sara Burgerhartstraat 25, 1055 KV Amsterdam, The Netherlands.]

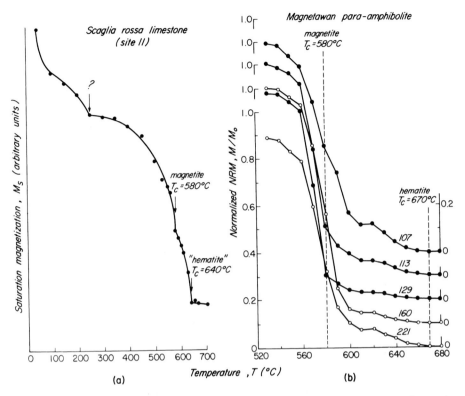

Figure 15.22 (a) Thermomagnetic curve of a late Cretaceous-early Tertiary red limestone showing major magnetic contributions from hematite, magnetite and a phase with a critical temperature around 150 °C (goethite dehydration?). (b) In a late Proterozoic metasedimentary rock from Ontario, thermal demagnetization in 10 °C steps was required to resolve the blocking temperatures of hematite, which are concentrated in an ≈ 50 °C interval below the Curie point. [After Lowrie and Alvarez (1975) © American Geophysical Union and Buchan *et al.* (1977), with the permission of the authors and the publishers, The American Geophysical Union, Washington, DC and Journal of Geomagnetism and Geoelectricity, Tokyo, Japan.]

Gose and Helsley (1972) found that CRM and/or partial TRM of this magnetite obscured the NRM unless residual fields during heating and cooling were < 10 nT. Field cancellation to this level requires a carefully shielded environment.

Roy and Park (1972) showed that if suitably porous rocks are chemically leached with HCl, NRM components not seen after thermal demagnetization were revealed by the leaching. A comparison of results of the two methods is given in Fig. 15.23.

Collinson (1965b) originally developed the leaching technique as a means of estimating the relative importance of pigment and specularite, and by inference, of CRM and DRM, in the NRM of red sandstones. Leaching first attacks the finest pigment particles. Only after most of the pigmentation has disappeared are

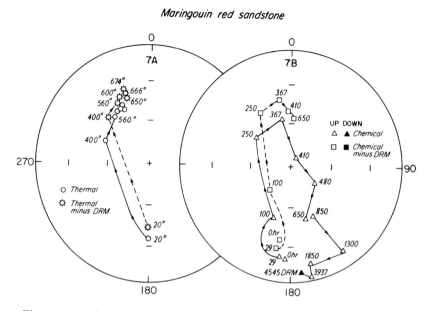

Figure 15.23   Comparison of thermal cleaning (left) and chemical leaching (right) of the NRM of a fine-grained Carboniferous red sandstone. Leaching revealed an underlying DRM that was undetected in thermal cleaning. [After Roy and Park (1974), with the permission of the authors and the publisher, The National Research Council of Canada, Ottawa, Canada.]

detrital grains affected to any great extent. Collinson (1966) concluded that rather unstable pigment CRM is predominant in the NRM, but that DRM of detrital hematite particles is also present and is the most stable part of the NRM (Fig. 15.24). This conclusion would have to be modified in the event that authigenic grains carrying CRM have grown to sizes approaching those of detrital grains, or if the detrital grains were in part oxidized after deposition and do not carry pure DRM (Steiner, 1983).

The disadvantages of leaching are the long times required and the uncertainty as to what range of grain sizes is attacked by the acid at any stage. Different rocks leach at different rates, depending on their porosity and grain size, and apart from colour, there are no quantitative guidelines like coercive force and blocking temperature to monitor the process. Given these limitations, thermal demagnetization is the recommended standard method for cleaning red beds.

15.5.5   The reliability of red bed results

There are good reasons for anticipating dispersion of the primary DRM, overprinting by secondary CRM, and prominent viscous or other noise components in the NRM of red beds. Hematite is notably stress-sensitive. The defect moment is directly affected through mobility of dislocations in the lattice, and the spin-

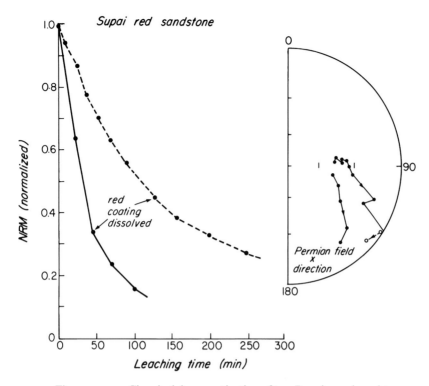

Figure 15.24 Chemical demagnetization of two Permian red sandstones, revealing primary DRM underlying later CRM. [After Collinson (1965b), with the permission of the author and the publisher, Blackwell Science, Oxford, UK.]

canted moment is indirectly affected, since much of the basal plane anisotropy that pins its remanence is magnetoelastic. Temperature has an equally important effect on both lattice and chemical defect moments, as well as increasing the intensity and coercivity of hard VRM (§16.3.4).

Is defect ferromagnetism more common in the pigment or the detrital fractions? Residual Fe_3O_4 and γFe_2O_3 are more likely to persist in large grains generally and in detrital hematite in particular. But lattice defect moments probably are more important in small grains. All theories relating coercivity of MD grains to wall-pinning by lattice defects predict larger effects when the grains are small. The defect moment seems an inescapable component of the remanence in hematites of all grain sizes.

Whatever the relative importance of spin-canted and defect moments, pigmentary hematite with its ultrafine crystallite size is a potential source of incoherent magnetization if the crystallites remain essentially superparamagnetic in size ($\approx 0.03\,\mu$m); of hard VRM if the crystallites grow to 0.05–0.1 μm; and of stable CRM if larger crystallites are produced. The *timing* of grain growth is crucial. Growth is promoted by mild heating and by circulation of oxidizing ground water solutions. Should the grain growth occur long after deposition, a strong and stable secondary NRM is inescapable. If the thermal or chemical event(s) that led to growth can be dated (Channell *et al.*, 1982), the CRM overprint is perfectly usable, although it is unlikely to retain the fine time resolution of the primary DRM (Channell, 1978; Larson *et al.*, 1982).

There is some experimental support for these speculations. Red beds from tectonically disturbed locales where beds have been buried to considerable depths do generally possess a prominent secondary NRM (e.g., Chamalaun, 1964; Irving and Opdyke, 1965; Kent, 1985; Kent and Opdyke, 1985). On the other hand, red bed sequences from stable continental areas, the Colorado plateau being the classic example, sometimes yield apparently faithful records of polarity transitions (Helsley and Steiner, 1969) and even of paleosecular variation (Baag and Helsley, 1974). In these formations, hematite must have formed very rapidly, probably soon after deposition. Indeed Elston and Purucker (1979) present evidence for DRM dating from initial deposition of specularite.

In apparently contradictory results from the same general area, Larson and Walker (1982) report NRM variations and reversals (presumably due to chemical remagnetization) on a fine scale, even within single specimens. Obviously generalizations about red beds are not possible. The magnetization history of each individual red bed sequence must be studied as exhaustively as possible, and the paleomagnetic results judged accordingly.

Chapter 16

Magnetism of metamorphic rocks

In Chapters 14 and 15, we considered several ways in which rocks are altered. Basalts in the upper 1–2 km of the oceanic crust were penetrated by seawater and underwent low-temperature oxidation. Gabbros and peridotites intruded at depth were subject to autometamorphism at moderate hydrostatic pressures. Subaerial lavas were deuterically oxidized, and sediments could be altered during or after diagenesis. All these processes occurred, geologically speaking, soon after the rocks formed.

In the present chapter, we will consider the effect on magnetism of metamorphic processes occuring tens, hundreds or even thousands of millions of years after formation. Metamorphic changes in the NRM and in the magnetic minerals themselves are dictated by four factors: temperature, determined by depth, geothermal gradient and tectonic setting; stress, both hydrostatic and directed, and the deformational response of the rock to stress; the presence of hydrothermal fluids; and time. We shall examine how primary NRM is partially or completely 'overprinted' by metamorphic processes in which these factors operate separately or in combination.

16.1 Metamorphic regimes

16.1.1 Contact metamorphism

Chemical and/or thermal remagnetization of the country rock is usually total at the contact with an intrusion but inappreciable at distances comparable to the dimensions of the intrusion. Figure 16.1 illustrates progressive resetting of

461

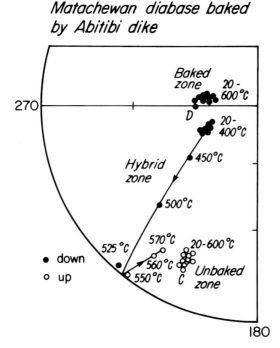

Figure 16.1 Thermal overprinting of NRM in Matachewan diabase country rock within and beyond the baked contact zone with an intruding Abitibi diabase dike. [Unpublished data courtesy of L. D. Schutts.]

primary NRM of 2450-Ma-old Matachewan diabase in the zone reheated by an intruding 2150-Ma-old Abitibi dike. At and near the dike contact, the Matachewan NRM direction has been reset parallel to the NRM of the dike. At large distances, the primary NRM is unaffected. At intermediate distances, in the so-called 'hybrid zone' (Schwarz, 1977; Schwarz and Buchan, 1989), overprinting is partial and the younger NRM is superimposed on the older. The two NRM's can be resolved by thermal cleaning because the blocking temperatures of the domains that have been remagnetized are lower than those of domains whose magnetizations have remained pinned. As a further illustration, subtracted vectors at each step of thermal cleaning of a Siluro-Devonian lava reheated by a Tertiary dike (Fig. 16.2) cluster around the dike NRM direction at low temperatures but approach the primary NRM direction of the lava at high temperatures.

A baked-contact test compares the direction of NRM in an intrusion with NRM directions in the country rock both near and far from the contact. The result is positive if the intrusion and contact rock have a common NRM direction, which is significantly different from that of the distant country rock. The examples shown in Figs. 16.1 and 16.2 are positive contact tests, demonstrating in each case a significant difference between the ages of magnetization of the country rocks and the intruding dikes. The NRM's of dike and country rock have not been mutually overprinted by a subsequent metamorphic event, as frequently occurs in regionally metamorphosed terrains.

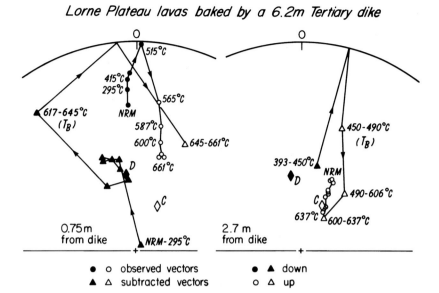

Figure 16.2   Thermal demagnetization of Lorne Plateau lavas of Siluro-Devonian age sampled at two different distances away from an intruding Tertiary dike. C, D are country rock and dike directions, respectively. [After McClelland Brown (1981), with the permission of the author and the publisher, The Geological Society of America, Boulder, CO.]

The baked contact test can be used even with quite young rocks where the time between formation and intrusion is too brief for significant apparent polar wander to have taken place, provided the earth's field has reversed polarity in the interval. A positive contact test of this kind – intrusion and contact rock having opposite NRM polarity to that of the distant country rock – demonstrates the reality of field reversal. The intruding rock could possess self-reversed TRM, but it is very improbable that the contact rock does too, particularly since the country rock (if it is igneous) did not exhibit self-reversing behaviour in its original cooling.

16.1.2   Hydrothermal alteration during low-grade metamorphism

An abundance of water (groundwater or seawater), pressures below 500 Mpa (5 kb) and temperatures below 250–300 °C, characterize the environment of low-grade hydrothermal metamorphism, e.g., in thick submarine or subaerial lava piles or deep sedimentary basins. The alteration may be local, e.g., surrounding a single volcanic source or in the circulation system above a single intrusion, or regional. The main secondary mineral zones that delineate increasing hydrothermal metamorphism are given by Ade-Hall et al. (1971).

The temperatures achieved in low-grade metamorphism are sufficient to unblock low-Curie-point titanomagnetites, but as we saw in §14.2.4 and will expand upon in the present chapter, the titanomagnetites themselves tend to oxidize under these conditions. The NRM then may become a pseudomorph of the primary NRM.

16.1.3   Medium-grade and high-grade regional metamorphism

Typical of hydrothermal alteration are zeolite, prehnite and pumpellyite (low temperature, low pressure) grades of metamorphism (Fig. 16.3). At temperatures of more than 250–300 °C, implying burial (generally of regional extent) to >5–10 km, hydrothermal solutions are no longer an important agent of alteration. This is not to say that magnetic minerals undergo no further changes, but under high greenschist-grade or low amphibolite-grade conditions, purely thermal overprinting of the NRM of equilibrium high-Curie-point minerals like magnetite and hematite begins to predominate. In high amphibolite and granulite facies metamorphism, thermal overprinting is paramount.

16.1.4   Stress and deformation

The response of rocks and their NRM's to stress is extremely varied. At this point, we shall single out folding because of its importance in the paleomagnetic fold test. Folding requires either brittle fracture or plastic flow at moderate temperatures and relieves compressive tectonic stress. The fold test depends for its

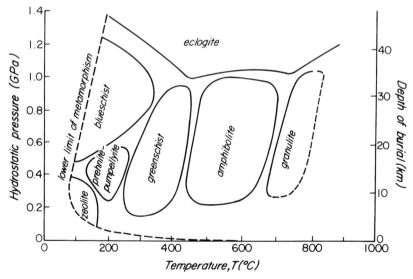

Figure 16.3   Approximate pressure–temperature limits for medium- and high-grade metamorphism.

success on the rather surprising ability of pre-folding NRM to survive (in direction anyway) the temperatures, stresses and strains experienced before and during folding.

The fold test as usually applied is positive if the initially scattered NRM vectors of a folded stratum tend to converge when each vector is 'unfolded', i.e., corrected for the local angle of bedding (Fig. 16.4). A positive fold test demonstrates that the NRM predated folding, and has not been overprinted by stress and heating during folding nor by subsequent CRM or VRM. Partial AF or thermal cleaning may reveal coexisting pre-folding and post-folding components of NRM.

A practical difficulty in applying the test (apart from finding good exposures of both limbs of a fold) is the lack of a three-dimensional view of the fold and consequent uncertainty as to the plunge and fold correction angles (MacDonald, 1980). There is frequently similar uncertainty in correcting for the regional tilt of non-folded formations. In fact, since it is difficult to test whether NRM predates or postdates tilting, it is often unclear whether a tilt correction should be applied at all to any but the most lightly metamophosed strata.

A serious objection to the usual version of the fold test is the naive assumption of rigid-body rotation during folding. The development of folds involves large internal strains. Successful 'unstraining' of NRM vectors in penetratively deformed rocks (whether in folds or not) requires detailed knowledge of both the mechanism of strain and the response of magnetic grains and their individual remanence vectors to the strain (for reviews, see Cogné and Perroud, 1987 and

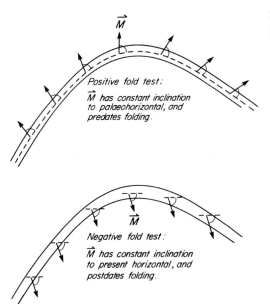

$\vec{M}$

Positive fold test:
$\vec{M}$ has constant inclination
to palaeohorizontal, and
predates folding.

Negative fold test:
$\vec{M}$ has constant inclination
to present horizontal, and
postdates folding.

Figure 16.4 Sketch of the paleomagnetic fold test.

Borradaile, 1993a). For example, flexural shear produces rotations of pre-fold-ing NRM vectors which are less than the apparent bed rotations (Kodama, 1988). The fold test would misidentify these NRM vectors as synkinematic (i.e., acquired when the beds were partly folded), a common interpretation of fold tests in Appalachian rocks. Stamatakos and Kodama (1991a,b) have reported examples in which the primary NRM appears synkinematic because of flexural slip or flow, while the secondary NRM is truly synfolding. Folds can also form without any flexure, by simple lateral shortening of beds. In this regime, all grains rotate by the same amount. NRM vectors remain parallel (although rotated), leading to a false negative fold test.

If the strain can be measured quantitatively, for example from the strain of reduction spots in red beds (Kligfield *et al.*, 1983; Cogné and Perroud, 1985; Cogné, 1988a), and is not the product of successive non-coaxial strains and rotations (Borradaile, 1993a), the rock matrix can be 'destrained'. However oxide and sulphide minerals are expected to deform very little at low tempera-tures compared to the matrix. Thus the usual model of rotation of remanence vectors as passive lines in the rock (Kligfield *et al.*, 1983; Cogné and Perroud, 1985; Hirt *et al.*, 1986; Lowrie *et al.*, 1986) is less realistic than active rotation of the long axes (to which NRM vectors are assumed to be pinned by shape anisotropy) of rigid grains in a passive matrix (Kodama, 1988; Kodama and Goldstein, 1991; Stamatakos and Kodama, 1991b). However, theoretically (Cogné and Gapais, 1986; Cogné *et al.*, 1986; Borradaile, 1993b) and experi-mentally (Cogné, 1987a; Borradaile and Mothersill, 1991; Borradaile, 1991, 1992a, Jackson *et al.*, 1993b), passive line and rigid marker models are almost equally good predictors of NRM rotations for homogeneous strains up to 40% (see Borradaile, 1993a).

Not considered in any of the 'destraining' models are rotations of NRM vec-tors relative to the magnetic grains themselves (i.e., remagnetization) as a result of stress. Piezomagnetization, the reversible and irreversible responses of SD moments and of domain walls in MD grains to stress, will be treated in §16.4.

16.2   Chemical remagnetization during metamorphism

16.2.1   Low-grade metamorphism of basaltic rocks

One conspicuous magnetic mineralogical change in low-temperature hydrother-mal environments is the formation and eventual inversion of titanomaghemite. In §14.2.4, we studied maghemitization of homogeneous titanomagnetites in sub-marine basalts of low deuteric oxidation state. Figure 16.5 shows that the low-temperature alteration of deuterically oxidized subaerial basalts gives rise to some familiar and some novel features.

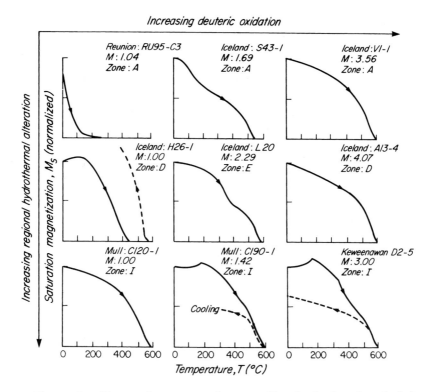

Figure 16.5   Vacuum thermomagnetic curves of basalts that have been both deuterically and hydrothermally altered. [After Ade-Hall *et al.* (1971), with the permission of the authors and the publisher, Blackwell Science, Oxford, UK.]

The top three curves for unmaghemitized lavas merely reproduce the low-,
double- and high-Curie point vacuum thermomagnetic curves of class-I, -II and
-III basalts (cf. Fig. 14.13). The three left-hand curves are also familiar. They
characterize, respectively, homogeneous titanomagnetite with $x \approx 0.6$, titanomaghemite with the same x value (cf. Fig. 14.4), and magnetite, the principal
magnetic inversion product of titanomaghemite of this composition. CRM
accompanying both these changes tends to reproduce the original NRM in direction, although not in intensity (§13.3.1 and 14.2.7).

A class-II titanomagnetite is little affected magnetically by middle-zeolite-
grade metamorphism, except that the lower of the two Curie points rises, but
under prehnite or epidote grade conditions, the thermomagnetic curve changes
markedly and becomes irreversible (Fig. 16.5, middle three curves, vertical).
The heating curve develops a cusp that is suppressed by sufficiently high fields,
reflecting increased coercivity in an existing mineral rather than a new mineral
of low Curie point. The cooling curve indicates a single magnetic phase less magnetic than the starting material. Class-III and higher basalts develop similar
characteristics (Fig. 16.5, lower right).

In an accompanying study of opaque minerals, Ade-Hall *et al.* (1971) were unable to identify the phase or phases responsible for this characteristic behaviour of highly altered, deuterically oxidized basalts. The irreversible decrease in saturation magnetization after heating suggests that maghemite was generated during hydrothermal alteration and then partly inverted to hematite during heating in vacuum. If this is so, we can expect a CRM that remembers the NRM direction if the maghemite is produced from SD or pseudo-SD magnetite or a field-controlled CRM if the grains are outside this size range (Johnson and Merrill, 1972, 1974). However, the principal Curie point indicated (Fig. 16.5, bottom middle and bottom right) is that of magnetite (580 °C), not maghemite (≈ 645 °C: §3.3).

Independent of deuteric oxidation state, Ade-Hall *et al.* (1971) describe three diagnostic effects of hydrothermal alteration on magnetic minerals. At and above 150 °C, finely-divided ferri-rutile produces a granulation texture in titanomagnetite. When temperatures approach 250 °C, sphene begins to replace ilmenite lamellae in titanomagnetite. Finally, around 300 °C, titanomagnetite is completely pseudomorphed by hematite. The likely magnetic consequences of these alterations are increased coercivity, as a result of granulation, and CRM accompanying hematite production.

Hall (1985) and Beske-Diehl and Li (1993) examined the effect of hydrothermal metamorphism at temperatures up to ≈ 300 °C on both deuteric hematite and titanomagnetite in subaerial basalts from the Iceland Research Drilling Project core. Titanium is lost by titanohematite between 1.8 and 2.4 km crustal depth, with an accompanying rise in unblocking temperatures from 610–650 °C to 650–690 °C. Neither magnetite nor hematite acquired a field-controlled CRM during Ti migration out of the lattice but in the epidote zone (≥ 250 °C approximately), magnetite began to acquire thermoviscous remanence (see §10.6, 16.3). Over most of the depth of the core, magnetite and hematite NRM directions were very similar and appeared to be inherited from primary TRM.

It is remarkable that metamorphism to almost greenschist facies, despite its profound mineralogical consequences, apparently has so little effect on the fidelity or resolution of the paleomagnetic record in Icelandic and Keweenawan (≈ 1100 Ma old) lavas (Ade-Hall *et al.*, 1971; Beske-Diehl and Li, 1993). If this conclusion is valid, it is intriguing to speculate at what level of metamorphism chemical changes are so profound that CRM loses all ability to track the original NRM.

Of course, the various mineralogical changes just described relate to specific subaerially erupted volcanic suites characterized by high oxidation state throughout. Submarine basaltic extrusive and intrusive rocks form under less oxidizing conditions, but because of the deep fissuring, high geothermal gradient and resulting large-scale convection cells near mid-ocean rifts, hydrothermal alteration may occur at great depths. Alteration of the opaque minerals and magnetic properties of basalts, diabases, gabbros and peridotites from hydrothermal fields in mid-ocean rift valleys have been studied by Wooldridge *et al.* (1990). They concluded that the degree of hydrothermal alteration increases with depth in the oceanic crust and upper mantle, although their samples were dredged and lacked stratigraphic control. Kikawa and Ozawa (1992), examining a stratigraphic succession of drilled gabbros from ODP hole 735B, found an inversion of the alteration gradient with depth: the upper gabbros were hydrothermally altered but the lower gabbros were fresh. Thus there appears to be a limiting depth for hydrothermal alteration of in-situ oceanic crust.

16.2.2   Moderate- and high-grade metamorphism

Magnetic mineralogical changes in rocks other than basalts, and under high temperature conditions generally, are poorly known. Rochette (1987) has demonstrated metamorphic control of the magnetic minerals in Alpine black shales. As the metamorphism increases from zeolite to amphibolite facies, susceptibility first decreases as magnetite breaks down, then recovers as pyrite transforms to pyrrhotite. Metamorphic isograds can be mapped, quickly and precisely, by magnetic measurements.

Volcanic rocks metamorphosed to greenschist facies ('greenstones') may be virtually non-magnetic (e.g., Fig. 14.11). Magnetite and hematite, the end products of deuteric and hydrothermal alteration, seem to be largely replaced by chlorite and sphene. On the other hand, diabases and gabbros metamorphosed under similar conditions may remain quite magnetic. There are several possible explanations. The coarser opaques in these rocks may resist chloritization, a new generation of magnetic minerals may have appeared, or the NRM may reside mainly in magnetic inclusions which are encased in silicate host minerals (§14.6.3; Geissman *et al.*, 1988) and therefore protected against later alteration.

In amphibolite-grade diorites, Buchan (1979) has shown that two distinct ancient NRM's are localized in different mineral fractions (Fig. 16.6). The younger NRM is carried by coarse hornblende–biotite aggregates containing infrequent titanomagnetite grains. The older NRM resides in the groundmass of feldspars and finer-grained (genetically distinct?) dark minerals, where both magnetite and hematite are also seen as discrete phases in thin section. Overprinting of NRM definitely has occurred in this rock, but we cannot say whether it is chemical overprinting or thermal overprinting of magnetic mineral fractions that differ in grain size and blocking temperature. If the NRM's are CRM overprints, it is tempting to associate them with magnetic inclusions in two generations of dark minerals, but there are many other potential NRM carriers.

Even at the highest grades of metamorphism, it is quite possible that primary magnetite and hematite inclusions in plagioclase and pyroxene (§14.4.1, 14.6.3) survive intact. The high temperatures ($>800\,°C$) at which these phases

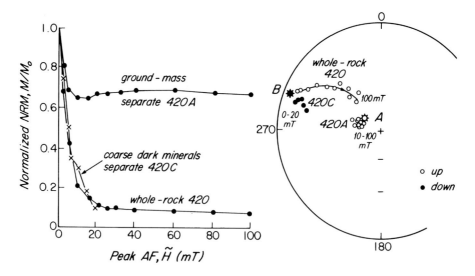

Figure 16.6 Microanalysis of NRM in a Precambrian diorite, using oriented microsamples. A and B are NRM components with approximate ages of 1000 and 800 Ma respectively. [After Buchan *et al.* (1977), with the permission of the authors and the publisher, Journal of Geomagnetism and Geoelectricity, Tokyo, Japan.]

equilibrate during initial cooling, combined with their chemical isolation within the host mineral, must serve to protect them from subsequent alteration at lower temperatures. If this is so, the primary NRM's of the magnetic silicates will be replaced by secondary TRM's rather than CRM's, and their high magnetic intensities will be preserved. Purely thermal remagnetization of this kind is the subject of the next section.

16.3   Thermoviscous remagnetization during metamorphism

Reheating during contact or burial metamorphism unpins the NRM of mineral grains at the lower end of the spectrum of unblocking temperatures, replacing it with a partial TRM or a viscous pTRM dating from the time of cooling of the intrusion or the region. If mineralogical changes continue throughout the heating, the unblocking temperatures themselves will change with changing Curie temperature, coercivity and effective grain size. But if the temperature is either low enough that the minerals do not alter appreciably (in addition oxygen and water must be lacking) or high enough that all reactions reach equilibrium, the extent of thermoviscous overprinting is readily predicted using Néel's (1949) SD theory (Chapters 8 and 10).

16.3.1   Metamorphic blocking temperature

At first glance, it seems self-evident that SD grains with blocking temperatures T_B less than the peak reheating temperature T_r will be overprinted by a new partial TRM, while those having $T_B > T_r$ will retain their preheating NRM. The difficulty is that the blocking temperature in a natural setting is lower than the blocking temperature we are able to measure in brief laboratory experiments, since T_B is time-dependent (eqn. (8.19), Fig. 8.10b). As we saw in Chapter 8, only order-of-magnitude differences in time t produce significant changes in T_B. Blocking temperatures in a cooling lava flow, for example, do not differ markedly from those measured on a time scale of hours in laboratory heating. But large bodies intruded at shallow depths may remain warm for hundreds or thousands of years, and deeply buried terrains may take millions or tens of millions of years to cool if erosion and isostatic uplift are very slow. In these situations, natural blocking temperatures may be as much as $200\,^\circ$C lower than laboratory blocking temperatures.

The simplest calculation follows directly from eqns. (8.18) or (10.2) (see §10.6.1 for a discussion). Consider an SD grain whose remanence is just unblocked when the grain is held at temperature T_r for a time t_r of geological length. It satisfies the blocking condition $\tau(T_r) \approx t_r$. This same grain, heated for a short time t in the laboratory has a different blocking temperature T_B (usually referred to as *the* blocking temperature), determined by the condition $\tau(T_B) \approx t$. Since $t \ll t_r$, $\tau(T_B) \ll \tau(T_r)$ and so $T_B > T_r$. In fact, T_B, t, T_r and t_r are related, for weak fields by

$$\frac{T_B \ln (2t/\tau_0)}{\beta_B \beta'_B} = \frac{T_r \ln(2t_r/\tau_0)}{\beta_r \beta'_r} = \frac{\mu_0 V M_{s0} H_{K0}}{2k}, \qquad (16.1)$$

which is equivalent to eqn. (10.13). In eqn. (16.1), $\beta \equiv M_s(T)/M_{s0}$, $\beta' \equiv H_K(T)/H_{K0}$, and subscripts B and r denote values of β or β' at T_B and T_r, respectively. All grains with T_B's less than the value given by (16.1) are thermally overprinted by the ambient field in time t_r at reheating temperature T_r.

That part of the partial TRM with (short-term) blocking temperatures between T_r and T_B can be viewed as a VRM produced at T_r in time t_r and added to the partial TRM that would be acquired in rapid cooling from T_r. The overprinting is therefore *thermoviscous* and the total remanence is often referred to as a viscous partial TRM or VpTRM (Chamalaun, 1964; Briden, 1965).

16.3.2   Thermoviscous overprinting viewed graphically

The only difficulty in applying eqn. (16.1) is that $\beta'(T)$, i.e., $H_K(T)$, is poorly known except for SD magnetite and hematite. Fortunately, these are the high-temperature equilibrium minerals commonly encountered in metamorphosed rocks, and the most stable fraction of NRM resides in SD (or PSD) grains of these minerals.

Figure 16.7 reproduces the graphical results of Pullaiah *et al.* (1975), based on eqn. (16.1). In the calculations, the $M_s(T)$ curves of Fig. 3.5(a) and 3.20 were

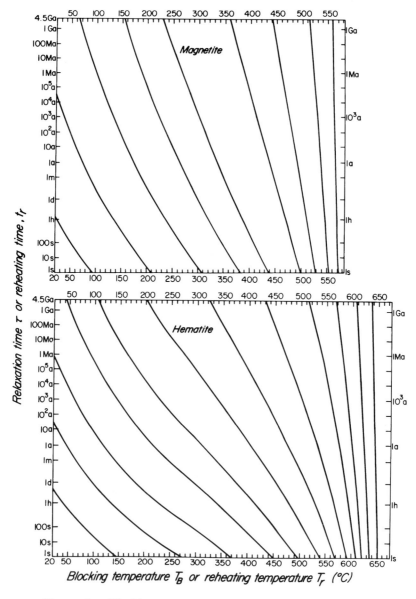

Figure 16.7 Blocking temperature as a function of time, for SD magnetite and hematite, contoured from eqn. (16.1) of the text. [Reprinted from Pullaiah *et al.* (1975), with kind permission of the authors and Elsevier Science – NL, Sara Burgerhartstraat 25, 1055 KV Amsterdam, The Netherlands.]

used for magnetite and hematite, respectively. For magnetite, $\beta' = \beta$ (i.e., $H_K(T) \propto M_s(T)$: shape anisotropy) was assumed, and for hematite, $\beta' = \beta^3$ (i.e., $H_K(T) \propto M_s^3(T)$; Flanders and Schuele, 1964) was assumed. The quantity contoured is $T_B \ln(2t/\tau_0)\beta_B\beta'_B$, which according to eqn. (16.1) is invariant for a particular grain ensemble (V, H_{K0}), indeed for all grain ensembles with the same value of VH_{K0}. Each contour traces the possible combinations of blocking temperature T_B and time t (or of T_r, and t_r, the temperature and duration of reheating for which overprinting is just possible) for grains lying on a common blocking curve $VH_{K0} = C$ (eqn. (8.33)) in a Néel diagram like Figs. 8.15 or 10.8.

It is immediately obvious that there are two general regions in the overprinting diagrams of Fig. 16.7. Within 50 °C or so of the Curie point of either hematite or magnetite, the contours are near-vertical and blocking temperature is insensitive to time. In this region, small changes in T result in large changes in β and β' (Figs. 3.5a, 3.20) that offset order-of-magnitude changes of t or t_r in eqn. (16.1). At lower temperatures, β and β' change more slowly and any change in $\ln t$ (or $\ln t_r$) must be balanced by a similar percentage change in T (or T_r). In this second region, the blocking temperature is ill-defined unless the time interval over which relaxation of the magnetization occurs is specified.

There is, of course, an intrinsic fuzziness in the contours in Fig. 16.7 at both low and high temperatures, since magnetization relaxes exponentially over a span of time that includes the relaxation time τ (eqns. (8.1) or (8.14)), not abruptly at τ (eqn. (10.8)). This fuzziness was illustrated in Fig. 8.10(b), where contours for 5% and 95% blocking and unblocking were plotted. Blocking is nevertheless relatively sharp, the blocking interval being typically 10–20 °C.

Overprinting diagrams like Fig. 16.7 answer two questions.

1. What fraction, if any, of the primary NRM can survive the thermal effect of a given metamorphic event?
2. Is a given secondary NRM likely to be a thermal overprint (as opposed to a chemical or other overprint) and as such amenable, for example, to paleointensity determination?

The blocking temperatures observed in a Thellier or other double-heating paleointensity determination do not need to be corrected for the effect of slow cooling, since NRM is demagnetized and laboratory partial TRM acquired under identical (T, t) conditions. However the equivalence does not extend to the *intensities* of NRM lost and partial TRM gained over the same temperature interval. Theoretically (Dodson and McClelland-Brown, 1980; Halgedahl *et al.*, 1980) and experimentally (Fox and Aitken, 1980; Fig. 10.14), the intensity of TRM increases with increased cooling time. Paleofield intensity estimates must be corrected accordingly.

16.3.3   Survival of primary NRM in thermal metamorphism

Table 16.1 gives an approximate scale of the 'survival potential' of primary NRM during thermal metamorphism, according to Fig. 16.7. For a nominal reheating time $t_r = 1$ Ma and a reheating temperature T_r corresponding to various grades of metamorphism (Fig. 16.3), only NRM with short-term blocking temperatures $\geq$ those tabulated can be primary. Laboratory blocking temperatures taken at face value may grossly overestimate resistance to thermal overprinting. Magnetite with $T_B \approx 400\,°C$ (for $t = 100$ s) is actually overprinted by prolonged heating at 250–300 °C. Hematite with $T_B \approx 450\,°C$ is remagnetized at only 150–200 °C. In fact, blocking temperatures below 350 °C in hematite are due entirely to VRM produced at T_0 by the present geomagnetic field over the past 0.78 Ma.

Blocking temperatures near the Curie point, in either magnetite or hematite, are time insensitive and approximate to the temperature the NRM could have survived in nature. Hematite grains with $T_B \geq 650\,°C$ or magnetite grains with $T_B \geq 565\,°C$ can, in the absence of chemical changes, preserve their remanence in the face of 550 °C (amphibolite-grade) metamorphism. Rocks metamorphosed above middle-amphibolite facies lose any primary NRM carried by magnetite but may preserve a fraction of the primary NRM carried by hematite. For this reason, hematite takes on a special importance in metamorphic rocks.

Table 16.1   An approximate scale of the survival potential of primary NRM during regional thermal metamorphism. [After Fig. 16.7 and Dunlop and Buchan (1977), with kind permission of the authors and Elsevier Science – NL, Sara Burgerhartstraat 25, 1055 KV Amsterdam, The Netherlands.]

Metamorphic facies	Approximate reheating temperature T_r (°C)[a]	Minimum blocking temperature of primary NRM, T_B (°C)[a]	
		Magnetite	Hematite
Zeolite	150–200	340	455
Prehnite/pumpellyite	250–300	415	510
Greenschist	400–450	500	560
Amphibolite	550	565	605
Granulite	700	No surviving primary NRM	

[a] Duration of T_r: 1 Ma; duration of T_B: 100 s

Although we argued earlier that high-temperature metamorphism inevitably transforms the primary magnetic minerals, except perhaps inclusions in 'magnetic silicates', the discussion of the last paragraph, confined as it is to purely thermal considerations, is not academic. A rock may experience two metamorphic events (this appears to have happened frequently in the Precambrian), the second more severe than the first and perhaps occuring much later. If the first metamorphism succeeds in producing an equilibrium mineral assemblage on which a CRM is imposed, Table 16.1 gives the survival potential of this CRM to the second, purely thermal, metamorphism.

Another possibility is that the CRM is not field-controlled, but a directional pseudomorph of the primary NRM. We will then observe two directionally distinct NRM components even if the two metamorphic episodes are not significantly separated in time, provided mineralogical changes are complete well below the eventual annealing temperature. Table 16.1 then estimates the survival potential of the pseudomorph primary NRM to thermal overprinting.

The minimum possible blocking temperatures of primary NRM in Table 16.1 are the maximum possible blocking temperatures of a secondary partial TRM for a given regional temperature T_r (or for a rock of given metamorphic grade, if T_r is imprecisely known). Observed T_B's significantly higher than those tabulated imply chemical or other non-thermal overprinting.

16.3.4 Prediction of thermoviscous effects

Figure 16.7 and Table 16.1 are useful for rocks from high-grade terrains, but not helpful with intermediate-grade (greenschist facies) rocks, where thermal and chemical effects have occurred concurrently. In low-grade rocks that have not altered appreciably, Fig. 16.7 again becomes useful as a predictor of the extent of thermoviscous overprinting. The contours of T_B as a function of t can just as well be viewed as contours of τ as a function of T (as was done in Fig. 8.10b). Figure 16.7 illustrates graphically the enhancement of VRM at higher-than-ambient temperatures discussed in §10.3. The enhancement is pronounced. In a single day's experiment at 150–200 °C, relaxation times up to 1 Ma are activated in both magnetite and hematite. The application to testing for viscous effects over the duration of a geomagnetic polarity epoch is obvious (see Dunlop and Hale (1977) for an example, including a set of t–T contours for TM60).

By the same token, Fig. 16.7 at once predicts the temperature to which any VRM must be heated to thermally demagnetize it. VRM's produced at higher temperatures are of course correspondingly more resistant to thermal cleaning. 'Hard' VRM acquired by hematite at room temperature (§15.5.3) is seen to be erased, whatever the coercivity of the grains, by heating to at most 360 °C, this figure corresponding to an exposure time of 1 Ma. The efficiency of thermal cleaning in erasing 'hard' VRM is obvious.

'Hard' VRM produced at above-ambient temperatures is undoubtedly a frequent contaminant of metamorphic rocks. It is more resistant to thermal cleaning than room-temperature VRM by an amount that is readily calculated from Fig. 16.7. It is likewise harder to AF demagnetize. Figure 16.8 illustrates the manner in which the blocking curve for AF demagnetization sweeps through the regions of the Néel diagram occupied by ordinary and higher-temperature 'hard' VRM's.

It is a straightforward matter to modify eqns. (10.12) or (15.8) for $T > T_0$. The values of H_c predicted for VRM acquired by hematite 100–200 °C above T_0 for the most part exceed the 100 mT (1 kOe) limit of the most commonly used commercial AF demagnetizers. For this reason, hematite-bearing rocks that have been mildly metamorphosed, e.g., red beds buried in a sedimentary pile, must be thermally demagnetized, a fact that has long been recognized in practice.

16.3.5   Experimental findings

Equations (8.18) and (8.19), from which eqn. (16.1) derives, have quite firm experimental backing (Everitt, 1961, 1962b; Dunlop and West, 1969; Dunlop, 1973b; see also Figs. 8.2, 8.3, 8.13b). Nonetheless, before drawing wide-ranging conclusions from Fig. 16.7 and Table 16.1, direct laboratory and field tests are highly desirable.

Field and laboratory tests involving the thermal demagnetization of low- and high-temperature VRM's were described in §10.6.1. In all cases where the VRM was carried by SD magnetite and hematite, the measured T_{UB}'s confirmed eqn. (16.1) (Fig. 10.10; Kent and Miller, 1987; Williams and Walton, 1988; Dunlop and Özdemir, 1993; Tyson Smith and Verosub, 1994). In some field studies of VRM and pTRM overprinting, even though it was not demonstrated that the rocks contained only SD grains, the theory was also verified. For example, unblocking temperatures of thermally overprinted secondary NRM in a deep drill core from the Michigan Basin (Fig. 16.9a) were related to temperatures measured in the borehole in the general way Fig. 16.7 predicts. Minimum blocking temperatures of surviving primary NRM at different distances from a Tertiary dike, when corrected for time according to Néel's SD theory, matched the expected profile of peak temperatures due to reheating by the dike (Fig. 16.9b).

Other experiments and field tests have given results that disagree with eqn. (16.1). The Matachewan diabase sample of Fig. 16.10 was exposed 10 days to the earth's field at 400 °C. According to Fig. 16.7, assuming magnetite as the carrier, the resulting VRM should have had laboratory unblocking temperatures no higher than 450 °C. In reality, more than 50% of the VRM remained at 450 °C and unblocking temperatures as high as 550 °C were observed. There are three possible explanations that would reconcile this result with SD theory:

(1) the rock may have altered during heating;

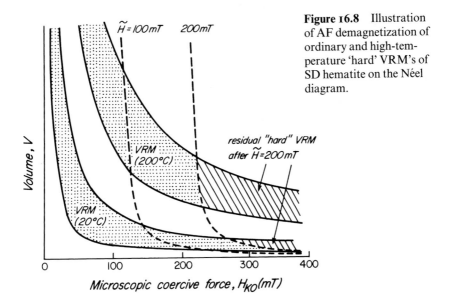

Figure 16.8 Illustration of AF demagnetization of ordinary and high-temperature 'hard' VRM's of SD hematite on the Néel diagram.

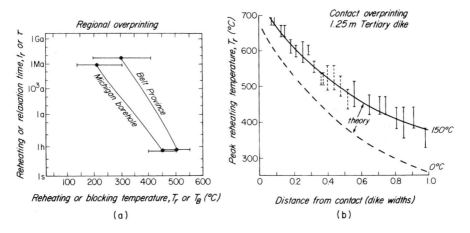

(a)

(b)

Figure 16.9 (a) Comparison of measured or inferred temperatures T_r of thermal remagnetization of red beds of Keweenawan (≈ 1100 Ma) and Beltian (≈ 1300 Ma) ages compared with laboratory blocking temperatures T_B of the thermal overprints predicted using the hematite curves of Fig. 16.7. [Reprinted from Van der Voo *et al.* (1978), with kind permission of the authors and Elsevier Science – NL, Sara Burgerhartstraat 25, 1055 KV Amsterdam, The Netherlands.] (b) Maximum reheating temperature required to demagnetize thermal overprints in the baked contact zone of a dike as a function of distance from the contact. The data match the expected profile of peak temperatures reached after dike intrusion if the country rock was at $\approx 150\,^{\circ}$C when the dike was intruded. [After McClelland Brown (1981), with the permission of the author and the publisher, The Geological Society of America, Boulder, CO.]

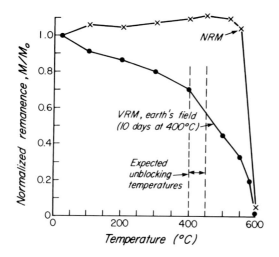

Figure 16.10 Thermal demagnetization of NRM and of VRM acquired in the earth's field during 10 days at 400 °C for a sample of Matachewan diabase. [Reprinted from Pullaiah *et al.* (1975), with kind permission of the authors and Elsevier Science – NL, Sara Burgerhartstraat 25, 1055 KV Amsterdam, The Netherlands.]

(2) hematite may be a carrier of VRM, originally or after alteration;

(3) the VRM may be carried in whole or part by MD magnetite.

MD grains carrying VRM or partial TRM give rise to pronounced 'tails' of high unblocking temperatures (Figs. 9.7b, 10.11b). These are suspected to be the cause of the 'anomalously' high unblocking temperatures observed by Kent (1985) for low- and high-temperature thermoviscous overprints of Appalachian limestones (Fig. 10.10, solid circles and triangles). In the case of similar overprints of the Milton monzonite (Middleton and Schmidt, 1982; Dunlop *et al.*, 1995a), MD carriers of remanence were demonstrably the cause of very high T_{UB}'s at two sites, while one site with only SD magnetite had lower T_{UB}'s that agreed with eqn. (16.1) (Fig. 10.10, open circle and triangles).

The fact that, when MD grains are the remanence carriers, T_{UB} during thermal cleaning can be $\gg T_B$ at which pTRM or VRM was acquired has more serious repercussions than the inability to determine T_r with certainty. A laboratory simulation of the overprinting of primary TRM by pTRM(400 °C, T_0) in large (135 μm) MD magnetite grains is illustrated in Fig. 16.11. The two remanences were produced along perpendicular directions so that the thermal demagnetization behaviour of each could be monitored separately. About 80% of the pTRM overprint demagnetizes below its maximum T_B of 400 °C, but the other 20% must be heated to T_C before it is completely erased (Fig. 16.11a). The simulated surviving 'primary TRM' behaves in a reciprocal manner: 85% is erased over its original T_B range above 400 °C, but 15% of its T_{UB}'s are low and overlap the T_B range of the overprint (see also Fig. 12.12).

A Zijderveld (1967) vector plot of the thermal demagnetization data (Fig. 16.11b) consists of two fairly linear sections, mimicking the ideal SD plot (dashed lines), but the break between the linear sections is at 450 °C, not 400 °C. T_r is

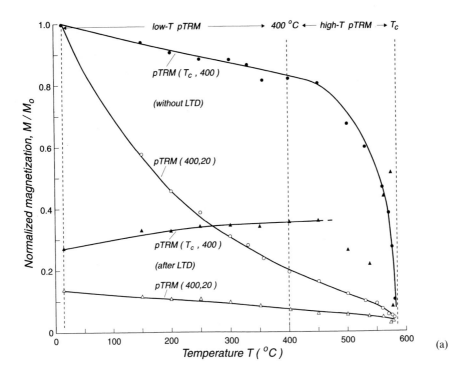

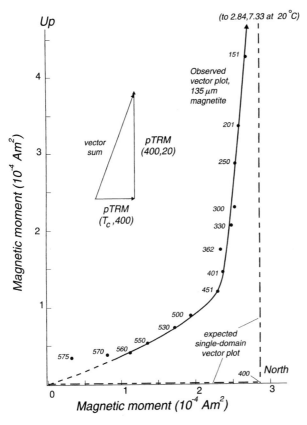

Figure 16.11 Thermal demagnetization of orthogonal partial TRM's carried by 135 μm magnetite grains, simulating overprinting of primary TRM by a later pTRM with a reheating temperature T_r of 400 °C. (a) Intensities of the pTRM's as a function of temperature. (b) Vector plot of the same pTRM data. [After Dunlop and Özdemir (1995) © American Geophysical Union, with the permission of the authors and the publisher.]

overestimated. More seriously, the slopes of the linear segments give incorrect directions for the pTRM overprint *and for the primary TRM*. The pTRM direction is 8° in error and the primary TRM direction is almost 20° in error. If MD grains are a major remanence carrier, paleomagnetists should be aware that thermal overprinting of primary TRM can produce such an overlap of T_{UB}'s that neither the overprint direction, nor the primary TRM direction, is trustworthy.

16.3.6   Blocking temperatures in continuous cooling

The calculation leading to eqn. (16.1) assumed constant temperature over extremely long times. In nature, a more frequent situation is very slow cooling at a rate dT/dt. Stacey and Banerjee (1974, p.97) and York (1978a) show that under these conditions, the blocking temperature is

$$T_{\mathrm{B}} = \frac{\Delta E}{k} \ln \left[\frac{\alpha k T_{\mathrm{B}}^2}{\tau_0 \Delta E(T_{\mathrm{B}})(-dT/dt)_{\mathrm{B}}} \right], \tag{16.2}$$

in which α is a number between 1 and about 4. ΔE is the principal energy barrier opposing magnetization relaxation. For SD grains, $\Delta E = \frac{1}{2}\mu_0 V H_{\mathrm{K}} M_{\mathrm{s}} = V K_{\mathrm{u}}$ (eqns. (8.12), (8.18)). For MD grains, $\Delta E = c\mu_0 V_{\mathrm{Bark}} H_{\mathrm{c}} M_{\mathrm{s}}$ (eqn. (9.20)). The temperature dependence of ΔE has been ignored in deriving eqn. (16.2) but it can be included in numerical calculations (York, 1978b; Dodson and McClelland-Brown, 1980).

Blocking temperatures from eqn. (16.2) or from numerical calculations, plotted as a function of cooling rate dT/dt, constitute an overprinting diagram. Alternatively one can utilize Fig. 16.7, using the following procedure. For a given constant cooling rate dT/dt and a chosen laboratory blocking temperature T_{B} (defined for a nominal time t of 60 s to 1 hr, say) calculate $\Delta E(T_{\mathrm{B}})$ (see the preceding paragraph) and substitute in

$$(t_{\mathrm{r}})_{\mathrm{eff}} = \frac{\alpha k T_{\mathrm{B}}^2}{\Delta E(T_{\mathrm{B}})(-dT/dt)}. \tag{16.3}$$

The 'characteristic cooling time' corresponding to dT/dt is $(t_{\mathrm{r}})_{\mathrm{eff}}$, which can be used to locate, on the contour passing through (T_{B}, t) in Fig. 16.7, the blocking temperature T_{r} during slow cooling at a rate dT/dt.

16.3.7   Relative ages of thermally overprinted NRM components

If one magnetization is thermally overprinted by another, the magnetization possessing the higher blocking temperature is clearly the older. The difficulty comes in proving that the overprint is purely thermal, i.e., that it is a partial TRM rather than CRM or TCRM. A hematite CRM of recent origin, for example, will possess higher blocking temperatures than a Precambrian NRM carried by magnetite.

One criterion for a thermal overprint is a maximum T_B compatible with Table 16.1. Two other criteria are helpful if two NRM components coexist.

1. The (un)blocking-temperature spectra of the two components must be mutually exclusive (McClelland Brown, 1982). Non-overlapping spectra cannot be demonstrated from directional plots alone (e.g., Fig. 16.1). It is necessary to show that the vector plot or Zijderveld diagram (Zijderveld, 1967, 1975; Dunlop, 1979) of the thermal demagnetization data consists of two straight-line segments (cf. Fig. 16.11b) or, equivalently, that the vector erased in successive temperature steps remains constant in direction, then changes abruptly as the high-temperature component is isolated (e.g., Fig. 16.2).

2. The unblocking temperature at which one component is eliminated and the other begins to be attacked must be the same for all samples from any site and must vary in a regular fashion (e.g., Fig. 16.9b), if at all, from site to site in a formation.

Figure 16.12(a) is an example of thermal overprinting in a Precambrian intrusive rock. The younger 'B' NRM direction is mainly associated with partial TRM carried by magnetite having blocking temperatures below 450°C. In some samples, a small 'B' NRM is carried also by hematite with much higher blocking temperatures: this part of the 'B' NRM must be a chemical overprint.

In Fig. 16.12(b), another formation, the Thanet gabbro, from the same area exhibits the same 'A' and 'B' NRM components, but here the 'B' NRM has no low blocking temperatures. Instead the 'B' blocking-temperature spectrum lies above and slightly overlaps the 'A' spectrum. The 'B' component in this rock must be a CRM, and no inference about the relative age of 'A' and 'B' NRMs can be drawn from the thermal results.

None of the criteria above serve to demonstrate that the NRM component with the higher blocking temperatures is of thermal origin. The only reliable criterion is a linear Thellier paleointensity plot over the appropriate blocking temperature range.

16.3.8   **Thermal overprinting of radiometric dates and absolute ages of NRM components**

Any component or components of NRM in a metamorphic rock that are demonstrably thermal overprints can ultimately be assigned absolute ages in the following manner. Radiometric clocks, like magnetizations, are reset by sufficiently severe heating because daughter (and parent) isotopes diffuse from one mineral to another. ^{40}Ar, the daughter product of ^{40}K, diffuses out of various minerals at different, but relatively low, temperatures. It is for this reason that K/Ar

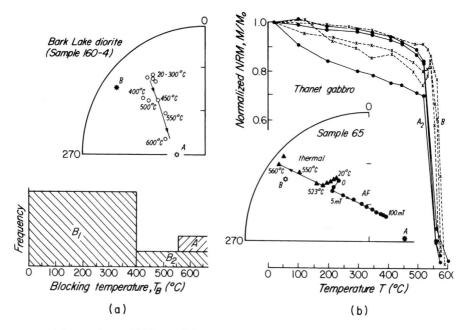

Figure 16.12   (a) Thermal demagnetization of rocks with multicomponent NRM's and a block model of the blocking temperature spectrum. B_1 and B_2 are thermal and chemical overprints respectively. (b) Thermal and AF demagnetization of multicomponent NRM in which a younger chemical overprint (B) has higher unblocking temperatures than component A. Note that A is reversed compared to Fig. 16.12(a). [After Buchan and Dunlop (1976) © American Geophysical Union and Buchan (1978), with the permission of the publishers, The American Geophysical Union, Washington, DC and The National Research Council of Canada, Ottawa, Canada.]

mineral ages, taken at face value, are notoriously poor age estimators for metamorphic rocks.

However, York (1978a) has demonstrated that the thermal activation equation for Ar diffusion is formally identical to the equations for SD and MD magnetization relaxation. It follows that there are well-defined 'blocking temperatures' for the diffusion of ^{40}Ar out of various minerals, i.e., for the resetting of K/Ar ages in these minerals. Field studies (Berger, 1975; Schutts et al., 1976; Buchan et al., 1977) confirm this expectation. Since argon diffusion depends on pressure (i.e., depth of burial), crystal size and shape, and the assemblage of neighbouring minerals, as well as on temperature, blocking temperatures for particular minerals are not hard and fast numbers, but must be determined for each rock (see McDougall and Harrison (1988) for details).

Isotope diffusion in the Rb/Sr and U/Pb systems is less well studied but it is clear that Rb/Sr whole-rock ages are reset at considerably higher temperatures than K/Ar ages. U/Pb ages of zircons and baddeleyites seem to be practically immune to metamorphism. They can be used to date any identifiable surviving primary NRM. U/Pb sphene or monazite ages, on the other hand, can be reset by metamorphism, again at higher temperatures than K/Ar ages.

Rb/Sr and K/Ar mineral ages, if the associated blocking temperatures are well determined, can be used to reconstruct the cooling curve for a metamorphosed region. The ^{40}Ar/^{39}Ar variant of K/Ar dating is particularly useful because it yields both an age and a blocking temperature for each mineral (Berger et al., 1979). The blocking temperature ranges of overprinted partial TRM's (corrected for slow cooling using Fig. 16.7) can then be translated into corresponding absolute ages for these NRM components (thermochronometry). The application of these principles is illustrated in Fig. 16.13.

A more direct method is to date K-bearing inclusions in magnetite crystals. This approach has been shown to be feasible using a high-sensitivity laser ^{40}Ar/^{39}Ar system by Özdemir and York (1990, 1992). In both studies, dates for the magnetite crystals (≈ 1150 Ma and 323 ± 5 Ma, respectively) were in reasonable agreement with independently determined metamorphic ages from micas.

Rochette et al. (1992a) have described an independent, non-radiometric method of paleomagnetic thermochronometry using pyrrhotite-bearing metamorphic rocks from the French Alps. These rocks cooled slowly enough during Alpine uplift that individual samples preserved a record of successive polarity reversals in their TRM. When the TRM was thermally demagnetized, the times of reversal (from the geomagnetic polarity time scale) versus (corrected) blocking temperature yielded an average cooling rate of $\approx 50\,^{\circ}C/Ma$. It is of course this very slow cooling that makes the method successful; more rapid cooling would not have permitted enough reversals to be recorded.

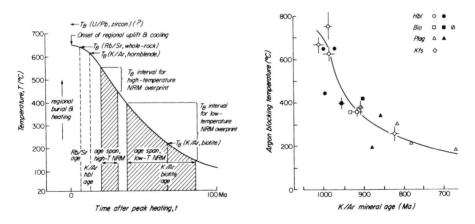

Figure 16.13 Hypothetical regional cooling curve, with blocking temperatures and ages determined by various radiometric systems and for thermally overprinted NRM's compared with an experimental cooling curve (for part of the Grenville orogen in Ontario) determined from ^{40}Ar/^{39}Ar mineral ages. [Experimental curve after Berger et al. (1979), with the permission of the authors. Reprinted with permission from Nature. Copyright © 1979 Macmillan Magazines Ltd.]

16.4   Stress, deformation and anisotropy effects

Deeply buried rocks experience, in addition to elevated temperatures, stresses up to 1 GPa (10 kb or 10^{10} dyne/cm^2). In the main these stresses are hydrostatic, but in tectonically active areas, the stress will have compressive and shear components as well.

16.4.1   Anisotropy of petrofabric and its effect on TRM

We discussed in §15.2.2 and 15.2.4 the magnetic anisotropy introduced by the tendency for magnetite, titanomagnetite or hematite grains to settle with their long axes in the bedding plane of sediments and the additional anisotropy produced by compaction of the sediment. As a result of post-depositional reorientation, the anisotropic alignment of magnetic grains may be less than that of non-magnetic crystals. In this case, distinct fabrics of the matrix and oxides may be measurable using the anisotropies of paramagnetic susceptibility and of ferrimagnetic susceptibility or remanence, respectively (Borradaile, 1988; Jackson et al., 1989). Composite magnetic fabrics may be due to distinct sedimentary and tectonic fabrics (Housen et al., 1993a) or to different magnetic minerals in different domain states (Rochette et al., 1992b).

In metamorphic rocks, anisotropy may be pronounced. Lineation, foliation, schistosity and gneissicity are the aftermath of plastic deformation or of recrystallization of minerals under directed stresses. The quantitative relation between strain and degree of anisotropy is not necessarily simple (Rochette et al., 1992b). For example, Cambrian slates with large finite strains (shortening ratio X/Z of ≈ 4) were found to have weak (ferrimagnetic) susceptibility anisotropies ($\chi_{max}/\chi_{min} = 1.1$–$1.2$) (Jackson and Borradaile, 1991). Leaching experiments (cf. §15.5.4) and measurement of IRM anisotropy showed that coarse detrital hematite grains had weak magnetic fabric while fine authigenic or metamorphic hematite grains gave rise to a stronger fabric. Most of the strain was accommodated by quartz grains ($X/Z \approx 3$); fine hematite grains behaved more rigidly ($X/Z = 1.7$). In less deformed rocks (the Flamanville granite), on the other hand, Cogné and Perroud (1988) reported good correlation between the strain and susceptibility ellipsoids, in both directions and lengths of principal axes.

Remanence anisotropy is a more sensitive and reliable measure of magnetic fabric for paleomagnetic purposes than susceptibility anisotropy (Jackson, 1991; see also discussion below). Anhysteretic remanence (ARM) anisotropy is generally used. Fabrics due to different parts of the coercivity spectrum (e.g., SD and MD grains) can be examined by using partial ARM's (Jackson et al., 1988a) or different fractions of IRM (§11.3.2; Jackson and Borradaile, 1991), or by AF demagnetizing the remanence to different levels. A final advantage is that ARM anisotropy is more directly correlated with strain and strain gradients than is susceptibility anisotropy. Housen and van der Pluijm (1991) studied the transition from undeformed shale to slate in the Martinsburg formation, and found that susceptibility anisotropy was controlled by chlorite fabric and was unrelated to strain. The ARM anisotropy, on the other hand, progressively changed from bedding-normal to cleavage-normal, and correlated closely with the strain gradient.

A direct relation between tectonic stress and remagnetization of apparently stable Paleozoic carbonate rocks (§15.3.4) has been demonstrated by Sun *et al.* (1993). Remagnetized and unremagnetized sites have much different degrees of ARM anisotropy, and the magnetic anisotropy ellipsoid has the same orientation as layer-parallel shortening during Appalachian orogenesis (indicated by calcite twinning). The diagenetic/authigenic magnetite responsible for remagnetization formed under at least the partial control of a tectonically produced fabric. In the Martinsburg formation, within the Appalachian orogen itself, shale and slate anisotropies (see above) were used to determine the timing of magnetite dissolution and neocrystallization (Housen *et al.*, 1993b). This source of remagnetization is in addition to, and independent of, strain-induced grain rotations (§16.1.4).

Despite the advantages of remanence anisotropy, susceptibility is the most frequently used measure of magnetic anisotropy. For reviews, see Borradaile (1988), Jackson and Tauxe (1991), Rochette *et al.* (1992b) or Tarling and Hrouda (1993). Anisotropy of magnetic susceptibility (AMS) is defined either as $A = \chi_{max}/\chi_{min}$ or as percentage anisotropy, $\Delta\chi/\chi_{av} = (\chi_{max} - \chi_{min})/\chi_{av}$. In 'isotropic' rocks, the grains are almost randomly oriented and $\Delta\chi/\chi_{av}$ is generally only a few per cent. In lineated or foliated rocks, AMS may be much more pronounced.

For strongly magnetic minerals like magnetite, titanomagnetite or titanomaghemite, AMS usually reflects preferential linear or planar alignment of the long axes of grains. (Planar or other non-isotropic clusters of grains without preferred orientations will also produce AMS ('distribution anisotropy': Hargraves *et al.*, 1991; Stephenson, 1994)). Average grain elongation is about 1.5 : 1 (Nagata, 1961, p. 131), strongly favouring magnetization parallel to the long axis (§4.2.4). Susceptibility will be maximum in this direction for MD grains, which have $\chi_0 \approx 1/N$ (Fig. 5.16, eqns. (5.30), (9.4)), but *minimum* for SD grains, whose maximum susceptibility is *perpendicular* to the long axis (Fig. 8.6). This effect results in an inverse AMS fabric (i.e., χ_{max} perpendicular instead of parallel to the lineation or foliation) if SD grains of a strongly magnetic mineral are responsible, although the ARM or other remanence anisotropy will be normal (Stephenson *et al.*, 1986; Rochette *et al.*, 1992b).

Hematite tends to occur as plate-like crystals that are short along the c-axis and elongated in the basal plane, which above the Morin transition is the plane of easy magnetization. Alignment of these platelets in the plane of foliation by either rotation (Cogné and Canot-Laurent, 1992) or (re)crystallization during deformation results in a pronounced planar anisotropy. Susceptibility and remanence are greatly reduced perpendicular to the foliation. Figure 16.14 is an example of AMS data from which petrofabric and paleostrain of a metamorphic rock can be estimated.

Anisotropy of remanence and AMS due to the same grains are related tensors. McElhinny (1973, pp. 66–67) and Stacey and Banerjee (1974, pp. 118–120) give equivalent approximate theories for multidomain TRM. All grains are taken to have the same shape, and attention is restricted to grains so oriented that the TRM deviates a maximum amount from the applied field H_0. Following eqns. (9.9) and (9.10), the TRM components $M_\parallel$, $M_\perp$, parallel and perpendicular to the long axis, considered to be independently induced by the field components $H_0 \cos \phi$ and $H_0 \sin \phi$, are

$$M_{\parallel,\perp} = H_0 \frac{\cos \phi, \ \sin \phi}{N_{\parallel,\perp} \beta(T_B)(1 + N_{\parallel,\perp}\chi_i)}, \tag{16.4}$$

where ϕ is the angle between H_0 and the long axis, and $N_\parallel$, $N_\perp$ are counterparts to N_a, N_b used previously for demagnetizing factors parallel to the major and minor axes, respectively, of spheroidal grains (§4.2.4).

The angle θ between the TRM M_{tr} and the long axis is then

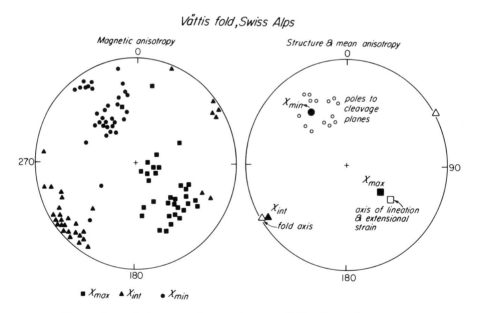

Figure 16.14 Anisotropy of magnetic susceptibility (AMS) for folded and strained Alpine rocks. Maximum susceptibility is in the direction of petrofabric lineation, while minimum susceptibility is measured perpendicular to the plane of foliation. [After Kligfield *et al.* (1982), with the permission of the authors and Eclogae Geologicae Helvetiae, Zürich, Switzerland.]

$$\tan\theta = \frac{M_\perp}{M_\parallel} = \frac{N_\parallel}{N_\perp}\frac{1+N_\parallel\chi_i}{1+N_\perp\chi_i} \qquad \tan\phi = \frac{1}{A_{tr}}\tan\phi. \qquad (16.5)$$

The 'anisotropy of TRM', A_{tr}, is defined by analogy with the anisotropy of susceptibility, as the ratio of maximum and minimum TRM's, induced parallel and perpendicular, respectively, to the long axis. For maximum deflection $\theta - \phi$,

$$\tan(\theta - \phi) = \frac{1}{2}\frac{A_{tr} - 1}{\sqrt{A_{tr}}}. \qquad (16.6)$$

Since for magnetite and other strongly magnetic minerals $\chi_0 = \chi_i/(1 + N\chi_i) \to 1/N$ (eqn. (9.4)), A_{tr} and A are related by

$$A_{tr} = \frac{N_\perp}{N_\parallel}\frac{\chi_{0\parallel}}{\chi_{0\perp}} \approx \left(\frac{\chi_{max}}{\chi_{min}}\right)^2 = A^2 \approx 1 + \frac{2\Delta\chi}{\chi_{av}}. \qquad (16.7)$$

The final approximation assumes small anisotropy, in which case multidomain TRM has twice the percentage anisotropy of susceptibility.

Table 16.2 gives a few values of maximum $\theta - \phi$ with the corresponding anisotropies, A_{tr} and A, of remanence and susceptibility. Quoting corresponding values of $\theta - \phi$ and A_{tr} but not A (McElhinny, 1973, p. 67) is a little misleading. A maximum deflection of $\approx 3°$ (comparable to the error in orienting a paleomagnetic sample and determining its NRM direction) requires $\approx 10\%$ anisotropy of TRM but only $\approx 5\%$ anisotropy of susceptibility. The latter criterion is more meaningful, since A, not A_{tr}, is normally measured.

The experimental evidence is sparse. Cogné (1987b) found that $A_{tr} \approx A^2$ within experimental error for two highly anisotropic rocks (a gabbro and a granite, the latter with 15–35% AMS). Irving and Park (1973) found the deflection of saturation IRM carried by high-coercivity grains in a number of metamorphic rocks to be only a few degrees and not systematic, although the remanence of low-coercivity MD grains was more anisotropic. Vetter et al. (1989), on the other hand, documented systematic deflections of remanence from the expected

Table 16.2 Maximum deflection of TRM from H_0 in an anisotropic rock according to MD theory (eqn. (16.6))

Susceptibility anisotropy, A	Anisotropy of TRM, A_{tr}	Maximum deflection of TRM (°)
1.05	1.10	2.7
1.095	1.20	5.2
1.225	1.50	11.6

Cambrian field direction as a result of tectonic fabric in Appalachian rocks. NRM deviations correlated well with strain and AMS magnitudes.

It is sometimes possible to correct TRM directions of igneous and high-grade metamorphic rocks using either the AMS or TRM anisotropy tensor. It is assumed that TRM was acquired after deformation and fabric development were complete. In addition, if the AMS tensor is used as a proxy for the TRM tensor, it must be assumed that the eigenvalues of the two tensors are quadratically related (Cogné, 1987b), in the manner of eqn (16.7). Cogné (1988b) was successful in using the measured AMS tensor of the Flamanville granite to restore dispersed NRM vectors to the orientations they would have had in an isotropic rock. The restored NRM vectors clustered well and agreed with the expected Carboniferous field direction. Hyodo and Dunlop (1993) used the directly measured TRM anisotropy tensor to correct deflected NRM directions in a Grenvillian (≈ 1000 Ma old) tonalitic gneiss and obtain a more reasonable paleofield direction.

A general theory relating AMS and anisotropy of magnetic remanence (AMR) has been advanced by Stephenson *et al.* (1986). They point out that even when (ferrimagnetic) susceptibilities are too weak to measure accurately, anisotropy can still be determined using remanence. They recommend weak-field IRM, which is 'magnetically non-destructive' in the sense that it has a negligible effect on NRM (cf. §11.2, 11.3.2, Fig. 11.1). (By the same token, of course, a weak-field IRM samples the part of the coercivity spectrum which is least relevant to stable NRM.) Their theory is too intricate to reproduce here, but it predicts relationships between AMS and AMR which are distinctively different for MD grains, for SD grains with a prolate AMR ellipsoid (i.e., an easy remanence axis, like magnetite), and for SD grains with an oblate AMR ellipsoid (an easy plane of magnetization, like hematite). The expected inverse (oblate) AMS fabric for SD grains with a prolate AMR ellipsoid is correctly predicted: susceptibility is maximum in a plane perpendicular to the easy axis. Experiments on rocks, dispersed PSD-MD magnetites, and SD magnetic tape particles gave results compatible with the theory.

16.4.2   Reversible piezomagnetic effects

We now turn to a consideration of **piezomagnetization**: reversible and irreversible changes in the magnetization of individual grains, e.g., rotations of SD moments, in response to stress. These changes may occur even in rocks with no permanent strain. In strained rocks, piezomagnetic effects are largely independent of strain-induced changes like rotations of the grains themselves. Partial remagnetization by permanent piezomagnetic changes, such as pressure demagnetization (§16.4.3) and piezoremanent magnetization (PRM) (§16.4.4, 16.4.5), causes deflections of the NRM vector which add to any strain-induced deflections due to (re)crystallization or rotation of grains.

A compressive stress σ with its axis at angle ψ to the magnetization M of a domain in a cubic ferromagnet gives rise to a magnetostrictive strain energy of

$$E_\sigma = \tfrac{3}{2}\lambda_s \sigma V \cos^2 \psi, \tag{16.8}$$

as in eqn. (2.66). The isotropic magnetostriction λ_s is appropriate for the randomly oriented crystals in a rock. In individual crystals, the energy depends on the angle between the stress axis and $\langle 100 \rangle$ and $\langle 111 \rangle$ axes. In magnetite, for instance, $\lambda_{100} = -19.5 \times 10^{-6}$ but $\lambda_{111} = +72.6 \times 10^{-6}$ (§3.2.3), so that compressive stress deflects magnetization away from $\langle 111 \rangle$ easy axes and toward $\langle 100 \rangle$ hard axes. Since $\lambda_s = +35.8 \times 10^{-6}$, the net effect overall is deflection of M away from the axis of compression in magnetite.

From thermodynamic considerations (for a detailed theory, see Hodych, 1976, 1977), small reversible magnetization changes parallel and perpendicular to a directed stress σ are related by

$$\frac{\partial M_\perp}{\partial \sigma} = -\frac{1}{2}\frac{\partial M_\parallel}{\partial \sigma} \tag{16.9}$$

Since in magnetite and other spinels with positive λ_s, $\partial M_\parallel/\partial\sigma$ is negative (for compression), the component of magnetization measured normal to the stress axis in these minerals should increase with compression. Kean *et al.* (1976) and Kapicka (1992) have found results generally compatible with these predictions in studies of reversible piezomagnetic changes in susceptibility for magnetite and titanomagnetites of different compositions and grain sizes.

Domain structure of MD magnetite compressed along a $\langle 111 \rangle$ axis was observed by Bogdanov and Vlasov (1966b). Below 0.5 kb, 180° walls of domains aligned with the stress were displaced, but around 0.85 kb, the entire domain structure was reconstituted along another $\langle 111 \rangle$ axis, at 71° to the stress. In principle, 180° walls should not be mobilized by stress since E_σ is identical in either domain. (σ defines an axis, not a vector direction.) 71° and 109° walls *are* moved by stress. Such motions, or direct rotations of the domains, explain the domain structure observations above 0.85 kb. Similar observations on polycrystalline high-Ti titanomagnetite were shown in Fig. 6.10.

Soffel (1966) also studied the effect of compressive stress on domain structure in magnetite. Observations were made on a $\{111\}$ plane with $\langle 112 \rangle$ as the axis of compression. The domains reformed with their M_s directions perpendicular to the stress axis when $E_\sigma > E_K$. In addition, a PRM was acquired perpendicular to the axis of compression.

It is worth emphasizing that only a field H_0 can create new magnetization. Stress alone cannot do so. Displacement of any 71° or 109° wall, in the absence of H_0, is balanced by an equal displacement of a wall bounding a domain of opposite magnetization.

Figure 16.15 shows large, purely reversible changes in the TRM of a gabbro compressed far beyond the elastic range. (The fifth compression cycle is illustrated. In the first cycle, where most of the plastic strain occurred, about 20% of the initial TRM was permanently demagnetized.) The axial component of TRM decreased > 20%, the perpendicular component increased about 5% (less than expected from eqn. (16.9)), the remanent intensity decreased 10% and the M_r vector rotated about 10° away from the compression axis. Initial values were recovered in all cases when the stress was released.

16.4.3   Effects of hydrostatic pressure

Hydrostatic pressures of 5 kb or more accompany deep (≥ 20 km) burial. The Curie point increases 2.3 °C/kb in magnetite, and to a lesser extent in titanomagnetite with $x = 0.5$–0.6 (Schult, 1970). Blocking temperatures at depth are

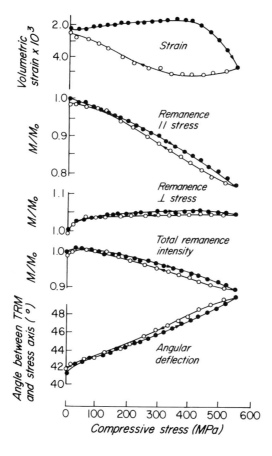

Figure 16.15   Reversible directional and intensity changes in the TRM of a gabbro during the fifth cycle of uniaxial compression. [After Martin *et al.* (1978) © American Geophysical Union, with the permission of the authors and the publisher.]

increased at most 10–20 °C. Magnetostriction and anisotropy constants change 5–15%/kb (Nagata and Kinoshita, 1967) but changes resulting from the increased temperature at depth are more important. VRM acquisition by MD magnetite is scarcely affected by hydrostatic pressure up to 1 kb (Kelso and Banerjee, 1995). Again, the effect of increased temperature is potentially more important (Fig. 10.3).

One would anticipate minor changes in magnetization due to hydrostatic pressure. Experimentally the changes are less for hydrostatic than for non-hydrostatic loading (Martin and Noel, 1988), but they can be substantial. 20 kb pressures have been found to permanently demagnetize 50% of the saturation IRM of dispersions of fine magnetite grains, much of the demagnetization occurring below 2 kb (Pearce and Karson, 1981). Magnetite-bearing rocks were affected less and hematite-bearing rocks hardly at all. On the other hand, Borradaile (1993a) reported 65% demagnetization of SIRM by 1–2 kb pressure for both magnetite and hematite-bearing rocks and for artificial dispersions of hematite. The reduction was greatest for the lowest coercivity fraction, and ARM was reduced less than IRM (Jackson *et al.*, 1993b). Girdler (1963), Avchyan (1967) and Nagata (1970) found only small (10–20%) permanent changes in the TRM of magnetite- and hematite-bearing rocks stressed to 5–10 kb, although weak-field IRM was largely demagnetized (Fig. 16.16).

These results suggest that hydrostatic stress merely aids internal stresses in effecting changes towards an equilibrium (lower magnetization) configuration. These changes are confined mainly to low-coercivity remanence, of which weak-field IRM is largely composed. Borradaile and Jackson (1993) demonstrated this association very directly in hydrostatic compaction up to 220 MPa (2.2 kb) of two-component IRM's carried by fine SD-PSD (0.02–2 μm) and MD (40 μm mean size) magnetites. The IRM with coercivities < 30 mT was preferentially demagnetized in both cases. Pressure demagnetization is thus akin in its effect to AF demagnetization in low-to-moderate fields of a few tens of mT.

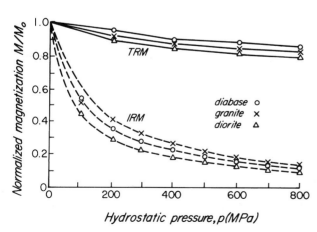

Figure 16.16 Demagnetization of weak-field IRM and TRM by hydrostatic pressure. [After Avchyan (1967) © American Geophysical Union, with the permission of the author and the publisher.]

16.4.4   Uniaxial stress and piezoremanent effects: single-domain theory

Small fields applied during hydrostatic pressurization produce insignificant remanence (Borradaile and Mothersill, 1991). A field applied during non-hydrostatic compression, however, creates a **piezoremanent magnetization (PRM)**. Directed stress is effective in reducing the anisotropy of an SD grain or the barriers to wall motion in MD grains. The applied field can then rotate domains or displace walls that were previously pinned.

There have been numerous demonstrations of the acquisition of PRM in the laboratory. Of particular interest are experiments by Borradaile (1994b), in which PRM was added to, and partly replaced, a pre-existing remanence, simulating tectonic overprinting of NRM in nature. Remagnetization of synthetic dispersions of PSD size magnetites by differential stresses up to 30 MPa (0.3 kb) was evidenced by remanence directions intermediate between those of the pre-existing ARM and of the 35 μT field applied during stressing. The newly acquired PRM's were mainly of lower coercivity (< 35 mT), but their effect completely masked that of strain-induced changes in remanence due to grain rotation (the shortening was up to 25%). Stress-induced remagnetization may thus be the main cause of syntectonic remagnetization in naturally deformed rocks.

Theoretically, the SD piezoremanent effect is easily demonstrated (Dunlop *et al.*, 1969). Taking the simplest possible case, in which a field H_0 and a uniaxial compression σ are applied parallel to the axis of a spheroidal SD grain, eqn. (8.5) for the total angle-dependent energy becomes:

$$E(\theta, \phi) = -\mu_0 V \boldsymbol{M}_s \cdot \boldsymbol{H}_0 + \tfrac{1}{2}\mu_0 V M_s [(N_b - N_a)M_s - 3\lambda_s\sigma/\mu_0 M_s]\sin^2\theta.$$

$$(16.10)$$

The analysis from this point is identical to that of eqns. (8.6) and (8.7) except that microscopic coercive force H_K is everywhere replaced by

$$H_K' = (N_b - N_a)M_s - \frac{3\lambda_s\sigma}{\mu_0 M_s} = H_K - \frac{3\lambda_s\sigma}{\mu_0 M_s}. \qquad (16.11)$$

As far as its effect on the shape anisotropy barrier is concerned, σ is equivalent to a reduction in coercivity of $3\lambda_s/\mu_0 M_s \approx 25$ mT/kb (250 Oe/kb) for magnetite. In reality, the reduction in coercivity is extremely variable because of the many possible relative orientations of H_0, σ, and the shape and crystalline easy axes. The coercivity may even be raised if a grain is elongated along a $\langle 100 \rangle$ axis, since λ_{100} is negative in Fe_3O_4. The experimentally observed average reduction in coercivity turns out to be much less than 25 mT/kb for magnetite.

Despite the numerical uncertainty of eqn. (16.11), the principles on which it is based are sound and lead to a simple picture for irreversible stress effects. The magnetization relaxation of an SD grain ensemble (V, H_{K0}) under compressive stress in a weak field, following eqn. (8.18), is

$$\frac{1}{\tau} = \frac{2}{\tau_0} \exp\left[-\frac{\mu_0 V M_s (H_K - m\lambda_s \sigma / \mu_0 M_s)}{2kT} \right],$$ (16.12)

m being a number that, experimentally, is ≤ 3. Compression of an assembly of SD grains with distributed (V, H_{K0}) is then described, at room temperature, by the following blocking curve on a Néel diagram:

$$V\left(H_{K0} - \frac{m\lambda_s \sigma}{\mu_0 M_{s0}}\right) = \frac{2kT_0}{\mu_0 M_{s0}} \ln\left(\frac{2t}{\tau_0}\right) = C(T_0, t, \sigma).$$ (16.13)

The blocking curves (Fig. 16.17) are identical to those for field-induced magnetization changes (IRM, ARM and AF demagnetization) and in marked contrast to those for thermally activated changes (TRM, VRM and thermal demagnetization).

In theory, PRM for a given stress σ affects the same grains as IRM produced in a field $H_0 = m\lambda_s \sigma / \mu_0 M_{s0}$. An even closer analogy is between partial ARM and PRM, $\tilde{H}$ in the former playing the role of σ in PRM. In the same manner, zero-field compression of SD grains should have an effect equivalent to AF demagnetization to $\tilde{H} = m\lambda_s \sigma / \mu_0 M_{s0}$, for any type of initial remanence.

Demagnetization by compression is well-documented. The curves resemble, in aspect and quantitatively, those for hydrostatic stress (Fig. 16.16). This resemblance implies that the hydrostatic effect is actually due to differential stresses at the grain scale, as the externally applied load is transmitted through grain contacts (Borradaile and Jackson, 1993). According to Ohnaka and Kinoshita (1969), about 70% of an initial TRM survives uniaxial compression to 5 kb. If m were equal to 3 in eqn. (16.13), 70% of the TRM would have coercivities in excess of 125 mT (1250 Oe). The actual coercivities are more like 25 mT (250 Oe), indi-

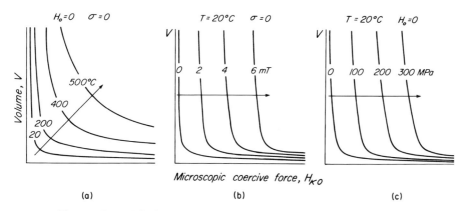

Microscopic coercive force, H_{K0}

(a)   (b)   (c)

Figure 16.17 Displacement of SD blocking curves due to (a) temperature, (b) applied field, and (c) uniaxial compression, for a material with positive magnetostriction. [After Dunlop *et al.* (1969), with the permission of the authors and the publisher, Journal of Geomagnetism and Geoelectricity, Tokyo, Japan.]

cating an average reduction in coercivity of ≈ 5 mT/kb, five times less than the 25 mT/kb predicted by (16.11).

Hamano (1983) made detailed comparisons of AF and uniaxial stress demagnetization curves for magnetite-bearing volcanic and pyroclastic rocks. The two types of demagnetization curves were strikingly similar and gave equivalent fields ranging from 2.3 mT/kb to 37.5 mT/kb. The lower figures were shown to be compatible with Brown's (1949) theory of domain wall motion in response to uniaxial stress. The higher figures, which are actually in good agreement with the SD theory embodied in (16.11), were considered by Hamano to result from the focusing of stress on MD magnetite grains in porous rocks. In the next section, we shall analyse companion stress and AF demagnetization data for hematite, which certainly has SD behaviour.

Figure 16.18 models the acquisition of remanence when σ is removed before H_0 and when field is removed before stress. The former remanence is the one usually referred to as PRM. It is considerably more intense than the latter (Fig. 16.19).

16.4.5   Piezoremanent effects: experimental results

Contrary to theoretical prediction, PRM does not increase with σ (Fig. 16.19) in a manner reminiscent of the increase of IRM with H_0 (Fig. 11.5). Instead there are two distinct regions of approximately linear trend below and above 200 bars. The initial slope is probably controlled by internal stresses, the slope above 200 bars being intrinsic.

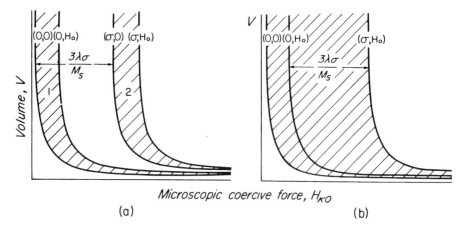

Figure 16.18   Néel diagrams illustrating (a) IRM (H_0) and PRM $(\sigma, H_0, H_0 \to 0$ before $\sigma \to 0)$ and (b) PRM $(\sigma, H_0, \sigma \to 0$ before $H_0 \to 0)$. [After Dunlop *et al.* (1969), with the permission of the authors and the publisher, Journal of Geomagnetism and Geoelectricity, Tokyo, Japan.]

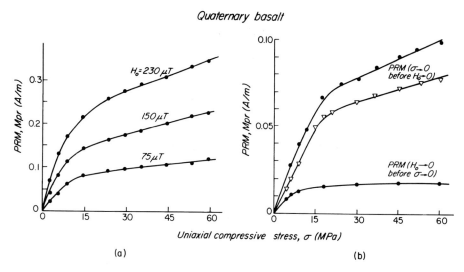

Figure 16.19 Dependence of PRM on (a) H_0 and σ, and (b) the order in which H_0 and σ are removed, for a Quaternary basalt. In (b), the open triangles denote results when σ was applied before H_0; these measurements are approximately equal to the difference between the other two curves. [After Pozzi (1973), with the permission of the author.]

Figure 16.19 shows that PRM intensity is proportional to H_0 in the weak field region. The intensity is low, only 1.5–4 times that of IRM produced in the same weak field, over the pressure range 0.5–4 kb (Nagata, 1966). The resistance to AF demagnetization is low. PRM is a relatively weak and soft remanence, in marked contrast to TRM.

PRM, unlike most other remanences, does not exactly parallel H_0. The deflection produced by σ does not entirely disappear when stress is released. The deflection is maximum when H_0 is at 45° to the stress axis (Fig. 16.20) and, for the basalt tested, was at most 4° for stresses of 0.3–0.6 kb.

Despite the shortcomings of the SD theory, PRM does appear to be above all a property of SD grains. Magnetically hard rocks acquire PRM more readily than soft rocks (Nagata and Carleton, 1969) and PRM is about 10 times as intense in very fine magnetite grains as it is in coarse MD grains (Kinoshita, 1968).

TM60 has a positive λ_s 3–4 times larger than that of magnetite and M_s about 4 times smaller. The reduction of SD microcoercivity by uniaxial stress (eqn. (16.11)) should therefore be an order of magnitude larger than in magnetite, but on the other hand, the microcoercivities are themselves much higher. Nagata and Carleton (1969) noted that among their samples, the largest PRM was exhibited by a basalt containing homogeneous titanomagnetite.

A direct calibration of the coercivity equivalence of stress is possible using Pozzi's (1973) data for hematite (Fig. 16.21). Despite the high coercivities of

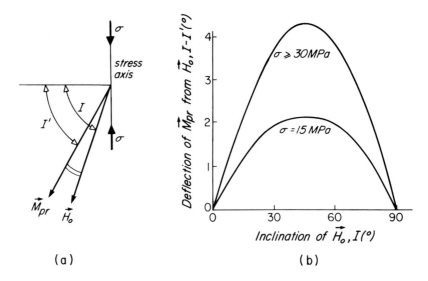

(a)   (b)

Figure 16.20   Deflection of PRM from the direction of H_0 during uniaxial compression of a basalt. [After Pozzi (1973), with the permission of the author.]

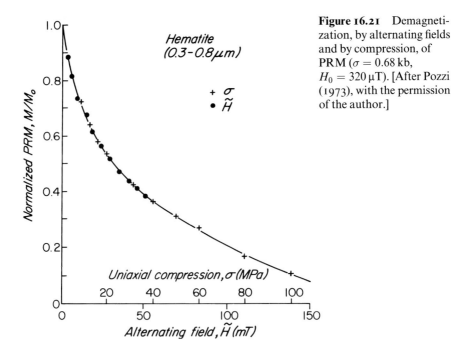

Figure 16.21   Demagnetization, by alternating fields and by compression, of PRM ($\sigma = 0.68$ kb, $H_0 = 320\,\mu$T). [After Pozzi (1973), with the permission of the author.]

hematite, the PRM illustrated is relatively soft. The H and σ axes have been appropriately scaled so as to show the equivalence of AF and pressure demagnetization. A coercivity of ≈ 150 mT (1.5 kOe) would be equivalent to a 1 kb compression. Although the theory of eqns. (16.8) and (16.11) is not appropriate for a rhombohedral mineral, $\lambda/\mu_0 M_{s0}$ for hematite is ≈ 300 mT/kb (3 kOe/kb), in order-of-magnitude accord with the experimental figure.

16.4.6 Overprinting by combined stress and temperature

At ordinary temperatures, overprinting due to directed stress is of minor importance. PRM produced by many kb of compression is weak and soft compared to the TRM that would be produced in thermal overprinting (or the CRM in chemical overprinting). But deep burial subjects minerals to simultaneous high temperature and high pressure. Although there are few experimental data to guide us, we can anticipate a number of effects peculiar to high (T, σ) conditions.

Both the defect moment and the intrinsic (spin-canted) moment of hematite are stress-sensitive, the latter because its remanence is pinned by magnetoelastic basal-plane anisotropy. Annealing substantially modifies both internal stress concentrations and basal-plane anisotropy (Fig. 15.18). Overprinting of NRM by external stress may therefore be important at metamorphic temperatures, the overprinting being stabilized after cooling and stress release by the lower mobility of dislocations at ordinary temperatures.

The blocking temperatures for thermal overprinting are lowered by a net directed stress. For weak fields, eqn. (8.19) becomes

$$T_B = \frac{\mu_0 M_{sB} V}{2k \ln(2t/\tau_0)} \left(H_{KB} - \frac{m\lambda_{sB}\sigma}{\mu_0 M_{sB}} \right). \tag{16.14}$$

Numerical values can be calculated for magnetite or TM60 using the $\lambda_s(T)$ data of Figs. 3.8(b) and 3.13. T_B decreases in proportion to the decrease in H_{KB}, e.g., by $\approx 80\,°$C for a (zero-compression) T_B of 500 °C (≈ 800 K), if stress lowers the coercivity by 10%. Overprinting in deep orogens may thus be rather easier than Fig. 16.7 predicts.

Uniaxial stress has another potentially serious effect in thermal overprinting: deflection of TRM or partial TRM in a cooling orogen from the direction of H_0. Figure 16.22 is a qualitative model of this syntectonic effect in SD grains. For simplicity, the shape axes are taken to be randomly oriented in a plane and the field is assumed to saturate the TRM, i.e., each moment is aligned with the stress and shape-defined easy direction nearest H_0, then relaxes into the corresponding shape-controlled easy direction when compression is released. Under stress, the deflection is, of course, substantial. When σ is removed, most of the deflection disappears, but not all. The residual deflection is due to moments trapped in orientations at $>90°$ to H_0. These moments cannot cross the shape-anisotropy barrier to reach the field-favoured orientation when $T < T_B$. The same reasoning can be applied to the permanent deflection of PRM or any other remanence by stress.

Experimentally, Hall and Neale (1960) reported TRM deflections of several degrees in the expected direction (away from both H_0 and σ), but Kern (1961) and Stott and Stacey (1960) found no systematic deflections except in anisotropic rocks that would show deflections even in the absence of stress. TCRM produced by oxidizing a stressed class-I basalt, initially containing homogeneous titano-magnetite, in the course of cooling also exhibited no significant deflection. Some of Stott and Stacey's data are reproduced in Table 16.3.

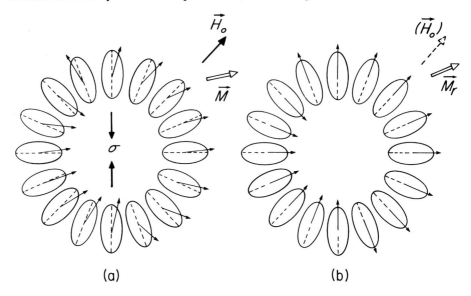

(a)   (b)

Figure 16.22 Magnetization of a uniform angular distribution of uniaxial SD grains (a) before and (b) after the release of compressive stress σ and field H_0. Shape-defined easy axes are dashed lines. H_0 is assumed large enough to turn all moments into the nearest easy axis defined by stress and shape.

Table 16.3 Experimental directions of TRM in rocks cooled under uniaxial compression σ. [After Stacey and Banerjee (1974) Table 11.2, with the permission of the authors.]

Rock type	No. of specimens (average or range)	Compression, (bar)	Angle between H_0 and σ	Angle between M and σ
Various igneous rocks	20	500	66°	65.4°
Basalt	20	500	46°	47.0°
Dolerite	13	1000	46°	45.5°
Porphyry	5	500	46°	45.8°
Serpentinite	4	0	66°	58–68°
	4	500	66°	10–75°

A systematic study of the deflection of TRM by uniaxial stress as a function of mineral composition and grain size remains to be done, but gross deflections from the applied field direction, such as were feared in the early days of paleomagnetism (e.g., Graham, 1956), seem unlikely in reasonably isotropic rocks.

16.4.7 Shock metamorphism and shock remanence

Large meteorite impacts, on earth and other planetary bodies (see Chapter 17), are accompanied by high pressures (tens of kb) of brief duration. Hargraves and Perkins (1969), Pohl *et al.* (1975) and Cisowski and Fuller (1978) have shown that laboratory shock remanent magnetization (SRM) in lunar and terrestrial rocks and synthetic analogs is of only moderate intensity and that like PRM, it affects only the lower part of the coercivity spectrum. The nature of the experiments did not permit a check on the collinearity of SRM and H_0. Experiments by Wasilewski (1977) suggest that SRM can be produced even in zero field, the SRM direction being dictated by the shock front vector. Although seemingly a negation of thermodynamic principles, zero-field SRM deserves investigation.

Shock textures in rocks from impact craters testify to complex physico-chemical changes that distinguish the SRM environment from the environment of elastic or even plastic deformation in which PRM forms. In natural situations, however, the strain energy of a large impact is partly dissipated as heat, and although the stresses are transitory, the heating is not. Its effect is to moderate the magnetic (and other) effects of shock. Textural and chemical changes can reach a thermal equilibrium. At the same time, annealing reduces internal stress buildups to near-normal levels. Finally, the long annealing and relatively slow cooling probably result in thermal overprinting of much of the SRM. It is quite possible that thermal overprinting of NRM in highly shocked but subsequently annealed minerals is the principal outcome of meteorite impact.

Figure 16.23 illustrates overprinting of the primary NRM of Keweenawan volcanic rocks from the Slate Islands crater by a later remanence dating from the time of a major meteorite impact (Halls, 1975, 1979). The overprint is not as hard as the primary NRM (presumed to be a TRM or TCRM) but it is harder than a typical PRM. Other paleomagnetic studies of impact structures include Pohl and Soffel (1971), Hargraves and Roy (1974), Cisowski (1977) and Robertson and Roy (1979).

16.4.8 Summary of stress effects

Both hydrostatic and directed stresses erase a fraction of pre-metamorphic NRM. This fraction has low coercivities and would be erased in the low-field steps of laboratory AF cleaning. Pressure demagnetization is considerable below 2 kb, even in the elastic range, where pressure may have a triggering effect on changes favoured by internal stresses. Stresses of many kb, accompanied by

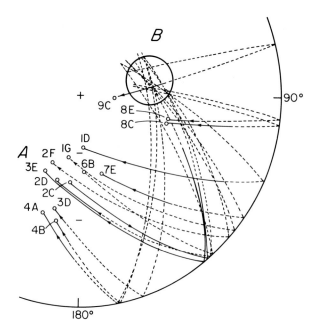

Figure 16.23 An example of shock overprinting in volcanic rocks from the Slate Islands (Lake Superior) meteorite crater. A is the approximate direction of primary NRM for Keweenawan (≈1100 Ma) rocks. B is believed to be a shock overprint (≈ 1000 Ma). The shock overprint is softer than the primary TRM but has some coercivities as high as 100 mT. [After Halls (1975), with the permission of the author. Reprinted with permission from *Nature*. Copyright © Macmillan Magazines Ltd.]

plastic deformation, have a proportionately smaller effect. A large fraction (70% or more) of TRM survives repeated stressing in the plastic range, although there are large reversible changes in its intensity and direction during stressing (Fig. 16.15). The success of the paleomagnetic fold test (§16.1.4) depends on the ability of NRM to survive the stresses – and the strains – that accompany folding. In tectonically deformed terrains, 'unstraining' paleomagnetic vectors in order to recover pre-tectonic NRM directions (e.g., Kligfield *et al.*, 1983; Cogné and Perroud, 1985; Hirt *et al.*, 1986; Lowrie *et al.*, 1986; Vetter *et al.*, 1989), using AMS or AMR as a guide, is probably only successful because tectonic stresses consistently demagnetize the lowest-coercivity remanence (Borradaile, 1992b). (However, Borradaile (1992a) presents evidence that strain-induced rotations of hematite grains, unlike those of magnetite grains, preferentially affect the *high-coercivity* grains. If this effect outweighs stress demagnetization, tectonic deformation of hematite-bearing rocks may selectively preserve low-coercivity NRM components.)

In the case of syntectonic magnetization, both strain-induced remagnetization, by rotation of existing grains and crystallization of new ones, and acquisition of PRM are important processes. In the presence of an applied field H_0, PRM overprints and replaces the low-coercivity fraction of NRM erased by uniaxial compression. The PRM is approximately parallel to H_0, but it is weak compared to TRM or CRM produced by the same field, even for stresses of several kb. PRM has maximum coercivities of 10–20 mT (100–200 Oe) in magnetite and ≈ 100 mT (1 kOe) in hematite. Fine-grained rocks containing SD (or nearly SD) grains acquire the largest PRM overprints. The main difficulties in using syntectonic PRM's or CRM's paleomagnetically are in distinguishing them from strain-rotated pre-tectonic NRM's and in dating the time of remanence acquisition.

Where the effects of stress and temperature are combined, as in deep burial, tectonism, or meteorite impact, temperature is the controlling factor. Stress can be regarded as simply lowering the blocking temperature for thermal overprinting. Permanent deflection of TRM or partial TRM in rocks cooled under (subsequently released) stress seems not to exceed a few degrees, unless the rock is highly anisotropic, in which case deflection occurs whether the rock is stressed or unstressed. Post-tectonic TRM's of anisotropic rocks, acquired in the uplift of deeply buried orogens for example, can be corrected by using AMS or AMR data (Cogné, 1988b; Hyodo and Dunlop, 1993).

16.5 The reliability of paleomagnetic results from metamorphic rocks

Most paleomagnetists regard metamorphic rocks, unless very mildly metamorphosed, as dubious recorders of primary NRM. Only very detailed study, often at the grain level, can separate grain rotation and crystallization effects (e.g., Housen *et al.*, 1993b). Although there have been successes in 'unstraining' NRM vectors (see previous section) in even quite strongly deformed rocks, improved grouping of initially scattered directions does not by itself prove that the corrected direction is primary. Only in cases where the age of pre-tectonic magnetization and the corresponding geomagnetic field direction are known in advance can the case be proven.

On the other hand, metamorphic rocks may be excellent recorders of one or more generations of secondary NRM, overprinted via physico-chemical changes, temperature, stress, or a combination of these agents. Directionally distinct NRM components may have been imprinted during different orogenic events or at different times in a single prolonged metamorphism.

The major difficulty comes in dating these NRM overprints. Much more detailed investigation of the nature of the NRM components is needed than in paleomagnetic work on 'fresh' igneous and sedimentary rocks (with the exception of red beds). Careful thermal demagnetization is indispensable to resolve remanences carried by magnetite and hematite and to establish, if possible, the thermal (i.e., partial TRM) origin of one or more components.

One is not limited in this quest to a single formation. Regional metamorphism produces pervasive overprints of regional extent, and if relative ages cannot be established in one study, they may in another (Figs. 16.12, 16.13). A regional cooling curve established by radiometric mineral ages and blocking temperatures in principle allows secondary NRM's of demonstrated thermal origin to be dated absolutely (Fig. 16.13). We are only beginning to reach this level of sophistication in studying metamorphosed terrains (e.g., Dallmeyer and Sutter, 1980; Dunlop *et al.*, 1980).

Fears that stress would destroy the directional fidelity of NRM in virtually any metamorphic rock have proven to be unfounded. Systematic deflection of remanence, either reflected in or produced by the anisotropic fabric of metamorphic rocks, remains a serious problem, however. A test of remanence anisotropy, e.g., the anisotropy of ARM (Jackson, 1991), is highly desirable in all studies.

Chapter 17

Magnetism of extraterrestrial rocks

17.1   Introduction

Extraterrestrial rocks formed in quite different environments from those on earth. In practically all cases, oxygen fugacity was very low during crystallization and later shock or thermal metamorphism. Iron–nickel alloys, rather than iron–titanium oxides, are the principal magnetic minerals.

Erosion and sedimentation occur to a very limited extent on the other terrestrial planets and their satellites. Lack of a shielding atmosphere has favoured another secondary process, bombardment of surface rocks by meteorites, principally 4000 to 3300 Ma ago.

The global fields of Mercury, Venus and the moon are orders of magnitude less than the present earth's field. The moon does have magnetic anomalies of regional extent, bearing witness to underlying magnetized crust, but none have a lineated or other regular pattern that would suggest plate tectonic processes. Except on Venus (Solomon *et al.*, 1992), there is no surface evidence for vertical or horizontal tectonic movements except those related to volcanism or meteorite impacts.

Of course, crustal spreading from centres of igneous activity on the moon might not leave the distinctive magnetic signature it does on earth. There is no compelling evidence that the moon ever possessed a substantial global field, although comparatively strong fields must have existed locally to produce the observed anomalies, and no evidence at all for reversal of either global or local fields. If a global field existed, it may not have been dipolar. For these reasons, it is doubtful whether paleodirectional and linear magnetic anomaly studies of the terrestrial variety would be successful on the moon, even if conditions permitted such studies. As it is, we are of necessity limited to attempting to determine paleo-field intensity.

The sampling problems in extraterrestrial paleomagnetic studies are enormous. Lunar samples, for example, are scarce and selected more or less at random. The few sites sampled are unlikely to provide a representative picture of the lunar surface. Bedrock on the moon is largely buried under a regolith of rock fragments, and even where bedrock is accessible rocks that originated at depth are not exposed by erosion or tectonic uplift. Very few lunar samples are oriented, even relative to other samples from the same site.

Weak sample moments and soft magnetic behaviour pose additional problems. Lunar samples are small, many contain only 0.1% iron, and the NRM seems to have been imprinted by fields $<10\%$ of the present earth's field (§17.3.2). With the measurement of chips and subsamples, one is in the realm of magnetic microanalysis (§14.6.3, 16.2.2). Unfortunately, lunar rocks are very susceptible to overprinting by IRM or VRM acquired in the spacecraft or on earth. These secondary remanences are amplified relative to the primary NRM because they were acquired in fields $\geq 50\,\mu\mathrm{T}$ (0.5 Oe).

Paleointensity determination is accompanied by serious physical and chemical alteration of the samples. Samples must be heated to 765 °C, the Curie temperature of iron. Above 700 °C, iron (especially the extremely fine-grained iron in some lunar breccias) oxidizes, sinters and even evaporates. Even high-vacuum systems cannot prevent oxidation above 700 °C. Only heating in a regulated H_2–N_2 mixture will do so. Unfortunately methods of reducing alteration of lunar samples during heating have developed slowly, being based on experience gained in previous paleointensity runs. The reliability of earlier results must be continually reassessed.

Because of the unparalleled experimental problems and the sparse and heterogeneous sample collections, which largely preclude intersample comparisons and statistical tests, lunar and meteorite results cannot be judged by the same standards as terrestrial studies. Nevertheless the existence of primary TRM or partial TRM overprints in many samples is now beyond doubt and we have approximate estimates of the magnetic fields that existed in the early solar system.

17.2   Magnetic properties of iron and of lunar rocks

17.2.1   Lunar rock types

Lunar rocks include primary igneous rocks, soils (igneous rocks fragmented by repeated meteorite impacts), and breccias (shock-welded fragments that have been metamorphosed to varying degrees). The oldest rocks (≈ 4000 Ma in age) are gabbros, norites, anorthosites and basalts. These are exposed in the highlands and were sampled during the Apollo 14, 16 and 17 missions. Some are perhaps not truly primary but derived from local impact melts. Among the youngest rocks (≥ 3300 Ma) are the mare basalts sampled during the Apollo 11, 12, 15 and 17 missions. They fill giant impact craters. The highlands rocks have all been metamorphosed to some extent through long exposure to meteorite bombardment.

Breccias, sampled by all missions but especially by Apollo 14, vary from loosely welded soil fragments to rocks with a recrystallized matrix that resemble primary igneous rocks. A complete range of metamorphic grade is seen in ejecta blankets surrounding giant impact craters. Even in quite high-grade breccias, metamorphic overprinting of NRM may be incomplete in large clasts that were only slightly heated. Large clasts from an Apollo 16 anorthosite (samples 68415, 68416) preserve an ^{40}Ar/^{39}Ar plagioclase age of 4100 Ma, while whole-rock Rb/Sr and ^{40}Ar/^{39}Ar ages have been reset to 3850 Ma (Huneke *et al.*, 1973).

17.2.2   Magnetic properties of iron and iron–nickel

The thermomagnetic curve of igneous sample 14053 (Fig. 17.1a) exhibits a single Curie point of 765 °C due to body-centred cubic kamacite (αFe). The heating and cooling curves for high-grade breccia 14303 (Fig. 17.1b) have a thermal hysteresis that is not related to irreversible chemical changes. It reflects the sluggish transformation, during cooling, from face-centred cubic taenite (γFe), which is

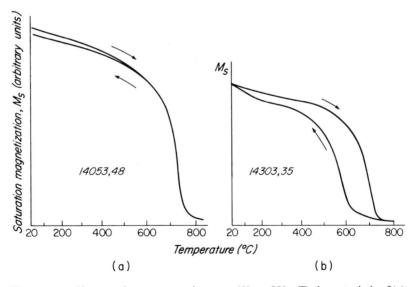

Figure 17.1   Vacuum thermomagnetic curves ($H_0 = 550$ mT) characteristic of (a) iron (kamacite or αFe), and (b) iron–nickel containing 5–10% Ni, showing the sluggish kamacite $\rightarrow$ taenite ($\alpha \rightarrow \gamma$) transformation. [After Nagata *et al.* (1972). Reprinted by permission of Kluwer Academic Publishers.]

stable at high temperatures, to kamacite (Fig. 17.2). The $\gamma \to \alpha$ transformation occurs below the α-phase Curie point if the iron is alloyed with $>5\%$ Ni. A sample like 14303 acquires phase-transformation CRM rather than true TRM if it cools in a field (assuming the transformation occurs below the lowest TRM blocking temperature of the α-phase).

For $>10\%$ Ni, both $\alpha \to \gamma$ (heating) and $\gamma \to \alpha$ (cooling) transformations occur below the kamacite Curie temperature. For $>30\%$ Ni, kamacite is not stable but taenite becomes ferromagnetic (Fig. 17.2). Ni-rich phases are more common in meteorites than in lunar rocks. The range of magnetic properties as a function of Ni content in different classes of meteorites and in synthetic analogs of meteoritic Fe–Ni particles are described by Nagata (1979a) and Wasilewski (1981a).

Néel *et al.* (1964) discovered that the iron and nickel atoms, which have a random or disordered arrangement in taenite, can be ordered by neutron irradiation in the presence of a strong magnetic field applied parallel to a $\langle 100 \rangle$ axis if the composition is 50–55% Ni (NiFe or close to it). Ordering occurs in nature below a critical temperature of 320 °C over long periods of time. The ordered phase, called tetrataenite (Nagata, 1983; Wasilewski, 1988a,b), has distinctly different magnetic properties from disordered taenite. Its coercivities are high (30–600 mT) and the $M_s(T)$ curve is 'blocky' like that of hematite (Fig. 3.20), dropping steeply around the Curie temperature of ≈ 550 °C. This is also the Curie point for taenite of this composition, to which the tetrataenite transforms during cooling. The thermal hysteresis in measured thermomagnetic curves is similar to that seen in plessite ($\alpha + \gamma$ intergrowths) of lower Ni content (Fig. 17.1b, 17.2). However, the heating and cooling curves generally cross: $M_s(T)$ increases more gradually with cooling below T_C in taenite than in tetrataenite, but M_{s0} at room temperature is somewhat higher for taenite

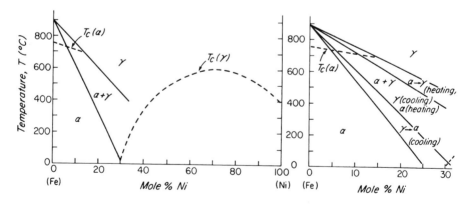

Figure 17.2 Part of the phase diagram of Fe–Ni. The kamacite region (0–30% Ni) is enlarged to illustrate the sluggish $\alpha \to \gamma$ and $\gamma \to \alpha$ transformations at typical thermomagnetic heating rates.

(see Nagata, 1983; Wasilewski, 1988a). Tetrataenite is common in chondritic meteorites (Nagata and Funaki, 1982), from which it was first described as a mineral phase by Clarke and Scott (1980).

In both tetrataenite and plessite, the NRM acquired during cooling is not a simple TRM but at least partly a phase-transformation CRM (§13.4). It is currently not known whether the CRM can 'remember' the original taenite TRM direction or is completely reset during transformation, which in the case of the ordering of tetrataenite occurs very slowly and only below 320 °C. Because of these uncertainties, meteorites containing mainly kamacite (E-chondrites and some classes of achondrites) or magnetite (carbonaceous or C-chondrites) are preferred for paleointensity determinations (Nagata, 1983).

In the case of lunar rocks, iron originated in at least four different ways.

1. Primary igneous iron, usually Ni-poor, often as kamacite–taenite (i.e., plessite) or kamacite–troilite (FeS) intergrowths.

2. Meteoritic iron, often nickel-rich and coarse-grained.

3. Secondary iron from the breakdown of troilite, a result of impact melting.

4. Ultra-fine-grained, nickel-free iron spheres in impact-melted glass. In higher-grade breccias, either shock causes larger particles to precipitate or the fine particles grow by subsolidus reduction of the glass during thermal annealing.

Mare basalts contain ≈0.1% igneous iron. Excess iron ($\geq 0.5\%$) in soils and breccias is partly meteoritic but most is due to impact melting. Repeatedly shocked highlands rocks are sometimes very iron rich (up to 4% Fe: e.g., basalt 60315).

17.2.3   Domain structure, hysteresis, and magnetic viscosity of lunar rocks

Domain structure and magnetic hysteresis in iron change very rapidly with grain size. Spherical grains 200–350 Å (0.02–0.035 μm) in size have hysteresis that combines SD and superparamagnetic (SP) features (Fig. 17.3a; cf. Figs. 5.19, 5.20). Saturation remanence, coercivity and initial susceptibility are all high. Lunar soils and low-grade breccias have just such properties (Fig. 17.3b). Judging by their hysteresis and magnetic viscosity, the predominant size range of the iron they contain is 100–200 Å. SP grains are volumetrically about ten times as abundant as SD grains.

Slightly larger synthetic grains, 800–1100 Å in diameter, have the low coercivity and saturation remanence and ramp-like hysteresis of true MD grains (Fig. 17.3). The saturation field, $H_{sat} = NM_s$ (cf. Fig. 5.16), is about equal to the predicted $\frac{1}{3}M_s = 570 \, \text{kA/m} = 710 \, \text{mT}$ $((4\pi/3)M_s = 7 \, \text{kOe})$ for iron

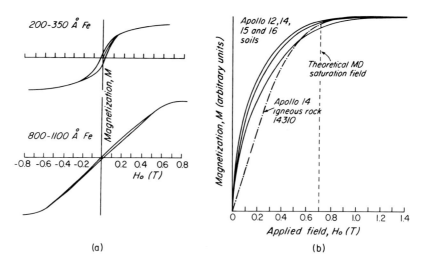

Figure 17.3   Isothermal magnetization curves for (a) synthetic iron spheres pre-cipitated in copper, and (b) lunar soils, and igneous rock 14310. (a) After Wasi-lewski (1977), with the permission of the author and the publisher, Journal of Geomagnetism and Geoelectricity, Tokyo, Japan. (b) Reprinted from Gose *et al.* (1972) and Pearce *et al.* (1973), with kind permission from the authors and Elsevier Science Ltd., The Boulevard, Langford Lane, Kidlington OX5 1GB, UK.

($M_s = 1715\,\text{kA/m}$ or emu/cm^3 for iron; Table 3.1). Lunar igneous rocks and high-grade breccias have this behaviour (Fig. 17.3b).

A slight contrast in initial grain size, or a small amount of grain growth, make the difference between a hard, strongly magnetized lunar rock (e.g., a low-grade breccia) and a soft, weakly magnetized one (e.g., a high-grade breccia). The transition from SD to MD hysteresis is quite sharp. There is no indication in lunar samples of an appreciable PSD size range, like that seen for magnetite and titano-magnetite (Chapter 12), although there is other evidence that a PSD range does exist in iron and iron–cobalt (e.g., Figs. 12.1, 13.1). The situation is quite different from that in terrestrial rocks, where truly MD behaviour is a property of $> 10\,\mu\text{m}$ grains, never of submicron grains.

Assuming coercivities $< 10\,\text{mT}$ (100 Oe) for domain-wall pinning in MD iron, eqn. (5.32) predicts $M_{rs}/M_s = H_c/NM_s < 0.015$. SD iron has $M_{rs}/M_s \geq 0.5$ (eqns. (5.41), (11.16)) and $25\,\text{mT} < H_c < 1\,\text{T}$. The minimum figure for H_c is determined by crystalline anisotropy in equant grains, the maximum value by shape anisotropy in long needles. Igneous and recrystallized rocks have the expected low values of M_{rs}/M_s, while soils and low-grade breccias have distinctly higher values that remain well below 0.5 because of the large SP fraction.

An experimentally determined grain distribution for SD iron in low-grade breccia 14313 (Fig. 17.4) illustrates the predicted maximum size for SD grains around $d_0 = 200\,\text{Å}$ (Table 5.1). The $25\,\text{mT}$ (250 Oe) cutoff coercivity is also

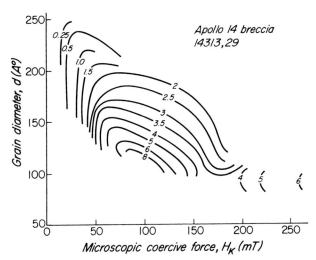

Figure 17.4 Experimentally determined grain distribution $f(V, H_{K0})$ for SD iron in low-grade breccia 14313. Contour values are in arbitrary units. [Reprinted from Dunlop *et al.* (1973b), with kind permission from the authors and Elsevier Science Ltd., The Boulevard, Langford Lane, Kidlington OX5 1GB, UK.]

obvious, as is the presence of elongated (perhaps shock-deformed) grains with very high coercivities.

The unusual concentration of SD grains just above SP size in lunar soils and soil breccias leads to pronounced short-term VRM, sometimes comparable to the stable NRM (e.g., Nagata and Carleton, 1970). Rocks containing mostly MD iron exhibit weaker but very long-lasting viscous effects due to domain-wall activation (Gose *et al.*, 1972).

Table 17.1 traces the simple progression of grain size and magnetic properties in Apollo 14 breccias as the metamorphic grade increases. The Warner (1972) grades of metamorphism given in the table, although they have been nominally temperature calibrated by Williams (1972), are a doubtful basis for calibrating thermal overprinting of NRM (in the manner of Table 16.1, for instance). As noted earlier for samples 68415 and 68416, shock heating is inhomogeneous and may scarcely penetrate large clasts. Furthermore, magnetic overprinting in breccias is not necessarily purely thermal. Subsolidus growth of iron in simulated lunar glasses occurs at temperatures as low as 700 °C (Usselman and Pearce, 1974) and will impart CRM to the highest blocking-temperature fraction of the NRM. Overprinting of slightly heated clasts as a result of shock alone is also a possibility (Cisowski *et al.*, 1973).

17.3 Lunar paleomagnetism

17.3.1 The NRM of lunar rocks

Fuller (1974, 1987) has reviewed NRM measurements for lunar samples. We shall consider only a few representative examples (Fig. 17.5). Sample 12063 has

Table 17.1 Variation of predominant grain size of Fe and of magnetic properties with metamorphic grade of lunar breccias

Metamorphic grade	Sample	Mean Fe grain size	VRM behaviour	Hysteresis loop shape	Remanence ratio, M_{rs}/M_s	Remanent coercive force, H_{cr} (mT)[a]
1	14313		Short-term, strong	Rounded	0.066	47
	14049	<500 Å			0.058	51
	14047				0.152	35
2	14301				0.009	45
3	14063		Long-term, weak	Ramp-like		
4	14321	<1 μm			0.019	
5	14311				0.006	14
6	14303				0.016	18
7	14312	>1 μm			0.010	

[a] H_{cr} is the reverse field required to reduce the remanence to zero after saturation.

[Data from Gose et al. (1972) and Fuller (1974) Table 2, with the permission of the authors and the publishers, Elsevier Science Ltd., The Boulevard, Langford Lane, Kidlington OX5 1GB, UK and the American Geophysical Union, Washington, DC.]

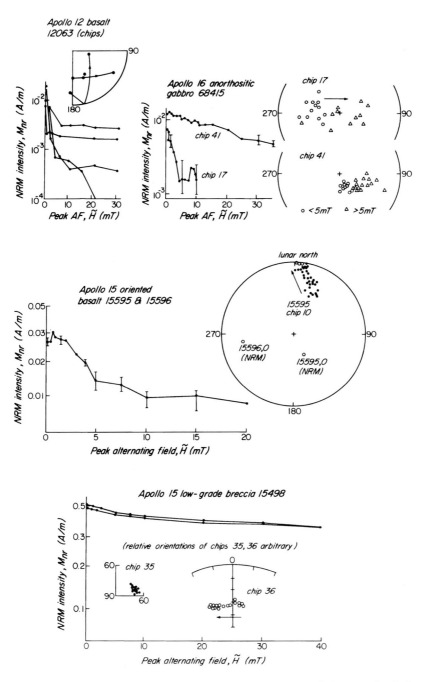

Figure 17.5 Examples of AF demagnetization of NRM in lunar rocks. Only 15595 is oriented in lunar co-ordinates. Open, closed symbols are up, down directions, respectively. [Reprinted from Pearce *et al.* (1973), with kind permission from the authors and Elsevier Science Ltd., The Boulevard, Langford Lane, Kidlington OX5 1GB, UK, and from Fuller (1974) © American Geophysical Union, with the permission of the author and the publisher.]

the typical features of a mare basalt. The NRM is weak: $1\text{--}5 \times 10^{-2}\,\mathrm{A/m}$ ($1\text{--}5 \times 10^{-5}\,\mathrm{emu/cm^3}$), decreasing to $2\text{--}10 \times 10^{-3}\,\mathrm{A/m}$ after $10\,\mathrm{mT}$ ($100\,\mathrm{Oe}$) AF cleaning. There is a prominent soft remanence, largely eliminated by $10\,\mathrm{mT}$ AF cleaning, and considerable dispersion of stable NRM directions between specimens of the same sample. The directional scatter is not all intrinsic to the NRM. Relative orientations of small subsamples cut from a single sample may be uncertain by as much as $30°$.

Highlands anorthosite 68415 has stronger and more stable NRM. Directions from two chips agree well after AF cleaning. The possibility of partial overprinting in this rock was suggested earlier, but directions of hard and soft components in chip 17 are not greatly different.

Low-grade breccia 15498 has a very stable and unusually intense NRM. Soils and soil breccias usually have NRM's of $0.5\text{--}5 \times 10^{-1}\,\mathrm{A/m}$, decreasing to $0.2\text{--}1 \times 10^{-1}\,\mathrm{A/m}$ after $10\,\mathrm{mT}$ AF cleaning. 15498 lacks any soft NRM component, as do also some Apollo 17 breccias.

Rille-edge basalt 15595 is of special interest because it is oriented with respect to bedrock. The AF-cleaned magnetization of chip 10 is toward lunar north and is horizontal. Since the latitude of the Apollo 15 site is $25°\mathrm{N}$, the inclination should have been about $45°$ if 15595 acquired its primary remanence in a dipolar lunar field. NRM directions of two other chips from this sample and neighbouring sample 15596 diverge considerably from the results for chip 10.

The NRM of most lunar rocks consists of a relatively hard component of variable directional stability and a soft component as much as ten times more intense. Sometimes the soft remanence can be shown or inferred to be IRM or VRM acquired in the spacecraft or on earth, but sometimes it has a similar direction to the hard remanence, attesting to a lunar origin. In nearly all cases, the soft remanence is carried by MD iron and the hard remanence by SD iron, the latter evidently present even in igneous rocks.

TRM (or phase transformation CRM) and SRM are the likely mechanisms of 'primary' (i.e., pre-thermal-overprinting) NRM. SRM has hardness approaching that of medium-temperature partial TRM, although less than that of total TRM (§16.4.7; Cisowski et al., 1974). Similarity between the AF or thermal demagnetization characteristics of NRM and TRM has been demonstrated for about 20 lunar rocks (Fuller, 1974, Fig. 25), but only four of these have been used for Thellier-type paleointensity determinations (see next section).

Figure 17.6(a) represents a 'conglomerate test' (§15.2.1) for semi-oriented lunar samples, whose top surfaces could be determined by cosmic ray exposure. The inclinations I, after AF cleaning, follow a $\cos I$ distribution, showing that the NRM has remained randomly oriented since the fragments tumbled into their present positions. The stable NRM is thus 'primary' in the sense that it predates the shattering of the rock.

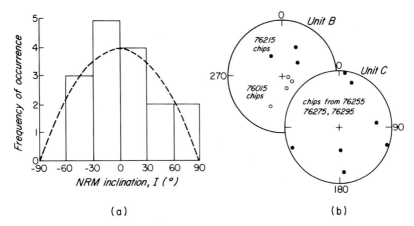

Figure 17.6 (a) Histogram of NRM inclinations for lunar surface oriented samples. The dashed curve is the expected random distribution. (b) AF-cleaned stable magnetization directions for subsamples of two lithographic units of a layered Apollo 17 breccia boulder. Open, closed symbols are up, down directions, respectively. [Reprinted from Pearce *et al.* (1973) and Gose *et al.* (1978), with kind permission from the authors, Elsevier Science Ltd., The Boulevard, Langford Lane, Kidlington OX5 1GB, UK, and Elsevier Science – NL, Sara Burgerhartstraat 25, 1055 KV Amsterdam, The Netherlands.]

A similar argument implies survivial of pre-brecciation remanence in subsamples from unit C of an Apollo 17 boulder (Fig. 17.6b). Unit C contains large clasts that escaped total overprinting when the matrix was impact melted. Unit B, containing small clasts, seems to have been entirely overprinted. Stable magnetizations of seven subsamples of 76015 and 76215 are reasonably well grouped (the two samples are not oriented relative to each other). The pre-impact and overprinted remanences are tentatively dated at 4100 Ma and 4000 Ma respectively (Turner and Cadogan, 1975).

17.3.2 Lunar paleofield intensity

Examples of Thellier-type paleointensity studies on lunar rocks appear in Fig. 17.7. The stable NRM of low-grade breccia 15498 (cf. Fig. 17.5) is probably a partial TRM acquired below 650 °C. Linearity is reasonable between 650 and 500 °C, the range of most blocking temperatures. A paleofield of 2.1 μT (0.021 Oe) is indicated.

Recrystallized anorthosite breccia 62235 yields a very high paleofield value of 1.2 Oe (Collinson *et al.*, 1973) and the Thellier plot is linear over the entire 500–20 °C range. A replicate determination using another subsample of 62235 gives essentially the same results (Sugiura and Strangway, 1983). These results are astonishing, since the sample has been reheated each lunar day (i.e., monthly) to 100–150 °C in the present tiny lunar field. Possibly, despite the convincing

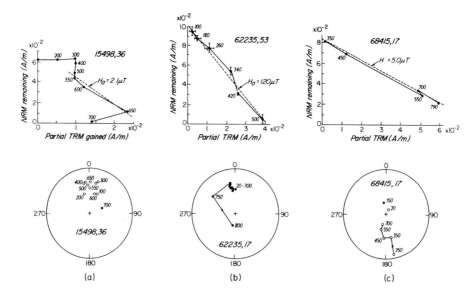

Figure 17.7 Thellier-type paleointensity determinations for lunar rocks. Labels on points are temperatures in °C. Open, closed symbols on stereonets are up, down directions, respectively. [Reprinted from Pearce *et al.* (1976), with kind permission from the authors and Elsevier Science Ltd., The Boulevard, Langford Lane, Kidlington OX5 1GB, UK.]

linearity of the Thellier plot, all or part of the NRM is a non-thermal spacecraft or terrestrial overprint.

Anorthosite or recrystallized breccia 68415 (cf. Fig. 17.5) has a soft remanence component with low blocking temperatures. The magnetization direction stabilizes and the Thellier plot is linear between 350 and 760 °C, giving a paleofield of 5 μT (0.05 Oe).

Many approximate paleointensity estimates have been published (see, e.g., Fuller, 1974, Fig. 28), most falling between 1 and 10 μT (0.01 and 0.1 Oe). However, Banerjee and Mellema (1976) obtained lunar paleofield intensities (based on the comparison of NRM and TRM AF demagnetization curves) ranging from 20 to 70 μT (0.2–0.7 Oe) for three subsamples of recrystallized breccia 72215.

Although estimated paleofields tend to be larger for the older rocks, there are exceptions. The NRM of 68415 is certainly as old as the NRM's of 62235 or 72215, but the paleointensity recorded is only 5 μT (0.05 Oe). Sugiura and Strangway (1980) applied the Thellier and several other techniques intended to minimize or eliminate heating to breccia sample 60225. High apparent paleointensities resulted from spurious but stable viscous overprints, which were only recognizable as such in the Thellier method.

It is disturbing that chips of the same sample give incompatible or widely dispersed results. 62235,53 and 68415,17 gave apparently reliable results but

62235,17 and 68415,41 failed the Thellier test (Collinson *et al.*, 1973; Pearce *et al.*, 1976). The dispersion in replicate determinations on 72215 has already been noted. Perhaps the lunar paleofield, far from being global, was extremely local and variable over geologically brief times. The other possibility is that lunar rocks are misleading recorders of paleofield intensity.

A problem in lunar paleointensity determination is self-reversal of partial TRM's acquired in certain temperature ranges. Figure 17.8(a) shows that partial TRM in 15498,36 decreases between the 250 °C and 350 °C in-field cooling steps and again, more strikingly, between the 550 °C and 600 °C cooling steps. Since no partial self-reversal is manifested in heating of the NRM (Fig. 17.7a) or of a laboratory TRM (Fig. 17.8a), the Thellier plot takes on a jagged aspect.

A synthetic lunar glass containing Fe–FeS eutectic intergrowths also shows jagged behaviour, with self-reversed partial TRM being acquired between the 250 °C and 325 °C cooling steps, although not between the 550 °C and 600 °C steps (Fig. 17.8b). The 325–250 °C partial self-reversal occurs just below the 320 °C Néel point of troilite and is probably caused by exchange interaction in the Fe–FeS intergrowths (Pearce *et al.*, 1976). The 600–550 °C self-reversal of Fig. 17.8(a) was tentatively attributed by the same authors to intergrown, exchange-coupled kamacite and nickel-rich, ferromagnetic taenite with $T_C < 600$ °C (see Fig. 17.2).

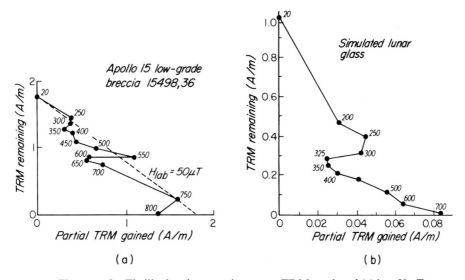

Figure 17.8 Thellier heating experiments on TRM produced (a) in a 50 μT (0.5Oe) laboratory field in breccia 15498, and (b) during synthesis of a simulated lunar glass. Self-reversal of partial TRM is seen in cooling from 350–250 °C in both samples, and also between 600 and 550 °C in the breccia. [Reprinted from Pearce *et al.* (1976), with kind permission from the authors and Elsevier Science Ltd., The Boulevard, Langford Lane, Kidlington OX5 1GB, UK.]

Chowdhary *et al.* (1987) produced similar partial self-reversals of pTRM using synthetic intergrowths in which iron was surrounded by troilite. Mechanical mixtures of Fe and FeS behaved normally. The partial self-reversal was attributed to reduction of the magnetizing field seen by the iron by the troilite demagnetizing field, which reaches its maximum value around the FeS Néel point (magnetic screening: §11.5), rather than to exchange coupling of the phases.

The underlying phenomena are complex and need further study, but they seem to be widespread. Samples 62235,17 and 68415,41 (see above) behaved in the same fashion as shown for 15498,36 in Fig. 17.8(a). So also did samples 10072,153 and 10020,209 (Apollo 11 basalts), 12018,224 and 12022,237 (Apollo 12 basalts) and 15597,68 (Apollo 15 porphyritic basalt) (Chowdhary *et al.*, 1987).

It is interesting that a line joining the points at 20 °C and at the highest temperature for which no chemical changes occurred (750 °C in Fig. 17.8a) gave a good estimate of the known field in which TRM was produced for 15498,36, and also for 62235,17 and 68415,41. Lines fitted to various quasi-linear segments of the plots gave very poor estimates. Non-ideal behaviour in the Thellier experiment may, after all, be the source of disagreement among various apparently reliable lunar paleointensity estimates.

17.4   Paleomagnetism of meteorites

17.4.1   Origin and types of meteorites

Meteorites as paleofield records possess the unique advantages of being the oldest available rocks, predating even the lunar samples, and of sampling all parts of their parent planet or planets.

The following is a tentative history of the formation of meteorites.

1. Primitive condensates from the undifferentiated solar nebula froze into chondrules, preserved today as inclusions in chondritic meteorites.

2. About 4550 Ma ago, an asteroidal planet or planets accreted, incorporating chondrules and also dust grains that crystallized later from the outer nebula.

3. The planetary interior melted and differentiated into an iron–nickel core, the source of iron meteorites, and an iron-rich silicate mantle, the source of achondrites.

4. The outer part of the planet, to a depth of a few hundred km, heated up and was thermally metamorphosed but did not totally melt (Wood, 1967). Differentiation was partial at depth (ordinary chondrites), inappreciable at the surface (carbonaceous chondrites).

5. Breakup of the planet(s) occurred about 4000 Ma ago and meteorite bombardment of the moon and earth began.

6. Collision and welding of planetary fragments produced multiply brecciated rocks. Shock features dating from breakup or subsequent collisions are prominent in some achondrites.

Ordinary chondrites and achondrites have compositional and textural similarities to the lunar rocks, but iron meteorites and carbonaceous chondrites are unique. They are also our sole samples of planetary cores and of the primitive solar nebula.

17.4.2　The NRM of meteorites

Meteorites are usually strongly magnetized and even small subsamples are readily measurable. Most contain $> 1\%$ magnetic material, the extreme case being hexahedrites, iron meteorites composed of 100% kamacite. NRM intensities are extremely variable, even within a single sample, but average 10^{-1}–10^{-2} A/m (10^{-4}–10^{-5} emu/cm^3) for carbonaceous chondrites, 1–100 A/m for stony meteorites generally and 50 to > 1000 A/m for iron meteorites. Iron–nickel alloys exsolved into coarse α–γ (plessite) intergrowths (bulk composition 5–20% Ni) are the norm but magnetite predominates in many carbonaceous chondrites. For reviews of meteorite magnetic properties and paleointensity results, see Stacey (1976), Sugiura (1977) and Nagata (1979a, 1983).

NRM originating as TRM during initial cooling in the planet's own field (Gus'kova and Pochtarev, 1967) seems unlikely (see however Collinson and Morden, 1994). All meteorites except carbonaceous chondrites formed at sufficient depths that they could not have been cooled below 765 °C while a core large enough to support a self-sustaining dynamo remained molten (Stacey, 1976). Carbonaceous chondrites, on the other hand, were never heated above 200 °C after accretion and could not have acquired a total TRM in a planetary or any other field.

Individual chondrules may preserve TRM acquired in the field of the early solar nebula. 'Detrital' remanence due to partial alignment of TRM vectors of chondrules during cold accretion forms part of the chondrite's primary NRM. Low-temperature partial TRM, imprinted in post-accretion heating to < 200 °C, is a more recent component of 'primary' NRM.

The mechanism of primary NRM in meteorites of other types and in the matrix of chondrites is more problematic (Funaki et al., 1981; Sugiura and Strangway, 1982; Collinson, 1987; Collinson and Morden, 1994). It is probably not either simple TRM or phase-transformation CRM. Grain growth almost certainly continued below 765 °C in the slowly cooling planetary interior, generating some form of TCRM. Shock or thermal overprinting erased most primary

NRM in brecciated achondrites, during planetary breakup or shortly thereafter. The extent of overprinting in other meteorite types is unclear.

The surface layer of a meteorite must be avoided in measurements because it bears a recent thermal overprint resulting from frictional heating by the earth's atmosphere. Only the outer 1 mm or so of a stony meteorite is reheated significantly (Nagata, 1979a). In iron meteorites, the outer 1 cm is heated. The surface overprint is softer than TRM in the meteorite's interior because, except in the surface fusion skin, the overprint is a partial, not a total, TRM.

17.4.3   Anisotropic and spontaneous remanence in iron meteorites

Iron meteorites are giant crystals of Fe–Ni that exsolved during extremely slow (1–100 °C/Ma) cooling into an octahedral or Widmanstatten structure. Ni-poor kamacite plates grew to mm size, with their {110} planes in {111} planes of a Ni-rich taenite host. Since exsolution occurred mainly between 760 °C and 600 °C (Scott and Wasson, 1975), any primary NRM is principally CRM or TCRM.

The NRM is surprisingly resistant to AF demagnetization (Fig. 17.9). AF coercivity correlates with kamacite bandwidth, being lowest in coarse octahedrites and mono-mineralic hexahedrites. Potential sites for high-coercivity NRM are micro-kamacite grains in fine-grained plessite ($\alpha + \gamma$) intergrowths (Brecher and Cutrera, 1976).

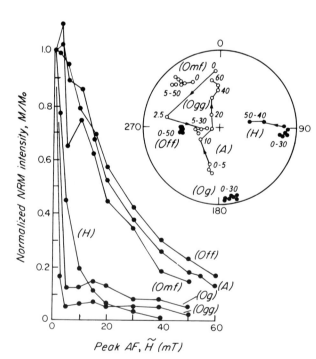

Figure 17.9 AF demagnetization of NRM in iron meteorites. Open, closed symbols in stereonet are up, down directions, respectively. Octahedrites, with kamacite band widths, are Odessa (Ogg, 1.7 mm), Cosby's Creek (Og, 2.5 mm), Smith's Mountain (Omf, 630 μm), and Butler (Off, 150 μm). Ataxite: Babb's Mill (A, 30 μm). Hexahedrite: Coahuila (H, Kamacite only). [After Brecher and Albright (1977), with the permission of the authors and the publisher, Journal of Geomagnetism and Geoelectricity, Tokyo, Japan.]

Laboratory-induced TRM is softer than NRM in all iron meteorites examined by Brecher and Albright (1977) and TRM intensity is not proportional to field strength. In fact, cooling in zero field imparts a remanence of similar intensity to $50\,\mu T$ (0.5 Oe) TRM! This 'spontaneous remanence' behaves erratically during AF cleaning but is not soft.

Iron meteorites, being single crystals, understandably have pronounced susceptibility anisotropy. Figure 17.10 shows that NRM, spontaneous remanence, and parts of the TRM are just as anisotropic. NRM and zero-field 'TRM's' are pinned near the intersection of taenite $\{111\}$ exsolution planes (i.e., kamacite $\{110\}$ planes), revealed by etching. TRM's are acquired approximately parallel to H_0 but upon AF cleaning migrate to and remain near exsolution planes.

Weak-field TRM thus contains two components, one parallel to H_0 and perhaps residing in micro-kamacite grains, the other a harder spontaneous remanence crystallographically pinned to, and presumably residing in, the coarse single-crystal plates of kamacite. Shape anisotropy strongly favours magnetization in the plane of a kamacite plate (§4.2.5). (Magnetic refraction or deflection of magnetization towards the plane of a dike, sill, lava flow, or kiln floor (§14.6.2, 14.7) is a similar but weaker effect.) The $\langle 100 \rangle$ crystalline easy axis within the $\{110\}$ plane provides additional pinning. Shape or crystalline pinning

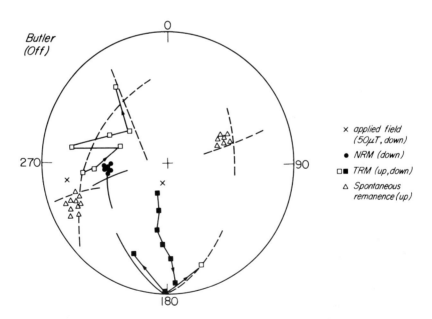

Figure 17.10 Pinning of remanence in kamacite $\{110\}$ planes (solid, dashed lines: lower, upper hemispheres, respectively) during AF demagnetization. [After Brecher and Albright (1977), with the permission of the authors and the publisher, Journal of Geomagnetism and Geoelectricity, Tokyo, Japan.]

of remanence occurs in all rocks but it is usually undetectable macroscopically because of the large numbers and random orientations of the crystals.

The spontaneous remanence is only apparently hard. It is analogous to the intrinsically soft random moments, due for example to loosely pinned domain walls, that align with stray fields when any rock is AF or thermally demagnetized and that ultimately determine the magnetic noise level. The giant random moments of iron meteorites simply reflect the tremendous amount of soft kamacite these rocks contain. The spontaneous remanence imposes a noise level so high that stable weak-field TRM cannot be resolved with confidence. The reliability of published paleointensities from iron meteorites is therefore doubtful.

17.4.4   Overprinting of NRM in achondrites

The three main classes of achondrites, diogenite, eucrite and howardite, and ureilite have distinctive proportions of α and γ phases and are therefore recognizable by their magnetic properties (Nagata, 1983). Kamacite is dominant in diogenites, eucrites and howardites, whose magnetic properties resemble those of lunar basalts (Nagata, 1979b). However, whereas the lunar rocks have rather unstable NRM's, these achondrites typically have NRM's which are directionally consistent at the subsample level. The NRM's are multicomponent, consisting of a heterogeneous NRM which varies in direction from one subsample to another and has somewhat erratic AF and thermal demagnetization behaviour, plus some secondary post-accretionary NRM (Collinson and Morden, 1994). The pre-accretionary remanence surviving in individual breccia fragments may be in part primary TRM acquired during cooling of the parent planet (see Fig. 17.6b for a breccia test on a lunar rock).

A stable, single-component overprint characterizes the heavily shocked ureilites (Fig. 17.11). The strong coherent NRM may be due to impact-generated iron, carrying TRM, TCRM or SRM, as in lunar soils.

17.4.5   NRM of ordinary chondrites

Kamacite containing $>5\%$ Ni is the dominant magnetic mineral in ordinary chondrites (Stacey et al., 1961). The directional stability and hardness of NRM improve with increasing Ni content (probably because of the microstructure of plessite $\alpha + \gamma$ intergrowths) and decreasing grain size, measured directly or indirectly through M_{rs}/M_s (Sugiura, 1977).

At least some ordinary chondrites of all classes have given apparently reliable paleointensities (see tabulations in Stacey and Banerjee, 1974, p. 173 and Brecher and Ranganayaki, 1975; values summarized in Fig. 17.14). In some low-grade chondrites, Sugiura (1977) has shown that pTRM with blocking temperatures below 200–300 °C behaves almost identically to NRM. In others, particularly those containing plessite, the NRM must be at least partly phase-transformation

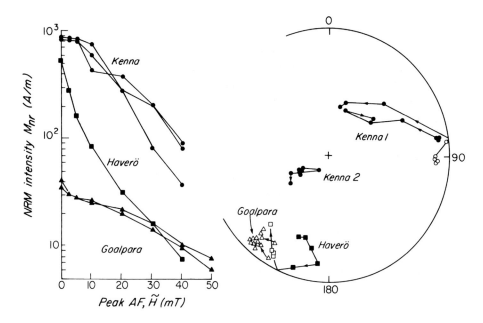

Figure 17.11 AF demagnetization of NRM of ureilites. Relative orientation of Kenna 1 and 2 is unknown. [After Brecher and Fuhrman (1979), with the permission of the authors. Reprinted by permission of Kluwer Academic Publishers.]

CRM and paleointensities can only be inferred from heatings below $\approx 550\,^{\circ}$C (Sugiura and Strangway, 1982).

Many low-grade chondrites have randomly distributed subsample NRM directions, suggesting a pre-accretionary NRM (Sugiura and Strangway, 1982). Funaki *et al.* (1981) have shown that large metal grains in several chondrites are unstably magnetized, the matrix NRM is directionally stable up to $\approx 15\,\text{mT}$ AF demagnetization and clusters within a hemisphere, while individual chondrules have quite stable NRM's which are randomly oriented from one chondrule to the next. Thus the chondrules may preserve pre-accretionary TRM recording an early magnetic field in the solar nebula (see next section).

Because of the variety of magnetic phases in chondrites, Funaki (1993) recommends using the temperature dependence of coercive force (H_c, H_{cr}) or saturation remanence (M_{rs}), rather than standard thermomagnetic curves ($M_s(T)$ or $\chi_0(T)$; e.g., Larson *et al.*, 1974), to identify the major contributors to NRM. In at least two cases, tetrataenite was found to be the phase responsible for reliable NRM (Collinson, 1987; Funaki, 1993), although as explained earlier kamacite is normally preferred for paleointensity determination.

High-grade breccias are less useful as magnetic field recorders. Most shock-reheated and brecciated chondrites are chaotic field recorders (e.g., Larson *et al.*, 1973).

17.4.6   Paleomagnetism of carbonaceous chondrites and their chondrules

Table 17.2 illustrates the varied magnetic mineralogy of carbonaceous chondrites. Curie points, memory of remanence after cycling through the magnetite Verwey transition (§3.2.2, 12.4.1), and values of M_{rs}/M_s have been used in conjunction to infer the relative amounts of Fe–Ni and of SD and MD Fe_3O_4. Class-I (undifferentiated) carbonaceous chondrites contain abundant Fe_3O_4. The importance of Fe–Ni (usually kamacite) increases with differentiation and, within class III, with increasing thermal metamorphism (C2 to C4).

The NRM of most carbonaceous chondrites is quite directionally stable during AF demagnetization (Fig. 17.12), whether carried by SD or MD grains of Fe_3O_4 or Fe–Ni. This stability is all the more unexpected since low-temperature partial TRM is a substantial component of the NRM (Fig. 17.13).

Individual chondrules separated from the Allende chondrite show confused behaviour when heated in the range of thermal overprinting ($< 200\,^\circ$C) but behave ideally in the Thellier sense at higher temperatures (Fig. 17.13b). They seem to have preserved a primordial TRM acquired in a field of 0.2–0.3 mT (2–3 Oe) (Sugiura et al., 1979). Wasilewski (1981b) has reported that the chondrules yielding these high paleofield values are anomalous in their magnetic properties compared to other chondrules from Allende, and probably contain interacting magnetic phases. Nagata and Funaki (1983) detected three magnetic

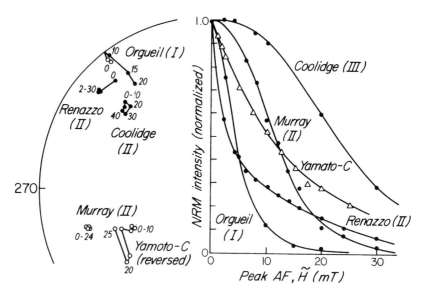

Figure 17.12   AF demagnetization of some carbonaceous chondrites (class given in parentheses). [Reprinted from Banerjee and Hargraves (1971) and Brecher and Arrhenius (1974) © American Geophysical Union, with kind permission from the authors, Elsevier Science – NL, Sara Burgerhartstraat 25, 1055 KV Amsterdam, The Netherlands, and The American Geophysical Union, Washington, DC.]

Table 17.2 Magnetic properties and paleointensity estimates for some carbonaceous chondrites

Meteorite	Type	Curie point, T_C (°C)	Memory of initial remanence	Remanence ratio, M_{rs}/M_s	Magnetic minerals	Paleofield(s), H_a (μT)
Orgueil	I,C1	590	1.0	0.08	$Fe_3O_4 (\approx 10\%)$	67
Ivuna	I,C1		0.63			≈ 50
Haripura	II,C2		0.55		Fe_3O_4	
Murray	II,C2		0.94	0.31–0.65	Fe_3O_4	33–135
Cold Bokkeveld	II,C2	590, 765	0.79		Fe_3O_4, Fe–Ni	
Mighei	II,C2		0.82	0.07–0.13	Fe–Ni, Fe_3O_4	
Murchison	II,C2		0.98			18, 40
Renazzo	II,C2		0.98	0.06	Fe–Ni	230
Mokoia	III–V,C2		0.23	0.17	Fe_3O_4 ($\approx 8\%$)	
Vigarano	III–V,C3		0.60		Fe_3O_4 (Fe–Ni?)	
Coolidge	III–V,C4		1.0			
Allende	III–V,C4	320, 600	0.98	0.03–0.12	γFe–Ni, $Fe_{1-x}S$	95–130, 200–300 (chondrules)

[Data from Brecher and Arrhenius (1974) © American Geophysical Union, and from Brecher (1977), reprinted with permission from *Nature*, copyright © 1977 Macmillan Magazines Ltd.]

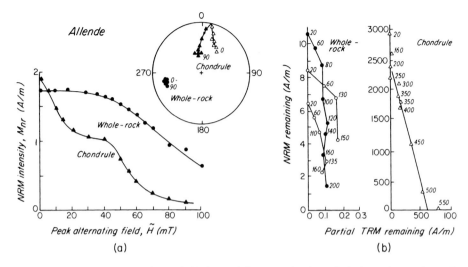

Figure 17.13 (a) AF and (b) thermal demagnetization of whole-rock samples and (unoriented) chondrules from the Allende (III) carbonaceous chondrite. Partial TRM's in (b) acquired in 50 μT (0.5 Oe). [Reprinted from Banerjee and Hargraves (1972), Lanoix *et al.* (1977), and Lanoix *et al.* (1978) © American Geophysical Union, with kind permission from the authors, Elsevier Science – NL, Sara Burgerhartstraat 25, 1055 KV Amsterdam, The Netherlands, and The American Geophysical Union, Washington, DC.]

phases in Allende: Ni-rich taenite, pyrrhotite, and magnetite, taenite being dominant. They found only weak unstable NRM's in individual chondrules, but bulk specimens, matrix material, and one stable chondrule all yielded high Thellier paleointensities of 0.1–0.3 mT.

17.5   Paleofields in the early solar system

Figure 17.14 sums up the scanty data from lunar and meteorite paleointensity studies. Only Thellier–Thellier or similar lunar determinations are shown (for a summary of other lunar paleointensity data, see Cisowski *et al.*, 1983). There are fewer meteorite determinations of comparable reliability, and in many cases, the thermal nature of the NRM was not adequately tested. Ages have very large uncertainties in most cases. The age (and nature) of overprinted NRM is particularly ill-defined.

There may be a real trend from relatively intense paleofields before and during planetary accretion to rather weak fields 4.0–3.0 Ga ago. The Alfvén (1954) model of solar-system formation postulates transfer of angular momentum from the sun to the protoplanets via a strong magnetic field (~10 mT or 100 Oe at the sun's surface). It is hard to believe that this early T-Tauri phase of mass

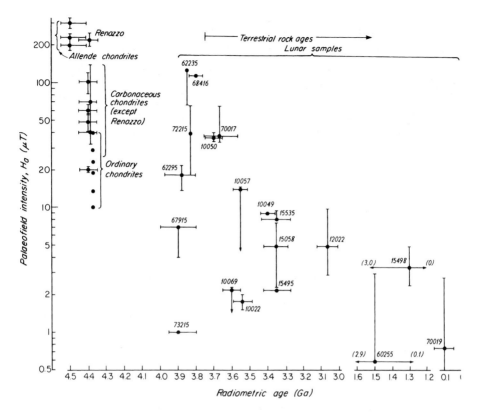

Figure 17.14 Summary of paleointensity determinations for lunar rocks and meteorites. Error bars on ages include uncertainty of age of magnetic overprinting (dashed lines). [Data from Collinson *et al.* (1973), Stacey and Banerjee (1974) Table 13.1, Pearce *et al.* (1976), Brecher (1977) © Macmillan Magazines Ltd., Lanoix *et al.* (1977) and Cisowski *et al.* (1983) © American Geophysical Union, with kind permission of the authors and the publishers, Elsevier Science Ltd., The Boulevard, Langford Lane, Kidlington OX5 1GB, UK, Macmillan Magazines Ltd., and the American Geophysical Union.]

and magnetic field transfer via a strong solar wind endured 400 Ma or more and was instrumental also in magnetizing igneous rocks on the moon. Another objection (Brecher, 1977) is that the moon condensed and accreted much closer to the sun than did the asteroids. It should have experienced a much larger T-Tauri field in consequence.

Models of global-scale lunar fields are of two sorts: dynamo and fossil-field models. Dynamo models (e.g., Runcorn *et al.*, 1971; Runcorn, 1994) face the difficulty that the moon's core must have been small and marginally able to support dynamo action even immediately after accretion and differentiation, let alone 500 Ma later when the highlands rocks formed.

The fossil-field model (Strangway *et al.*, 1974) postulates an intense early solar-wind field to produce uniform IRM in the still cold moon, just after accretion. Most of this IRM has today been lost through internal heating, but at the time the igneous surface rocks cooled, they did so in the surface dipole field of whatever uniformly magnetized spherical shell still remained below 765 °C. (The dipole field of a uniformly magnetized sphere was described in §4.3.2.) Since lunar igneous rocks contain only $\approx 0.1\%$ iron, a solar-wind field of $\approx 5\,\text{mT}$ (50 Oe) at 1 AU from the sun is required to give enough IRM for a $2\,\mu\text{T}$ lunar field. An implausibly large solar-wind field is needed to give a $0.1\,\text{mT}$ (1 Oe) surface field.

A general difficulty faced by all models that require an external field, whether solar, terrestrial, or interplanetary, is that the moon rotates with respect to the sun and, although it does not do so today, must have rotated with respect to the earth as well 4 Ga ago. No TRM or thermal overprint is acquired rapidly enough to be coherent under these conditions.

Local fields acting during meteorite impacts are a possible source of NRM's of soils and breccias (Cisowski *et al.*, 1973). Indeed impacts may possibly generate their own fields, or SRM may be acquired even in the absence of a field (Wasilewski, 1977; cf. §16.4.7). The relative merits of local shock-generated magnetizations and TRM's due to planetary-scale fields as sources of lunar magnetism are reviewed by Collinson (1993).

Although planetary field models are uncertain and planetary magnetic studies are still in their infancy, other solar system bodies probably possessed magnetic fields of their own or were exposed to fairly strong interplanetary fields before any record exists of a terrestrial magnetic field. The earth is not the sole arena of paleomagnetism.

References

Ade-Hall, J.M. & Johnson, H.P. (1976).
Paleomagnetism of basalts, Leg 34. *Init. Rep. Deep Sea Drill. Proj.*, **34**, 513–532.

Ade-Hall, J.M., Palmer, H.C. & Hubbard, T.P. (1971). The magnetic and opaque petrological response of basalts to regional hydrothermal alteration. *Geophys. J. Roy. Astron. Soc.*, **24**, 137–174.

Aharoni, A., Frei, E. H. & Schieber, M. (1962). Curie point and origin of weak ferromagnetism in hematite. *Phys. Rev.*, **127**, 439–441.

Aitken, M.J. & Hawley, H.N. (1971). Archaeomagnetism: Evidence for magnetic refraction in kiln structures. *Archaeometry*, **13**, 83–85.

Akimoto, S. (1962). Magnetic properties of FeO-$Fe_2O_3 - TiO_2$ system as a basis of rock magnetism. *J. Phys. Soc. Japan*, **17**, suppl. B1, 706–710.

Akulov, N.S. (1936). Zur Quantentheorie der Temperaturabhändigkeit der Magnetisierungskurve. *Z. Phys.*, **100**, 197–202.

Alfvén, H. (1954). *The Origin of the Solar System*. Clarendon Press, Oxford.

Amar, H. (1958). Magnetization mechanism and domain structure of multidomain particles. *Phys. Rev.*, **111**, 149–153.

Ambatiello, A., Fabian, K. & Hoffmann, V. (1995). Changes in the magnetic domain structures of magnetite during heating up to the Curie temperature (abstract). *Int. Un. Geod. Geophys. XXI Gen. Ass., Abstracts*, p. B96.

Amin, N., Arajs, S. & Matijevic, E. (1987). Magnetic properties of uniform spherical magnetic particles prepared from ferrous hydroxide gels. *Phys. Stat. Sol.*, (a), **101**, 233–238.

Anson, G.L. & Kodama, K. P. (1987). Compaction-induced inclination shallowing of the post-depositional remanent magnetization in a synthetic sediment. *Geophys. J. Roy. Astron. Soc.*, **88**, 673–692.

Appel, E. (1987). Stress anisotropy in Ti-rich titanomagnetites. *Phys. Earth Planet. Inter.*, **46**, 233–240.

Appel, E. & Soffel, H.C. (1984). Model for the domain state of Ti-rich titanomagnetites. *Geophys. Res. Lett.*, **11**, 189–192.

Appel, E. & Soffel, H.C. (1985). Domain state of Ti-rich titanomagnetites deduced from domain structure observations and susceptibility measurements. *J. Geophys.*, **56**, 121–132.

Arason, P. & Levi, S. (1990a). Models of inclination shallowing during sediment compaction. *J. Geophys. Res.*, **95**, 4481–4500.

Arason, P. & Levi, S. (1990b). Compaction and inclination shallowing in deep-sea sediments from the Pacific Ocean. *J. Geophys. Res.*, **95**, 4501–4510.

Argyle, K.S. & Dunlop, D.J. (1984). Theoretical domain structure in multidomain magnetite particles. *Geophys. Res. Lett.*, **11**, 185–188.

Argyle, K.S. & Dunlop, D.J. (1990). Low-temperature and high-temperature hysteresis of small multidomain magnetites (215–540 nm). *J. Geophys. Res.*, **95**, 7069–7083.

Argyle, K.S., Dunlop, D.J. & Xu, S. (1994). Single-domain behaviour of multidomain magnetite grains (abstract). *Eos (Trans. Am. Geophys. Un.)*, **75**, Fall Meeting suppl., 196.

Arkani-Hamed, J. (1988). Remanent magnetization of the oceanic upper mantle. *Geophys. Res. Lett.*, **15**, 48–51.

Arkani-Hamed, J. (1989). Thermoviscous remanent magnetization of oceanic lithosphere inferred from its thermal evolution. *J. Geophys. Res.*, **94**, 17,421–17,436.

Audunsson, H. & Levi, S. (1989). Drilling-induced remanent magnetization in basalt drill cores. *Geophys. J. Roy. Astron. Soc.*, **98**, 613–622.

Avchyan, G.M. (1967). Effect of hydrostatic pressure up to 8000 kg/cm^2 on various types of remanent magnetization of rocks (transl. from Russian). *Izv. Phys. Solid Earth*, no. 7, 465–469.

Avchyan, G.M. & Faustov, S.S. (1966). On the stability of viscous magnetization in variable magnetic fields (transl. from Russian). *Izv. Phys. Solid Earth*, no. 5, 336–341.

Baag, C-G. & Helsley, C.E. (1974). Evidence for penecontemporaneous magnetization of the Moenkopi Formation. *J. Geophys. Res.*, **79**, 3308–3320.

Baag, C., Helsley, C.E., Xu, S-Z. & Lienert, B.R. (1995). Deflection of paleomagnetic directions due to magnetization of the underlying terrain. *J. Geophys. Res.*, **100**, 10,013–10,027.

Bacri, J.C., Perzynski, R. & Salin, D. (1988). Magnetic and thermal behaviour of γ-Fe$_2$O$_3$ fine grains. *J. Magn. Magn. Mat.*, **71**, 246–254.

Bailey, M.E. & Dunlop, D.J. (1977). On the use of anhysteretic remanent magnetization in paleointensity determination. *Phys. Earth Planet. Inter.*, **13**, 360–362.

Bailey, M.E. & Dunlop, D.J. (1983). Alternating field characteristics of pseudo-single-domain (2–14 μm) and multidomain magnetite. *Earth Planet. Sci. Lett.*, **63**, 335–352.

Bailey, M.E. & Hale, C.J. (1981). Anomalous magnetic directions recorded by laboratory-induced chemical remanent magnetization. *Nature*, **294**, 739–741.

Bando, Y., Kiyama, M., Yamamoto, N., Takada, T., Shinjo, T. & Takaki, H. (1965). Magnetic properties of α-Fe$_2$O$_3$ fine particles. *J. Phys. Soc. Japan*, **20**, 2086.

Banerjee, S.K. (1970). Origin of thermoremanence in goethite. *Earth Planet. Sci. Lett.*, **8**, 197–201.

Banerjee, S.K. (1971). New grain size limits for palaeomagnetic stability in hematite. *Nature Phys. Sci.*, **232**, 15–16.

Banerjee, S.K. (1977). On the origin of stable remanence in pseudo-single domain grains. *J. Geomag. Geoelec.*, **29**, 319–329.

Banerjee, S.K. (1980). Magnetism of the oceanic crust: evidence from ophiolite complexes. *J. Geophys. Res.*, **85**, 3557–3566.

Banerjee, S.K. (1984). The magnetic layer of the ocean crust – how thick is it? *Tectonophysics*, **105**, 15–27.

Banerjee, S.K. (1991). Magnetic properties of Fe-Ti oxides. In *Oxide Minerals: Petrologic and Magnetic Significance*, 2nd edn., ed. D.H. Lindsley, pp. 107–128. Mineralogical Society of America, Washington, D.C.

Banerjee, S.K. & Hargraves, R.B. (1971). Natural remanent magnetization of carbonaceous chondrites. *Earth Planet. Sci. Lett.*, **10**, 392–396.

Banerjee, S.K. & Hargraves, R.B. (1972). Natural remanent magnetizations of carbonaceous chondrites and the magnetic field in the early solar system. *Earth Planet. Sci. Lett.*, **17**, 110–119.

Banerjee, S.K. & Mellema, J.P. (1974). A new method for the determination of paleointensity from the A.R.M. properties of rocks. *Earth Planet. Sci. Lett.*, **23**, 177–184.

Banerjee, S.K. & Mellema, J.P. (1976). Early lunar magnetism. *Nature*, **260**, 230–231.

Banerjee, S.K. & Moskowitz, B.M. (1985). Ferrimagnetic properties of magnetite. In *Magnetite Biomineralization and Magnetoreception in Organisms*, ed. J.L. Kirschvink, D.S. Jones & B.J. MacFadden, pp. 17–41. Plenum, New York.

Banerjee, S.K., Hunt, C.P. & Liu, X-M. (1993). Separation of local signals from the regional paleomonsoon record of the Chinese loess plateau: A rock magnetic approach. *Geophys. Res. Lett.*, **20**, 842–846.

Banerjee, S.K., King, J. & Marvin, J. (1981). A rapid method of magnetic granulometry with applications to environmental studies. *Geophys. Res. Lett.*, **8**, 333–336.

Barbetti, M.F., McElhinny, M.W., Edwards, D.J. & Schmidt, P.W. (1977). Weathering processes in baked sediments and their effects on archaeomagnetic field-intensity measurements. *Phys. Earth Planet. Inter.*, **13**, 346–354.

Barton, C.E. & McElhinny, M.W. (1979). Detrital remanent magnetization in five slowly redeposited long cores of sediment. *Geophys. Res. Lett.*, **6**, 229–232.

Barton, C.E., McElhinny, M.W. & Edwards, D.J. (1980). Laboratory studies of depositional DRM. *Geophys. J. Roy. Astron. Soc.*, **61**, 355–377.

Bate, G. (1980). Recording materials. In *Ferromagnetic Materials*, vol. 2, ed. E.P. Wohlfarth, pp. 381–507. North-Holland, Amsterdam and New York.

Bean, C.P. & Livingston, J.D. (1959). Superparamagnetism. *J. Appl. Phys.*, **30**, 120S–129S.

Becker, J.J. (1973). A model for the field dependence of magnetization discontinuities in high-anisotropy materials. *IEEE Trans. Magnetics*, **MAG9**, 161–167.

Berger, G.W. (1975). $^{40}Ar/^{39}Ar$ step heating of thermally overprinted biotite, hornblende and potassium feldspar from Eldora, Colorado. *Earth Planet. Sci. Lett.*, **26**, 387–408.

Berger, G.W., York, D. & Dunlop, D.J. (1979). Calibration of Grenvillian palaeopoles by $^{40}Ar/^{39}Ar$ dating. *Nature*, **277**, 46–47.

Berkov, D.V., Ramstöck, K. & Hubert, A. (1993). Solving micromagnetic problems: Towards an optimal numerical method. *Phys. Stat. Sol.*, (a), **137**, 207–225.

Bernal, J.D., Dasgupta, D.R. & Mackay, A.L. (1957). Oriented transformation in iron oxides and hydroxides. *Nature*, **180**, 645–647.

Bernal, J.D., Dasgupta, D.R. & Mackay, A.L. (1959). The oxides and hydroxides of iron and their structural inter-relationshps. *Clay Min. Bull.*, **4**, 15–30.

Beske-Diehl, S.J. (1988). Ilmenite lamellae and stability of magnetization. *Geophys. Res. Lett.*, **15**, 483–486.

Beske-Diehl, S.J. (1990). Magnetization during low-temperature oxidation of seafloor basalts: No large scale chemical remagnetization. *J. Geophys. Res.*, **95**, 21,413–21,432.

Beske-Diehl, S.J. & Li, H. (1993). Magnetic properties of hematite in lava flows from Iceland: Response to hydrothermal alteration. *J. Geophys. Res.*, **98**, 403–417.

Beske-Diehl, S. & Soroka, W.L. (1984). Magnetic properties of variably oxidized pillow basalt. *Geophys. Res. Lett.*, **11**, 217–220.

Bickford, L.R. (1950). Ferromagnetic resonance absorption in magnetite single crystal. *Phys. Rev.*, **78**, 449–457.

Bickford, L.R., Pappis, J. & Stull, J.L. (1955). Magnetostriction and permeability of magnetite and cobalt-substituted magnetite. *Phys. Rev.*, **99**, 1210–1214.

Bina, M. & Daly, L. (1994). Mineralogical change and self-reversed magnetizations in pyrrhotite resulting from partial oxidation; geophysical implications. *Phys. Earth Planet. Inter.*, **85**, 83–99.

Bina, M.M. & Henry, B. (1990). Magnetic properties, opaque mineralogy and magnetic anisotropies of serpentinized peridotites from ODP Hole 670A near the Mid-Atlantic Ridge. *Phys. Earth Planet. Inter.*, **65**, 88–103.

Bina, M.M. & Prévot, M. (1994). Thermally activated magnetic viscosity in natural multidomain titanomagnetite. *Geophys. J. Int.*, **117**, 495–510.

Biquand, D. (1971). Etude de quelques problèmes méthodologiques et techniques du paléomagnétisme des roches sédimentaires: Application à l'étude d'une formation lacustre d'âge mio-pliocène du Velay (Massif Central, France). *Ann. Géophys.*, **27**, 311–327.

Biquand, D. & Prévot, M. (1970). Sur la surprenante résistance à la destruction par champs magnétiques alternatifs de l'aimantation rémanente visqueuse acquise par certaines roches sédimentaires au cours d'un séjour, même bref, dans le champ magnétique terrestre. *C. R. Acad. Sci. Paris*, **270**, 362–365.

Biquand, D. & Prévot, M. (1971). A.F. demagnetization of viscous remanent magnetization in rocks. *Z. Geophys.*, **37**, 471–485.

Biquand, D. & Prévot, M. (1972). Comparaison des directions d'aimantation rémanente observées dans des roches volcaniques et sédimentaires mio-pliocènes du Velay. *Mem. Bur. Ressources Geol. Miner.*, **77**, 885–889.

Birks, J.B. (1950). The properties of ferromagnetic compounds at centimetre wavelengths. *Proc. Phys. Soc. (London)*, **B63**, 65–74.

Bitter, F. (1931). On inhomogeneities in the magnetization of ferromagnetic materials. *Phys. Rev.*, **38**, 1903–1905.

Blackman, M., Haigh, G. & Lisgarten, N.D. (1957). A new method of observing magnetic transformations. *Nature*, **179**, 1288–1290.

Blackman, M., Haigh, G. & Lisgarten, N.D. (1959). Investigations on the domain structure in hematite by means of the electron beam method. *Proc. Roy. Soc. London*, **A251**, 117–126.

Blakemore, R.P. (1975). Magnetotactic bacteria. *Science*, **190**, 377–379.

Blasse, G. (1964). Crystal chemistry and some magnetic properties of mixed metal oxides with spinel structure. *Philips Res. Rep.*, Suppl., **3**, 1–138.

Bleil, U. (1971). Cation distribution in titanomagnetites. *Z. Geophys.*, **37**, 305–319.

Bleil, U. & Petersen, N. (1983). Variations in magnetization intensity and low-temperature oxidation of ocean floor basalts. *Nature*, **301**, 384–388.

Bloemendal, J., King, J.W., Hall, F.R. & Doh, S-J. (1992). Rock magnetism of Late Neogene and Pleistocene deep-sea sediments: Relationship to sediment source, diagenetic processes and sediment lithology. *J. Geophys. Res.*, **97**, 4361–4375.

Blow, R.A. & Hamilton, N. (1978). Effect of compaction on the acquisition of a detrital remanent magnetization in fine-grained sediments. *Geophys. J. Roy. Astron. Soc.*, **52**, 13–23.

Bochirol, L. & Pauthenet, R. (1951). Aimantation spontanée des ferrites. *J. Physique*, **12**, 249–251.

Bogdanov, A.A. & Vlasov, A. Ya. (1965). Domain structure in a single crystal of magnetite: Changes of domain structure induced by an external magnetic field (transl. from Russian). *Izv., Phys. Solid Earth*, no. 1, 28–32.

Bogdanov, A.A. & Vlasov, A.Ya. (1966a). The domain structure on magnetite particles (transl. from Russian). *Izv., Phys. Solid Earth*, no. 9, 577–581.

Bogdanov, A.A. & Vlasov, A.Ya. (1966b). On the effect of elastic stresses on the domain structure of magnetite (transl. from Russian). *Izv., Phys. Solid Earth*, no. 1, 24–26.

Bol'shakov, A.S. & Shcherbakova, V.V. (1979). A thermomagnetic criterion for determining the domain structure of ferrimagnetics (transl. from Russian). *Izv., Phys. Solid Earth*, **15**, 111–117.

Bol'shakov, A.S., Gapeev, A.K., Dashevskaja, D.M., Melnikov, B.N. & Shcherbakov, V.P. (1978). Anomalous properties of the thermoremanence of synthetic ferrites (transl. from Russian). *Izv., Phys. Solid Earth*, **14**, 209–215.

Borradaile, G.J. (1988). Magnetic susceptibility, petrofabrics and strain – a review. *Tectonophys.*, **156**, 1–20.

Borradaile, G.J. (1991). Remanent magnetism and ductile deformation in an experimentally deformed magnetite-bearing limestone. *Phys. Earth Planet. Inter.*, **67**, 362–373.

Borradaile, G.J. (1992a). Deformation of remanent magnetism in a synthetic aggregate with hematite. *Tectonophys.*, **206**, 203–218.

Borradaile, G.J. (1992b). Experimental deformation of two-component IRM in magnetite-bearing limestone: A model for the behaviour of NRM during natural deformation. *Phys. Earth Planet. Inter.*, **70**, 64–77.

Borradaile, G.J. (1993a). Strain and magnetic remanence. *J. Struct. Geol.*, **15**, 383–390.

Borradaile, G. J. (1993b). The rotation of magnetic grains. *Tectonophys.*, **221**, 381–384.

Borradaile, G.J. (1994a). Paleomagnetism carried by crystal inclusions: The effect of preferred crystallographic orientations. *Earth Planet. Sci. Lett.*, **126**, 171–182.

Borradaile, G.J. (1994b). Remagnetization of a rock analogue during experimental triaxial deformation. *Phys. Earth Planet. Inter.*, **83**, 147–163.

Borradaile, G.J. (1996). A 1800-year archeological experiment in viscous remagnetization. *Geophys. Res. Lett.*, **23**, 1585–1588.

Borradaile, G.J & Jackson, M. (1993). Changes in magnetic remanence during simulated deep sedimentary burial. *Phys. Earth Planet. Inter.*, **77**, 315–327.

Borradaile, G.J. & Mothersill, J.S. (1991). Experimental strain of isothermal remanent magnetization in ductile sandstone. *Phys. Earth Planet. Inter.*, **65**, 308–318.

Borradaile, G.J., Chow, N. & Werner, T. (1993). Magnetic hysteresis of limestones: Facies control? *Phys. Earth Planet. Inter*, **76**, 241–252.

Boyd, J.R., Fuller, M. & Halgedahl, S.L. (1984). Domain wall nucleation as a controlling factor in the behaviour of fine magnetic particles in rocks. *Geophys. Res. Lett.*, **11**, 193–196.

Bozorth, R.M. (1951). *Ferromagnetism*. Van Nostrand, Princeton, NJ, 968 pp.

Brecher, A. (1977). Lunar and meteoritic palaeomagnetism: Common origin contested. *Nature*, **266**, 381–382.

Brecher, A. & Albright, L. (1977). The thermoremanence hypothesis and the origin of magnetization in iron meteorites. *J. Geomag. Geoelec.*, **29**, 379–400.

Brecher, A. & Arrhenius, G. (1974). The paleomagnetic record in carbonaceous chondrites: Natural remanence and magnetic properties. *J. Geophys. Res.*, **79**, 2081–2106.

Brecher, A. & Cutrera, M. (1976). A scanning electron microscope (SEM) study of the magnetic domain structure of iron meteorites and their synthetic analogs. *J. Geomag. Geoelec.*, **28**, 31–45.

Brecher, A. & Fuhrman, M. (1979). The magnetic effects of brecciation and shock in meteorites: II. The ureilites and evidence for strong nebular magnetic fields. *The Moon & Planets*, **20**, 251–263.

Brecher, A. & Ranganayaki, R.P. (1975). Paleomagnetic systematics of ordinary chondrites. *Earth Planet. Sci. Lett.*, **25**, 57–67.

Brewster, D. & O'Reilly, W. (1988). Magnetic properties of synthetic analogues of the altered olivines of igneous rocks. *Geophys. J.*, **95**, 421–432.

Briden, J.C. (1965). Ancient secondary magnetizations in rocks. *J. Geophys. Res.*, **70**, 5205–5221.

Brooks, P.J. & O'Reilly, W. (1970). Magnetic rotational hysteresis characteristics of red sandstones. *Earth Planet. Sci. Lett.*, **9**, 71–76.

Brown, K. & O'Reilly, W. (1987). Aqueous maghemitization of titanomagnetite. *J. Geophys.*, **61**, 82–89.

Brown, K. & O'Reilly, W. (1988). The effect of low - temperature oxidation on the remanence of TRM-carrying titanomagnetite $Fe_{2.4}Ti_{0.6}O_4$. *Phys. Earth Planet. Int.*, **52**, 108–116.

Brown, W.F. (1949). Irreversible magnetic effects of stress. *Phys. Rev.*, **75**, 147–154.

Brown, W.F. (1959). Relaxational behaviour of fine magnetic particles. *J. Appl. Phys.*, **30**, 130S–132S.

Brown, W.F. (1962). *Magnetostatic Principles in Ferromagnetism*. North-Holland, Amsterdam, 202 pp.

Brown, W.F. (1963a). *Micromagnetics*. Wiley-Interscience, New York, 143 pp. Reprinted in 1978 by Krieger, Huntingdon, NY.

Brown, W.F. (1963b). Thermal fluctuations of a single-domain particle. *Phys. Rev.*, **130**, 1677–1686.

Brunhes, B. (1906). Recherches sur la direction d'aimantation des roches volcaniques. *J. Physique*, **5**, 705–724.

Buchan, K.L. (1978). Magnetic overprinting in the Thanet gabbro complex, Ontario. *Can. J. Earth Sci.*, **15**, 1407–1421.

Buchan, K.L. (1979). Paleomagnetic studies of bulk mineral separates from the Bark Lake diorite, Ontario. *Can. J. Earth Sci.*, **16**, 1558–1565.

Buchan, K.L. & Dunlop, D.J. (1976). Paleomagnetism of the Haliburton intrusions: superimposed magnetizations, metamorphism, and tectonics in the late Precambrian. *J. Geophys. Res.*, **81**, 2951–2967.

Buchan, K.L. & Halls, H.C. (1990). Paleomagnetism of Proterozoic mafic dyke swarms of the Canadian Shield. In *Mafic Dykes and Emplacement Mechanisms*, ed A.J. Parker, P.C. Rickwood & D.H. Tucker, pp. 209–230. A.A. Balkema, Rotterdam.

Buchan, K.L., Berger, G.W., McWilliams, M.O., York, D. & Dunlop, D.J. (1977). Thermal overprinting of natural remanent magnetization and K/Ar ages in metamorphic rocks. *J. Geomag. Geoelec.*, **29**, 401–410.

Bucur, I. (1978). Experimental study of the origin and properties of the defect moment in single domain hematite. *Geophys. J. Roy. Astron. Soc.*, **55**, 589–604.

Buddington, A.F. & Lindsley, D.H. (1964). Iron-titanium oxide minerals and synthetic equivalents. *J. Petrol.*, **5**, 310–357.

Burmester, R.F. (1977). Origin and stability of drilling induced remanence. *Geophys. J. Roy. Astron. Soc.*, **48**, 1–14.

Butler, R.F. (1992). *Paleomagnetism: Magnetic Domains to Geologic Terranes*. Blackwell, Oxford, 319 pp.

Butler, R.F. & Banerjee, S.K. (1975a). Single-domain grain-size limits for metallic iron. *J. Geophys. Res.*, **80**, 252–259.

Butler, R. F. & Banerjee, S.K. (1975b). Theoretical single-domain grain-size range in magnetite and titanomagnetite. *J. Geophys. Res.*, **80**, 4049–4058.

Callen, H.B. & Callen, E.R. (1966). The present status of the temperature dependence of magnetocrystalline anisotropy, and the $l(l+1)/2$ power law. *J. Phys. Chem. Solids.*, **27**, 1271–1285.

Cann, J.R. (1979). Metamorphism in the oceanic crust. In *Deep Drilling Results in the Atlantic Ocean Crust*, ed. M. Talwani, C.G.A. Harrison & D.E. Hayes, pp. 230–238. American Geophysical Union, Washington, DC.

Carmichael, C.M. (1961). The magnetic properties of ilmenite-hematite crystals. *Proc. Roy. Soc. London*, **A263**, 508–530.

Carmichael, I.S.E. & Nicholls, J. (1967). Iron-titanium oxides and oxygen fugacities in volcanic rocks. *J. Geophys. Res.*, **72**, 4665–4687.

Celaya, M.A. & Clement, B.M. (1988). Inclination shallowing in deep-sea sediments from the North Atlantic. *Geophys. Res. Lett.*, **15**, 52–55.

Chamalaun, F.H. (1964). Origin of the secondary magnetization of the Old Red Sandstone of the Anglo-Welsh cuvette. *J. Geophys. Res.*, **69**, 4327–4337.

Chamalaun, F.H. & Creer, K.M. (1964). Thermal demagnetization studies of the Old Red Sandstone of the Anglo-Welsh cuvette. *J. Geophys. Res.*, **69**, 1607–1616.

Chang, S-B.R. & Kirschvink, J.L. (1989). Magnetofossils, the magnetization of sediments, and the evolution of magnetite biomineralization. *Ann. Rev. Earth Planet. Sci.*, **17**, 169–195.

Chang, S-B.R., Kirschvink, J.L. & Stolz, J.F. (1987). Biogenic magnetite as a primary remanence carrier in limestone. *Phys. Earth Planet. Inter.*, **46**, 289–303.

Channell, J.E.T. (1978). Dual magnetic polarity measured in a single bed of Cretaceous pelagic limestone from Sicily, *J. Geophys.*, **44**, 613–622.

Channell, J.E.T. & McCabe, C. (1994). Comparison of magnetic hysteresis parameters of unremagnetized and remagnetized limestones. *J. Geophys. Res.*, **99**, 4613–4623.

Channell, J.E.T., Freeman, R., Heller, F. & Lowrie, W. (1982). Timing of diagenetic haematite growth in red pelagic limestones from Gubbio (Italy). *Earth Planet. Sci. Lett.*, **58**, 189–201.

Channell, J.E.T., Heller, F. & van Stuijvenberg, J. (1979). Magnetic susceptibility anisotropy as an indicator of sedimentary fabric in the Gurnigel Flysch. *Eclog. Geol. Helv.*, **72**, 781–787.

Chevallier, R. (1951). Propriétés magnétiques de l'oxyde ferrique rhomboédrique ($Fe_2O_3\alpha$). *J. Physique.*, **12**, 172–188.

Chevallier, R. & Mathieu, S. (1943). Propriétés magnétiques des poudres d'hématite – influence des dimensions des grains. *Annales Phys.*, **18**, 258–288.

Chikazumi, S. (1964). *Physics of Magnetism*. John Wiley, New York, 554 pp.

Chowdhary, S.K., Collinson, D.W., Stephenson, A. & Runcorn, S.K. (1987). Further investigations into lunar palaeointensity determinations. *Phys. Earth Planet. Inter.*, **49**, 133–141.

Cisowski, S. (1977). Shock remanent magnetization in Deccan basalts due to the Lonar crater. Ph.D thesis, Univ. of Calif., Santa Barbara, CA.

Cisowski, S. (1981). Interacting vs. non-interacting single domain behavior in natural and synthetic samples. *Phys. Earth Planet. Inter.*, **26**, 56–62.

Cisowski, S. & Fuller, M. (1978). The effect of shock on the magnetism of terrestrial rocks. *J. Geophys. Res.*, **83**, 3441–3458.

Cisowski, S., Collinson, D.W., Runcorn, S.K., Stephenson, A. & Fuller, M. (1983). A review of lunar paleointensity data and implications for the origin of lunar magnetism. *J. Geophys Res.*, **88**, suppl. (Proc. 13th Lunar Sci. Conf.), pp. A691–A704.

Cisowski, S., Dunn, J.R., Fuller, M., Rose, M.F. & Wasilewski, P.J. (1974). Impact processes and lunar magnetism. *Geochim. Cosmochim. Acta.*, **38**, suppl. 5, 2841–2858.

Cisowski, S., Fuller, M., Rose, M.F. & Wasilewski, P.J. (1973). Magnetic effects of explosive shocking of lunar soil. *Geochim. Cosmochim. Acta.*, **37**, suppl. 4, 3003–3017.

Clark, D.A. (1984). Hysteresis properties of sized dispersed monoclinic pyrrhotite grains. *Geophys. Res. Lett.*, **11**, 173–176.

Clark, D.A. & Schmidt, P.W. (1982). Theoretical analysis of thermomagnetic properties, low-temperature hyseresis and domain structure of titanomagnetites. *Phys. Earth Planet. Inter.*, **30**, 300–316.

Clarke, R.S. & Scott, E.R.D. (1980). Tetrataenite – ordered FeNi, a new mineral in meteorites. *Am. Mineral.*, **65**, 624–630.

Clauter, D.A. & Schmidt, V.A. (1981). Shifts in blocking temperature spectra for magnetite powders as a function of grain size and applied magnetic field. *Phys. Earth Planet. Inter.*, **26**, 81–92.

Coe, R.S. (1967). The determination of paleointensities of the earth's magnetic field with emphasis on mechanisms which could cause non-ideal behavior in Thellier's method. *J. Geomag. Geoelec.*, **19**, 157–179.

Coe, R.S. (1979). The effect of shape anisotropy on TRM direction. *Geophys. J. Roy. Astron. Soc.*, **56**, 369–383.

Cogné, J-P. (1987a). Experimental and numerical modeling of IRM rotation in deformed synthetic samples. *Earth Planet. Sci. Lett.*, **86**, 39–45.

Cogné, J-P. (1987b). TRM deviations in anisotropic assemblages of multidomain magnetite. *Geophys. J. Roy. Astron. Soc.*, **91**, 1013–1023.

Cogné, J-P. (1988a). Strain, magnetic fabric, and paleomagnetism of the deformed red beds of the Pont-Réan formation, Brittany, France. *J. Geophys. Res.*, **93**, 13,673–13,687.

Cogné, J-P. (1988b). Strain-induced AMS in the granite of Flamanville and its effects upon TRM acquisition. *Geophys. J.*, **92**, 445–453.

Cogné, J-P. & Canot-Laurent, S. (1992). Simple shear experiments on magnetized wax-hematite samples. *Earth Planet. Sci. Lett.*, **112**, 147–159.

Cogné, J-P. & Gapais, D. (1986). Passive rotation of hematite during deformation: A comparison of simulated and natural redbeds fabrics. *Tectonophys.*, **121**, 365–372.

Cogné, J-P. & Perroud, H. (1985). Strain removal applied to paleomagnetic directions in an orogenic belt: The Permian red slates of the Alpes Maritimes, France. *Earth Planet. Sci. Lett.*, **72**, 125–140.

Cogné, J-P. & Perroud, H. (1987). Unstraining paleomagnetic vectors: The current state of debate. *Eos (Trans. Am. Geophys. Un.)*, **68**, 711–712.

Cogné, J-P. & Perroud, H. (1988). Anisotropy of magnetic susceptibility as a strain gauge in the Flamanville granite, NW France. *Phys. Earth Planet. Inter.*, **51**, 264–270.

Cogné, J-P., Perroud, H., Texier, M.P. & Bonhommet, N. (1986). Strain reorientation of hematite and its bearing upon remanent magnetization. *Tectonics*, **5**, 753–767.

Collinson, D.W. (1965a). Depositional remanent magnetization in sediments. *J. Geophys. Res.*, **70**, 4663–4668.

Collinson, D.W. (1965b). Origin of remanent magnetization and initial suceptibility of certain red sandstones. *Geophys. J. Roy. Astron. Soc.*, **9**, 203–217.

Collinson, D.W. (1966). Carrier of remanent magnetization in certain red sandstones. *Nature*, **210**, 516–517.

Collinson, D.W. (1969). Investigations into the stable remanent magnetization of sediments. *Geophys. J. Roy. Astron. Soc.*, **18**, 211–222.

Collinson, D.W. (1983). *Methods in Rock Magnetism and Palaeomagnetism: Techniques and Instrumentation*. Chapman and Hall, London & New York, 503 pp.

Collinson, D.W. (1987). Magnetic properties of the Olivenza meteorite – possible implications for its evolution and an early Solar System magnetic field. *Earth Planet. Sci. Lett.*, **84**, 369–380.

Collinson, D.W. (1993). Magnetism of the Moon – a lunar core dynamo or impact magnetization? *Surv. Geophys.*, **14**, 89–118.

Collinson, D.W. & Morden, S.J. (1994). Magnetic properties of howardite, eucrite and diogenite (HED) meteorites: Ancient magnetizing fields and meteorite evolution. *Earth Planet. Sci. Lett.*, **126**, 421–434.

Collinson, D.W., Stephenson, A. & Runcorn, S.K. (1973). Magnetic properties of Apollo 15 and 16 rocks. *Geochim. Cosmochim. Acta.*, **37**, suppl. 4, 2963–2976.

Collombat, H., Rochette, P. & Kent, D.V. (1993). Detection and correction of inclination shallowing in deep sea sediments using the anisotropy of anhysteretic remanence. *Bull. Soc. Géol. France*, **164**, 103–111.

Colombo, U., Fagherazzi, G., Gazzarrini, F., Lanzavecchia, G. & Sironi, G. (1964). Mechanisms in the first stage of oxidation of magnetites. *Nature*, **202**, 175–176.

Cox, A., ed. (1973). *Plate Tectonics and Geomagnetic Reversals*. Freeman, San Francisco, 702 pp.

Craik, D.J. (1995). *Magnetism: Principles and Applications*. John Wiley, New York and London, 459 pp.

Craik, D.J. & McIntyre, D.A. (1969). Anhysteretic magnetization process in multidomain crystals and polycrystals. *Proc. Roy. Soc. London.*, **A313**, 97–116.

Craik, D.J. & Tebble, R.S. (1965). *Ferromagnetism and Ferromagnetic Domains*. North-Holland, Amsterdam, 337 pp.

Creer, K.M. (1961). Superparamagnetism in red sandstones. *Geophys. J. Roy. Astron. Soc.*, **5**, 16–28.

Creer, K.M. (1974). Geomagnetic variations for the interval 7000–25,000 yr B.P. as recorded in a core of sediment from station 1474 of the Black Sea cruise of "Atlantis II". *Earth Planet. Sci. Lett.*, **23**, 34–42.

Creer, K.M., Molyneux, L., Vernet, J.P. & Wagner, J.J. (1975). Palaeomagnetic dating of 1-metre cores of sediment from Lake Geneva. *Earth Planet. Sci. Lett.*, **28**, 127–132.

Creer, K.M., Petersen, N. & Petherbridge, J. (1970). Partial self-reversal of remanent magnetization and anisotropy of viscous magnetization in basalts. *Geophys. J. Roy. Astron. Soc*, **21**, 471–483.

Creer, K.M., Tucholka, P. & Barton, C.E., eds. (1983). *Geomagnetism of Baked Clays and Recent Sediments*. Elsevier, Amsterdam, 324 pp.

Cui, Y., Verosub, K.L. & Roberts, A.P. (1994). The effect of low-temperature oxidation on large multi-domain magnetite. *Geophys. Res. Lett.*, **21**, 757–760.

Cullity, B.D. (1972). *Introduction to Magnetic Materials*. Addison-Wesley, Reading, MA, 666 pp.

Dahlberg, E.D. & Zhu, J-G. (1995). Micromagnetic microscopy and modeling. *Physics Today*, **48**, no. 4, 34–40.

Dallmeyer, R.D. & Sutter, J.F. (1980). Acquisitional chronology of remanent magnetization along the 'Grenville polar path': Evidence from $^{40}Ar/^{39}Ar$ ages of hornblende and biotite from the Whitestone diorite, Ontario. *J. Geophys. Res*, **85**, 3177–3186.

Dankers, P. (1981). Relationship between median destructive field and remanent coercive forces for dispersed natural magnetite, titanomagnetite and hematite. *Geophys. J. Roy. Astron. Soc.*, **64**, 447–461.

Dankers, P. & Sugiura, N. (1981). The effects of annealing and concentration on the hysteresis properties of magnetite around the PSD-MD transition. *Earth Planet. Sci. Lett.*, **56**, 422–428.

Dankers, P.H.M. & Zijderveld, J.D.A. (1981). Alternating field demagnetization of rocks, and the problem of gyromagnetic remanence. *Earth Planet. Sci. Lett.*, **53**, 89–92.

David, P. (1904). Sur la stabilité de la direction d'aimantation dans quelques roches volcaniques. *C. R. Acad. Sci. Paris*, **138**, 41–42.

Davis, K.E. (1981). Magnetite rods in plagioclase as the primary carrier of stable NRM in ocean floor gabbros. *Earth Planet. Sci. Lett.*, **55**, 190–198.

Davis, P.M. & Evans, M.E. (1976). Interacting single-domain properties of magnetite intergrowths. *J. Geophys. Res.*, **81**, 989–994.

Day, R. (1973). The effect of grain size on the magnetic properties of the magnetite-ulvospinel solid solution series. Ph.D. thesis, Univ. of Pittsburgh, Pittsburgh, PA.

Day, R. (1977). TRM and its variation with grain size. *J. Geomag. Geoelec.*, **29**, 233–265.

Day, R., Fuller, M.D. & Schmidt, V.A. (1977). Hysteresis properties of titanomagnetites: Grain size and composition dependence. *Phys. Earth Planet. Inter.*, **13**, 260–267.

Deamer, G.A. & Kodama, K.P. (1990). Compaction-induced inclination shallowing in synthetic and natural clay-rich sediments. *J. Geophys. Res.*, **95**, 4511–4529.

DeBlois, R.W. & Graham, C.D. (1958). Domain observations on iron whiskers. *J. Appl. Phys.*, **29**, 931–939.

Dekkers, M.J. (1988). Magnetic properties of natural pyrrhotite. I: Behaviour of initial susceptibility and saturation-magnetization-related rock-magnetic parameters in a grain-size dependent framework. *Phys. Earth Planet. Inter.*, **52**, 376–393.

Dekkers, M.J. (1989a). Magnetic properties of natural goethite – II. TRM behaviour during thermal and alternating field demagnetization and low-temperature treatment. *Geophys. J.*, **97**, 341–355.

Dekkers, M.J. (1989b). Magnetic properties of natural pyrrhotite. II. High- and low-temperature behaviour of J_{rs} and TRM as a function of grain size. *Phys. Earth Planet. Inter.*, **57**, 266–283.

Dekkers, M.J. (1990). Magnetic monitoring of pyrrhotite alteration during thermal demagnetization. *Geophys. Res. Lett.*, **17**, 779–782.

Dekkers, M.J., Mattéi, J-L., Fillion, G. & Rochette, P. (1989). Grain-size dependence of the magnetic behaviour of pyrrhotite during its low-temperature transition at 34 K. *Geophys. Res. Lett.*, **16**, 855–858.

deMenocal, P.B., Ruddiman, W.F. & Kent, D.V. (1990). Depth of post-depositional remanence acquisition in deep-sea sediments: A case study of the Brunhes-Matuyama reversal and oxygen isotopic stage 19.1. *Earth Planet. Sci. Lett.*, **99**, 1–13.

Denham, C.R. (1979). Oblique anhysteretic remanent magnetization. *J. Geophys. Res.*, **84**, 6286–6290.

Diaz Ricci, J.C. & Kirschvink, J.L. (1992). Magnetic domain state and coercivity predictions for biogenic greigite (Fe_3S_4): A comparison of theory with magnetosome observations. *J. Geophys. Res.*, **97**, 17,309–17,315.

Dodson, M.H. & McClelland Brown, E. (1980). Magnetic blocking temperatures of single domain grains during slow cooling. *J. Geophys. Res.*, **85**, 2625–2637.

Doraiswami, M.S. (1947). Elastic constants of magnetite, pyrite and chromite. *Proc. Indian Acad. Sci.*, **25**, 413–416.

Dubuisson, G., Pozzi, J-P. & Galdeano, A. (1991). Aimantation à haute température de roches profondes du forage continental de Couy (Bassin de Paris): Discussion d'une contribution aux anomalies magnétiques. *C. R. Acad. Sci. (Paris)*, **312**, 385–392.

Dunlop, D.J. (1969). Hysteretic properties of synthetic and natural monodomain grains. *Phil. Mag.*, **19**, 329–338.

Dunlop, D.J. (1971). Magnetic properties of fine-particle hematite. *Ann. Géophys.*, **27**, 269–293.

Dunlop, D.J. (1972). Magnetic mineralogy of unheated and heated red sediments by coercivity spectrum analysis. *Geophys. J. Roy. Astron. Soc.*, **27**, 37–55.

Dunlop, D.J. (1973a). Superparamagnetic and single-domain threshold sizes in magnetite. *J. Geophys. Res.*, **78**, 1780–1793.

Dunlop, D.J. (1973b). Thermoremanent magnetization in submicroscopic magnetite. *J. Geophys. Res.*, **78**, 7602–7613.

Dunlop, D.J. (1973c). Theory of the magnetic viscosity of lunar and terrestrial rocks. *Rev. Geophys. Space Phys.*, **11**, 855–901.

Dunlop, D.J. (1974). Thermal enhancement of magnetic susceptibility. *J. Geophys.*, **40**, 439–451.

Dunlop, D.J. (1976). Thermal fluctuation analysis: A new technique in rock magnetism. *J. Geophys. Res.*, **81**, 3511–3517.

Dunlop, D.J. (1977). The hunting of the 'psark'. *J. Geomag. Geoelec.*, **29**, 293–318.

Dunlop, D.J. (1979). On the use of Zijderveld vector diagrams in multicomponent paleomagnetic studies. *Phys. Earth Planet. Inter.*, **20**, 12–24.

Dunlop, D.J. (1981). The rock magnetism of fine particles. *Phys. Earth Planet. Inter.*, **26**, 1–26.

Dunlop, D.J. (1982). Field dependence of magnetic blocking temperature: Analog tests using coercive force data. *J. Geophys. Res.*, **87**, 1121–1126.

Dunlop, D.J. (1983a). On the demagnetizing energy and demagnetizing factor of a multidomain ferromagnetic cube. *Geophys. Res. Lett.*, **10**, 79–82.

Dunlop, D.J. (1983b). Viscous magnetization of 0.04–100 μm magnetites. *Geophys. J. Roy. Astron. Soc.*, **74**, 667–687.

Dunlop, D.J. (1983c). Determination of domain structure in igneous rocks by alternating field and other methods. *Earth Planet. Sci. Lett.*, **63**, 353–367.

Dunlop, D.J. (1984). A method of determining demagnetizing factor from multidomain hysteresis. *J. Geophys. Res.*, **89**, 553–558.

Dunlop, D.J. (1985). Paleomagnetism of Archean rocks from northwestern Ontario: V. Poohbah Lake alkaline complex, Quetico Subprovince. *Can. J. Earth Sci.*, **22**, 27–38.

Dunlop, D.J. (1986a). Coercive forces and coercivity spectra of submicron magnetites. *Earth Planet. Sci. Lett.*, **78**, 288–295.

Dunlop, D.J. (1986b). Hysteresis properties of magnetite and their dependence on particle size: A test of pseudo-single-domain remanence models. *J. Geophys. Res.*, **91**, 9569–9584.

Dunlop, D.J. (1987). Temperature dependence of hysteresis in 0.04–0.22 µm magnetites and implications for domain structure. *Phys. Earth Planet. Inter.*, **46**, 100–119.

Dunlop, D.J. (1990). Developments in rock magnetism. *Rep. Prog. Phys.*, **53**, 707–792.

Dunlop, D.J. (1995a). Magnetism in rocks. *J. Geophys. Res.*, **100**, 2161–2174.

Dunlop, D.J. (1995b). Using magnetic properties to determine grain size and domain state (abstract). *Ann. Geophysicae*, **13**, suppl. 1, C63.

Dunlop, D.J. & Argyle, K.S. (1991). Separating multidomain and single-domain-like remanences in pseudo-single-domain magnetites (215–540 nm) by low temperature demagnetization. *J. Geophys. Res.*, **96**, 2007–2017.

Dunlop, D.J. & Argyle, K.S. (1996). Thermoremanence and anhysteretic remanence of pseudo-single-domain magnetites (215–540 nm). *J. Geophys. Res.*, **101**, in press

Dunlop, D.J. & Bina, M-M. (1977). The coercive force spectrum of magnetite at high temperatures: Evidence for thermal activation below the blocking temperature. *Geophys. J. Roy. Astron. Soc.*, **51**, 121–147.

Dunlop, D.J. & Buchan, K.L. (1977). Thermal remagnetization and the paleointensity record of metamorphic rocks. *Phys. Earth Planet. Inter.*, **13**, 325–331.

Dunlop, D.J. & Hale, C.J. (1976). A determination of paleomagnetic field intensity using submarine basalts drilled near the Mid-Atlantic ridge. *J. Geophys. Res.*, **81**, 4166–4172.

Dunlop, D.J. & Hale, C.J. (1977). Simulation of long-term changes in the magnetic signal of the oceanic crust. *Can. J. Earth Sci.*, **14**, 716–744.

Dunlop, D.J. & Özdemir, Ö. (1990). Alternating field stability of high-temperature viscous remanent magnetization. *Phys. Earth Planet. Inter.*, **65**, 188–196.

Dunlop, D.J. & Özdemir, Ö. (1993). Thermal demagnetization of VRM and pTRM of single domain magnetite: No evidence for anomalously high unblocking temperatures. *Geophys. Res. Lett.*, **20**, 1939–1942.

Dunlop, D.J. & Özdemir, Ö. (1995). Thermal and low-temperature demagnetization of thermoviscous overprints carried by multidomain magnetite (abstract). *Eos (Trans. Am. Geophys. Un.)*, **76**, Fall Meeting suppl., F160.

Dunlop, D.J. & Prévot, M. (1982). Magnetic properties and opaque mineralogy of drilled submarine intrusive rocks. *Geophys. J. Roy. Astron. Soc.*, **69**, 763–802.

Dunlop, D.J. & Stirling, J.M. (1977). "Hard" viscous remanent magnetization (VRM) in fine-grained hematite. *Geophys. Res. Lett.*, **4**, 163–166.

Dunlop, D.J. & Waddington, E.D. (1975). The field dependence of thermoremanent magnetization in igneous rocks. *Earth Planet. Sci. Lett.*, **25**, 11–25.

Dunlop, D.J. & West, G.F. (1969). An experimental evaluation of single domain theories. *Rev. Geophys.*, **7**, 709–757.

Dunlop, D.J. & Xu, S. (1993). A comparison of methods of granulometry and domain structure determination (abstract). *Eos (Trans. Am. Geophys. Un.)*, **74**, Fall Meeting suppl., 203.

Dunlop, D.J. & Xu, S. (1994). Theory of partial thermoremanent magnetization in multidomain grains. 1. Repeated identical barriers to wall motion (single microcoercivity). *J. Geophys. Res.*, **99**, 9005–9023.

Dunlop, D.J. & Zinn, M.B. (1980). Archeomagnetism of a 19th century pottery kiln near Jordan, Ontario. *Can. J. Earth Sci.*, **17**, 1275–1285.

Dunlop, D.J., Argyle, K.S. & Bailey, M.E. (1990c). High-temperature techniques for measuring microcoercivity, crystallite size, and domain state. *Geophys. Res. Lett.*, **17**, 767–770.

Dunlop, D.J., Bailey, M.E. & Westcott-Lewis, M.F. (1975). Lunar paleointensity determination using anhysteretic remanence (ARM): A critique. *Geochim. Cosmochim. Acta*, **39**, suppl. 6, 3063–3069.

Dunlop, D.J., Enkin, R.J. & Tjan, E. (1990a). Internal field mapping in single-domain and multidomain grains. *J. Geophys. Res.*, **95**, 4561–4577.

Dunlop, D.J., Gose, W.A., Pearce, G.W. & Strangway, D.W. (1973b). Magnetic properties and granulometry of metallic iron in lunar breccia 14313. *Geochim. Cosmochim. Acta,* suppl. 4, vol. 3, pp. 2977–2990.

Dunlop, D.J., Hanes, J.A. & Buchan, K.L. (1973a). Indices of multidomain magnetic behaviour in basic igneous rocks: Alternating-field demagnetization, hysteresis, and oxide petrology. *J. Geophys. Res.,* **78**, 1387–1393.

Dunlop, D.J., Heider, F. & Xu, S. (1995b). Single domain moments in multidomain magnetite (abstract). *Int. Un. Geod. Geophys. XXI Gen. Ass., Abstracts Vol.,* B106.

Dunlop, D.J., Newell, A.J. & Enkin, R.J. (1994). Transdomain thermoremanent magnetization. *J. Geophys. Res.,* **99**, 19,741–19,755.

Dunlop, D.J., Özdemir, Ö. & Enkin, R.J. (1987). Multidomain and single domain relations between susceptibility and coercive force. *Phys. Earth Planet. Inter.,* **49**, 181–191.

Dunlop, D.J., Özdemir, Ö. & Schmidt, P.W. (1995a). Thermoviscous remagnetization of multidomain magnetite (abstract). *Int. Un. Geod. Geophys. XXI Gen. Ass., Abstracts Vol.,* B136.

Dunlop, D.J., Ozima, Minoru, & Kinoshita, H. (1969). Piezomagnetization of single-domain grains: A graphical approach. *J. Geomag. Geoelec.,* **21**, 513–518.

Dunlop, D.J., Schutts, L.D. & Hale, C.J. (1984). Paleomagnetism of Archean rocks from northwestern Ontario: III. Rock magnetism of the Shelley Lake granite, Quetico Subprovince. *Can. J. Earth Sci.,* **21**, 879–886.

Dunlop, D.J., Stacey, F.D. & Gillingham, D.E.W. (1974). The origin of thermoremanent magnetization: Contribution of pseudo-single-domain magnetic moments. *Earth Planet. Sci. Lett.,* **21**, 288–294.

Dunlop, D.J., Westcott-Lewis, M.F. & Bailey, M. E. (1990b). Preisach diagrams and anhysteresis: Do they measure interactions? *Phys. Earth Planet. Inter.,* **65**, 62–77.

Dunlop, D.J., York, D., Berger, G.W., Buchan, K.L. & Stirling, J.M. (1980). The Grenville Province: A case-study of Precambrian continental drift. In *The Continental Crust and its Mineral Deposits,* ed. D.W. Strangway, pp. 487–502. Geological Association of Canada, St. John's, Nfld., Special paper 20.

Dzyaloshinsky, I. (1958). A thermodynamic theory of "weak" ferromagnetism of antiferromagnetics. *J. Phys. Chem. Solids,* **4**, 241–255.

Eaton, J.A. & Morrish, A.H. (1969). Magnetic domains in hematite at and above the Morin transition. *J. Appl. Phys.,* **40**, 3180–3185.

Edwards, J. (1980). An experiment relating to rotational remanent magnetization and frequency of demagnetizing field. *Geophys. J. Roy. Astron. Soc.,* **60**, 283–288.

Elder, T. (1965). Particle size effect in oxidation of natural magnetite. *J. Appl. Phys.,* **36**, 1012–1013.

Ellwood, B.B., Balsam, W., Burkart, B., Long, G.J. & Buhl, M.L. (1986). Anomalous magnetic properties in rocks containing the mineral siderite: Paleomagnetic implications. *J. Geophys. Res.,* **91**, 12,779–12,790.

Ellwood, B.B., Chrzanowski, T.J., Long, G.J. & Buhl, M.L. (1988). Siderite formation in anoxic deep-sea sediments: A synergistic bacterially controlled process with important implications in paleomagnetism. *Geology,* **16**, 980–982.

Elmore, R.D. & Leach, M.C. (1990). Remagnetization of the Rush Springs Formation, Cement, Oklahoma: Implications for dating hydrocarbon migration and aeromagnetic exploration. *Geology,* **18**, 124–127.

Elmore, R.D. & McCabe, C. (1991). The occurrence and origin of remagnetization in the sedimentary rocks of North America. *Rev. Geophys.,* **29**, suppl. (IUGG report), 377–383.

Elmore, R.D., Engel, M.H., Crawford, L., Nick, K., Imbus, S. & Sofer, Z. (1987). Evidence for a relationship between hydrocarbons and authigenic magnetite. *Nature,* **325**, 428–430.

Elmore, R.D., Imbus, S.W., Engel, M.H. & Fruit, D. (1993b). Hydrocarbons and magnetizations in magnetite. In *Applications of Paleomagnetism to Sedimentary Geology,* pp. 181–191. Society for Sedimentary Geology, Spec. Publ. 49, ISBN 1-56567-002-6.

Elmore, R.D., London, D., Bagley, D. & Fruit, D. (1993a). Remagnetization by basinal fluids: Testing the hypothesis in the Viola Limestone, southern Oklahoma. *J. Geophys. Res.,* **98**, 6237–6254.

Elston, D.P. & Purucker, M.E. (1979). Detrital magnetization in red beds of the Moenkopi Formation (Triassic), Gray Mountain, Arizona. *J. Geophys. Res.,* **84**, 1653–1665.

Enkin, R.J. & Dunlop, D.J. (1987). A micromagnetic study of pseudo-single-domain remanence in magnetite. *J. Geophys. Res.,* **92**, 12,726–12,740.

Enkin, R.J. & Dunlop, D.J. (1988). The demagnetization temperature necessary to remove viscous remanent magnetization. *Geophys. Res. Lett.,* **15**, 514–517.

Enkin, R.J. & Williams, W. (1994). Three-dimensional micromagnetic analysis of stability in fine magnetic grains. *J. Geophys. Res.*, **99**, 611–618.

Evans, M.E. (1968). Magnetization of dikes: A study of the paleomagnetism of the Widgiemooltha dike suite, Western Australia. *J. Geophys. Res.*, **73**, 3261–3270.

Evans, M.E. (1977). Single domain oxide particles as a source of thermoremanent magnetization. *J. Geomag. Geoelec.*, **29**, 267–275.

Evans, M.E. & Heller, F. (1994). Magnetic enhancement and palaeoclimate: Study of a loess/palaeosol couplet across the Loess Plateau of China. *Geophys. J. Int.*, **117**, 257–264.

Evans, M.E. & McElhinny, M.W. (1969). An investigation of the origin of stable remanence in magnetite-bearing rocks. *J. Geomag. Geoelec.*, **21**, 757–773.

Evans, M.E. & Wayman, M.L. (1970). An investigation of small magnetic particles by means of electron microscopy. *Earth Planet. Sci. Lett.*, **9**, 365–370.

Evans, M.E. & Wayman, M.L. (1974). An investigation of the role of ultrafine titanomagnetite intergrowths in paleomagnetism. *Geophys. J. Roy. Astron. Soc.*, **36**, 1–10.

Everitt, C.W.F. (1961). Thermoremanent magnetization. I. Experiments on single-domain grains. *Phil. Mag.*, **6**, 713–726.

Everitt, C.W.F. (1962a). Thermoremanent magnetization. III. Theory of multidomain grains. *Phil. Mag.*, **7**, 599–616.

Everitt, C.W.F. (1962b). Thermoremanent magnetization. II. Experiments on multidomain grains. *Phil. Mag.*, **7**, 583–598.

Fabian, K., Kirchner, A., Williams, W., Heider, F., Hubert, A. & Leibl, T. (1996). Three-dimensional micromagnetic calculations for magnetite using FFT. *Geophys. J. Int.*, **124**, 89–104.

Fassbinder, J.W.E. & Stanjek, H. (1994). Magnetic properties of biogenic soil greigite (Fe_3S_4). *Geophys. Res. Lett.*, **21**, 2349–2352.

Fassbinder, J.W.E., Stanjek, H. & Vali, H. (1990). Occurrence of magnetic bacteria in soil. *Nature*, **343**, 161–163.

Fine, P., Singer, M.J., La Ven, R., Verosub, K.L. & Southard, R.J. (1989). Role of pedogenesis in distribution of magnetic susceptibility in two California chronosequences. *Geoderma*, **44**, 287–306.

Fine, P., Verosub, K.L. & Singer, M.J. (1995). Pedogenic and lithogenic contributions to the magnetic suceptibility record of the Chinese loess/palaeosol sequence. *Geophys. J. Int.*, **122**, 97–107.

Flanders, P.J. (1988). An alternating-gradient magnetometer. *J. Appl. Phys.*, **63**, 3940–3945.

Flanders, P.J. & Remeika, J.P. (1965). Magnetic properties of hematite single crystals. *Phil. Mag.*, **11**, 1271–1288.

Flanders, P.J. & Schuele, W.J. (1964). Anisotropy in the basal plane of hematite single crystals. *Phil. Mag.*, **9**, 485–490.

Fletcher, E.J. (1975). The temperature dependence of the magnetocrystalline anisotropy constants of titanium-doped magnetite. Ph.D. thesis, University of Newcastle upon Tyne.

Fletcher, E.J. & O'Reilly, W. (1974). Contribution of Fe^{2+} ions to the magnetocrystalline anisotropy constant K_1 of $Fe_{3-x}Ti_xO_4 (0 < x < 0.1)$. *J. Phys. C: Sol. State Phys.*, **7**, 171–178.

Fox, J.M.W. & Aitken, M.J. (1980). Cooling-rate dependence of thermoremanent magnetisation. *Nature*, **283**, 462–463.

Freeman, R. (1986). Magnetic mineralogy of pelagic limestones. *Geophys. J. Roy. Astron. Soc.*, **85**, 433–452.

Frei, E.J., Shtrikman, S. & Treves, D. (1957). Critical size and nucleation field of ideal ferromagnetic particles. *Phys. Rev.*, **106**, 446–455.

Fröhlich, F. & Vollstädt, H. (1967). Untersuchungen zur Bestimmung der Curietemperatur von Maghemit (γ-Fe_2O_3). *Monatsber. Deutsch. Akad. Wiss., Berlin*, **9**, 180–186.

Fukuma, K. (1992). A numerical simulation of magnetostatic coagulation in a fluid. *Geophys. J. Int.*, **111**, 357–362.

Fukuma, K. & Dunlop, D.J. (1995). Monte Carlo simulation of two-dimensional domain structures in magnetite (abstract). *Eos (Trans. Am. Geophys. Un.)*, **76**, Fall Meeting suppl., p. F160.

Fuller, M. (1970). Geophysical aspects of paleomagnetism. *Crit Rev. Solid State Phys.*, **1**, 137–219.

Fuller, M. (1974). Lunar magnetism. *Rev. Geophys. Space Phys.*, **12**, 23–70.

Fuller, M. (1984). On the grain size dependence of the behaviour of fine magnetic particles in rocks. *Geophys. Surv.* **7**, 75–87.

Fuller, M. (1987). Lunar palaeomagnetism. In *Geomagnetism.*, vol. 2, ed. J.A. Jacobs, pp. 307–448. Academic Press, London & New York.

Fuller, M.D. & Kobayashi, K. (1964). Identification of the magnetic phases carrying natural remanent magnetization in certain rocks. *J. Geophys. Res.*, **69**, 4409–4413.

Funaki, M. (1993). Temperature dependence of coercivity for chondrites: Allende, Allan Hills-769, and Nuevo Mercurio. In *Proc. Sympos. Antarct. Meteorites*, no. 6, *Nat. Inst. Polar Res., Tokyo*, pp. 391–400.

Funaki, M., Nagata, T. & Momose, K. (1981). Natural remanent magnetizations of chondrules, metallic grains and matrix of an Antarctic chondrite, ALH-769. In *Proc. 6th Sympos. Antarct. Meteorites. Mem. Nat. Inst. Polar Res., Tokyo*, spec. iss. **30**, pp. 300–315.

Funaki, M., Sakai, H., Matsunaga, T. & Hirose, S. (1992). The S pole distribution on magnetic grains in pyroxenite determined by magnetotactic bacteria. *Phys. Earth Planet. Inter.*, **70**, 253–260.

Futschik, K., Pfützner, H., Doblander, A., Schönhuber, P., Dobeneck, T., Petersen, N. & Vali, H. (1989). Why not use magnetotactic bacteria for domain analyses? *Physica Scripta*, **40**, 518–521.

Gallagher, K.J., Feitknecht, W. & Mannweiler, U. (1968). Mechanism of oxidation of magnetite to γ-Fe_2O_3. *Nature*, **217**, 1118–1121.

Gallon, T.E. (1968). The remanent magnetization of haematite single crystals. *Proc. Roy. Soc. London*, **A303**, 511–524.

Gapeev, A.K., Gribov, S.K., Dunlop, D.J., Özdemir, Ö. & Shcherbakov, V.P. (1991). A direct comparison of the properties of CRM and VRM in the low temperature oxidation of magnetite. *Geophys. J. Int.*, **105**, 407–418.

Gapeyev, A.K. & Tsel'movich, V.A. (1983). Microtexture of synthetic titanomagnetite oxidized at high partial pressures of oxygen (transl. from Russian). *Izv., Phys. Solid Earth*, **19**, 983–986.

Gapeyev, A.K. & Tsel'movich, V.A. (1988). Stages of oxidation of titanomagnetite grains in igneous rocks (in Russian). Viniti N, Moscow, 1331-B89, pp.3–8.

Gaunt, P.J. (1960). A magnetic study of precipitation in a gold-cobalt alloy. *Phil. Mag.*, **5**, 1127–1145.

Gaunt, P.J. (1977). The frequency constant for thermal activation of a ferromagnetic domain wall. *J. Appl. Phys.*, **48**, 4370–4374.

Gaunt, P.J. & Mylvaganam, C.K. (1979). The thermal activation of magnetic domain walls from continuous and discontinuous planar pinning sites. *Phil. Mag*, **39**, 313–320.

Gauss, C.F. (1839). *Allgemeine Theorie des Erdmagnetismus*, Leipzig. Reprinted in *C.F. Gauss, Werke*, vol. 5, pp. 121–193. König. Gesellsch. Wissen., Göttingen, 1877.

Gee, J. & Kent, D.V. (1994). Variations in Layer 2A thickness and the origin of the central anomaly magnetic high. *Geophys. Res. Lett.*, **21**, 297–300.

Gee, J. & Kent, D.V. (1995). Magnetic hysteresis in young mid-ocean ridge basalts: Dominant cubic anisotropy? *Geophys. Res. Lett.*, **22**, 551–554.

Geiss, C.E., Heider, F. & Soffel, H.C. (1996). Magnetic domain observations on magnetite and titanomagnetite grains (0.5 µm–10 µm). *Geophys. J. Int.*, **124**, 75–88.

Geissman, J.W., Harlan, S.S. & Brearley, A.J. (1988). The physical isolation and identification of carriers of geologically stable remanent magnetization: Paleomagnetic and rock magnetic microanalysis and electron microscopy. *Geophys. Res. Lett.*, **15**, 479–482.

Gie, T.I. & Biquand, D. (1988). Etude expérimentale de l'aimantation rémanente chimique acquise au cours des transformations réciproques hématite-magnétite. *J. Geomag. Geoelec.*, **40**, 177–206.

Gilbert, W. (1600). *De Magnete*. Reprinted by Dover, New York, 1958.

Gillingham, D.E.W. & Stacey, F.D. (1971). Anhysteretic remanent magnetization (A.R.M.) in magnetite grains. *Pure Appl. Geophys.*, **91**, 160–165.

Girdler, R.W. (1963). Sur l'application de pressions hydrostatiques à des aimantations thermorémanentes. *Ann. Géophys.*, **19**, 118–121.

Gordon, R.G. (1990). Test for bias in paleomagnetically determined paleolatitudes from Pacific plate Deep Sea Drilling Project sediments. *J. Geophys. Res.*, **95**, 8397–8404.

Gorter, E.W. (1954). Saturation magnetization and crystal chemistry of ferrimagnetic oxides. *Philips Res. Rep.*, **9**, 295–320 & 321–355.

Gose, W.A. & Helsley, C.E. (1972). Paleomagnetic and rock magnetic studies of the Permian Cutler and Elephant Canyon formations. *J. Geophys. Res.*, **77**, 1534–1548.

Gose, W.A., Pearce, G.W., Strangway, D.W. & Larson, E.E. (1972). Magnetic properties of Apollo 14 breccias and their correlation with metamorphism. *Geochim. Cosmochim. Acta*, **36**, suppl. 3, 2387–2395.

Gose, W.A., Strangway, D.W. & Pearce, G.W. (1978). Origin of magnetization in lunar breccias: An example of thermal overprinting. *Earth Planet. Sci. Lett.*, **38**, 373–384.

Goto, K. & Sakurai, T. (1977). A colloid-SEM method for the study of fine magnetic domain structures. *Appl. Phys. Lett.*, **30**, 355–356.

Gottschalk, V.H. (1935). The coercive force of magnetite powders. *Physics*, **6**, 127–132.

Graham, J.W. (1956). Paleomagnetism and magnetostriction. *J. Geophys. Res.*, **61**, 735–739.

Graham, K.W.T. (1961). The re-magnetization of a surface outcrop by lightning currents. *Geophys. J. Roy. Astron. Soc.*, **6**, 85–102.

Graham, S. (1974). Remanent magnetization of modern tidal flat sediments from San Francisco Bay, California. *Geology*, **2**, 223–226.

Granar, L. (1958). Magnetic measurements on Swedish varved sediments. *Ark. Geofys.*, **3**, 1–40.

Griffiths, D.H., King, R.F., Rees, A.I. & Wright, A.E. (1960). The remanent magnetism of some recent varved sediments. *Proc. Roy. Soc. London*, **A256**, 359–383.

Grommé, S., Mankinen, E.A., Marshall, M. & Coe, R.S. (1979). Geomagnetic paleointensities by the Thelliers' method from submarine pillow basalts: Effects of seafloor weathering. *J. Geophys. Res.*, **84**, 3553–3575.

Grommé, C.S., Wright, T.L. & Peck, D.L. (1969). Magnetic properties and oxidation of iron-titanium oxide minerals in Alae and Makaopuhi lava lakes, Hawaii. *J. Geophys. Res.*, **74**, 5277–5293.

Gus'kova, Ye.G. & Pochtarev, V.I. (1967). Magnetic fields in space according to a study of the magnetic properties of meteorites (transl. from Russian). *Geomag. Aeron.*, **7**, 245–250.

Gustard, B. (1967). The ferromagnetic domain structure in haematite. *Proc. Roy. Soc. London*, **A297**, 269–274.

Gustard, B. & Schuele, W.J. (1966). Anomalously high remanence in $(\gamma\text{-Fe}_2\text{O}_3)_{1-x}(\alpha\text{-Fe}_2\text{O}_3)_x$ particles. *J. Appl. Phys.*, **37**, 1168–1169.

Haag, M. & Allenspach, R. (1993). A novel approach to domain imaging in natural Fe/Ti oxides by spin-polarized scanning electron microscopy. *Geophys. Res. Lett.*, **20**, 1943–1946.

Haag, M., Heller, F. & Allenspach, R. (1988). Magnetic interaction in self-reversing andesitic pumice in relation to iron alloys. *J. Physique*, **49**, Coll. C8, 2065–2066.

Haag, M., Heller, F., Allenspach, R. & Roche, K. (1990). Self-reversal of remanent magnetization in andesitic pumice. *Phys. Earth Planet. Inter.*, **65**, 104–108.

Haag, M., Heller, F., Lutz, M. & Reusser, E. (1993). Domain observations of the magnetic phases in volcanics with self-reversed magnetization. *Geophys. Res. Lett.*, **20**, 675–678.

Haggerty, S.E. (1970). Magnetic minerals in pelagic sediments. *Yearb. Carnegie Inst. Washington.*, **68**, 332–336.

Haggerty, S.E. (1976). Oxidation of opaque mineral oxides in basalts, and Opaque mineral oxides in terrestrial igneous rocks. In *Oxide Minerals*, 1st edn., ed. D. Rumble, pp. Hg1–177.

Mineralogical Society of America, Washington, DC.

Haggerty, S.E. & Baker, I. (1967). The alteration of olivine in basaltic and associated lavas. Part I. High temperature alteration. *Contrib. Miner. Petrol.*, **16**, 233–257.

Haggerty, S.E. & Toft, P.B. (1985). Native iron in the continental lower crust: Petrological and geophysical implications. *Science*, **229**, 647–649.

Hagstrum, J.T. & Johnson, C.M. (1986). A paleomagnetic and stable isotope study of the pluton at Rio Hondo near Questa, New Mexico: Implications for CRM related to hydrothermal alteration. *Earth Planet. Sci. Lett.*, **78**, 296–314.

Haigh, G. (1957). Observations on the magnetic transition in hematite at $-15°C$. *Phil. Mag.*, **2**, 877–890.

Haigh, G. (1958). The process of magnetization by chemical change. *Phil. Mag.*, **3**, 267–286.

Hale, C.J. & Dunlop, D.J. (1984). Evidence for an early Archean geomagnetic field: A paleomagnetic study of the Komati Formation, Barberton greenstone belt, South Africa. *Geophys. Res. Lett.*, **11**, 97–100.

Halgedahl, S.L. (1987). Domain pattern observations in rock magnetism: Progress and problems. *Phys. Earth Planet. Inter.*, **46**, 127–163.

Halgedahl, S.L. (1991). Magnetic domain patterns observed on synthetic Ti-rich titanomagnetite as a function of temperature and in states of thermoremanent magnetization. *J. Geophys. Res.*, **96**, 3943–3972.

Halgedahl, S.L. (1993). Experiments to investigate the origin of anomalously elevated unblocking temperatures. *J. Geophys. Res.*, **98**, 22,443–22,460.

Halgedahl, S.L. (1995). Bitter patterns versus hysteresis behavior in small single particles of hematite. *J. Geophys. Res.*, **100**, 353–364.

Halgedahl, S.L. & Fuller, M. (1980). Magnetic domain observations of nucleation processes in fine particles of intermediate titanomagnetite. *Nature*, **288**, 70–72.

Halgedahl, S.L. & Fuller, M. (1981). The dependence of magnetic domain structure upon magnetization state in polycrystalline pyrrhotite. *Phys. Earth Planet. Inter.*, **26**, 93–97.

Halgedahl, S.L. & Fuller, M. (1983). The dependence of magnetic domain structure upon magnetization state with emphasis on nucleation as a mechanism for pseudo-single-domain behavior. *J. Geophys. Res.*, **88**, 6505–6522.

Halgedahl, S.L. & Jarrard, R.D. (1995). Low temperature behavior of single-domain through multidomain magnetite. *Earth Planet. Sci. Lett.*, **130**, 127–139.

Halgedahl, S.L., Day, R. & Fuller, M. (1980). The effect of cooling rate on the intensity of weak field TRM in a single-domain magnetite. *J. Geophys. Res.*, **85**, 3690–3698.

Hall, J.M. (1976). Major problems regarding the magnetization of oceanic crustal layer 2. *J. Geophys. Res.*, **81**, 4223–4230.

Hall, J.M. (1985). The Iceland Research Drilling Project crustal section: Variation of magnetic properties with depth in Icelandic-type oceanic crust. *Can. J. Earth Sci.*, **22**, 85–101.

Hall, J.M. & Fisher, B.E. (1987). The characteristics and significance of secondary magnetite in a profile through the dike component of the Troodos, Cyprus, ophiolite. *Can. J. Earth Sci.*, **24**, 2141–2159.

Hall, J.M. & Neale, R.N. (1960). Stress effects on thermoremanent magnetization. *Nature*, **188**, 805–806.

Hall, J.M., Walls, C.C. & Hall, S.L. (1995). Viscous magnetization at 300 K in a profile through Troodos type oceanic crust. *Phys. Earth Planet. Inter.*, **88**, 101–116.

Hall, J.M., Walls, C.C., Yang, J-S., Hall, S.L. & Bakor, A.R. (1991). The magnetization of oceanic crust: Contribution to knowledge from the Troodos, Cyprus, ophiolite. *Can. J. Earth Sci.*, **28**, 1812–1826.

Halls, H.C. (1975). Shock-induced remanent magnetisation in late Precambrian rocks from Lake Superior. *Nature*, **255**, 692–695.

Halls, H.C. (1979). The Slate Islands meteorite impact site: A study of shock remanent magnetization. *Geophys. J. Roy. Astron. Soc.*, **59**, 553–591.

Hamano, Y. (1980). An experiment on the post-depositional remanent magnetization in artificial and natural sediments. *Earth Planet. Sci. Lett.*, **51**, 221–232.

Hamano, Y. (1983). Experiments on the stress sensitivity of natural remanent magnetization. *J. Geomag. Geoelec.*, **35**, 155–172.

Haneda, K. & Morrish, A.H. (1977). Magnetite to maghemite transformation in ultrafine particles. *J. Physique.*, **38**, suppl. C1, 321–323.

Hanss, R.E. (1964). Thermochemical etching reveals domain structure in magnetite. *Science*, **146**, 398–399.

Hargraves, R.B. & Perkins, W.E. (1969). Investigations of the effect of shock on remanent magnetism. *J. Geophys. Res.*, **74**, 2576–2589.

Hargraves, R.B. & Roy, J.L. (1974). Paleomagnetism of anorthosite in and around

the Charlevoix cryptoexplosion structure, Quebec. *Can. J. Earth Sci.*, **11**, 854–859.

Hargraves, R.B. & Young, W.M. (1969). Source of stable remanent magnetism in Lambertville diabase. *Am. J. Sci.*, **267**, 1161–1167.

Hargraves, R.B., Johnson, D. & Chan, C.Y. (1991). Distribution anisotropy: the cause of AMS in igneous rocks? *Geophys. Res. Lett.*, **18**, 2193–2196.

Harrison, C.G.A. (1976). Magnetization of the oceanic crust. *Geophys. J. Roy. Astron. Soc.*, **47**, 257–283.

Hartmann, U. (1987). A theoretical analysis of Bitter-pattern evolution. *J. Magn. Magn. Mater.*, **68**, 298–304.

Hartmann, U. (1994). Fundamentals and special applications of non-contact scanning force microscopy. *Adv. Electron. Electron Phys.*, **87**, 49–200.

Hartmann, U. & Mende, H.H. (1985). Observation of Bloch wall fine structures on iron whiskers by a high-resolution interference contrast technique. *J. Phys. D: Appl. Phys.*, **18**, 2285–2291.

Hartstra, R.L. (1982a). A comparative study of the ARM and I_{sr} of some natural magnetites of MD and PSD grain size. *Geophys. J. Roy. Astron. Soc.*, **71**, 497–518.

Hartstra, R.L. (1982b). Grain-size dependence of initial susceptibility and saturation magnetization-related parameters of four natural magnetites in the PSD-MD range. *Geophys. J. Roy. Astron. Soc.*, **71**, 477–495.

Hartstra, R.L. (1982c). High-temperature characteristics of a natural titanomagnetite. *Geophys. J. Roy. Astron. Soc.*, **71**, 455–476.

Hartstra, R.L. (1983). TRM, ARM and I_{sr} of two natural magnetites of MD and PSD grain size. *Geophys. J. Roy. Astron. Soc.*, **73**, 719–737.

Hauptmann, Z. (1974). High-temperature oxidation, range of nonstoichiometry and Curie point variation of cation-deficient titanomagnetite $Fe_{2.4}Ti_{0.6}O_4$. *Geophys. J. Roy. Astron. Soc.*, **38**, 29–47.

Hedley, I.G. (1968). Chemical remanent magnetization of the FeOOH, Fe_2O_3 system. *Phys. Earth Planet. Inter.*, **1**, 103–121.

Hedley, I.G. (1971). The weak ferromagnetism of goethite (α-FeOOH). *Z. Geophys.*, **37**, 409–420.

Heider, F. (1990). Temperature dependence of domain structure in natural magnetite and its significance for multi-domain TRM models. *Phys. Earth Planet. Inter.*, **65**, 54–61.

Heider, F. & Bryndzia, L.T. (1987). Hydrothermal growth of magnetite crystals (1 µm to 1 mm). *J. Cryst. Growth*, **84**, 50–56.

Heider, F. & Dunlop, D.J. (1987). Two types of chemical remanent magnetization during oxidation of magnetite. *Phys. Earth Planet. Inter.*, **46**, 24–45.

Heider, F. & Hoffmann, V. (1992). Magneto-optical Kerr effect on magnetite crystals with externally applied magnetic fields. *Earth Planet. Sci. Lett.*, **108**, 131–138.

Heider, F. & Williams, W. (1988). Note on temperature dependence of exchange constant in magnetite. *Geophys. Res. Lett.*, **15**, 184–187.

Heider, F., Dunlop, D.J. & Soffel, H.C. (1992). Low-temperature and alternating field demagnetization of saturation remanence and thermoremanence in magnetite grains (0.037 μm to 5 mm). *J. Geophy. Res.*, **97**, 9371–9381.

Heider, F., Dunlop, D.J. & Sugiura, N. (1987). Magnetic properties of hydrothermally recrystalized magnetite crystals. *Science*, **236**, 1287–1290.

Heider, F., Halgedahl, S.L. & Dunlop, D.J. (1988). Temperature dependence of magnetic domains in magnetite crystals. *Geophys. Res. Lett.*, **15**, 499–502.

Heisenberg, W. (1928). Zur Theorie des Ferromagnetismus. *Z. Phys.*, **49**, 619–636.

Hejda, P. (1985). Simple representation of magnetic characteristics by smooth functions. *Czech J. Phys.*, **B35**, 442–458.

Hejda, P. & Zelinka, T. (1990). Modelling of hysteresis processes in magnetic rock samples using the Preisach diagram. *Phys. Earth Planet. Inter.*, **63**, 32–40.

Hejda, P., Kropaček, V., Petrovsky, E., Zelinka, T. & Zatecky, J. (1992). Some magnetic properties of synthetic and natural haematite of different grain size. *Phys. Earth Planet. Inter.*, **70**, 261–272.

Hejda, P., Kropaček, V. & Zelinka, T. (1990). Modelling magnetization processes in single- and multi-domain rock samples using the Preisach diagram. *Stud. Geophys. Geod. Cesk. Akad. Ved.*, **34**, 220–229.

Hejda, P., Petrovsky, E. & Zelinka, T. (1994). The Preisach diagram, Wohlfarth's remanence formula and magnetic interactions. *IEEE Trans. Magn.*, **30**, 896–898.

Heller, F. (1978). Rock magnetic studies of Upper Jurassic limestones from southern Germany. *J. Geophys.*, **44**, 525–543.

Heller, F. & Evans, M.E. (1995). Loess magnetism. *Rev. Geophys.*, **33**, 211–240.

Heller, F. & Markert, H. (1973). The age of viscous remanent magnetization of Hadrian's Wall (Northern England). *Geophys. J. Roy. Astron. Soc.*, **31**, 395–406.

Heller, F., Carracedo, J.C. & Soler, V. (1986). Reversed magnetization in pyroclastics from the 1985 erruption of Nevado del Ruiz, Colombia. *Nature*, **324**, 241–242.

Heller, F., Shen, C.D., Beer, J., Liu, X-M., Liu, T-S, Bronger, A., Suter, M. & Bonani, G. (1993). Quantitative estimates of pedogenic ferromagnetic mineral formation in Chinese loess and palaeoclimatic implications. *Earth Planet. Sci. Lett.*, **114**, 385–390.

Helsley, C.E. & Steiner, M.B. (1969). Evidence for long intervals of normal polarity during the Cretaceous period. *Earth Planet. Sci. Lett.*, **5**, 325–332.

Henkel, O. (1964). Remanenzverhalten und Wechselwirkungen in hartmagnetischen Teilchenkollektiven. *Phys. Stat. Sol.*, **7**, 919–924.

Henry, W.E. & Boehm, M.J. (1956). Intradomain magnetic saturation and magnetic structure of γ-Fe$_2$O$_3$. *Phys. Rev.*, **101**, 1253–1254.

Henshaw, P.C. & Merrill, R.T. (1980). Magnetic and chemical changes in marine sediments. *Rev. Geophys. Space Phys.*, **18**, 483–504.

Hillhouse, J.W. (1977). A method for the removal of rotational remanent magnetization acquired during alternating field demagnetization. *Geophys. J. Roy. Astron. Soc.*, **50**, 29–34.

Hirt, A.M., Lowrie, W. & Pfiffner, O.A. (1986). A paleomagnetic study of tectonically deformed red beds of the Lower Glarus Nappe Complex, Eastern Switzerland. *Tectonics*, **5**, 723–731.

Hodych, J.P. (1976). Single-domain theory for the reversible effect of small uniaxial stress upon the initial magnetic susceptibility of rock. *Can. J. Earth Sci.*, **13**, 1186-1200.

Hodych, J.P. (1977). Single-domain theory for the reversible effect of small uniaxial stress upon the remanent magnetization of rock. *Can. J. Earth Sci.*, **14**, 2047–2061.

Hodych, J.P. (1982). Magnetostrictive control of coercive force in multidomain magnetite. *Nature*, **298**, 542–544.

Hodych, J.P. (1986). Determination of self-demagnetizing factor N for multidomain magnetite grains in rock. *Phys. Earth Planet. Inter.*, **41**, 283–291.

Hodych, J.P. (1990). Magnetic hysteresis as a function of low temperature in rocks: Evidence for internal stress control of remanence in multi-domain and pseudo-single domain magnetite. *Phys. Earth Planet. Inter.*, **64**, 21–36.

Hodych, J.P. (1991). Low-temperature demagnetization of saturation remanence in rocks bearing multidomain magnetite. *Phys. Earth Planet. Inter.*, **66**, 144–152.

Hodych, J.P. & Bijaksana, S. (1993). Can remanence anisotropy detect paleomagnetic inclination shallowing due to compaction? A case study using Cretaceous deep-sea limestones. *J. Geophys. Res.*, **98**, 22,429–22,441.

Hodych, J.P., Pätzold, R.R. & Buchan, K.L. (1985). Chemical remanent magnetization due to deep-burial diagenesis in oolitic hematite-bearing ironstones of Alabama. *Phys. Earth Planet. Inter.*, **37**, 261–284.

Hoffman, K.A. (1975). Cation diffusion processes and self-reversal of thermoremanent magnetization in the ilmenite-haematite solid solution series. *Geophys. J. Roy. Astron. Soc.*, **41**, 65–80.

Hoffman, K.A. (1992). Self-reversal of thermoremanent magnetization in the ilmenite-hematite system: Order-disorder, symmetry, and spin alignment. *J. Geophys. Res.*, **97**, 10,883–10,895.

Hoffmann, V. (1992). Greigite (Fe_3S_4): Magnetic properties and first domain observations. *Phys. Earth Planet. Inter.*, **70**, 288–301.

Hoffmann, V. & Fehr, K.Th. (1996). (Micro-) magnetic and mineralogical studies on dacitic pumice from the Pinatubo eruption (1991, Phillipines) showing self-reversed TRM. *Geophys. Res. Lett.*, **23**, 2835–2838.

Hoffmann, V., Schäfer, R., Appel, E., Hubert, A. & Soffel, H. (1987). First domain observations with the magneto-optical Kerr effect on Ti-ferrites in rocks and their synthetic equivalents. *J. Magn. Magn. Mat.*, **71**, 90–94.

Holm, E.J. & Verosub, K.L. (1988). An analysis of the effects of thermal demagnetization on magnetic carriers. *Geophys. Res. Lett.*, **15**, 487–490.

Horen, H. & Dubuisson, G. (1996). Rôle de l'hydrothermalisme sur les propriétés magnétiques d'une lithosphère océanique formée à une ride lente: L'ophiolite de Xigaze (Tibet). *C.R. Acad. Sci. (Paris)*, in press.

Housen, B.A. & van der Pluijm, B.A. (1991). Slaty cleavage development and magnetic anisotropy fabrics. *J. Geophys. Res.*, **96**, 9937–9946.

Housen, B.A., Richter, C. & van der Pluijm, B.A. (1993a). Composite magnetic anisotropy fabrics: experiments, numerical models, and implications for the quantification of rock fabrics. *Tectonophys.*, **220**, 1–12.

Housen, B.A., van der Pluijm, B.A. & Van der Voo, R. (1993b). Magnetite dissolution and neocrystallization during cleavage formation: Paleomagnetic study of the Martinsburg Formation, Lehigh Gap, Pennsylvania. *J. Geophys. Res.*, **98**, 13,799–13,813.

Hoye, G.S. & Evans, M.E. (1975). Remanent magnetizations in oxidized olivine. *Geophys. J. Roy. Astron. Soc.*, **41**, 139–151.

Huneke, J.C., Jessberger, E.K., Podosek, F.A. & Wasserburg, G.J. (1973). $^{40}Ar/^{39}Ar$ measurements in Apollo 15 and 17 samples and the chronology of metamorphic and volcanic activity in the Taurus-Littrow region. *Geochim. Cosmochim. Acta.*, **37**, suppl. 4, 1725–1756.

Hunt, C.P., Moskowitz, B.M. & Banerjee, S.K. (1995a). Magnetic properties of rocks and minerals. In *Rock Physics and Phase Relations: A Handbook of Physical Constants*, vol. 3, ed. T.J. Ahrens, pp. 189–204. American Geophysical Union, Washington, DC.

Hunt, C.P., Singer, M.J., Kletetschka, G., TenPas, J. & Verosub, K.L. (1995b). Effect of citrate-bicarbonate-dithionite treatment on fine-grained magnetite and maghemite. *Earth Planet. Sci. Lett.*, **130**, 87–94.

Hus, J.J. (1990). The magnetic properties of siderite concretions and the CRM of their oxidation products. *Phys. Earth Planet. Inter.*, **63**, 41–57.

Hyodo, H. & Dunlop, D.J. (1993). Effect of anisotropy on the paleomagnetic contact test for a Grenville dike. *J. Geophys. Res.*, **98**, 7997–8017.

Hyodo, M. (1984). Possibility of reconstruction of the past geomagnetic field from homogeneous sediments. *J. Geomag. Geoelec.*, **36**, 45–62.

Hyodo, M. & Yaskawa, K. (1986). Geomagnetic secular variation recorded in remanent magnetization of silty sediments from the Inland Sea, Japan (Seto Naikai). *J. Geomag. Geoelec.*, **38**, 11–26.

Hyodo, M., Itota, C. & Yaskawa, K. (1993). Geomagnetic secular variation reconstructed from magnetizations of wide-diameter cores of Holocene sediments in Japan. *J. Geomag. Geoelec.*, **45**, 669–696.

Imaoka, Y. (1968). On the coercive force of magnetic iron oxides. *J. Electrochem. Soc. Japan.*, **36**, 15–22.

Irving, E. (1957). Origin of the palaeomagnetism of the Torridonian sandstone of north-west Scotland. *Phil. Trans. Roy. Soc. London.*, **A250**, 100–110.

Irving, E. (1964). *Palaeomagnetism and Its Applications to Geological and Geophysical Problems*. John Wiley, New York, 399 pp.

Irving, E. (1970). The Mid-Atlantic Ridge at 45°N. XVI. Oxidation and magnetic properties of basalts; review and discussion. *Can. J. Earth Sci.*, **7**, 1528–1538.

Irving, E. & Major, A. (1964). Post-depositional remanent magnetization in a synthetic sediment. *Sedimentology*, **3**, 135–143.

Irving, E. & Opdyke, N.D. (1965). The paleomagnetism of the Bloomsburg redbeds and its possible application to the tectonic history of the Appalachians. *Geophys. J. Roy. Astron. Soc.*, **9**, 153–167.

Irving, E. & Park, J.K. (1973). Palaeomagnetism of metamorphic rocks: Errors owing to intrinsic anisotropy. *Geophys. J. Roy. Astron. Soc.*, **34**, 489–493.

Irving, E., Robertson, W.A. & Aumento, F. (1970). The Mid-Atlantic Ridge at 45°N. VI. Remanent intensity, susceptibility, and iron content of dredged samples. *Can. J. Earth Sci.*, **7**, 226–238.

Irving, E., Robertson, W.A., Stott, P.M., Tarling, D.H. & Ward, M.A. (1961). Treatment of partially stable sedimentary rocks showing planar distribution of directions of magnetization. *J. Geophys. Res.*, **66**, 1927–1933.

Ishikawa, Y. & Syono, Y. (1963). Order–disorder transformation and reverse thermo-remanent magnetization in the $FeTiO_3$–Fe_2O_3 system. *J. Phys. Chem. Solids.*, **24**, 517–528.

Jackson, M. (1990). Diagenetic source of stable remanence in remagnetized Paleozoic cratonic carbonates: A rock magnetic study. *J. Geophys. Res.*, **95**, 2753–2762.

Jackson, M. (1991). Anisotropy of magnetic remanence: A brief review of mineralogical sources, physical origins, and geological applications, and comparison with susceptibility anisotropy. *Pure Appl. Geophys.*, **136**, 1–28.

Jackson, M. & Borradaile, G.J. (1991). On the origin of the magnetic fabric in purple Cambrian slates of North Wales. *Tectonophys.*, **194**, 49–58.

Jackson, M. & Tauxe, L. (1991). Anisotropy of magnetic susceptibility and remanence: Developments in the characterization of tectonic, sedimentary and igneous fabric. *Rev. Geophys.*, **29**, suppl. (IUGG Report), 371–376.

Jackson, M. & Van der Voo, R. (1985). Drilling-induced remanence in carbonate rocks: Occurrence, stability and grain-size dependence. *Geophys. J. Roy. Astron. Soc.*, **81**, 75–87.

Jackson, M. & Van der Voo, R. (1986). Thermally activated viscous remanence in some magnetite- and hematite-bearing dolomites. *Geophys. Res. Lett.*, **13**, 1434–1437.

Jackson, M., Banerjee, S.K., Marvin, J.A., Lu, R. & Gruber, W. (1991). Detrital remanence, inclination errors and anhysteretic remanence anisotropy: Quantitative model and experimental results. *Geophys. J. Int.*, **104**, 95–103.

Jackson, M., Borradaile, G., Hudleston, P. & Banerjee, S. (1993b). Experimental deformation of synthetic magnetite-bearing calcite sandstones: Effects on remanence, bulk magnetic properties, and magnetic anisotropy. *J. Geophys. Res.*, **98**, 383–401.

Jackson, M., Gruber, W., Marvin, J. & Banerjee, S.K. (1988a). Partial anhysteretic remanence and its anisotropy: Applications and grain size dependence. *Geophys. Res. Lett.*, **15**, 440–443.

Jackson, M., McCabe, C., Ballard, M. & Van der Voo, R. (1988b). Magnetite authigenesis and diagenetic paleotemperatures across the northern Appalachian basin. *Geology*, **16**, 592–595.

Jackson, M., Rochette, P., Fillion, G., Banerjee, S.K. & Marvin, J. (1993a). Rock magnetism of remagnetized Paleozoic carbonates: Low-temperature behavior and susceptibility characteristics. *J. Geophys. Res.*, **98**, 6217–6225.

Jackson, M., Sprowl, D. & Ellwood, B. (1989). Anisotropies of partial anhysteretic remanence and susceptibility in compacted black shales: Grain size- and composition-dependent magnetic fabric. *Geophys. Res. Lett.*, **16**, 1063–1066.

Jackson, M., Sun, W-W. & Craddock, J.P. (1992). The rock magnetic fingerprint of chemical remagnetization in midcontinental Paleozoic carbonates. *Geophys. Res. Lett.*, **19**, 781–784.

Jackson, M., Worm, H-U. & Banerjee, S.K. (1990). Fourier analysis of digital hysteresis data: Rock magnetic applications. *Phys. Earth Planet. Inter.*, **65**, 78–87.

Jacobs, I.S. & Bean, C.P. (1955). An approach to elongated fine-particle magnets. *Phys. Rev.*, **100**, 1060–1067.

Jacobs, I.S., Beyerlein, R.A., Foner, S. & Remeika, J.P. (1971). Field induced magnetic phase transitions in antiferromagnetic hematite (α-Fe_2O_3). *Int. J. Magn.*, **1**, 193–208.

Jaep, W.F. (1969). Anhysteretic magnetization of an assembly of single-domain particles. *J. Appl. Phys.*, **40**, 1297–1298.

Jaep, W.F. (1971). Role of interactions in magnetic tapes. *J. Appl. Phys.*, **42**, 2790–2794.

Jiles, D. (1991). *Introduction to Magnetism and Magnetic Materials*. Chapman and Hall, London and New York, 440 pp.

Joffe, I. & Heuberger, R. (1974). Hysteresis properties of distributions of cubic single-domain ferromagnetic particles. *Phil. Mag.*, **29**, 1051–1059.

Johnson, E.A., Murphy, T. & Torreson, O.W. (1948). Pre-history of the earth's magnetic field. *Terr. Magn. Atm. Elec.*, **53**, 349–372.

Johnson, H.P. & Atwater, T. (1977). Magnetic study of basalts from the Mid-Atlantic Ridge, lat. 37°N. *Geol. Soc. Am. Bull.*, **88**, 637–647.

Johnson, H.P. & Hall, J.M. (1978). A detailed rock magnetic and opaque mineralogy study of the basalts from the Nazca Plate. *Geophys. J. Roy. Astron. Soc.*, **52**, 45–64.

Johnson, H.P. & Merrill, R.T. (1972). Magnetic and mineralogical changes associated with low-temperature oxidation of magnetite. *J. Geophys. Res.*, **77**, 334–341.

Johnson, H.P. & Merrill, R.T. (1973). Low-temperature oxidation of a titanomagnetite and the implications for paleomagnetism. *J. Geophys. Res.*, **78**, 4938–4949.

Johnson, H.P. & Merrill, R.T. (1974). Low-temperature oxidation of a single-domain magnetite. *J. Geophys. Res.*, **79**, 5533–5534.

Johnson, H.P. & Pariso, J.E. (1993). Variations in oceanic crustal magnetization: Systematic changes in the last 160 million years. *J. Geophys, Res.*, **98**, 435–445.

Johnson, H.P. & Tivey, M.A. (1995). Magnetic properties of zero-age oceanic crust: A new submarine lava flow on the Juan de Fuca Ridge. *Geophys. Res. Lett.*, **22**, 175–178.

Johnson, H.P., Kinoshita, H. & Merrill, R.T. (1975b). Rock magnetism and paleomagnetism of some North Pacific deep-sea sediments. *Geol. Soc. Am. Bull.*, **86**, 412–420.

Johnson, H.P., Lowrie, W. & Kent, D.V. (1975a). Stability of anhysteretic remanent magnetization in fine and coarse magnetite and maghemite particles. *Geophys. J. Roy. Astron. Soc.*, **41**, 1–10.

Joseph, R.I. (1976). Demagnetizing factors in nonellipsoidal samples – a review. *Geophysics*, **41**, 1052–1054.

Kachi, S., Momiyama, K. & Shimizu, S. (1963). An electron diffraction study and a theory of the transformation from γ-Fe$_2$O$_3$ to α-Fe$_2$O$_3$. *J. Phys. Soc. Japan*, **18**, 106–116.

Kachi, S., Nakanishi, N., Kosuge, K. & Hiramatsu, H. (1971). Electron microscopic observations on the transformation of Fe$_2$O$_3$. In *Ferrites: Proceedings of the International Conference*, ed. Y. Hoshino, S. Iida & M. Sugimoto, pp. 141–143. University of Tokyo Press, Tokyo.

Kaneoka, M. (1980). Change in coercive force of γ-Fe$_2$O$_3$–Fe$_3$O$_4$ solid solutions with magnetostrictive anisotropy. *IEEE Trans. Magn.*, **MAG16**, 1319–1322.

Kapicka, A. (1992). Magnetic susceptibility under hydrostatic pressure of synthetic magnetite samples. *Phys. Earth. Planet. Inter.*, **70**, 248–252.

Karlin, R. (1990). Magnetite diagenesis in marine sediments from the Oregon continental margin. *J. Geophys. Res.*, **95**, 4405–4419.

Karlin, R. & Levi, S. (1985). Geochemical and sedimentological control of the magnetic properties of hemipelagic sediments. *J. Geophys. Res.*, **90**, 10,373–10,392.

Karlin, R., Lyle, M. & Heath, G.R. (1987). Authigenic magnetite formation in suboxic marine sediments. *Nature*, **326**, 490–493.

Kean, W.F., Day, R., Fuller, M. & Schmidt, V.A. (1976). The effect of uniaxial compression on the initial susceptibility of rocks as a function of the grain size and composition of their constituent titanomagnetites. *J. Geophys. Res.*, **81**, 861–872.

Kellogg, K., Larson, E.E. & Watson, D.E. (1970). Thermochemical remanent magnetization and thermal remanent magnetization: A comparison in a basalt. *Science*, **170**, 628–630.

Kelso, P.R. & Banerjee, S.K. (1994). Elevated temperature viscous remanent magnetization of natural and synthetic multidomain magnetite. *Earth Planet. Sci. Lett.*, **122**, 43–56.

Kelso, P.R. & Banerjee, S.K. (1995). Effect of hydrostatic pressure on viscous remanent magnetization in magnetite-bearing specimens. *Geophys. Res. Lett.*, **22**, 1953–1956.

Kelso, P.R., Banerjee, S.K. & Worm, H-U. (1991). The effect of low-temperature hydrothermal alteration on the remanent magnetization of synthetic titanomagnetites: A case for acquisition of chemical remanent magnetization. *J. Geophys. Res.*, **96**, 19,545–19,553.

Kennedy, L.P. (1981). Self-reversed thermoremanent magnetization in a late Brunhes dacite pumice. *J. Geomag. Geoelec.*, **33**, 429–448.

Kent, D.V. (1973). Post-depositional remanent magnetization in deep-sea sediment. *Nature*, **246**, 32–34.

Kent, D.V. (1985). Thermoviscous remagnetization in some Appalachian limestones. *Geophys. Res. Lett.*, **12**, 805–808.

Kent, D.V. & Gee, J. (1994). Grain size-dependent alteration and the magnetization of oceanic basalts. *Science*, **265**, 1561–1563.

Kent, D.V. & Lowrie, W. (1974). Origin of magnetic instability in sediment cores from the central North Pacific. *J. Geophys. Res.*, **79**, 2987–3000.

Kent, D.V. & Miller, J.D. (1987). Redbeds and thermoviscous magnetization theory for hematite. *Geophys. Res. Lett.*, **14**, 327–330.

Kent, D.V. & Opdyke, N.D. (1985). Multicomponent magnetizations from the Mississippian Mauch Chunk formation of the central Appalachians and their tectonic implications. *J. Geophys. Res.*, **90**, 5371–5383.

Kent, D.V., Honnorez, B.M., Opdyke, N.D. & Fox, P.J. (1978). Magnetic properties of dredged oceanic gabbros and the source of marine magnetic anomalies. *Geophys. J. Roy. Astron. Soc.*, **55**, 513–537.

Kern, J.W. (1961). Effects of moderate stresses on directions of thermoremanent magnetization. *J. Geophys. Res.*, **66**, 3801–3806.

Kikawa, E. & Ozawa, K. (1992). Contribution of oceanic gabbros to sea-floor spreading magnetic anomalies. *Science*, **258**, 796–799.

Kilgore, B. & Elmore, R.D. (1989). A study of the relationship between hydrocarbon migration and the precipitation of authigenic magnetic minerals in the Triassic Chugwater Formation, southern Montana. *Geol. Soc. Am. Bull.*, **101**, 1280–1288.

King, J.W. & Channell, J.E.T. (1991). Sedimentary magnetism, environmental magnetism, and magnetostratigraphy. *Rev. Geophys.*, **29**, suppl. (IUGG report), 358–370.

King, J.W., Banerjee, S.K. & Marvin, J. (1983). A new rock-magnetic approach to selecting sediments for geomagnetic paleointensity studies: Application to paleointensity for the last 4000 years. *J. Geophys. Res.*, **88**, 5911–5921.

King, J., Banerjee, S.K., Marvin, J. & Özdemir, Ö. (1982). A comparison of different magnetic methods for determining the relative grain size of magnetite in natural materials: Some results from lake sediments. *Earth Planet. Sci. Lett.*, **59**, 404–419.

King, R.F. (1955). The remanent magnetism of artificially deposited sediments. *Mon. Notices Roy. Astron. Soc., Geophys. Suppl.*, **7**, 115–134.

King, R.F. & Rees, A.I. (1966). Detrital magnetism in sediments: An examination of some theoretical models. *J. Geophys. Res.*, **71**, 561–571.

Kinoshita, H. (1968). Studies on piezo-magnetization (III) – PRM and related phenomena. *J. Geomag. Geoelec.*, **20**, 155–167.

Kirschvink, J.L. (1983). Biogenic ferrimagnetism: A new biomagnetism. In *Biomagnetism*, ed. S.J. Williamson, G-L. Romani, L. Kaufman, & I. Modena, pp. 501–531. Plenum, New York.

Kirschvink, J.L. & Lowenstam, H.A. (1979). Mineralization and magnetization of chiton teeth: paleomagnetic, sedimentologic, and biologic implications of organic magnetite. *Earth Planet. Sci. Lett.*, **44**, 193–204.

Kirschvink, J.L., Jones, D.S. & MacFadden, B.J., eds. (1985). *Magnetite Biomineralization and Magnetoreception in Organisms: A New Biomagnetism*. Plenum, New York, 682 pp.

Kittel, C. (1949). Physical theory of ferromagnetic domains. *Rev. Mod. Phys.*, **21**, 541–583.

Kittel, C. (1976). *Introduction to Solid State Physics*, 5th edn. John Wiley, New York, 608 pp.

Klapel, G.D. & Shive, P.N. (1974). High-temperature magnetostriction of magnetite. *J. Geophys. Res.*, **79**, 2629–2633.

Klerk, J., Brabers, V.A.M. & Kuipers, A.J.M. (1977). Magnetostriction of the mixed series $Fe_{3-x}Ti_xO_4$. *J. Physique*, **38**, C1, 187–189.

Kligfield, R., Lowrie, W. & Pfiffner, O.A. (1982). Magnetic properties of deformed oolitic limestones from the Swiss Alps: The correlation of magnetic anisotropy and strain. *Eclog. Geol. Helv.*, **75**, 127–157.

Kligfield, R., Lowrie, W., Hirt, A. & Siddans, A.W.B. (1983). Effect of progressive deformation on remanent magnetization of Permian redbeds from the Alpes Maritimes (France). *Tectonophys.*, **97**, 59–85.

Kneller, E. (1962). *Ferromagnetismus*. Springer-Verlag, Berlin, 792 pp.

Kneller, E. (1969). Fine particle theory. In *Magnetism and Metallurgy*, vol. 1, ed. A.E. Berkowitz & E. Kneller, pp. 365–471. Academic Press, London and New York.

Kneller, E.F. & Luborsky, F.E. (1963). Particle size dependence of coercivity and remanence of single-domain particles. *J. Appl. Phys.*, **34**, 656–658.

Kneller, E.F. & Wohlfarth, E.P. (1966). Effect of thermal fluctuations on the anhysteretic process in ferromagnetic fine-particle assemblies. *J. Appl. Phys.*, **37**, 4816–4818.

Kobayashi, K. (1959). Chemical remanent magnetization of ferromagnetic minerals and its application to rock magnetism. *J. Geomag. Geoelec.*, **10**, 99–117.

Kobayashi, K. (1961). An experimental demonstration of the production of chemical remanent magnetization with Cu-Co alloy. *J. Geomag. Geoelec.*, **12**, 148–164.

Kobayashi, K. (1962a). Magnetization-blocking process by volume development of ferromagnetic fine particles. *J. Phys. Soc. Japan*, **17**, suppl. B1, 695–698.

Kobayashi, K. (1962b). Crystallization or chemical remanent magnetization. In *Proc. Benedum Earth Magnetism Symp.*, University of Pittsburgh, pp. 107–112.

Kobayashi, K. & Fuller, M. (1968). Stable remanence and memory of multi-domain materials with special reference to magnetite. *Phil. Mag.*, **18**, 601–624.

Kodama, K.P. (1984). Paleomagnetism of granitic intrusives from the Precambrian basement under eastern Kansas: Orienting drill cores using secondary magnetization components. *Geophys. J. Roy. Astron. Soc.*, **76**, 273–287.

Kodama, K.P. (1988). Remanence rotation due to rock strain during folding and the stepwise application of the fold test. *J. Geophys. Res.*, **93**, 3357–3371.

Kodama, K.P. & Goldstein, A.G. (1991). Experimental simple shear deformation of magnetic remanence. *Earth Planet. Sci. Lett.*, **104**, 80–88.

Kodama, K.P. & Sun, W.W. (1992). Magnetic anisotropy as a correction for compaction-caused inclination shallowing. *Geophys. J. Int.*, **111**, 465–469.

Koenigsberger, J.G. (1938). Natural residual magnetism of eruptive rocks. *Terr. Magn. Atmos. Elec.*, **43**, 119–130 & 299–320.

Kok, Y.S. & Tauxe, L. (1996). Saw-toothed pattern of relative paleointensity records and cumulative viscous remanence. *Earth Planet. Sci. Lett.*, **137**, 95–99.

Kono, M. (1978). Reliability of paleointensity methods using alternating field demagnetization and anhysteretic remanence. *Geophys. J. Roy. Astron. Soc.*, **54**, 241–261.

Kono, M. (1979). Palaeomagnetism and palaeointensity studies of Scottish Devonian volcanic rocks. *Geophys. J. Roy. Astron. Soc.*, **56**, 385–396.

Kono, M. (1985). Changes in magnetic hysteresis properties of a basalt induced by heating in air. *J. Geomag. Geoelec.*, **37**, 589–600.

Kono, M. (1987). Changes in TRM and ARM in a basalt due to laboratory heating. *Phys. Earth Planet. Inter.*, **46**, 1–8.

Kropaček, V. (1968). Self-reversal of spontaneous magnetization of natural cassiterite. *Stud. Geophys. Geod. Cesk. Akad. Ved.*, **12**, 108–110.

Kumar, A. & Bhalla, M.S. (1984). Source of stable remanence in chromite ores. *Geophys. Res. Lett.*, **11**, 177–180.

Landau, L.D. & Lifschitz, E.M. (1935). On the theory of the dispersion of magnetic permeability in ferromagnetic bodies (transl. from Russian). *Phys. Z. Sowjetunion*, **8**, 153–169.

Langereis, C.G., Van Hoof, A.A.M. & Rochette, P. (1992). Longitudinal confinement of geomagnetic reversal paths as a possible sedimentary artefact. *Nature*, **358**, 226–229.

Lanoix, M., Strangway, D.W. & Pearce, G.W. (1977). Anomalous acquisition of thermoremanence at 130°C in iron and paleointensity of the Allende meteorite. *Geochim. Cosmochim. Acta.*, **41**, suppl. 8, 689–701.

Lanoix, M., Strangway, D.W. & Pearce, G.W. (1978). The primordial magnetic field preserved in chondrules of the Allende meteorite. *Geophys. Res. Lett.*, **5**, 73–76.

Lanos, P. (1987). The effects of demagnetizing fields on the thermoremanent magnetization acquired by parallel-sided baked clay blocks. *Geophys. J. Roy. Astron. Soc.*, **91**, 985–1012.

Larson, E.E. & Walker, T.R. (1982). A rock magnetic study of the lower Massive Sandstone, Moenkopi Formation (Triassic), Gray Mountain area, Arizona. *J. Geophys. Res.*, **87**, 4819–4836.

Larson, E.E., Ozima, Mituko, Ozima, Minoru, Nagata, T. & Strangway, D.W. (1969). Stability of remanent magnetization of igneous rocks. *Geophys. J. Roy. Astron. Soc.*, **17**, 263–292.

Larson, E.E., Walker, T.R., Patterson, P.E., Hoblitt, R.P. & Rosenbaum, J.G. (1982). Paleomagnetism of the Moenkopi Formation, Colorado Plateau: Basis for long-term model of acquisition of chemical remanent magnetism in red beds. *J. Geophys. Res.*, **87**, 1081–1106.

Larson, E.E., Watson, D.E., Herndon, J.M. & Rowe, M.W. (1973). Partial A.F. demagnetization studies of 40 meteorites. *J. Geomag. Geoelec.*, **25**, 331–338.

Larson, E.E., Watson, D.E., Herndon, J.M. & Rowe, M.W. (1974). Thermomagnetic analysis of meteorites, 1. C1 chondrites. *Earth Planet. Sci. Lett.* **21**, 345–350.

Lawley, E.A. & Ade-Hall, J.M. (1971). A detailed magnetic and opaque petrological study of a thick Palaeogene tholeiite lava flow from Northern Ireland. *Earth Planet. Sci. Lett.*, **11**, 113–120.

Lawson, C.A., Nord, G.L. & Champion, D.E. (1987). Fe–Ti oxide mineralogy and the origin of normal and reverse remanent magnetization in dacitic pumice blocks from Mt. Shasta, California. *Phys. Earth Planet. Inter.*, **46**, 270–288.

Lawson, C.A., Nord, G.L., Dowty, E. & Hargraves, R.B. (1981). Antiphase domains and reverse thermoremanent magnetism in ilmenite–hematite minerals. *Science*, **213**, 1372–1374.

Le Borgne, E. (1955). Susceptibilité magnétique anormale du sol superficiel. *Ann. Géophys.*, **11**, 399–419.

Le Borgne, E. (1960a). Etude expérimentale du traînage magnétique dans le cas d'un ensemble de grains magnétiques très fins dispersés dans une substance non magnétique. *Ann. Géophys.*, **16**, 445–494.

Le Borgne, E. (1960b). Influence du feu sur les propriétés magnétiques du sol et sur celles du schiste et du granite. *Ann. Géophys.*, **16**, 159–195.

Lederman, M., Schultz, S. & Ozaki, M. (1994). Measurement of the dynamics of the magnetization reversal in individual single-domain ferromagnetic particles. *Phys. Rev. Lett.*, **73**, 1986–1989.

Leslie, B.W., Lund, S.P. & Hammond, D.E. (1990). Rock magnetic evidence for the dissolution and authigenic growth of magnetic minerals within anoxic marine sediments of the California continental borderland. *J. Geophys. Res.*, **95**, 4437–4452.

Levi, S. (1974). Some magnetic properties of magnetite as a function of grain size and their implications for paleomagnetism. Ph.D. thesis, Univ. of Washington, Seattle, WA.

Levi, S. (1977). The effect of magnetite particle size on paleointensity determinations of the geomagnetic field. *Phys. Earth Planet. Inter.*, **13**, 245–259.

Levi, S. (1979). The additivity of partial thermal remanent magnetization in magnetite. *Geophys. J. Roy. Astron. Soc.*, **59**, 205–218.

Levi, S. (1989). Chemical remanent magnetization (CRM). In *The Encyclopedia of Solid Earth Geophysics*, ed. D.E. James, pp. 49–58. Van Nostrand Reinhold, New York.

Levi, S. & Banerjee, S.K. (1976). On the possibility of obtaining relative paleointensities from lake sediments. *Earth Planet. Sci. Lett.*, **29**, 219–226.

Levi, S. & Banerjee, S.K. (1977). The effects of alteration on the natural remanent magnetization of three ophiolite complexes: Possible implications for the oceanic crust. *J. Geomag. Geoelec.* **29**, 421–439.

Levi, S. & Merrill, R.T. (1976). A comparison of ARM and TRM in magnetite. *Earth Planet. Sci. Lett.*, **32**, 171–184.

Levi, S. & Merrill, R.T. (1978). Properties of single-domain, pseudo-single-domain and multidomain magnetite. *J. Geophys. Res.*, **83**, 309–323.

Li, H. & Beske-Diehl, S.J. (1991). Magnetic properties of deuteric hematite in young lava flows from Iceland. *Geophys. Res. Lett.*, **18**, 597–600.

Liebermann, R.C. & Banerjee, S.K. (1971). Magnetoelastic interactions in hematite: Implications for geophysics. *J. Geophys. Res.*, **76**, 2735–2756.

Lilley, B.A. (1950). Energies and widths of domain boundaries in ferromagnetics. *Phil. Mag.*, **41**, 792–813.

Lin, S.T. (1960). Magnetic behaviour in the transition region of a hematite single crystal. *J. Appl. Phys.*, **31**, 2735–2745.

Lindsley, D.H. (1976). The crystal chemistry and structure of oxide minerals as exemplified by the Fe–Ti oxides, and Experimental studies of oxide minerals. In *Oxide Minerals*, 1st edn., ed. D. Rumble, pp. L1–84. Mineralogical Society of America, Washington, DC.

Lindsley, D.H. (1991). Experimental studies of oxide minerals. In *Oxide Minerals: Petrologic and Magnetic Significance*, 2nd edn., ed. D.H. Lindsley, pp. 69–106. Mineralogical Society of America, Washington, DC.

Liu, X-M., Shaw, J., Liu, T-S., Heller, F. & Cheng, M-Y. (1993). Rock magnetic properties and palaeoclimate of Chinese loess. *J. Geomag. Geoelec.*, **45**, 117–124.

Longworth, G., Becker, L.W., Thompson, R., Oldfield, F., Dearing, J.A. & Rummery, T.A. (1979). Mössbauer effect and magnetic studies of secondary iron oxides in soils. *J. Soil Sci.*, **30**, 93–110.

Lovley, D.R., Stolz, J.F., Nord, G.L. & Phillips, E.J.P. (1987). Anaerobic production of magnetite by a dissimilatory iron-reducing microorganism. *Nature*, **330**, 252–254.

Løvlie, R. (1976). The intensity pattern of post-depositional remanence acquired in some marine sediments deposited during a reversal of the external magnetic field. *Earth Planet. Sci. Lett.*, **30**, 209–214.

Lowenstam, H.A. & Kirschvink, J.L. (1985). Iron biomineralization: A geobiological perspective. In *Magnetite Biomineralization and Magnetoreception in Organisms: A New Biomagnetism*, ed. J.L. Kirschvink, D.S. Jones & B.J. MacFadden, pp. 3–15. Plenum, New York.

Lowrie, W. (1977). Intensity and direction of magnetization in oceanic basalts. *J. Geol. Soc. London*, **133**, 61–82.

Lowrie, W. (1990). Identification of ferromagnetic minerals in a rock by coercivity and unblocking temperature properties. *Geophys. Res. Lett.*, **17**, 159–162.

Lowrie, W. & Alvarez, W. (1975). Paleomagnetic evidence for rotation of the Italian peninsula. *J. Geophys. Res.*, **80**, 1579–1592.

Lowrie, W. & Alvarez, W. (1977). Late Cretaceous geomagnetic polarity sequence: detailed rock and paleomagnetic studies of the Scaglia Rossa limestone at Gubbio, Italy. *Geophys. J. Roy. Astron. Soc.*, **51**, 561–581.

Lowrie, W. & Alvarez, W. (1981). One hundred million years of geomagnetic polarity history. *Geology*, **9**, 392–397.

Lowrie, W. & Fuller, M. (1971). On the alternating field demagnetization characteristics of multidomain thermoremanent magnetization in magnetite. *J. Geophys. Res.*, **76**, 6339–6349.

Lowrie, W. & Heller, F. (1982). Magnetic properties of marine limestones. *Rev. Geophys. Space Phys.*, **20**, 171–192.

Lowrie, W. & Kent, D.V. (1978). Characteristics of VRM in oceanic basalts. *J. Geophys.*, **44**, 297–315.

Lowrie, W., Channell, J.E.T. & Alvarez, W. (1980). A review of magnetic stratigraphy investigations in Cretaceous pelagic carbonate rocks. *J. Geophys. Res.*, **85**, 3597–3605.

Lowrie, W., Hirt, A.M. & Kligfield, R. (1986). Effects of tectonic deformation on the remanent magnetization of rocks. *Tectonics*, **5**, 713–722.

Luborsky, F.E. (1961). Development of elongated particle magnets. *J. Appl. Phys.*, **32**, 171S–183S.

Lund, S.P. & Karlin, R. (1990). Introduction to the special section on "Physical and Biogeochemical Processes Responsible for the Magnetization of Sediments". *J. Geophys. Res.*, **95**, 4353–4354.

Lund, S.P. & Keigwin, L. (1994). Measurement of the degree of smoothing in sediment paleomagnetic secular variation records: An example from late Quaternary deep-sea sediments of the Bermuda Rise, western North Atlantic Ocean. *Earth Planet. Sci. Lett.*, **122**, 317–330.

Luo, Y. & Zhu, J-G. (1994). Switching field characteristics of individual iron particles by MFM. *IEEE Trans. Magn.*, **30**, 4080–4082.

MacDonald, W.D. (1980). Net tectonic rotation, apparent tectonic rotation and the structural tilt correction in paleomagnetic studies. *J. Geophys. Res.*, **85**, 3659–3669.

Maher, B.A. (1986). Characterization of soils by mineral magnetic measurements. *Phys. Earth Planet. Inter.*, **42**, 76–92.

Maher, B.A. (1988). Magnetic properties of some synthetic submicron magnetites. *Geophys. J.*, **94**, 83–96.

Maher, B.A. & Taylor, R.M. (1988). Formation of ultrafine-grained magnetite in soils. *Nature*, **336**, 368–370.

Mann, S. (1985). Structure, morphology, and crystal growth of bacterial magnetite. In *Magnetite Biomineralization and Magnetoreception in Organisms: A New Biomagnetism*, ed. J.L. Kirschvink, D.S. Jones & B.J. MacFadden, pp. 311–332. Plenum, New York.

Mann, S., Sparks, N.H.C., Frankel, R.B., Bazylinksi, D.A. & Jannasch, H.W. (1990). Biomineralization of ferrimagnetic greigite (Fe_3S_4) and iron pyrite (FeS_2) in a magnetotactic bacterium. *Nature*, **343**, 258–261.

Manson, A.J. & O'Reilly, W. (1976). Submicroscopic texture in titanomagnetite grains in basalt studied using the torque magnetometer and the electron microscope. *Phys. Earth Planet. Inter.*, **11**, 173–183.

Markov, G.P., Shcherbakov, V.P., Bol'shakov, A.S. & Vinogradov, Yu.K. (1983). On the temperature dependence of the partial thermoremanent magnetization of multidomain grains (transl. from Russian). *Izv., Phys. Solid Earth*, **19**, 625–630.

Marshall, M. & Cox, A. (1971). Effect of oxidation on the natural remanent magnetization of titanomagnetite in sub-oceanic basalt. *Nature*, **230**, 28–31.

Marshall, M. & Cox, A. (1972). Magnetic changes in pillow basalt due to sea floor weathering. *J. Geophys. Res.*, **77**, 6459–6469.

Martin, R.J. & Noel, J.S. (1988). The influence of stress path on thermoremanent magnetization. *Geophys. Res. Lett.*, **15**, 507–510.

Martin, R.J., Habermann, R.E. & Wyss, M. (1978). The effect of stress cycling and inelastic volumetric strain on remanent magnetization. *J. Geophys. Res.*, **83**, 3485–3496.

Matuyama, M. (1929). On the direction of magnetisation of basalt in Japan, Tyosen and Manchuria. *Proc. Imper. Acad. Japan.*, **5**, 203–205.

Maxwell, L.R., Smart, J.S. & Brunaver, S. (1949). Dependence of the intensity of magnetization and the Curie point of certain iron oxides upon the ratio of Fe^{2+}/Fe^{3+}. *Phys. Rev.*, **76**, 459–460.

McCabe, C. & Elmore, R.D. (1989). The occurrence and origin of Late Paleozoic remagnetization in the sedimentary rocks of North America. *Rev. Geophys.*, **27**, 471–494.

McCabe, C., Jackson, M. & Saffer, B. (1989). Regional pattern of magnetite authigenesis in the Appalachian Basin: Implications for the mechanism of Late Paleozoic remagnetization. *J. Geophys. Res.*, **94**, 10,429–10,443.

McCabe, C., Van der Voo, R., Peacor, D.R., Scotese, C.R. & Freeman, R. (1983). Diagenetic magnetite carries ancient yet secondary remanence in some Paleozoic sedimentary carbonates. *Geology*, **11**, 221–223.

McClay, K.R. (1974). Single-domain magnetite in the Jimberlana norite, Western Australia. *Earth Planet. Sci. Lett.*, **21**, 367–376.

McClelland Brown, E. (1981). Paleomagnetic estimates of temperatures reached in contact metamorphism. *Geology*, **9**, 112–116.

McClelland Brown, E. (1982). Discrimination of TRM and CRM by blocking temperature spectrum analysis. *Phys. Earth Planet. Inter.*, **30**, 405–414.

McClelland Brown, E. (1984). Experiments on TRM intensity dependence on cooling rate. *Geophys. Res. Lett.*, **11**, 205–208.

McClelland, E. (1987). Self-reversal of chemical remanent magnetization: A palaeomagnetic example. *Geophys. J. Roy. Astron. Soc.*, **90**, 615–625.

McClelland, E. & Goss, C. (1993). Self-reversal of chemical remanent magnetization on the transformation of maghemite to haematite. *Geophys. J. Int.*, **112**, 517–532.

McClelland, E. & Shcherbakov, V.P. (1995). Metastability of domain state in multidomain magnetite: Consequences for remanence acquisition. *J. Geophys. Res.*, **100**, 3841–3857.

McClelland, E. & Sugiura, N. (1987). A kinematic model of TRM acquisition in multidomain magnetite. *Phys. Earth Planet. Inter.*, **46**, 9–23.

McDougall, I. & Harrison, T.M. (1988). *Geochronology and Thermochronology by the $^{40}Ar/^{39}Ar$ Method*. Oxford Univ. Press, New York & Clarendon Press, Oxford, 208 pp.

McElhinny, M.W. (1973). *Palaeomagnetism and Plate Tectonics*. Cambridge University Press, 358 pp.

McElhinny, M.W. & Evans, M.E. (1968). An investigation of the strength of the geomagnetic field in the early Precambrian. *Phys. Earth Planet. Inter.*, **1**, 485–497.

McElhinny, M.W. & Opdyke, N.D. (1973). Remagnetization hypothesis discounted: A paleomagnetic study of the Trenton limestone, New York state. *Geol. Soc. Am. Bull.*, **84**, 3697–3708.

McNab, T.K., Fox, R.A. & Boyle, A.J.F. (1968). Some magnetic properties of magnetite (Fe_3O_4) microcrystals. *J. Appl. Phys.*, **39**, 5703–5711.

Mehra, O.P. & Jackson, M.L. (1960). Iron oxide removal from soils and clays by a dithionite-citrate system buffered with sodium bicarbonate. *Clays Clay Miner.*, **5**, 317–327.

Meiklejohn, W.E. (1953). Experimental study of the coercive force of fine particles. *Rev. Mod. Phys.*, **25**, 302–306.

Menyeh, A. & O'Reilly, W. (1991). The magnetization process in monoclinic pyrrhotite (Fe_7S_8) particles containing few domains. *Geophys. J. Int.*, **104**, 387–399.

Menyeh, A. & O'Reilly, W. (1995). The coercive force of fine particles of monoclinic pyrrhotite (Fe_7S_8) studied at elevated temperature. *Phys. Earth Planet. Inter.*, **89**, 51–62.

Merrill, R.T. (1970). Low-temperature treatment of magnetite and magnetite-bearing rocks. *J. Geophys. Res.*, **75**, 3343–3349.

Merrill, R.T. (1975). Magnetic effects associated with chemical changes in igneous rocks. *Geophys. Surv.*, **2**, 277–311.

Merrill, R.T. & Grommé, C.S. (1969). Non-reproducible self-reversal of magnetization in diorite. *J. Geophys. Res.*, **74**, 2014–2024.

Merrill, R.T. & McElhinny, M.W. (1983). *The Earth's Magnetic Field: Its History, Origin and Planetary Perspective*. Academic Press, London, and New York, 401 pp.

Metcalf, M. & Fuller, M. (1987a). Magnetic remanence measurements of single particles and the nature of domain patterns in titanomagnetites. *Geophys. Res. Lett.*, **14**, 1207–1210.

Metcalf, M. & Fuller, M. (1987b). Domain observations of titanomagnetites during hysteresis at elevated temperatures and thermal cycling. *Phys. Earth Planet. Inter.*, **46**, 120–126.

Metcalf, M. & Fuller, M. (1988). A synthetic TRM induction curve for fine particles generated from domain observations. *Geophys. Res. Lett.*, **15**, 503–506.

Meynadier, L., Valet, J.-P., Bassinot, R.C., Shackleton, N. J. & Guyodo, Y. (1994). Asymmetrical saw-tooth pattern of the geomagnetic field intensity from equatorial sediments in the Pacific and Indian Oceans. *Earth Planet. Sci. Lett.*, **126**, 109–127.

Meynadier, L., Valet, J-P. & Shackleton, N.J. (1995). Relative geomagnetic intensity during the last 4 million years from the equatorial Pacific. *Proc. Ocean Drill Program. Sci. Res.*, ed. N.G. Pisias et al., **138**, 779–793.

Meynadier, L., Valet, J-P., Weeks, R., Shackleton, N.J. & Hagee, V.L. (1992). Relative geomagnetic intensity of the field during the last 140 ka. *Earth Planet. Sci. Lett.*, **114**, 39–57.

Michel, A. & Chaudron, G. (1935). Etude du sesquioxyde de fer cubique stabilisé. *C. R. Acad. Sci. Paris*, **201**, 1191–1193.

Michel, A., Chaudron, G. & Bénard, J. (1951). Propriétés des composés ferromagnétiques non métalliques. *J. Physique*, **12**, 189–201.

Middleton, M.F. & Schmidt, P.W. (1982). Paleothermometry of the Sydney Basin. *J. Geophys. Res.*, **87**, 5351–5359.

Mollard, P., Rousset, A. & Dupré, G. (1977). Moment à saturation du sesquioxyde de fer cubique. *Mat. Res. Bull.*, **12**, 797–801.

Moon, T.S. (1991). Domain states in fine particle magnetite and titanomagnetite. *J. Geophys. Res.*, **96**, 9909–9923.

Moon, T.S. & Merrill, R.T. (1984). The magnetic moments of non-uniformly magnetized grains. *Phys. Earth Planet. Inter.*, **34**, 186–194.

Moon, T.S. & Merrill, R.T. (1985). Nucleation theory and domain states in multidomain magnetic material. *Phys. Earth Planet. Inter.*, **37**, 214–222.

Moon, T.S. & Merrill, R.T. (1986a). A new mechanism for stable viscous remanent magnetization and overprinting during long magnetic polarity intervals. *Geophys. Res. Lett.*, **13**, 737–740.

Moon, T.S. & Merrill, R.T. (1986b). Magnetic screening in multidomain material. *J. Geomag. Geoelec.*, **38**, 883–894.

Moon, T.S. & Merrill, R.T. (1988). Single-domain theory of remanent magnetization. *J. Geophys. Res.*, **93**, 9202–9210.

Morgan, G.E. & Smith, P.P.K. (1981). Transmission electron microscope and rock magnetic investigations of remanence carriers in a Precambrian metadolerite. *Earth Planet. Sci. Lett.*, **53**, 226–240.

Morrish, A.H. (1965). *The Physical Principles of Magnetism.* John Wiley, New York, 680 pp.

Morrish, A.H. & Yu, S.P. (1955). Dependence of the coercive force on the density of some iron oxide powders. *J. Appl. Phys.*, **26**, 1049–1055.

Moskowitz, B.M. (1980). Theoretical grain size limits for single-domain, pseudo-single-domain and multidomain behavior in titanomagnetite ($x = 0.6$) as a function of low-temperature oxidiation. *Earth Planet. Sci. Lett.*, **47**, 285–293.

Moskowitz, B.M. (1985). Magnetic viscosity, diffusion after-effect, and disaccommodation in natural and synthetic samples. *Geophys. J. Roy. Astron. Soc.*, **82**, 143–161.

Moskowitz, B.M. (1987). Towards resolving the inconsistencies in characteristic physical properties of synthetic titanomaghemites. *Phys. Earth Planet. Inter.*, **46**, 173–183.

Moskowitz, B.M. (1992). *Rock Magnetism Laboratory Notes.* University of Minnesota, Minneapolis, MN, 40 pp.

Moskowitz, B.M. (1993a). High-temperature magnetostriction of magnetite and titanomagnetites. *J. Geophys. Res.*, **98**, 359–371.

Moskowitz, B.M. (1993b). Micromagnetic study of the influence of crystal defects on coercivity in magnetite. *J. Geophys. Res.*, **98**, 18,011–18,026.

Moskowitz, B.M. (1995). Biomineralization of magnetic minerals. *Rev. Geophys.*, **33**, suppl. (IUGG Report), 123–128.

Moskowitz, B.M. & Banerjee, S.K. (1979). Grain size limits for pseudosingle domain behaviour in magnetite. *IEEE Trans. Magn.*, **MAG15**, 1241–1246.

Moskowitz, B.M. & Banerjee, S.K. (1981). A comparison of the magnetic properties of synthetic titanomaghemites and some oceanic basalts. *J. Geophys. Res.*, **86**, 11,869–11,882.

Moskowitz, B.M. & Halgedahl, S.L. (1987). Theoretical temperature and grain-size dependence of domain state in $x = 0.6$ titanomagnetite. *J. Geophys. Res.*, **92**, 10,667–10,682.

Moskowitz, B.M., Frankel, R. B. & Bazylinksi, D.A. (1993). Rock magnetic criteria for the detection of biogenic magnetite. *Earth Planet. Sci. Lett.*, **120**, 283–300.

Moskowitz, B.M., Frankel, R.B., Flanders, P.J., Blakemore, R.P. & Schwartz, B.B. (1988a). Magnetic properties of magnetotactic bacteria. *J. Magn. Magn. Mat.*, **73**, 273–288.

Moskowitz, B.M., Halgedahl, S.L. & Lawson, C.A. (1988b). Magnetic domains on unpolished and polished surfaces of titanium-rich titanomagnetite. *J. Geophys. Res.*, **93**, 3372–3386.

Muench, G.J., Arajs, S. & Matijevic, E. (1985). The Morin transition in small α-Fe_2O_3 particles. *Phys. Stat. Sol.*, (a), **92**, 187–192.

Mullins, C.E. (1977). Magnetic susceptibility of the soil and its significance in soil science – A review. *J. Soil. Sci.*, **28**, 223–246.

Mullins, C.E. & Tite, M.S. (1973). Magnetic viscosity, quadrature susceptibility, and frequency dependence of susceptibility in single-domain assemblies of magnetite and maghemite. *J. Geophys. Res.*, **78**, 804–809.

Murthy, G.S., Evans, M.E. & Gough, D.I. (1971). Evidence of single-domain magnetite in the Michikamau anorthosite. *Can. J. Earth Sci.*, **8**, 361–370.

Murthy, G.S., Pätzold, R. & Brown, C. (1981). Source of stable remanence in certain intrusive rocks. *Phys. Earth Planet. Inter.*, **26**, 72–80.

Nagata, T. (1943). The natural remanent magnetism of volcanic rocks and its relation to geomagnetic phenomena. *Bull. Earthquake Res. Inst., Univ. Tokyo*, **21**, 1–196.

Nagata, T. (1953). *Rock Magnetism*, 1st edn. Maruzen, Tokyo, 225 pp.

Nagata, T. (1961). *Rock Magnetism*, 2nd edn. Maruzen, Tokyo, 350 pp.

Nagata, T. (1966). Main characteristics of piezo-magnetization and their qualitative interpretation. *J. Geomag. Geoelec.*, **18**, 81–97.

Nagata, T. (1970). Basic magnetic properties of rocks under the effects of mechanical stresses. *Tectonophys.*, **9**, 167–195.

Nagata, T. (1979a). Meteorite magnetism and the early solar system magnetic field. *Phys. Earth Planet. Inter.*, **20**, 324–341.

Nagata, T. (1979b). Magnetic properties and paleointensity of achondrites in comparison with those of lunar rocks. *Proc. 10th Lunar Sci. Conf.*, pp. 2199–2210. Pergamon, New York,

Nagata, T. (1983). Meteorite magnetization and paleointensity. *Adv. Space Res.*, **2**, 55–63.

Nagata, T. & Carleton, B. J. (1969). Notes on piezo-remanent magnetization of igneous rocks II. *J. Geomag. Geoelec.*, **21**, 427–445.

Nagata, T. & Carleton, B.J. (1970). Natural remanent magnetization and viscous magnetization of Apollo 11 lunar materials. *J. Geomag. Geoelec.*, **22**, 491–506.

Nagata, T. & Funaki, M. (1982). Magnetic properties of tetrataenite-rich stony meteorites. In *Proc. 7th Sympos. Antarct. Meteorites, Mem. Nat. Inst. Polar Res., Tokyo*, spec. iss. **25**, pp. 222–250.

Nagata, T. & Funaki, M. (1983). Paleointensity of the Allende carbonaceous chondrite. In *Proc. 8th Sympos. Antarct. Meteorites. Mem. Nat. Inst. Polar Res. Tokyo*, spec. iss. **30**, pp. 403–434.

Nagata, T. & Kinoshita, H. (1967). Effect of hydrostatic pressure on magnetostriction and magnetocrystalline anisotropy of magnetite. *Phys. Earth Planet. Inter.*, **1**, 44–48.

Nagata, T., Arai, Y. & Momose, K. (1963). Secular variation of the geomagnetic total force during the last 5000 years. *J. Geophys. Res.*, **68**, 5277–5281.

Nagata, T., Fisher, R.M. & Schwerer, F.C. (1972). Lunar rock magnetism. *The Moon*, **4**, 160–186.

Nagata, T., Uyeda, S. & Akimoto, S. (1952). Self-reversal of thermo-remanent magnetization of igneous rocks. *J. Geomag. Geoelec.*, **4**, 22–38.

Néel, L. (1944a). Quelques propriétés des parois des domaines élémentaires ferromagnétiques. *Cahiers Phys.*, **25**, 1–20.

Néel, L. (1944b). Effet des cavités et des inclusions sur le champ coercitif. *Cahiers Phys.*, **25**, 21–44.

Néel, L. (1947). Propriétés d'un ferromagnétique cubique en grain fins. *C. R. Acad. Sci. Paris*, **224**, 1488–1490.

Néel, L. (1948). Propriétés magnétiques des ferrites: ferrimagnétisme et antiferromagnétisme. *Annales Phys.*, **3**, 137–198.

Néel, L. (1949). Théorie du traînage magnétique des ferromagnétiques en grains fins avec applications aux terres cuites. *Ann. Géophys.*, **5**, 99–136.

Néel, L. (1950). Théorie du traînage magnétique des substances massives dans le domaine de Rayleigh. *J. Physique.*, **11**, 49–61.

Néel, L. (1951). L'inversion de l'aimantation permanente des roches. *Ann. Géophys.*, **7**, 90–102.

Néel, L. (1954). Remarques sur la théorie des propriétés magnétiques des substances dures. *Appl. Sci. Res. (Hague)*, **B4**, 13–24.

Néel, L. (1955). Some theoretical aspects of rock magnetism. *Adv. Phys.*, **4**, 191–243.

Néel, L. & Pauthenet, R. (1952). Étude thermomagnétique d'un monocristal de $Fe_2O_3\alpha$. *C. R. Acad. Sci. (Paris)*, **234**, 2172–2174.

Néel, L. Pauleve, J., Pauthenet, R., Laugier, J. & Dautreppe, D. (1964). Magnetic properties of an iron-nickel single crystal ordered by neutron bombardment. *J. Appl. Phys.*, **35**, 873–876.

Newell, A.J., Dunlop, D.J. & Enkin, R.J. (1990). Temperature dependence of critical sizes, wall widths and moments in two-domain magnetite grains. *Phys. Earth Planet. Inter.*, **65**, 165–176.

Newell, A.J., Dunlop, D.J. & Williams, W. (1993b). A two-dimensional micromagnetic model of magnetization and fields in magnetite. *J. Geophys. Res.*, **98**, 9533–9549.

Newell, A.J., Williams, W. & Dunlop, D.J. (1993a). A generalization of the demagnetizing tensor for nonuniform magnetization. *J. Geophys. Res.*, **98**, 9551–9555.

Nguyen, T.K.T. & Pechersky, D.M. (1987a). Experimental study of chemical and crystallization remanent magnetizations in magnetite. *Phys. Earth Planet. Inter.*, **46**, 46–63.

Nguyen, T.K.T. & Pechersky, D.M. (1987b). Properties of chemical remanent magnetization of magnetite formed on decomposition of titanomaghemite (transl. from Russian). *Izv. Phys. Solid Earth.*, **23**, 413–419.

Nininger, R.C. & Schroeer, D. (1978). Mössbauer studies of the Morin transition in bulk and microcrystalline α-Fe_2O_3. *J. Phys. Chem. Solids.*, **39**, 137–144.

Nishitani, T. & Kono, M. (1982). Grain size effect on the low-temperature oxidation of titanomagnetite. *J. Geophys.*, **50**, 137–142.

Nishitani, T. & Kono, M. (1983). Curie temperature and lattice constant of oxidized titanomagnetite. *Geophys. J. Roy. Astron. Soc.*, **74**, 585–600.

Nishitani, T. & Kono, M. (1989). Effect of low-temperature oxidation on the remanence properties of titanomagnetites. *J. Geomag. Geoelec.*, **41**, 19–38.

Nord, G.L. & Lawson, C.A. (1989). Order–disorder transition-induced twin domains and magnetic properties in ilmenite–hematite. *Am. Mineral.*, **74**, 160–176.

Nord, G.L. & Lawson, C.A. (1992). Magnetic properties of ilmenite$_{70}$-hematite$_{30}$: Effect of transformation-induced twin boundaries. *J. Geophys. Res.*, **97**, 10,897–10,910.

Oades, J.M. & Townsend, W.N. (1963). The detection of ferromagnetic minerals in soils and clays. *J. Soil Sci.*, **14**, 180–187.

O'Donovan, J.B. & O'Reilly, W. (1977). Monodomain behaviour in multiphase oxidized titanomagnetite. *Earth Planet. Sci. Lett.*, **34**, 396–402.

O'Donovan, J.B., Facey, D. & O'Reilly, W. (1986). The magnetization process in titanomagnetite (Fe$_{2.4}$Ti$_{0.6}$O$_4$) in the 1–30 µm particle size range. *Geophys. J. Roy. Astron. Soc.*, **87**, 897–916.

Ohnaka, M. & Kinoshita, H. (1969). Effects of uniaxial compression on remanent magnetization. *J. Geomag. Geoelec.*, **20**, 93–99.

Oldfield, F. (1994). Toward the discrimination of fine-grained ferrimagnets by magnetic measurements in lake and near-shore marine sediments. *J. Geophys. Res.*, **99**, 9045–9050.

Opdyke, N.D. (1972). Paleomagnetism of deep-sea cores. *Rev. Geophys. Space Phys.*, **10**, 213–249.

Opdyke, N.D. & Henry, K.W. (1969). A test of the dipole hypothesis. *Earth Planet. Sci. Lett.*, **6**, 139–151.

Opdyke, N.D., Ninkovich, D., Lowrie, W. & Hays, J.D. (1972). The paleomagnetism of two Aegean deep-sea cores. *Earth Planet. Sci. Lett.*, **14**, 145–159.

O'Reilly, W. (1976). Magnetic minerals in the crust of the earth. *Rep. Prog. Phys.*, **39**, 857–908.

O'Reilly, W. (1983). The identification of titanomaghemites: Model mechanisms for the maghemitization and inversion processes and their magnetic consequences. *Phys. Earth Planet. Inter.*, **31**, 65–76.

O'Reilly, W. (1984). *Rock and Mineral Magnetism.* Blackie, Glasgow and London, & Chapman and Hall, New York, 220 pp.

O'Reilly, W. & Banerjee, S.K. (1965). Cation distribution in titanomagnetites $(1 - x)$Fe$_3$O$_4$–xFe$_2$TiO$_4$. *Phys. Lett.*, **17**, 237–238.

O'Reilly, W. & Banerjee, S.K. (1966). Oxidation of titanomagnetites and self-reversal. *Nature.*, **211**, 26–28.

Osborn, J.A. (1945). Demagnetizing factors of the general ellipsoid. *Phys. Rev.*, **67**, 351–357.

Ouliac, M.(1976). Removal of secondary magnetization from natural remanent magnetization of sedimentary rocks: Alternating field or thermal demagnetization technique? *Earth Planet. Sci. Lett.*, **29**, 65–70.

Özdemir, Ö. (1979). An experimental study of thermoremanent magnetization acquired by synthetic monodomain titanomagnetites and titanomaghemites. Ph.D. thesis, Univ. of Newcastle upon Tyne, U.K.

Özdemir, Ö. (1987). Inversion of titanomaghemites. *Phys. Earth Planet. Inter.*, **46**, 184–196.

Özdemir, Ö. (1990). High-temperature hysteresis and thermoremanence of single-domain maghemite. *Phys. Earth Planet. Inter.*, **65**, 125–136.

Özdemir, Ö. & Banerjee, S.K. (1981). An experimental study of magnetic viscosity in synthetic monodomain titanomaghemites: Implications for the magnetization of the ocean crust. *J. Geophys. Res.*, **86**, 11,864–11,868.

Özdemir, Ö. & Banerjee, S.K. (1982). A preliminary magnetic study of soil samples from west-central Minnesota. *Earth Planet. Sci. Lett.*, **59**, 393–403.

Özdemir, Ö. & Banerjee, S.K. (1984). High temperature stability of maghemite. *Geophys. Res. Lett.*, **11**, 161–164.

Özdemir, Ö. & Deutsch, E.R. (1984) Magnetic properties of oolitic iron ore on Bell Island, Newfoundland. *Earth Planet. Sci. Lett.*, **69**, 427–441.

Özdemir, Ö. & Dunlop, D.J. (1985). An experimental study of chemical remanent magnetizations of synthetic monodomain titanomaghemites with initial thermoremanent magnetizations. *J. Geophys. Res.*, **90**, 11,513–11,523.

Özdemir, Ö. & Dunlop, D.J. (1988). Crystallization remanent magnetization during the transformation of maghemite to hematite. *J. Geophys. Res.*, **93**, 6530–6544.

Özdemir, Ö. & Dunlop, D.J. (1989). Chemico-viscous remanent magnetization in the Fe$_3$O$_4$–γFe$_2$O$_3$ system. *Science*, **243**, 1043–1047.

Özdemir, Ö. & Dunlop, D.J. (1992). Domain structure observations in biotites and hornblendes (abstract). *Eos (Trans. Am. Geophys. Un.)*, **73**, Spring Meeting suppl., 93.

Özdemir, Ö. & Dunlop, D.J. (1993a). Magnetic domain structures on a natural single crystal of magnetite. *Geophys. Res. Lett.*, **20**, 1835–1838.

Özdemir, Ö. & Dunlop, D.J. (1993b). Chemical remanent magnetization during γFeOOH phase transformations. *J. Geophys. Res.*, **98**, 4191–4198.

Özdemir, Ö. & Dunlop, D.J. (1995). Single-domain state in a large (3 mm) multidomain magnetite crystal (abstract). *Eos (Trans. Am. Geophys. Un.)*, **76**, Fall Meeting suppl., F160.

Özdemir, Ö. & Dunlop, D.J. (1996a). Thermoremanence and Néel temperature of goethite. *Geophys. Res. Lett.*, **23**, 921–924.

Özdemir, Ö. & Dunlop, D.J. (1996b). Effect of crystal defects on the domain structure of magnetite. *J. Geophys. Res.*, **101**, in press.

Özdemir, Ö. & Moskowitz, B.M. (1992). Magnetostriction in aluminium-substituted titanomagnetites. *Geophys. Res. Lett.*, **19**, 2361–2364.

Özdemir, Ö. & O'Reilly, W. (1978). Magnetic properties of monodomain aluminium-substituted titanomagnetite. *Phys. Earth Planet. Inter.*, **16**, 190–195.

Özdemir, Ö. & O'Reilly, W. (1981). High-temperature hysteresis and other magnetic properties of synthetic monodomain titanomagnetites. *Phys. Earth Planet. Inter.*, **25**, 406–418.

Özdemir, Ö. & O'Reilly, W. (1982a). Magnetic hysteresis properties of synthetic monodomain titanomaghemites. *Earth Planet. Sci. Lett.*, **57**, 437–447.

Özdemir, Ö. & O'Reilly, W. (1982b). An experimental study of thermoremanent magnetization acquired by synthetic monodomain titanomaghemites. *J. Geomag. Geoelec.*, **34**, 467–478.

Özdemir, Ö. & O'Reilly, W. (1982c). An experimental study of the intensity and stability of thermoremanent magnetization acquired by synthetic monodomain titanomagnetite substituted by aluminium. *Geophys. J. Roy. Astron. Soc.*, **70**, 141–154.

Özdemir, Ö. & York, D. (1990). ^{40}Ar/^{39}Ar laser dating of a single grain of magnetite. *Tectonophys.*, **184**, 21–33.

Özdemir, Ö. & York, D. (1992). ^{40}Ar/^{39}Ar laser dating of biotite inclusions in a single crystal of magnetite. *Geophys. Res. Lett.*, **19**, 1799–1802.

Özdemir, Ö., Dunlop, D.J. & Moskowitz, B.M. (1993). The effect of oxidation on the Verwey transition in magnetite. *Geophys. Res. Lett.*, **20**, 1671–1674.

Özdemir, Ö., Dunlop, D.J., Reid, B.J. & Hyodo, H. (1988). An early Proterozoic VGP from an oriented drill core into the Precambrian basement of southern Alberta. *Geophys. J.*, **95**, 69–78.

Özdemir, Ö., Xu, S. & Dunlop, D.J. (1995). Closure domains in magnetite. *J. Geophys. Res.*, **100**, 2193–2209.

Ozima, Minoru (1971). Magnetic processes in oceanic ridge. *Earth Planet. Sci. Lett.*, **13**, 1–5.

Ozima, Minoru & Ozima, Mituko (1965). Origin of thermoremanent magnetization. *J. Geophys. Res.*, **70**, 1363–1369.

Ozima, Mituko & Sakamoto, N. (1971). Magnetic properties of synthesized titanomaghemite. *J. Geophys. Res.*, **76**, 7035–7046.

Ozima, Mituko, Funaki, M., Hamada, N., Aramaki, S. & Fujii, T. (1992). Self-reversal of thermo-remanent magnetization in pyroclastics from the 1991 eruption of Mt. Pinatubo, Philippines. *J. Geomag. Geoelec.*, **44**, 979–984.

Ozima, Mituko, Ozima, Minoru & Akimoto, S. (1964). Low temperature characteristics of remanent magnetization of magnetite: Self-reversal and recovery phenomena of remanent magnetization. *J. Geomag. Geoelec.*, **16**, 165–177.

Pan, H. & Symons, D.T.A. (1993). Paleomagnetism of the Mississippi Valley-type Newfoundland zinc deposit: Evidence for Devonian mineralization and host rock remagnetization in the northern Appalachians. *J. Geophys. Res.*, **98**, 22,415–22,427.

Pariso, J.E. & Johnson, H.P. (1989). Magnetic properties and oxide petrography of the sheeted dike complex in Hole 504B. In *Proc. Ocean Drill. Program. Sci. Res.*, ed. K. Becker, H. Sakai, et al., vol. **111**, pp. 159–167. Ocean Drilling Program, College Station, TX.

Pariso, J.E. & Johnson, H.P. (1991). Alteration processes at Deep Sea Drilling Project/Ocean Drilling Program Hole 504B at the Costa Rica Rift: Implications for magnetization of oceanic crust. *J. Geophys. Res.*, **96**, 11,703–11,722.

Pariso, J.E. & Johnson, H.P. (1993a). Do lower crustal rocks record reversals of the earth's magnetic field? Magnetic petrology of oceanic gabbros from Ocean Drilling Program Hole 735B. *J. Geophys. Res.*, **98**, 16,013–16,032.

Pariso, J.E. & Johnson, H.P. (1993b). Do lower crustal rocks record reversals of the earth's magnetic field? In situ magnetization of gabbros at Ocean Drilling Program Hole 735B. *J. Geophys. Res.*, **98**, 16,033–16,052.

Parry, L.G. (1965). Magnetic properties of dispersed magnetite powders. *Phil. Mag.*, **11**, 303–312.

Parry, L.G. (1979). Magnetization of multidomain particles of magnetite. *Phys. Earth Planet. Inter.*, **19**, 21–30.

Parry, L.G. (1980). Shape-related factors in the magnetization of immobilized magnetite particles. *Phys. Earth Planet. Inter.*, **22**, 144–154.

Parry, L.G. (1982). Magnetization of immobilized particle dispersions with two distinct particle sizes. *Phys. Earth Planet. Inter.*, **28**, 230–241.

Pearce, G.W. & Karson, J.A. (1981). On pressure demagnetization. *Geophys. Res. Lett.*, **8**, 725–728.

Pearce, G.W., Gose, W.A. & Strangway, D.W. (1973). Magnetic studies on Apollo 15 and 16 lunar samples. *Geochim. Cosmochim. Acta.*, **37**, suppl. 4, 3045–3076.

Pearce, G.W., Hoye, G.S., Strangway, D.W., Walker, B.M. & Taylor, L.A. (1976). Some complexities in the determination of lunar paleointensities. *Geochim. Cosmochim. Acta.*, **40**, suppl. 7, 3271–3297.

Peregrinus, Petrus (1269). *Epistola de Magnete*. See Smith, P.J. (1970). Petrus Peregrinus Epistola – The beginning of experimental studies of magnetism in Europe. *Atlas (News Suppl., Earth Sci. Rev.)*, **6**, A11.

Petersen, N. & Vali, H. (1987). Observation of shrinkage cracks in ocean floor titanomagnetites. *Phys. Earth Planet. Inter.*, **46**, 197–205.

Petersen, N., von Dobeneck, T. & Vali, H. (1986). Fossil bacterial magnetite in deep-sea sediments from the South Atlantic Ocean. *Nature*, **320**, 611–615.

Petersen, N., Weiss, D.G. & Vali, H. (1989). Magnetic bacteria in lake sediments. In *Geomagnetism and Palaeomagnetism*, ed. F.J. Lowes et al., pp. 231–241. Kluwer, Dordrecht.

Petherbridge, J. (1977). A magnetic coupling occurring in partial self-reversal of magnetism and its association with increased magnetic viscosity in basalts. *Geophys. J. Roy. Astron. Soc.*, **50**, 395–406.

Petrovsky, E., Hejda, P., Zelinka, T., Kropaček, V. & Šubrt, J. (1993). Experimental determination of magnetic interactions within a system of synthetic haematite particles. *Phys. Earth Planet. Inter.*, **76**, 123–130.

Pick, T. & Tauxe, L. (1991). Chemical remanent magnetization in synthetic magnetite. *J. Geophys. Res.*, **96**, 9925–9936.

Pick, T. & Tauxe, L. (1993). Holocene paleointensities: Thellier experiments on submarine basaltic glass from the East Pacific Rise. *J. Geophys. Res.*, **98**, 17,949–17,964.

Pinto, M.J. & McWilliams, M.O. (1990). Drilling-induced isothermal remanent magnetization. *Geophysics*, **55**, 111–115.

Plessard, C. (1971). Modification des propriétés magnétiques, en particulier du traînage, après réchauffement d'une roche préalablement stabilisée thermiquement. *C. R. Acad. Sci. Paris*, **273**, 97–100.

Plessard, C. & Prévot, M. (1977). Magnetic viscosity of submarine basalts. *Init. Rep. Deep Sea Drill. Proj.*, **37**, 503–506.

Pohl, J. & Soffel, H.C. (1971). Paleomagnetic age determinaton of the Rochechouart impact structure (France). *Z. Geophys.*, **37**, 857–866.

Pohl, J., Bleil, U. & Hornemann, U. (1975). Shock magnetization and demagnetization of basalt by transient stress up to 10 kbar. *J. Geophys.*, **41**, 23–41.

Pokhil, T.G. & Moskowitz, B.M. (1995). MFM study of domain walls in small magnetite particles (abstract). *Int. Un. Geod. Geophys. XXI Gen. Ass., Abstracts vol.*, p. B96.

Porath, H. (1968a). Magnetic studies on specimens of intergrown maghemite and hematite. *J. Geophys. Res.*, **73**, 5959–5965.

Porath, H. (1968b). Stress induced anisotropy in natural single crystals of hematite. *Phil. Mag.*, **17**, 603–608.

Porath, H. & Raleigh, C.B. (1967). An origin of the triaxial basal-plane anisotropy in hematite crystals. *J. Appl. Phys.*, **38**, 2401–2402.

Pozzi, J-P. (1973). Effets de pression en magnétisme des roches. D.ès Sc. thesis, Univ. de Paris.

Pozzi, J-P. & Dubuisson, G. (1992). High temperature viscous magnetization of oceanic deep crustal- and mantle-rocks as a partial source for Magsat magnetic anomalies. *Geophys. Res. Lett.*, **19**, 21–24.

Preisach, F. (1935). Über die magnetische Nachwirkung. *Z. Phys.*, **94**, 277–302.

Prévot, M. (1981). Some aspects of magnetic viscosity in subaerial and submarine volcanic rocks. *Geophys. J. Roy. Astron. Soc.*, **66**, 169–192.

Prévot, M. & Bina, M.M. (1993). Origin of magnetic viscosity and estimate of long-term induced magnetization in coarse-grained submarine basalts. *Geophys. Res. Lett.*, **20**, 2483–2486.

Prévot, M., Lecaille, A. & Hekinian, R. (1979). Magnetism of the Mid-Atlantic Ridge crest near 37°N from FAMOUS and DSDP results: A review. In *Deep Drilling Results in the Atlantic Ocean: Ocean Crust*, ed. M. Talwani, C.G.A. Harrison & D.E. Hayes, pp. 210–229. American Geophysical Union, Washington, DC.

Prévot, M., Mankinen, E.A., Coe, R.S. & Grommé, C.S. (1985). The Steens Mountain (Oregon) geomagnetic polarity transition, 2. Field intensity variations and discussion of reversal models. *J. Geophys. Res.*, **90**, 10,417–10,448.

Proksch, R.B. & Moskowitz, B.M. (1994). Interactions between single domain particles. *J. Appl. Phys.*, **75**, 5894–5896.

Proksch, R.B., Foss, S. & Dahlberg, E.D. (1994a). High resolution magnetic force microscopy of domain wall fine structures, *IEEE Trans. Magn.*, **30**, 4467–4472.

Proksch, R.B., Foss, S., Dahlberg, E.D. & Prinz, G. (1994b). Magnetic fine structure of domain walls in iron films observed with a magnetic force microscope. *J. Appl. Phys.*, **75**, 5776–5778.

Proksch, R.B., Schäffer, T.E., Moskowitz, B.M., Dahlberg, E.D., Bazylinski, D.A. & Frankel, R.B. (1995). Magnetic force microscopy of the submicron magnetic assembly in a magnetotactic bacterium. *Appl. Phys. Lett.*, **66**, 2582–2584.

Pucher, R. (1969). Relative stabilities of chemical and thermal remanence in synthetic ferrites. *Earth Planet. Sci. Lett.*, **6**, 107–111.

Pullaiah, G., Irving, E., Buchan, K.L. & Dunlop, D.J. (1975). Magnetization changes caused by burial and uplift. *Earth Planet. Sci. Lett.*, **28**, 133–143.

Purucker, M.E., Elston, D.P. & Shoemaker, E.M. (1980). Early acquisition of characteristic magnetization in red beds of the Moenkopi Formation (Triassic), Gray Mountain, Arizona. *J. Geophys. Res.*, **85**, 997–1012.

Quidelleur, X. & Valet, J-P. (1994). Paleomagnetic records of excursions and reversals: Possible biases caused by magnetization artefacts. *Phys. Earth Planet. Inter.*, **82**, 27–48.

Quidelleur, X., Valet, J-P., LeGoff, M. & Bouldoire, X. (1995). Field dependence on magnetization of laboratory-redeposited deep-sea sediments: First results. *Earth Planet. Sci. Lett.*, **133**, 311–325.

Rahman, A.A. & Parry, L.G. (1975). Self shielding of inclusions in titanomagnetite grains. *Phys. Earth Planet. Inter.*, **11**, 139–146.

Rahman, A.A., Duncan, A.D. & Parry, L.G. (1973). Magnetization of multidomain magnetite particles. *Riv. Ital. Geofis.*, **22**, 259–266.

Raymond, C. & LaBrecque, J.L. (1987). Magnetization of the oceanic crust: Thermoremanent magnetization or chemical remanent magnetization? *J. Geophys. Res.*, **92**, 8077–8088.

Readman, P.W. & O'Reilly, W. (1972). Magnetic properties of oxidized (cation-deficient) titanomagnetites, $(Fe,Ti,\square)O_4$. *J. Geomag. Geoelec.*, **24**, 69–90.

Rees, A.I. (1961). The effect of water currents on the magnetic remanence and anisotropy of susceptibility of some sediments. *Geophys. J. Roy. Astron. Soc.*, **5**, 235–251.

Reynolds, R.L. (1979). Comparison of the TRM of the Yellowstone Group and the DRM of some pearlette ashbeds. *J. Geophys. Res.*, **84**, 4525–4532.

Reynolds, R.L., Fishman, N.S., Wanty, R.B. & Goldhaber, M. (1990). Iron sulphide minerals at Cement Oil Field, Oklahoma: Implications for the magnetic detection of oil fields. *Geol. Soc. Am. Bull.*, **102**, 368–380.

Reynold, R.L., Tuttle, M.L., Rice, C.A., Fishman, N.S., Karachewski, J.A. & Sherman, D.M. (1994). Magnetization and geochemistry of greigite-bearing Cretaceous strata, North Slope Basin, Alaska. *Am. J. Sci.*, **294**, 485–528.

Rhodes, P. & Rowlands, G. (1954). Demagnetising energies of uniformly magnetised rectangular blocks. *Proc. Leeds Phil. Lit. Soc., Sci. Sect.*, **6**, 191–210.

Richter, G. (1937). Über die magnetische Nachwirkung am Carbonyleisen. *Ann. Physik*, **29**, 605–635.

Rimbert, F. (1959). Contribution a l'étude de l'action de champs alternatifs sur les aimantations rémanentes des roches. *Rev. Inst. Franç. Pétrole Ann. Combust. Liq.*, **14**, 17–54 & 123–155.

Riste, T. & Tenzer, L. (1961). A neutron diffraction study of the temperature variation of the spontaneous sublattice magnetization of ferrites and the Néel theory of ferrimagnetism. *J. Phys. Chem. Solids.*, **19**, 117–123.

Robbins, M., Wertheim, G.K., Sherwood, R.C. & Buchanan, D.N.E. (1971). Magnetic properties and site distributions in the system $FeCr_2O_4$–$Fe_3O_4(Fe^{2+}Cr_{2-x}Fe_x^{3+}O_4)$. *J. Phys. Chem. Solids.*, **32**, 717–729.

Roberts, A.P. (1995). Magnetic properties of sedimentary greigite (Fe_3S_4). *Earth Planet. Sci. Lett.*, **134**, 227–236.

Roberts, A.P. & Turner, G.M. (1993). Diagenetic formation of ferrimagnetic iron sulphide minerals in rapidly deposited marine sediments, South Island, New Zealand. *Earth Planet. Sci. Lett.*, **115**, 257–273.

Roberts, A.P., Cui, Y. & Verosub, K.L. (1995). Wasp-waisted hysteresis loops: Mineral magnetic characteristics and discrimination of components in mixed magnetic systems. *J. Geophys. Res.*, **100**, 17,909–17,924.

Robertson, D.J. & France, D.E. (1994). Discrimination of remanence-carrying minerals in mixtures, using isothermal remanent magnetisation acquisition curves. *Phys. Earth Planet. Inter.*, **82**, 223–234.

Robertson, P.B. & Roy, J.L. (1979). Shock-diminished paleomagnetic remanence at the Charlevoix impact structure, Quebec. *Can J. Earth Sci.*, **16**, 1842–1856.

Robins, B.W. (1972). Remanent magnetization in spinel iron oxides. Ph.D. thesis, Univ. of New South Wales, Sydney.

Rochette, P. (1987). Metamorphic control of the magnetic mineralogy of black shales in the Swiss Alps: Toward the use of "magnetic isogrades". *Earth Planet. Sci. Lett.*, **84**, 446–456.

Rochette, P., Fillion, G., Mattéi, L-L. & Dekkers, M.J. (1990). Magnetic transition at 30–34 Kelvin in pyrrhotite: Insight into a widespread occurrence of this mineral in rocks. *Earth Planet. Sci. Lett.*, **98**, 319–328.

Rochette, P., Jackson, M. & Aubourg, C. (1992b). Rock magnetism and the interpretation of anisotropy of magnetic susceptibility. *Rev. Geophys*, **30**, 209–226.

Rochette, P., Ménard, G. & Dunn, R. (1992a). Thermochronometry and cooling rates deduced from single sample records of successive magnetic polarities during uplift of metamorphic rocks in the Alps (France). *Geophys. J. Int.*, **108**, 491–501.

Rogers, J., Fox, J.M.W. & Aitken, M.J. (1979). Magnetic anisotropy in ancient pottery. *Nature*, **277**, 644–646.

Roquet, J. (1954). Sur les rémanences des oxydes de fer et leur intérêt en géomagnetisme. *Ann. Géophys.*, **10**, 226–247 & 282–325.

Roquet, J. & Thellier, E. (1946). Sur les lois numériques simples, relatives à l'aimantation thermorémanente du sesquioxyde de fer rhomboédrique. *C. R. Acad. Sci. Paris.*, **222**, 1288–1290.

Roy, J.L. & Park, J.K. (1972). Red beds: DRM or CRM? *Earth Planet. Sci. Lett.*, **17**, 211–216.

Roy, J.L. & Park, J.K. (1974). The magnetization process of certain red beds: Vector analysis of chemical and thermal results. *Can. J. Earth Sci.*, **11**, 437–471.

Runcorn, S.K. (1994). The early magnetic field and primeval satellite system of the Moon: Clues to planetary formation. *Phil. Trans. Roy. Soc. London.*, **349**, 181–196.

Runcorn, S.K., Collinson, D.W., O'Reilly, W., Stephenson, A., Battey, M.H., Manson, A.J. & Readman, P.W. (1971). Magnetic properties of Apollo 12 lunar samples. *Proc. Roy. Soc. London*, **A325**, 157–174.

Ryall, P.J.C. & Ade-Hall, J.M. (1975). Radial variation of magnetic properties in submarine pillow basalt. *Can. J. Earth Sci.*, **12**, 1959–1969.

Sahu, S. & Moskowitz, B.M. (1995). Thermal dependence of magnetocrystalline anisotropy and magnetostriction constants of single crystal. $Fe_{2.4}Ti_{0.6}O_4$. *Geophys. Res. Lett.*, **22**, 449–452.

Sakaguchi, T., Burges, J.G. & Matsunaga, T. (1993). Magnetite formation by a sulphate-reducing bacterium. *Nature*, **365**, 47–49.

Sakamoto, N., Ince, P.I. & O'Reilly, W. (1968). The effect of wet grinding on the oxidation of titanomagnetites. *Geophys. J. Roy. Astron. Soc.*, **15**, 509–515.

Sato, T., Nakatsuka, K., Toita, K. & Shimoizaka, J. (1967). Study on the artificial magnetite by wet method. *J. Jap. Soc. Powders Powder Metal.*, **14**, 17–25.

Schabes, M.E. (1991). Micromagnetic theory of non-uniform magnetization processes in magnetic recording particles. *J. Magn. Magn. Mat.*, **95**, 249–288.

Schabes, M.E. & Aharoni, A. (1987). Magnetostatic interaction fields for a three-dimensional array of ferromagnetic cubes. *IEEE Trans. Magn.*, **MAG23**, 3882–3888.

Schabes, M.E. & Bertram, H.N. (1988). Magnetization processes in ferromagnetic cubes. *J. Appl. Phys.*, **64**, 1347–1357.

Scheinfein, M.R., Unguris, J., Blue, J.L., Coakley, K.J., Pierce, D.T., Celotta, R.J. & Ryan, P.J. (1991). Micromagnetics of domain walls at surfaces. *Phys. Rev.*, **B43**, 3395–3422.

Schmidbauer, E. (1971). Magnetization of Fe–Cr spinels and its application for the identification of such ferrites in rocks. *Z. Geophys.*, **37**, 421–424.

Schmidbauer, E. & Schembera, N. (1987). Magnetic hysteresis properties and anhysteretic remanent magnetization of spherical Fe_3O_4 particles in the grain size range 60–160 nm. *Phys. Earth Planet. Inter.*, **46**, 77–83.

Schmidbauer, E. & Veitch, R.J. (1980). Anhysteretic remanent magnetization of small multidomain Fe_3O_4 particles in a non-magnetic matrix. *J. Geophys.*, **48**, 148–152.

Schmidt, P. W. (1993). Palaeomagnetic cleaning strategies. *Phys. Earth Planet. Inter.*, **76**, 169–178.

Schmidt, V.A. (1973). A multidomain model of thermoremanence. *Earth Planet. Sci. Lett.*, **20**, 440–446.

Schmidt, V.A. (1976). The variation of the blocking temperature in models of thermoremanence. *Earth Planet. Sci. Lett.*, **29**, 146–154.

Schneider, D.A. & Kent, D.V. (1990). The time-averaged paleomagnetic field. *Rev. Geophys.*, **28**, 71–96.

Schult, A. (1968). Self-reversal of magnetization and chemical composition of titanomagnetites in basalts. *Earth Planet. Sci. Lett.*, **4**, 57–63.

Schult, A. (1970). Effect of pressure on the Curie temperature of titanomagnetites $[(1 - x)Fe_3O_4 - xTiFe_2O_4]$. *Earth Planet. Sci. Lett.*, **10**, 81–86.

Schult, A. (1976). Self-reversal above room temperature due to N-type magnetization in basalt. *J. Geophys.*, **42**, 81–84.

Schutts, L.D., Brecher, A., Hurley, P.M., Montgomery, C.W. & Krueger, H.W. (1976). A case study of the time and nature of paleomagnetic resetting in a mafic complex in New England. *Can. J. Earth Sci.*, **13**, 898–907.

Schwarz, E.J. (1975). Magnetic properties of pyrrhotite and their use in applied geology and geophysics. *Geol. Surv. Canada. Prof. Paper*, **74-59**, pp. 1–24.

Schwarz, E.J. (1977). Depth of burial from remanent magnetization: The Sudbury irruptive at the time of diabase intrusion (1250 Ma). *Can. J. Earth Sci.*, **14**, 82–88.

Schwarz, E.J. & Buchan, K.L. (1989). Identifying types of remanent magnetization in igneous contact zones. *Phys. Earth Planet. Inter.*, **58**, 155–162.

Schwarz, E.J. & Vaughan, D.J. (1972). Magnetic phase relations of pyrrhotite. *J. Geomag. Geoelec.*, **24**, 441–458.

Schwertmann, U. (1971). Transformation of hematite to goethite in soils. *Nature*, **232**, 624–625.

Schwertmann, U. (1988). Occurrence and formation of iron oxides in various pedoenvironments. In *Iron in Soils and Clay Minerals*, ed. J.W. Stucki et al., pp. 267–308. D. Reidel, Norwell, MA.

Scott, E.R.D. & Wasson, J.T. (1975). Classification and properties of iron meteorites. *Rev. Geophys. Space Phys.*, **13**, 527–546.

Serway, R.A. (1986). *Physics for Scientists and Engineers*, 2nd edn. Saunders, Philadelphia, 1108 pp.

Shau, Y-H., Peacor, D.R. & Essene, E.J. (1993). Formation of magnetic single-domain magnetite in ocean ridge basalts with implications for sea-floor magnetism. *Science*, **261**, 343–345.

Shaw, J. (1974). A new method for determining the magnitude of the palaeomagnetic field. *Geophys. J. Roy. Astron. Soc.*, **39**, 133–141.

Shcherbakov, V.P. (1978). Theory concerning the magnetization properties of pseudo-single domain grains (transl. from Russian). *Izv., Phys. Solid Earth*, **14**, 356–362.

Shcherbakov, V.P. & Lamash, B.E. (1988). Metastability threshold sizes in single-domain magnetite particles. *Geophys. Res. Lett.*, **15**, 526–529.

Shcherbakov, V.P. & Markov, G.P. (1982). Theory of thermoremanent magnetization in a multidomain grain of inhomogeneous magnetic hardness (transl. from Russian). *Izv. Phys. Solid Earth*, **18**, 681–689.

Shcherbakov, V.P. & Shcherbakova, V.V. (1977). Calculation of thermoremanence and ideal magnetization of an ensemble of interacting single-domain grains (transl. from Russian). *Izv., Phys. Solid Earth*, **13**, 413–421.

Shcherbakov, V.P. & Shcherbakova, V.V. (1979). The concentration dependence of thermoremanent and ideal magnetization of an assembly of one-dimensional granules (transl. from Russian). *Izv., Phys. Solid Earth*, **15**, 933–936.

Shcherbakov, V.P. & Tarashchan, S.A. (1990). Domain structure of titanomagnetite grains with closure domains. *Phys. Earth Planet. Inter.*, **65**, 177–187.

Shcherbakov, V.P., Lamash, B.E., Schmidt, P.W. & Sycheva, N.K. (1990). Micromagnetic formulation for the personal computer. *Phys. Earth Planet. Inter.*, **65**, 15–27.

Shcherbakov, V.P., Lamash, B.E. & Sycheva, N.K. (1995). Monte Carlo modelling of thermoremanence acquisition in interacting single-domain grains. *Phys. Earth. Planet. Inter.*, **87**, 197–211.

Shcherbakov, V.P., McClelland, E. & Shcherbakova, V.V. (1993). A model of multidomain thermoremanent magnetization incorporating temperature-variable domain structure. *J. Geophys. Res.*, **98**, 6201–6216.

Shimizu, Y. (1960). Magnetic viscosity of magnetite. *J. Geomag. Geoelec.*, **11**, 125–138.

Shive, P.N. (1969). Dislocation control of magnetization. *J. Geomag. Geoelec.*, **21**, 519–529.

Shive, P.N. (1985). Alignment of magnetic grains in fluids. *Earth Planet. Sci. Lett.*, **72**, 117–124.

Shive, P.N. & Fountain, D.M. (1988). Magnetic mineralogy in an Archean crustal section: Implications for crustal magnetization. *J. Geophys. Res.*, **93**, 12,177–12,186.

Shtrikman, S. & Treves, D. (1960). Internal structure of Bloch walls. *J. Appl. Phys.*, **31**, 147S–148S.

Smit, J. & Wijn, H.P.J. (1959). *Ferrites*. John Wiley, New York, 369 pp.

Smith, B.M. (1984). Magnetic viscosity of some doleritic basalts in relation to the interpretation of the oceanic magnetic anomalies. *Geophys. Res. Lett.*, **11**, 213–216.

Smith, B.M. (1987). Consequences of the maghemitization on the magnetic properties of submarine basalts: Synthesis of previous works and results concerning basement rocks from mainly DSDP Legs 51 and 52. *Phys. Earth Planet. Inter.*, **46**, 206–226.

Smith, G.M. (1988). The growth of Fe–Ti oxides: Phase relations in the Fe–Ti–B–O system. *Geophys. Res. Lett.*, **15**, 495–498.

Smith, G.M. & Banerjee, S.K. (1985). Magnetic properties of basalts from Deep Sea Drilling Project Leg 83: The origin of remanence and its relation to tectonic and chemical evolution. *Init. Rep. Deep Sea Drill Proj.*, **83**, 347–357.

Smith, G.M. & Banerjee, S.K. (1986). Magnetic structure of the upper kilometer of the marine crust at Deep Sea Drilling Project Hole 504B, eastern Pacific Ocean. *J. Geophys. Res.*, **91**, 10,337–10,354.

Smith, G.M. & Merrill, R.T. (1982). The determination of the internal field in magnetic grains. *J. Geophys. Res.*, **87**, 9419–9423.

Smith, P.P.K. (1979). The identification of single-domain titanomagnetite particles by means of transmission electron microscopy. *Can. J. Earth Sci.*, **16**, 375–379.

Smith, P.P.K. (1980). The application of Lorentz electron microscopy to the study of rock magnetism. *Inst. Phys. Conf. Ser.*, No. 52, pp. 125–128. Institute of Physics, Bristol, U.K.

Smith, R.W. & Fuller, M. (1967). Alpha-hematite: Stable remanence and memory. *Science*, **156**, 1130–1133.

Snowball, I.F. (1994). Bacterial magnetite and the magnetic properties of sediments in a Swedish lake. *Earth Planet. Sci. Lett.*, **126**, 129–142.

Snowball, I.F. & Thompson, R. (1990). A stable chemical remanence in Holocene sediments. *J. Geophys. Res.*, **95**, 4471–4479.

Soffel, H. (1965). Magnetic domains of polycrystalline natural magnetite. *Z. Geophys.*, **31**, 345–361.

Soffel, H. (1966). Stress dependence of the domain structure of natural magnetite. *Z. Geophys.*, **32**, 63–77.

Soffel, H.C. (1971). The single domain–multidomain transition in natural intermediate titanomagnetites. *Z. Geophys.*, **37**, 451–470.

Soffel, H.C. (1977a). Pseudo-single-domain effects and single-domain multidomain transition in natural pyrrhotite deduced from domain structure observations. *J. Geophys.*, **42**, 351–359.

Soffel, H.C. (1977b). Domain structure of titanomagnetites and its variation with temperature. *J. Geomag. Geoelec.*, **29**, 277–284.

Soffel, H.C. & Appel, E. (1982). Domain structure of small synthetic titanomagnetite particles and experiments with IRM and TRM. *Phys. Earth Planet. Inter.*, **30**, 348–355.

Soffel, H.C., Aumüller, C., Hoffmann, V. & Appel, E. (1990) Three-dimensional domain observations of magnetite and titanomagnetites using the dried colloid SEM method. *Phys. Earth Planet. Inter.*, **65**, 43–53.

Soffel, H.C., Deutsch, E.R., Appel, E., Eisenach, P. & Petersen, N. (1982). The domain structure of synthetic stoichiometric TM10–TM75 and Al-, Mg-, Mn- and V-doped TM62 titanomagnetites. *Phys. Earth Planet. Inter.*, **30**, 336–346.

Solomon, S.C. and 10 co-authors (1992). Venus tectonics: An overview of Magellan observations. *J. Geophys. Res.*, **97**, 13,199–13,255.

Soroka, W. & Beske-Diehl, S. (1984). Variation of magnetic directions within pillow basalts. *Earth Planet. Sci. Lett.*, **69**, 215–223.

Stacey, F.D. (1958). Thermoremanent magnetization (TRM) of multidomain grains in igneous rocks. *Phil. Mag.*, **3**, 1391–1401.

Stacey, F.D. (1962). A generalized theory of thermoremanence, covering the transition from single domain to multi-domain magnetic grains. *Phil. Mag.*, **7**, 1887–1900.

Stacey, F.D. (1963). The physical theory of rock magnetism. *Adv. Phys.*, **12**, 45–133.

Stacey, F.D. (1967). The Koenigsberger ratio and the nature of thermoremanence in igneous rocks. *Earth Planet. Sci. Lett.*, **2**, 67–68.

Stacey, F.D. (1972). On the role of Brownian motion in the control of detrital remanent magnetization in sediments. *Pure Appl. Geophys.*, **98**, 139–145.

Stacey, F.D. (1976). Paleomagnetism of meteorites. *Ann. Rev. Earth Planet. Sci.*, **4**, 147–157.

Stacey, F.D. & Banerjee, S.K. (1974). *The Physical Principles of Rock Magnetism*. Elsevier, Amsterdam, 195 pp.

Stacey, F.D. & Wise, K.N. (1967). Crystal dislocations and coercivity in fine grained magnetite. *Austral. J. Phys.*, **20**, 507–513.

Stacey, F.D., Lovering, J.F. & Parry, L.G. (1961). Thermomagnetic properties, natural magnetic moments and magnetic anisotropies of some chondritic meteorites. *J. Geophys. Res.*, **66**, 1523–1534.

Stamatakos, J. & Kodama, K.P. (1991a). Flexural flow folding and the paleomagnetic fold test: An example of strain reorientation of remanence in the Mauch Chunk formation. *Tectonics*, **10**, 807–819.

Stamatakos, J. & Kodama, K.P. (1991b). The effects of grain-scale deformation on the Bloomsburg

Formation pole. *J. Geophys. Res.*, **96**, 17,919–17,933.

Stapper, C.H. (1969). Micromagnetic solutions for ferromagnetic spheres. *J. Appl. Phys.*, **40**, 798–803.

Steiner, M.B. (1983). Detrital remanent magnetization in hematite. *J. Geophys. Res.*, **88**, 6523–6539.

Stephenson, A. (1969). The temperature-dependent cation distribution in titanomagnetites. *Geophys. J. Roy. Astron. Soc.*, **18**, 199–210.

Stephenson, A. (1975). The observed moment of a magnetized inclusion of high Curie point within a titanomagnetite particle of lower Curie point. *Geophys. J. Roy Astron. Soc.*, **40**, 29–36.

Stephenson, A. (1980). Rotational remanent magnetization and the torque exerted on a rotating rock in an alternating magnetic field. *Geophys. J. Roy. Astron. Soc.*, **62**, 113–132.

Stephenson, A. (1981). Gyromagnetic remanence and anisotropy in single-domain particles, rocks, and magnetic recording tape. *Phil. Mag.*, **B44**, 635–664.

Stephenson, A. (1994). Distribution anisotropy: Two simple models for magnetic lineation and foliation. *Phys. Earth Planet. Inter.*, **82**, 49–53.

Stephenson, A. and Potter, D.K. (1995). Gyroremanent magnetization in rock magnetism and palaeomagnetism. *Int. Un. Geod. Geophys. XXI Gen. Ass., Abstracts vol.*, p. B107.

Stephenson, A., Sadikun, S. & Potter, D.K. (1986). A theoretical and experimental comparison of the anisotropies of magnetic susceptibility and remanence in rocks and minerals. *Geophys. J. Roy. Astron. Soc.*, **84**, 185–200.

Stephenson, R.W. (1976). A study of rotational remanent magnetization. *Geophys. J. Roy. Astron. Soc.*, **47**, 363–373.

Stokking, L. & Tauxe, L. (1987). Acquisition of chemical remanent magnetization by synthetic iron oxide. *Nature*, **327**, 610–612.

Stokking, L. & Tauxe, L. (1990). Properties of chemical remanence in synthetic hematite: Testing theoretical predictions. *J. Geophys. Res.*, **95**, 12,639–12,652.

Stolz, J.F., Chang, S-B.R & Kirschvink, J.L. (1986). Magnetotactic bacteria and single-domain magnetite in hemipelagic sediments. *Nature*, **321**, 849–851.

Stolz, J.F., Lovley, D.R. & Haggerty, S.E. (1990). Biogenic magnetite and the magnetization of sediments. *J. Geophys. Res.*, **95**, 4355–4361.

Stoner, E.C. (1945). The demagnetizing factors for ellipsoids *Phil. Mag.*, **36**, 803–821.

Stoner, E.C. & Wohlfarth, E.P. (1948). A mechanism of magnetic hysteresis in heterogeneous alloys. *Phil. Trans. Roy. Soc. London*, **A240**, 599–642.

Stott, P.M. & Stacey, F.D. (1960). Magnetostriction and paleomagnetism of igneous rocks. *J. Geophys. Res.*, **65**, 2419–2424.

Strangway, D.W. (1961). Magnetic properties of diabase dikes. *J. Geophys. Res.*, **66**, 3021–3031.

Strangway, D.W., Gose, W.A., Pearce, G.W. & Carnes, J.G. (1974). Magnetism and the history of the moon. *J. Appl. Phys.*, **45**, 1178–1196.

Strangway, D.W., Larson, E.E. & Goldstein, M. (1968). A possible cause of high magnetic stability in volcanic rocks. *J. Geophys. Res.*, **73**, 3787–3795.

Strangway, D.W., McMahon, B.E., Honea, R.M. & Larson, E.E. (1967). Superparamagnetism in hematite. *Earth Planet. Sci. Lett.*, **2**, 367–371.

Street, R. & Woolley, J.C. (1949). A study of magnetic viscosity. *Proc. Phys. Soc. (London)*, **A62**, 562–572.

Sugiura, N. (1977). Magnetic properties and remanent magnetization of stony meteorites. *J. Geomag. Geoelec.*, **29**, 519–539.

Sugiura, N. (1979). ARM, TRM and magnetic interactions: Concentration dependence. *Earth Planet. Sci. Lett.*, **42**, 451–455.

Sugiura, N. (1980). Field dependence of blocking temperature of single-domain magnetite. *Earth Planet. Sci. Lett*, **46**, 438–442.

Sugiura, N. (1981). A new model for the acquisition of thermoremanence by multidomain magnetite. *Can. J. Earth Sci.*, **18**, 789–794.

Sugiura, N. (1988). On the origin of PSD moment in magnetite. *Geophys. J.*, **92**, 479–485.

Sugiura, N. & Strangway, D.W. (1980). Comparisons of magnetic paleointensity methods using a lunar sample. *Proc. 11th Lunar Sci. Conf.*, pp. 1801–1813. Pergamon, New York.

Sugiura, N. & Strangway, D.W. (1982). Magnetic properties of low-petrologic grade non-carbonaceous chondrites. In *Proc. 7th Sympos. Antarct. Meteorites, Mem. Nat. Inst. Polar Res., Tokyo*, spec. iss. **25**, pp. 260–280.

Sugiura, N. & Strangway, D.W. (1983). Magnetic paleointensity determination on lunar sample 62235 *J. Geophys. Res.*, **88**, suppl. (Proc. 13th Lunar Sci. Conf.), pp. A684–A690.

Sugiura, N., Lanoix, M. & Strangway, D.W. (1979). Magnetic fields of the solar nebula as recorded in chondrules from the Allende meteorite. *Phys. Earth Planet. Inter.*, **20**, 342–349.

Suk, D. & Halgedahl, S.L. (1996). Hysteresis properties of magnetic spherules versus whole-rock specimens from some Paleozoic platform carbonates. *J. Geophys. Res.*, **101**, 25053–25075.

Suk, D., Peacor, D.R. & Van der Voo, R. (1990). Replacement of pyrite framboids by magnetite in limestone and implications for palaeomagnetism. *Nature*, **345**, 611–613.

Suk, D., Van der Voo, R. & Peacor, D.R. (1992). SEM/STEM observation of magnetic minerals in presumably unremagnetized Paleozoic carbonates from Indiana and Alabama. *Tectonophys.*, **215**, 255–272.

Sun, W. & Jackson, M. (1994). Scanning electron microscopy and rock magnetic studies of magnetic carriers in remagnetized early Paleozoic carbonates from Missouri. *J. Geophys. Res.*, **99**, 2935–2942.

Sun, W., Banerjee, S.K. & Hunt, C.P. (1995). The role of maghemite in the enhancement of magnetic signal in the Chinese loess–paleosol sequence: An extensive rock magnetic study combined with citrate-bicarbonate-dithionite treatment. *Earth Planet. Sci. Let.*, **133**, 493–505.

Sun, W., Jackson, M. & Craddock, J.P. (1993). Relationship between remagnetization, magnetic fabric and deformation in Paleozoic carbonates. *Tectonophys*, **221**, 361–366.

Swift, B.A. & Johnson, H.P. (1984). Magnetic properties of the Bay of Islands ophiolite suite and implications for the magnetization of oceanic crust. *J. Geophys. Res.*, **89**, 3291–3308.

Symons, D.T.A., Sangster, D.F. & Jowett, E.C. (1993). Paleomagnetism of the Pine Point Zn–Pb deposits. *Can. J. Earth Sci.*, **30**, 1028–1036.

Syono, Y. (1965). Magnetocrystalline anisotropy and magnetostriction of Fe_3O_4–Fe_2TiO_4 series – with special application to rock magnetism. *Jap. J. Geophys.*, **4**, 71–143.

Syono, Y. & Ishikawa, Y. (1963a). Magnetocrystalline anisotropy of $xFe_2TiO_4 \cdot (1-x)Fe_3O_4$. *J. Phys. Soc. Japan*, **18**, 1230–1231.

Syono, Y. & Ishikawa, Y. (1963b). Magnetostriction constants of $xFe_2TiO_4 \cdot (1-x)Fe_3O_4$. *J. Phys. Soc. Japan*, **18**, 1231–1232.

Szymczak, R. (1968). The magnetic structure of ferromagnetic materials of uniaxial symmetry. *Electron Techn.*, **1**, 5–43.

Takei, H. & Chiba, S. (1966). Vacancy ordering in epitaxially-grown single crystals of γFe_2O_3. *J. Phys. Soc. Japan*, **21**, 1255–1263.

Takeuchi, H., Uyeda, S. & Kanamori, H. (1970). *Debate about the Earth*, 2nd edn. Freeman & Cooper, San Francisco, 281 pp.

Talwani, M., Windich, C.C. & Langseth, M.G. (1971). Reykjanes ridge crest: A detailed geophysical study. *J. Geophys. Res.*, **76**, 473–517.

Tarduno, J.A. (1990). Absolute inclination value from deep sea sediments: A reexamination of the Cretaceous Pacific record. *Geophys. Res. Lett.*, **17**, 101–104.

Tarduno, J.A. (1994). Temporal trends of magnetic dissolution in the pelagic realm: Gauging paleoproductivity? *Earth Planet. Sci. Lett.*, **123** 39–48.

Tarduno, J.A. (1995). Superparamagnetism and reduction diagenesis in pelagic sediments: Enhancement or depletion? *Geophys. Res. Lett.*, **22**, 1337–1340.

Tarling, D.H. & Hrouda, F. (1993). *The Magnetic Anisotropy of Rocks*. Chapman & Hall, London, 217 pp.

Tauxe, L. (1993). Sedimentary records of relative paleointensity: Theory and practice. *Rev. Geophys.*, **31**, 319–354.

Tauxe, L. & Badgley, C. (1984). Transition stratigraphy and the problem of remanence lock-in times in the Siwalik red beds. *Geophys. Res. Lett.*, **11**, 611–613.

Tauxe, L. & Kent, D.V. (1984). Properties of a detrital remanence carried by haematite from study of modern river deposits and laboratory redeposition experiments. *Geophys. J. Roy. Astron. Soc.*, **76**, 543–561.

Tauxe, L. & Wu, G. (1990). Normalized remanence in sediments of the western equatorial Pacific: Relative paleointensity of the geomagnetic field. *J. Geophys. Res.*, **95**, 12,337–12,350.

Tauxe, L., Mullender, T.A.T. & Pick, T. (1996). Pot-bellies, wasp-waists and superparamegnetism in magnetic hysteresis. *J. Geophys. Res.*, **101**, 571–583.

Taylor, R.M., Maher, B.A. & Self, P.G. (1987). Magnetite in soils: I. The synthesis of single-domain and superparamagnetic magnetite. *Clay Minerals*, **22**, 411–422.

Thellier, E. (1938). Sur l'aimantation des terres cuites et ses applications géophysiques. *Ann. Inst. Phys. Globe Univ. Paris*, **16**, 157–302.

Thellier, E. (1941). Sur la vérification d'une méthode permettant de déterminer l'intensité du champ terrestre dans le passé. *C. R. Acad. Sci. Paris*, **212**, 281–283.

Thellier, E. & Thellier, O. (1959). Sur l'intensité du champ magnétique terrestre dans le passé historique et géologique. *Ann. Géophys.*, **15**, 285–376.

Thibal, J., Pozzi, J-P., Barthès, V. & Dubuisson, G. (1995). Continuous record of geomagnetic field intensity between 4.7 and 2.7 Ma from downhole measurements. *Earth Planet. Sci. Lett.*, **136** 541–550.

Thompson, R. (1975). Long period European geomagnetic secular variation confirmed. *Geophys. J. Roy. Astron. Soc.*, **43**, 847–859.

Thompson, R. & Oldfield, F. (1986). *Environmental Magnetism*. Allen and Unwin, London, 227 pp.

Thompson, R. & Peters, C. (1995). Unmixing natural magnetic mixtures (abstract). *Int. Un. Geod. Geophys. XXI Gen. Ass., Abstracts vol.*, p. B99.

Thomson, L.C., Enkin, R.J. & Williams, W. (1994). Simulated annealing of three-dimensional micromagnetic structures and simulated thermoremanent magnetization. *J. Geophys. Res.*, **99**, 603–609.

Tite, M.S. & Linington, R.E. (1975). Effect of climate on the magnetic susceptibility of soils. *Nature, **256**, 565–566.

Tivey, M. & Johnson, H.P. (1981). Characterization of viscous remanent magnetization in single- and multi-domain magnetite grains. *Geophys. Res. Lett.*, **8**, 217–220.

Tivey, M. & Johnson, H.P. (1984). The characterization of viscous remanent magnetization in large and small magnetite particles. *J. Geophys. Res.*, **89**, 543–552.

Toft, P.B. & Haggerty, S.E. (1988). Limiting depth of magnetization in cratonic lithosphere. *Geophys. Res. Lett.*, **15**, 530–533.

Tric, E., Laj, C., Jehanno, C., Valet, J-P., Kissel, C., Mazaud, A. & Iaccarino, S. (1991). High-resolution record of the Upper Olduvai transition from Po Valley (Italy) sediments: Support for dipolar transition geometry? *Phys. Earth Planet. Inter.*, **65**, 319–336.

Tric, E., Valet, J-P., Tucholka, P., Paterne, M., Labeyrie, L., Guichard, F., Tauxe, L. & Fontugne, M. (1992). Paleointensity of the geomagnetic field during the last 80, 000 years. *J. Geophys. Res.*, **97**, 9337–9351.

Tucker, P. (1980). A grain mobility model of post-depositional realignment. *Geophys. J. Roy. Astron. Soc.*, **63**, 149–163.

Tucker, P. (1981). Palaeointensities from sediments: Normalization by laboratory redepositions. *Earth Planet. Sci. Lett.*, **56**, 398–404.

Tucker, P. & O'Reilly, W. (1980a). The acquisition of thermoremanent magnetization by multidomain single-crystal titanomagnetite. *Geophys. J. Roy. Astron. Soc.*, **60**, 21–36.

Tucker, P. & O'Reilly, W. (1980b). The laboratory simulation of deuteric oxidation of titanomagnetites: Effect on magnetic properties and stability of thermoremanence. *Phys. Earth Planet. Inter.*, **23**, 112–133.

Tucker, P. & O'Reilly, W. (1980c). Reversed thermoremanent magnetization in synthetic titanomagnetites as a consequence of high temperature oxidation. *J. Geomag. Geoelec.*, **32**, 341–355.

Turner, G. & Cadogan, P.H. (1975). The history of lunar bombardment inferred from $^{40}Ar/^{39}Ar$ dating of highland rocks. *Geochim. Cosmochim. Acta*, **39**, suppl. 6, 1509–1538.

Tyson Smith, R. & Verosub, K.L. (1994). Thermoviscous remanent magnetism of Columbia River basalt blocks in the Cascade landslide. *Geophys. Res. Lett.*, **21**, 2661–2664.

Usselman, T.M. & Pearce, G.W. (1974). The grain growth of iron: Implications for the thermal conditions in a lunar ejecta blanket. *Geochim. Cosmochim. Acta*, **38**, suppl. 5, 597–603.

Uyeda, S. (1958). Thermo-remanent magnetism as a medium of paleomagnetism, with special reference to reverse thermo-remanent magnetism. *Jap. J. Geophys.*, **2**, 1–123.

Valet, J-P. & Meynadier, L. (1993). Geomagnetic field intensity and reversals during the past four million years. *Nature,* **366**, 91–95.

Vali, H., Förster, O., Amarantidis, G. & Petersen, N. (1987). Magnetotactic bacteria and their magnetofossils in sediments. *Earth Planet. Sci. Lett.*, **86**, 389–400.

Vali, H., von Dobeneck, T., Amarantidis, G., Förster, O., Morteani, G., Bachmann, L. & Petersen, N. (1989). Biogenic and lithogenic magnetic minerals in Atlantic and Pacific deep sea sediments and their paleomagnetic significance. *Geol. Rund.*, **78**, 753–764.

Van der Voo, R., Henry, S.G. & Pollack, H.N. (1978). On the significance and utilization of secondary magnetizations in red beds. *Phys. Earth Planet. Inter.*, **16**, 12–19.

van der Woude, F., Sawatzky, G.A. & Morrish, A.H. (1968). Relation between hyperfine magnetic fields and sublattice magnetizations in Fe_3O_4. *Phys. Rev.*, **167**, 533–535.

Van Hoof, A.A.M. & Langereis, C.G. (1992). The Upper Kaena sedimentary geomagnetic reversal record from southern Sicily. *J. Geophys. Res.*, **97**, 6941–6957.

Van Houten, F.B. (1968). Iron oxides in red beds. *Geol. Soc. Am. Bull.*, **79**, 399–416.

van Velzen, A.J. & Zijderveld, J.D.A. (1992). A method to study alterations of magnetic minerals during thermal demagnetization applied to a fine-grained marine marl (Trubi formation,. Sicily). *Geophys. J. Int.*, **110**, 79–90.

van Velzen, A.J., Dekkers, M.J. & Zijderveld, J.D.A. (1993). Magnetic iron–nickel sulphides in the Pliocene and Pleistocene marine marls from the Vrica section (Calabria, Italy). *Earth Planet. Sci. Lett.*, **115**, 43–55.

Van Vleck, J.H. (1937). On the anisotropy of cubic ferromagnetic crystals. *Phys. Rev.*, **52**, 1178–1198.

Van Zijl, J.S.V., Graham, K.W.T. & Hales, A.L. (1962). The palaeomagnetism of the Stormberg lavas, II: The behaviour of the magnetic field during a reversal. *Geophys. J. Roy. Astron. Soc.*, **7**, 169–182.

Veitch, R.J. (1983). Magnetostatic energy of spherical two-domain particles and the upper single-domain particle size in magnetite. *J. Geophys.*, **53**, 141–143.

Veitch, R.J. (1984). Calculated anhysteretic susceptibility due to domain wall motion in two-domain magnetite spheres. *Geophys. Res. Lett.*, **11**, 181–184.

Verhoogen, J. (1959). The origin of thermoremanent magnetization. *J. Geophys. Res.*, **64**, 2441–2449.

Verosub, K.L. (1975). Paleomagnetic excursions as magneto-stratigraphic horizons: A cautionary note. *Science*, **190**, 48–50.

Verosub, K.L. (1977). Depositional and post-depositional processes in the magnetization of sediments. *Rev. Geophys. Space Phys.*, **15**, 129–143.

Verosub, K.L. & Banerjee, S.K. (1977). Geomagnetic excursions and their paleomagnetic record. *Rev. Geophys. Space Phys.*, **15**, 145–155.

Verosub, K.L. & Roberts, A.P. (1995). Environmental magnetism: Past, present, and future. *J. Geophys. Res.*, **100**, 2175–2192.

Verosub, K.L., Ensley, R.A. & Ulrick, J.S. (1979). The role of water content in the magnetization of sediments. *Geophys. Res. Lett.*, **6**, 226–228.

Verwey, E.J.M. (1935). The crystal structure of γ-Fe_2O_3 and γ-Al_2O_3. *Z. Krist.*, **91**, 65–69.

Vetter, J.R. ,Kodama, K.P. & Goldstein, A. (1989). Reorientation of remanent magnetism during tectonic fabric development: An example from the Waynesboro Formation, Pennsylvania, U.S.A. *Tectonophys.*, **165**, 29–39.

Vine, F.J. & Matthews, D.H. (1963). Magnetic anomalies over ocean ridges. *Nature*, **199**, 947–949.

Walderhaug, H. (1992). Directional properties of alteration CRM in basic igneous rocks. *Geophys. J. Int.*, **111**, 335–347.

Walderhaug, H., Torsvik, T.H. & Løvlie, R. (1991). Experimental CRM production in a basaltic rock; evidence for stable, intermediate palaeomagnetic directions. *Geophys. J. Int.*, **105**, 747–756.

Walton, D. (1980). Time–temperature relations in the magnetization of assemblies of single domain grains. *Nature*, **286**, 245–247.

Walton, D. (1983). Viscous magnetization. *Nature*, **305**, 616–619.

Walton, D. (1984). Re-evaluation of Greek archaeomagnitudes. *Nature*, **310**, 740–743.

Warner, J. (1972). Metamorphism of Apollo 14 breccias. *Geochim. Cosmochim. Acta.*, **36**, suppl. 3, 623–643.

Wasilewski, P. (1977). Characteristics of first-order shock-induced magnetic transitions in iron and discrimination from TRM. *J. Geomag. Geoelec.*, **29**, 355–377.

Wasilewski, P. (1981a). Magnetization of small iron–nickel spheres. *Phys. Earth Planet. Inter.*, **26**, 149–161.

Wasilewski, P. (1981b). New magnetic results from Allende C3(V). *Phys. Earth Planet. Inter.*, **26**, 134–148.

Wasilewski, P. (1988a). Magnetic characterization of the new magnetic mineral tetrataenite and its contrast with isochemical taenite. *Phys. Earth Planet. Inter.*, **52**, 150–158.

Wasilewski, P. (1988b). A new class of natural magnetic materials: The ordering alloys. *Geophys. Res. Lett.*, **15**, 534–537.

Wasilewski, P. & Mayhew, M.A. (1982). Crustal xenolith magnetic properties and long wavelength anomaly source requirements. *Geophys. Res. Lett.*, **9**, 329–332.

Watkins, N.D. & Haggerty, S.E. (1967). Primary oxidation variation and petrogenesis in a single lava. *Contrib. Miner. Petrol.*, **15**, 251–271.

Weeks, R., Laj, C., Endignoux, L., Fuller, M., Roberts, A.P., Manganne, R., Blanchard, E. & Goree, W. (1993). Improvements in long-core measurement techniques: Applications in palaeomagnetism and palaeoceanography. *Geophys. J. Int.*, **114**, 651–662.

Weiss, P. (1907). L'hypothèse du champ moleculaire et la propriété ferromagnétique. *J. Physique.*, **6**, 661–690.

Weiss, P. & Forrer, R. (1929). Saturation absolue des ferromagnétiques et loi d'approche en fonction d'H et de T. *Ann. Physique.*, **12**, 279–324.

Westcott-Lewis, M.F. (1971). Grain size dependence of thermoremanence in ilmenite-haematites. *Earth Planet. Sci. Lett.*, **12**, 124–128.

Westcott-Lewis, M.F. & Parry, L.G. (1971). Magnetism in rhombohedral iron-titanium oxides. *Austral. J. Phys.*, **24**, 719–734.

Williams, C.D.H., Evans, D. & Thorp, J.S. (1988). Size dependent magnetometric demagnetisation tensors for single domain particles. *J. Magn. Magn. Mat.*, **73**, 123–128.

Williams, R.J. (1972). The lithification and metamorphism of lunar breccias. *Earth Planet. Sci. Lett.*, **16**, 256–260.

Williams, W. & Dunlop, D.J. (1989). Three-dimensional micromagnetic modelling of ferromagnetic domain structure. *Nature*, **337**, 634–637.

Williams, W. & Dunlop, D.J. (1990). Some effects of grain shape and varying external magnetic fields on the magnetic structure of small grains of magnetite. *Phys. Earth Planet. Inter.*, **65**, 1–14.

Williams, W. & Dunlop, D.J. (1995). Simulation of magnetic hysteresis in pseudo-single-domain grains of magnetite. *J. Geophys. Res.*, **100**, 3859–3871.

Williams, W. & Walton, D. (1988). Thermal cleaning of viscous magnetic moments. *Geophys. Res. Lett.*, **15**, 1089–1092.

Williams, W., Enkin, R.J. & Milne, G. (1992a). Magnetic domain wall visibility in Bitter pattern imaging. *J. Geophys. Res.*, **97**, 17,433–17,438.

Williams, W., Hoffmann, V., Heider, F., Göddenhenreich, T. & Heiden, C. (1992b). Magnetic force microscopy imaging of domain walls in magnetite. *Geophys. J. Int.*, **111**, 417–423.

Wilson, R.L. (1961). Palaeomagnetism in Northern Ireland. I. The thermal demagnetization of natural magnetic moments in rocks. *Geophys. J. Roy. Astron. Soc.*, **5**, 45–69.

Wilson, R.L. & Haggerty, S.E. (1966). Reversals of the earth's magnetic field. *Endeavour*, **25**, 104–109.

Wilson, R.L. & Lomax, R. (1972). Magnetic remanence related to slow rotation of ferromagnetic material in alternating magnetic fields. *Geophys. J. Roy. Astron. Soc.*, **30**, 295–303.

Wilson, R.L. & Smith, P.J. (1968). The nature of secondary magnetizations in some igneous and baked rocks. *J. Geomag. Geoelec.*, **20**, 367–380.

Wilson, R.L. & Watkins, N.D. (1967). Correlation of petrology and natural magnetic polarity in Columbia Plateau basalts. *Geophys. J. Roy. Astron. Soc.*, **12**, 405–424.

Wohlfarth, E.P. (1958). Relations between different modes of acquisition of the remanent magnetization of ferromagnetic particles. *J. Appl. Phys.*, **29**, 595–596.

Wood, J.A. (1967). Chondrites: Their metallic minerals, thermal histories, and parent planets. *Icarus*, **6**, 1–49.

Wooldridge, A.L., Haggerty, S.E., Rona, P.A. & Harrison, C.G.A. (1990). Magnetic properties and opaque mineralogy of rocks from selected seafloor hydrothermal sites at oceanic ridges. *J. Geophys. Res.*, **95**, 12,351–12,374.

Worm, H-U. (1986). Herstellung und magnetische Eigenschaften kleiner Titanomagnetit-Ausscheidungen in Silikaten. Doctoral thesis. Univ. Bayreuth, Germany.

Worm, H-U. & Banerjee, S.K. (1984). Aqueous low-temperature oxidation of titanomagnetite. *Geophys. Res. Lett.*, **11**, 169–172.

Worm, H-U. & Jackson, M. (1988). Theoretical time-temperature relationships of magnetization for distributions of single domain magnetite grains. *Geophys. Res. Lett.*, **15**, 1093–1096.

Worm, H-U. & Markert, H. (1987). Magnetic hysteresis properties of fine particle titanomagnetites precipitated in a silicate matrix. *Phys. Earth Planet. Inter.*, **46**, 84–92.

Worm, H-U., Jackson, M., Kelso, P. & Banerjee, S.K. (1988). Thermal demagnetization of partial thermoremanent magnetization. *J. Geophys. Res.*, **93**, 12,196–12,204.

Worm, H-U., Ryan, P.J. & Banerjee, S.K. (1991). Domain size, closure domains, and the importance of magnetostriction in magnetite. *Earth Planet. Sci. Lett.*, **102**, 71–78.

Wright, T.M. (1995). Three dimensional micromagnetic modeling of fine magnetite grains (abstract). *Int. Un. Geod. Geophys. XXI Gen. Ass., Abstracts vol.*, p. B96.

Wright, T.M., Williams, W. & Dunlop, D.J. (1996). An improved algorithm for micromagnetics. *J. Geophys. Res.*, **101**, in press.

Wu, Y.T., Fuller, M. & Schmidt, V.A. (1974). Microanalysis of N.R.M. in a granodiorite intrusion. *Earth Planet. Sci. Lett.*, **23**, 275–285.

Xu, S. & Dunlop, D.J. (1993). Theory of alternating field demagnetization of multidomain grains and implications for the origin of pseudo-single-domain remanence. *J. Geophys. Res.*, **98**, 4183–4190.

Xu, S. & Dunlop, D.J. (1994). The theory of partial thermoremanent magnetization in multidomain grains. 2. Effect of microcoercivity distribution and comparison with experiment. *J. Geophys. Res.*, **99**, 9025–9033.

Xu, S. & Dunlop, D.J. (1995a). Thellier paleointensity determination using PSD and MD grains (abstract). *Eos (Trans. Am. Geophys. Un.)*, **76**, Fall Meeting suppl., F170.

Xu, S. & Dunlop, D.J. (1995b). Towards a better understanding of the Lowrie–Fuller test. *J. Geophys. Res.*, **100**, 22,533–22,542.

Xu, S. & Dunlop, D.J. (1996). Micromagnetic modeling of Bloch walls with Néel caps in magnetite. *Geophys. Res. Lett.*, **23**, 2819–2822.

Xu, S. & Merrill, R.T. (1987). The demagnetizing factors in multidomain grains. *J. Geophys. Res.*, **92**, 10,657–10,665.

Xu, S. & Merrill, R.T. (1989). Microstress and microcoercivity in multidomain grains. *J. Geophys. Res.*, **94**, 10,627–10,636.

Xu, S. & Merrill, R.T. (1990a). Thermal variations of domain wall thickness and number of domains in magnetic rectangular grains. *J. Geophys. Res.*, **95**, 21,433–21,440.

Xu, S. & Merrill, R.T. (1990b). Microcoercivity, bulk coercivity and saturation remanence in multidomain materials. *J. Geophys. Res.*, **95**, 7083–7090.

Xu, S. & Merrill, R.T. (1990c). Toward a better understanding of magnetic screening in multidomain grains. *J. Geomagn. Geoelec.*, **42**, 637–652.

Xu, S. & Merrill, R.T. (1992). Stress, grain size and magnetic stability of magnetite. *J. Geophys. Res.*, **97**, 4321–4329.

Xu, S., Dunlop, D.J. & Newell, A.J. (1994). Micromagnetic modeling of two-dimensional domain structures in magnetite. *J. Geophys. Res.*, **99**, 9035–9044.

Yamazaki, T. (1986). Secondary remanent magnetization of pelagic clay in the South Pacific: Application of thermal demagnetization. *Geophys. Res. Lett.*, **13**, 1438–1441.

Yamazaki, T. & Katsura, I. (1990). Magnetic grain size and viscous remanent magnetization of pelagic clay. *J. Geophys. Res.*, **95**, 4373–4382.

Yamazaki, T., Katsura, I. & Marumo, K. (1991). Origin of stable remanent magnetization of siliceous sediments in the central equatorial Pacific. *Earth Planet. Sci. Lett.*, **105**, 81–93.

Ye, J. & Merrill, R.T. (1991). Differences between magnetic domain imaging observations and theory. *Geophys. Res. Lett.*, **18**, 593–596.

Ye, J. & Merrill, R.T. (1995). The use of renormalization group theory to explain the large variation of domain states observed in titanomagnetites and implications for paleomagnetism. *J. Geophys. Res.*, **100**, 17,899–17,907.

Ye, J., Newell, A.J. & Merrill, R.T. (1994). A re-evaluation of magnetocrystalline anisotropy and magnetostriction constants of single crystal $Fe_{2.4}Ti_{0.6}O_4$. *Geophys. Res. Lett.*, **21**, 25–28.

York, D. (1978a). A formula describing both magnetic and isotopic blocking temperatures. *Earth Planet. Sci. Lett.*, **39**, 89–93.

York, D. (1978b). Magnetic blocking temperature. *Earth Planet. Sci. Lett.*, **39**, 94–97.

Yoshida, S. & Katsura, I. (1985). Characterization of fine magnetic grains in sediments by the suspension method. *Geophys. J. Roy. Astron. Soc.*, **82**, 301–317.

Yuan, S.W. & Bertram, H.N. (1992). Fast adaptive algorithms for micromagnetics. *IEEE Trans. Magn.*, **28**, 2031–2036.

Zelinka, T., Hejda, P. & Kropaček, V. (1987). The vibrating-sample magnetometer and Preisach diagram. *Phys. Earth Planet. Inter.*, **46**, 241–246.

Zhang, B. & Halls, H.C. (1995). The origin and age of feldspar clouding in the Matachewan dyke swarm, Canada. In *Physics and Chemistry of Dykes*, ed. G. Baer & A. Heimann, pp. 171–176. Balkema, Rotterdam.

Zhilyaeva, V.A. & Minibaev, R.A. (1965). The relation of the parameters of magnetic stability and the coefficient of magnetic viscosity to the particle size of ferromagnetic minerals (transl. from Russian). *Izv., Phys. Solid Earth*, no. 4, 275–278.

Zijderveld, J.D.A. (1967). A.C. demagnetization of rocks: Analysis of results. In *Methods in Palaeomagnetism*, ed. D.W. Collinson, K.M. Creer & S.K. Runcorn, pp. 254–286. Elsevier, Amsterdam.

Zijderveld, J.D.A. (1975). Paleomagnetism of the Esterel rocks. Doctoral thesis, Univ. of Utrecht, 199 pp.